HANDBOOK OF STATISTICS
VOLUME 3

Handbook of Statistics

VOLUME 3

General Editor

P. R. Krishnaiah

NORTH-HOLLAND
AMSTERDAM • NEW YORK • OXFORD • TOKYO

Time Series in the Frequency Domain

Edited by

D. R. Brillinger
Department of Statistics
University of California, Berkeley, CA, U.S.A.

P. R. Krishnaiah
Center for Multivariate Analysis
University of Pittsburgh, Pittsburgh, PA, U.S.A.

NORTH-HOLLAND
AMSTERDAM • NEW YORK • OXFORD • TOKYO

ELSEVIER SCIENCE PUBLISHERS B.V.
Sara Burgerhartstraat 25
P.O. Box 1991, 1000 BZ Amsterdam
The Netherlands

Distributors for the United States and Canada:
ELSEVIER SCIENCE PUBLISHING COMPANY, INC.
655 Avenue of the Americas
New York, N.Y. 10010
U.S.A.

ISBN: 0 444 86726 0

First edition: 1983
Second impression: 1990

Library of Congress Cataloging in Publication Data
Main entry under title:
Time series in the frequency domain.

(Handbook of statistics ; v. 3)
Includes index.
1. time-series analysis--Addresses, essays, lectures.
I. Brillinger, David R. II. Krishnaiah, Paruchuri R.
III. Series.
QA280.T56 1983 519.5'5 83-20641
ISBN 0-444-86726-0 (U.S.)

PRINTED IN THE NETHERLANDS

Preface

Time series analysis is one of the most flourishing of the fields of present day statistics. Exciting developments are taking place: in pure theory and in practice, with broad relevance and with narrow intent, for large samples and for small samples. The flourishing results in part, from the dramatic increase in the availability of computing power for both number crunching and for graphical display and in part from a compounding of knowledge as more and more researchers involve themselves with the problems of the field.

This volume of the *Handbook of Statistics* is concerned particularly with the frequency side, or spectrum, approach to time series analysis. This approach involves essential use of sinusoids and bands of (angular) frequency, with Fourier transforms playing an important role. A principal activity is thinking of systems, their inputs, outputs, and behavior in sinusoidal terms. In many cases, the frequency side approach turns out to be simpler in each of computational, mathematical, and statistical respects. In the frequency approach, an assumption of stationarity is commonly made. However, the essential roles played by the techniques of complex demodulation and seasonal adjustment show that stationarity is far from a necessary condition. So too are assumptions of Gaussianity and linearity commonly made. As various of the papers in this Volume show, nor are these necessary assumptions.

The Volume is meant to represent the frequency approach to time series analysis as it is today. Readers working their way through the papers and references included will find themselves abreast of much of contemporary spectrum analysis.

We wish to express our deep appreciation to Professors E. J. Hannan and M. B. Priestley for serving as members of the editorial board. Thanks are due to Professors P. Guttorp, E. J. Hannan, T. Hasan, J. Lillestøl, and M. B. Priestley for refereeing various chapters in the volume. We are most grateful to the authors and North-Holland Publishing Company for their excellent cooperation in bringing out this volume.

D. R. Brillinger
P. R. Krishnaiah

Table of Contents

Contributors

R. J. Bhansali, *Statistics Department, University of Liverpool, P.O. Box 147, Liverpool L69 3BX, U.K.* (Ch. 1)

David R. Brillinger, *Department of Statistics, University of California, Berkeley, California 94720* (Ch. 2)

T. C. Chang, *Department of Mathematical Sciences, University of Cincinnati, Cincinnati, Ohio 45221* (Ch. 20)

William S. Cleveland, *Bell Laboratories, Murray Hill, New Jersey 07974* (Ch. 3)

Robert B. Davies, *Department of Scientific and Industrial Research, Applied Mathematics Division, P.O. Box 1335, Wellington, New Zealand* (Ch. 4)

Robert Engle, *Department of Economics, University of California, San Diego, La Jolla, California 92093* (Ch. 5)

C. W. J. Granger, *Department of Economics, University of California, San Diego, La Jolla, California 92093* (Ch. 5)

E. J. Hannan, *Department of Statistics, The Australian National University, Box 4, G.P.O., Canberra, A.C.T., Australia 2600* (Ch. 6)

T. Hasan, *Bell Laboratories, Murray Hill, New Jersey 07974* (Ch. 7)

Melvin J. Hinich, *Department of Government, University of Texas at Austin, Austin, Texas 78712* (Ch. 8)

D. Karavellas, *Statistics Department, University of Liverpool, P.O. Box 147, Liverpool L69 3BX, U.K.* (Ch. 1)

L. H. Koopmans, *Department of Mathematics, University of New Mexico, Albuquerque, New Mexico 87131* (Ch. 9)

P. R. Krishnaiah, *Center for Multivariate Analysis, University of Pittsburgh, Pittsburgh, Pennsylvania 15260* (Ch. 20)

J. C. Lee, *Bell Laboratories, Murray Hill, New Jersey 07974* (Ch. 20)

R. Douglas Martin, *Department of Statistics, University of Washington, Seattle, Washington 98195* (Ch. 10)

Emanuel Parzen, *Institute of Statistics, Texas A & M University, College Station, Texas 77843* (Ch. 11)

J. Pemberton, *Department of Mathematics, University of Salford, Salford, M5 4WT, U.K.* (Ch. 12)

M. B. Priestley, *Department of Mathematics, University of Manchester Institute of Science & Technology, P.O. Box 88, Manchester, M60 1QD, U.K.* (Ch. 13)

T. Subba Rao, *Department of Mathematics, University of Manchester Institute of Science & Technology, P.O. Box 88, Manchester, M60 1QD, U.K.* (Ch. 14)

Enders A. Robinson, *Department of Geophysics, University of Tulsa, Tulsa, Oklahoma 74104* (Ch. 15)

P. M. Robinson, *Department of Mathematics, University of Surrey, Guildford, Surrey, CU2 5XH, U.K.* (Ch. 16)

M. Rosenblatt, *Department of Mathematics, University of California, San Diego, La Jolla, California 92093* (Ch. 17)

R. H. Shumway, *Division of Statistics, University of California, Davis, California 95616* (Ch. 18)

Tony Thrall, *Systems Applications, Inc., 101 Lucas Valley Rd., San Rafael, California 94903* (Ch. 19)

H. Tong, *Department of Statistics, Chinese University of Hong Kong, New Territories, Hong Kong* (Ch. 12)

D. R. Brillinger and P. R. Krishnaiah, eds., *Handbook of Statistics, Vol. 3*
© Elsevier Science Publishers B.V. (1983) 1–19

1

Wiener Filtering
(with emphasis on frequency-domain approaches)

R. J. Bhansali and D. Karavellas

1. Introduction

Let $\{y_t, x_t\}$ $(t = 0, \pm 1, \ldots)$ be a bivariate process. An important class of problems considered in time-series analysis may be formulated in terms of the problem: How can we best predict y_t from $\{x_s, s \leq t\}$? If $y_t = x_{t+\nu}$, $\nu > 0$, then the problem is that of predicting the 'future' of x_t on the basis of its past. If $x_t = \xi_t + \zeta_t$, where ζ_t is 'noise' and ξ_t the 'signal' and $y_t = \xi_{t+\nu}$, then for $\nu = 0$ the problem is that of 'signal extraction', for $\nu > 0$ that of predicting the signal and for $\nu < 0$ that of interpolating the signal, in the presence of noise. If y_t and x_t are arbitrary, then the problem is simply that of predicting one series from another. This last problem is itself of interest in a number of disciplines: for example, in Economics, interest is often centred on obtaining a distributed lag relationship between two economic variables (see, e.g., Dhrymes [11]) such as level of unemployment and the rate of inflation.

A complete solution to the problem of predicting y_t from the past, $\{x_s, s \leq t\}$, of x_t would consist of giving the conditional probability distribution of the random variable y_t when the observed values of the random variables $\{x_s, s \leq t\}$ are given. However, this is seldom practicable as finding such a conditional distribution is usually a formidable problem. A simplifying procedure of taking the mean value of this conditional distribution as the predictor of y_t is also rarely feasible because this mean value is in general a very complicated function of the past x's. Progress may, however, be made if $\{y_t, x_t\}$ is assumed to be jointly stationary and attention is restricted to the consideration of the linear least-squares predictor of y_t, i.e. the best predictor, $\hat{y}_t$, say, of y_t is chosen from the comparatively narrow class of linear functions of $\{x_s, s \leq t\}$,

$$\hat{y}_t = \sum_{j=0}^{\infty} h(j) x_{t-j}, \tag{1.1}$$

the coefficients $h(j)$ being chosen on the criterion that the mean square error of prediction

$$\eta^2 = E(\hat{y}_t - y_t)^2 \tag{1.2}$$

be a minimum.

Formation of $\hat{y}_t$ from the $\{x_s, s \leq t\}$ may be viewed as a filtering operation applied to the past of x_t, and, especially in engineering literature, $\hat{y}_t$ is known as the Wiener filter.

It should be noted that if $\{y_t, x_t\}$ is Gaussian, then the linear least-squares predictor, $\hat{y}_t$, of y_t is also the best possible predictor in the sense that it minimises the mean square error of prediction within the class of all possible predictors of y_t; hence for the Gaussian case the consideration of only linear predictors is not a restriction.

2. Derivation of the filter transfer function and the filter coefficients

Suppose that $\{y_t, x_t\}$ $(t = 0, \pm 1, \ldots)$ is real-valued jointly stationary with zero means, i.e. $Ex_t = Ey_t = 0$. If the means are nonzero, then these may be subtracted out. Let $R_{xx}(u) = E(x_{t+u}x_t)$ and $R_{yy}(u) = E(y_{t+u}y_t)$ denote the autocovariance functions of x_t and y_t, respectively, and let $R_{yx}(u) = Ey_{t+u}x_t$ denote their cross-covariance function. Assume that

$$\sum_{u=-\infty}^{\infty} |R_{xx}(u)| < \infty, \qquad \sum_{u=-\infty}^{\infty} |R_{yy}(u)| < \infty, \qquad \sum_{u=-\infty}^{\infty} |R_{yx}(u)| < \infty$$

and let

$$f_{xx}(\lambda) = (2\pi)^{-1} \sum_{u=-\infty}^{\infty} R_{xx}(u) \exp(-iu\lambda),$$

$$f_{yy}(\lambda) = (2\pi)^{-1} \sum_{u=-\infty}^{\infty} R_{yy}(u) \exp(-iu\lambda)$$

denote the power spectral density functions of x_t and y_t, respectively, and

$$f_{yx}(\lambda) = (2\pi)^{-1} \sum_{u=-\infty}^{\infty} R_{yx}(u) \exp(-iu\lambda)$$

their cross-spectral density function. Assume also that $f_{xx}(\lambda) \neq 0$ $(-\infty < \lambda < \infty)$.

Under these conditions x_t has the one-sided moving average representation (see Billinger [9, p. 78])

$$x_t = \sum_{j=0}^{\infty} b(j)\varepsilon_{t-j}, \quad b(0) = 1, \tag{2.1}$$

and the autoregressive representation

$$\sum_{j=0}^{\infty} a(j)x_{t-j} = \varepsilon_t, \quad a(0) = 1. \tag{2.2}$$

Here ε_t is a sequence of uncorrelated random variables with 0 mean and finite variance σ^2, say, and the $\{b(j)\}$ and $\{a(j)\}$ are absolutely summable coefficients, i.e. they satisfy

$$\sum_{j=0}^{\infty} |b(j)| < \infty, \qquad \sum_{j=0}^{\infty} |a(j)| < \infty.$$

Also, if

$$B(z) = \sum_{j=0}^{\infty} b(j)z^j, \qquad A(z) = \sum_{j=0}^{\infty} a(j)z^j, \tag{2.3}$$

respectively, denote the characteristic polynomials of the $b(j)$ and the $a(j)$, then $B(z) \neq 0$, $A(z) \neq 0$, $|z| \leq 1$ and $A(z) = \{B(z)\}^{-1}$. The transfer functions $B(e^{-i\lambda})$ and $A(e^{-i\lambda})$ of the $b(j)$ and $a(j)$ are denoted by $B(\lambda)$ and $A(\lambda)$ respectively. We have $A(\lambda) = \{B(\lambda)\}^{-1}$ and $f_{xx}(\lambda) = \sigma^2(2\pi)^{-1}|B(\lambda)|^2$.

If $f_{xx}(\lambda)$ is known exactly, then the $\{b(j)\}$ and $\{a(j)\}$ may be determined, by the Wiener–Hopf spectral factorization procedure (Wiener [25, p. 78]). The assumptions made previously on $R_{xx}(u)$ and $f_{xx}(\lambda)$ ensure that $\log f_{xx}(\lambda)$ is integrable and hence has the Fourier series expansion

$$\log f_{xx}(\lambda) = \sum_{v=-\infty}^{\infty} c(v)\exp(-iv\lambda), \tag{2.4}$$

with

$$c(v) = (2\pi)^{-1}\int_{-\pi}^{\pi} \log f_{xx}(\lambda)\exp(iv\lambda)\,d\lambda \tag{2.5}$$

and

$$\sum_{v=-\infty}^{\infty} |c(v)| < \infty.$$

Set

$$B(\lambda) = \exp\left\{\sum_{v=1}^{\infty} c(v)\exp(-iv\lambda)\right\}, \tag{2.6}$$

$$A(\lambda) = \{B(\lambda)\}^{-1} \tag{2.7}$$

and

$$\sigma^2 = 2\pi\exp\{c(0)\}. \tag{2.8}$$

Then

$$b(j) = (2\pi)^{-1}\int_{-\pi}^{\pi} B(\lambda)\exp(ij\lambda)\,d\lambda, \tag{2.9}$$

$$a(j) = (2\pi)^{-1}\int_{-\pi}^{\pi} A(\lambda)\exp(ij\lambda)\,d\lambda, \tag{2.10}$$

and the $\{b(j)\}$ and $\{a(j)\}$ thus obtained are absolutely summable (Brillinger [9,

p. 79]); see also Doob [12, pp. 160–164] and Grenander and Rosenblatt [16, pp. 67–81] for related work.

Next, consider prediction of y_t from the past, $\{x_s, s \leqslant t\}$, of x_t, and in particular the determination of the filter coefficients $h(j)$ of the linear least-squares predictor $\hat{y}_t$ of y_t. The mean square error of prediction η^2 is given by

$$\eta^2 = R_{yy}(0) - 2\sum_{j=0}^{\infty} h(j)R_{yx}(j) + \sum_{j=0}^{\infty}\sum_{k=0}^{\infty} h(j)h(k)R_{xx}(k-j)\,. \tag{2.11}$$

If the $h(j)$ minimise η^2, then we must have $\partial\eta^2/\partial h(j) = 0$ $(j = 0, 1, \ldots)$. This requirement leads to the equations

$$\sum_{k=0}^{\infty} h(k)R_{xx}(k-j) = R_{yx}(j) \quad (j = 0, 1, \ldots)\,. \tag{2.12}$$

That the $h(k)$ satisfying (2.12) also minimise η^2 may be established by using an argument analogous to that given, for example, by Jenkins and Watts [18, pp. 204–205].

Equations (2.12) provide discrete analogues of the Wiener–Hopf integral equations (Wiener [25, p. 84]). As their left-hand side is of the form of a convolution, the use of Fourier series techniques is a natural approach to adopt for solving them. However, as discussed by N. Levinson (see [25, p. 153]) a direct use of the Fourier series techniques for obtaining the $h(j)$ is not feasible as well, because (2.12) is valid only for $j \geqslant 0$. Therefore, a somewhat indirect approach is adopted for expressing $h(j)$ in terms of $f_{yx}(\lambda)$ and $f_{xx}(\lambda)$.

The representation (2.1) implies that

$$R_{xx}(u) = \sigma^2 \sum_{s=0}^{\infty} b(s)b(s+u) \quad (u = 0, 1, \ldots)\,. \tag{2.13}$$

Put

$$D(\lambda) = f_{yx}(\lambda)\overline{A(\lambda)} = \sum_{u=-\infty}^{\infty} d(u)\, e^{-iu\lambda}\,, \tag{2.14}$$

and

$$[D(\lambda)]_+ = \sum_{u=0}^{\infty} d(u) \exp(-iu\lambda)\,, \tag{2.15}$$

where

$$d(u) = (2\pi)^{-1} \int_{-\pi}^{\pi} f_{yx}(\lambda)\overline{A(\lambda)} \exp(iu\lambda)\, d\lambda \tag{2.16}$$

and

$$\sum_{u=-\infty}^{\infty} |d(u)| < \infty\,.$$

Note that $2\pi d(u) = E(y_t\varepsilon_{t-u})$ and $D(\lambda)$ gives the cross-spectral density function of y_t and ε_t.

From (2.14), we get

$$R_{yx}(j) = 2\pi \sum_{s=0}^{\infty} b(s)d(j+s)\,. \tag{2.17}$$

Hence, (2.12) may be rewritten as

$$\sum_{s=0}^{\infty} b(s)d(j+s) = \frac{\sigma^2}{2\pi} \sum_{s=0}^{\infty} \sum_{k=0}^{\infty} h(k)b(s)b(s+j-k) \quad (j=0,1,\ldots)\,,$$

or, as

$$d(v) = \frac{\sigma^2}{2\pi} \sum_{k=0}^{\infty} h(k)b(v-k) \quad (v=0,1,\ldots)\,. \tag{2.18}$$

Since, $b(v)=0$, $v<0$, (2.18) may be solved by the Fourier series techniques. On multiplying both the sides of (2.18) by $e^{-iv\lambda}$ and summing for all $v \geqslant 0$, we get

$$\begin{aligned} H(\lambda) &= \sum_{k=0}^{\infty} h(k)\exp(-\mathrm{i}k\lambda) \\ &= \frac{2\pi}{\sigma^2} B(\lambda)^{-1}[D(\lambda)]_+ = \frac{2\pi}{\sigma^2} A(\lambda)[f_{yx}(\lambda)\overline{A(\lambda)}]_+ \end{aligned} \tag{2.19}$$

and

$$h(j) = (2\pi)^{-1} \int_{-\pi}^{\pi} H(\lambda)\exp(\mathrm{i}j\lambda)\,\mathrm{d}\lambda\,. \tag{2.20}$$

Since the $d(u)$ given by (2.16) and the $a(j)$ given by (2.9) are absolutely summable, so are the $h(j)$ (see, e.g., Fuller [14, p. 120]). Thus, the $h(j)$'s satisfy

$$\sum_{j=0}^{\infty} |h(j)| < \infty\,.$$

The mean square error of prediction η^2 is

$$\begin{aligned} \eta^2 &= E\{(y_t - \hat{y}_t)^2\} = R_{yy}(0) - \sum_{j=0}^{\infty} h(j)R_{yx}(j) \\ &= \int_{-\pi}^{\pi} \left[f_{yy}(\lambda) - \left\{\frac{2\pi}{\sigma^2} |[D(\lambda)]_+|^2\right\}\right] \mathrm{d}\lambda = R_{yy}(0) - \frac{4\pi^2}{\sigma^2} \sum_{j=0}^{\infty} d^2(j)\,. \end{aligned} \tag{2.21}$$

Equations (2.19) and (2.21) are consistent with the results of Whittle [24, pp. 66–68], but note that a dividing factor of σ^2 is missing in equation (3.7.2) of Whittle [24, p. 42]; see also Bhansali [3].

It is instructive to compare the 'one-sided' predictor (2.10) with the corresponding 'two-sided' predictor of y_t obtained by assuming that the complete past, present and the complete future of x_t is known. Let

$$\tilde{y}_t = \sum_{j=-\infty}^{\infty} g(j) x_{t-j} \tag{2.22}$$

be the 'two-sided' linear least-squares predictor of y_t. Then, as in (2.12), the $g(j)$ are the solutions of the equations

$$R_{yx}(u) = \sum_{j=-\infty}^{\infty} g(j) R_{xx}(u-j) \quad (u = 0, \pm 1, \ldots). \tag{2.23}$$

Since these equations are two-sided and are valid for all integral values of u, they may be solved by the Fourier series techniques. On multiplying both the sides of (2.23) by $(2\pi)^{-1} \mathrm{e}^{-\mathrm{i}u\lambda}$ and summing over u, we have

$$\Gamma(\lambda) = \sum_{j=-\infty}^{\infty} g(j) \exp(-\mathrm{i}j\lambda) = f_{yx}(\lambda)/f_{xx}(\lambda) \tag{2.24}$$

and

$$g(j) = \frac{1}{2\pi} \int_{-\pi}^{\pi} \Gamma(\lambda) \exp(\mathrm{i}j\lambda). \tag{2.25}$$

Let $\tau^2 = E\{(y_t - \tilde{y}_t)^2\}$ be the corresponding mean square error of prediction. We have

$$\tau^2 = \int_{-\pi}^{\pi} \left\{ f_{yy}(\lambda) - \frac{|f_{yx}(\lambda)|^2}{f_{xx}(\lambda)} \right\} \mathrm{d}\lambda = \int_{-\pi}^{\pi} \{1 - C_{yx}^2(\lambda)\} f_{yy}(\lambda)\, \mathrm{d}\lambda, \tag{2.26}$$

where $C_{yx}(\lambda) = |f_{yx}(\lambda)| / \{f_{yy}(\lambda) f_{xx}(\lambda)\}^{1/2}$ is called the coherence between y_t and x_t. Note that $0 \leq C_{yx}(\lambda) \leq 1$, all λ. Expression (2.26) therefore shows that if $C_{yx}(\lambda)$ is close to 1 at all frequencies, then τ^2 is close to 0, and one would expect to obtain a close linear fit between y_t and x_t. In this sense, $C_{yx}(\lambda)$ may be interpreted as a correlation coefficient 'in the frequency domain' (see, e.g., Priestley [22] and Granger and Hatanaka (15]).

On using (2.14)–(2.16), (2.21) may be rewritten as (see Whittle [24, p. 69])

$$\eta^2 = \tau^2 + \frac{4\pi^2}{\sigma^2} \sum_{j=-\infty}^{-1} d^2(j). \tag{2.27}$$

The second term to the right of this expression gives the increase in mean square error due to the restriction that only the 'past' of x_t may be used for predicting y_t. In general, therefore, $\eta^2 \geq \tau^2$.

There is, however, one important situation in which $\eta^2 \equiv \tau^2$. This occurs when x_t is the input to, and y_t the output of, a physically realizable linear time-invariant filter with uncorrelated noise, i.e. when,

$$y_t = \sum_{j=0}^{\infty} l(j) x_{t-j} + z_t, \tag{2.28}$$

$\{z_t\}$ is a stationary process uncorrelated with x_t and $\Sigma |l(j)| < \infty$.

We have

$$L(\lambda)=\sum_{j=0}^{\infty} l(j)\exp(-\mathrm{i}j\lambda)=\{f_{yx}(\lambda)/f_{xx}(\lambda)\}=\Gamma(\lambda).$$

Thus,

$$\begin{aligned}D(\lambda)&=f_{yx}(\lambda)\overline{A(\lambda)}=L(\lambda)f_{xx}(\lambda)\overline{A(\lambda)}\\&=\sigma^2(2\pi)^{-1}L(\lambda)A(\lambda)=[D(\lambda)]_+,\end{aligned}$$

the last equality following from the fact that the Fourier coefficients, $l(j)$ and $a(j)$, respectively, of $L(\lambda)$ and $A(\lambda)$ vanish for $j<0$. We, therefore, have

$$H(\lambda)=L(\lambda)=\Gamma(\lambda), \qquad \eta^2=\tau^2. \tag{2.29}$$

Let $f_{zz}(\lambda)$ denote the power spectral density function of the 'residual' process z_t of (2.28). We have

$$f_{zz}(\lambda)=\left\{f_{yy}(\lambda)-\frac{|f_{yx}(\lambda)|^2}{f_{xx}(\lambda)}\right\}. \tag{2.30}$$

3. Realization of the Wiener filter in some special cases

First consider the case of *pure prediction*, and, thus suppose that $y_t=x_{t+\nu}$, $\nu\geqslant 1$. We have

$$R_{yx}(u)=R_{xx}(u+\nu), \qquad f_{yx}(\lambda)=\mathrm{e}^{\mathrm{i}\nu\lambda}f_{xx}(\lambda).$$

Hence

$$H(\lambda)=A(\lambda)[\mathrm{e}^{\mathrm{i}\nu\lambda}B(\lambda)]_+=\{B(\lambda)\}^{-1}\sum_{j=0}^{\infty} b(j+\nu)\exp(-\mathrm{i}j\lambda) \tag{3.1}$$

gives the transfer function of the prediction constants, and

$$\eta^2=\sigma^2\sum_{j=0}^{\nu-1} b^2(j)$$

gives the mean square error of ν-step prediction. Note that if $\nu=1$, then $\eta^2=\sigma^2$, which may be determined using (2.8).

Let $\hat{x}_t(\nu)$ be the linear least-squares predictor of $x_{t+\nu}$ ($\nu\geqslant 1$) when $\{x_s, s\leqslant t\}$ is known. Then, an explicit expression for $\hat{x}_t(\nu)$ in terms of $\hat{x}_t(\nu-1),\ldots,\hat{x}_t(1)$ and $\{x_s, s\leqslant t\}$ may also be written down. We have

$$\hat{x}_t(\nu)=-\sum_{j=1}^{\nu-1} a(j)\hat{x}_t(\nu-j)-\sum_{j=\nu}^{\infty} a(j)x_{t+\nu-j}, \tag{3.2}$$

where the first sum to the right is 0 if $\nu=1$.

That (3.1) and (3.2) are mutually consistent is easily verified. Thus, for $\nu = 1$, (3.1) gives

$$\sum_{j=0}^{\infty} h(j)\,\mathrm{e}^{-\mathrm{i}j\lambda} = H(\lambda) = \mathrm{e}^{\mathrm{i}\lambda}\{1 - A(\lambda)\}$$
$$= -\sum_{j=0}^{\infty} a(j+1)\,\mathrm{e}^{-\mathrm{i}j\lambda}. \tag{3.3}$$

Hence on comparing the coefficient of $\mathrm{e}^{-\mathrm{i}j\lambda}$ on the right- and the left-hand sides of (3.3) we get

$$h(j) = -a(j+1),$$

which is immediately seen to be consistent with (3.2), see also Whittle [24, p. 33]. The argument may similarly be generalised for an arbitrary ν, though the algebra now is more complicated.

A comparison of (3.2) with (2.2) shows that the linear least-squares predictor of $x_{t+\nu}$ when the past $\{x_t, x_{t-1}, \ldots\}$ is known is obtained by (i) setting $\varepsilon_{t+\nu} \equiv 0$; and (ii) for $j < \nu$ replacing the unknown $x_{t+\nu-j}$ by their linear least-squares predictor $\hat{x}_t(\nu - j)$. A related reference is Box and Jenkins [8, pp. 130–131].

An alternative expression for $\hat{x}_t(\nu)$ is given by Bhansali [7].

We note from (3.3) that if x_t is a finite autoregressive process of order m, i.e. if in (2.2) for some finite $m \geq 1$, $a(u) \equiv 0$, $u > m$, then $\hat{x}_t(\nu)$ depends only on $x_{t-1}, \ldots, x_{t-m}$. Similarly, if x_t is a finite moving average process of order p, i.e. if in (2.1), for some finite $p \geq 1$, $b(u) \equiv 0$, $u > p$, then (3.1) shows that $\hat{x}_t(\nu) = 0$ if $\nu > p$; and for this particular class of processes, the knowledge of the complete history of the process does not help, in the linear least-squares sense, for prediction more than p steps ahead. Related reference is Akaike [2], who studies some of the properties of the 'predictor space' spanned by $\{\hat{x}_t(\nu), \nu \geq 1\}$, when x_t is a mixed autoregressive-moving average process.

Second, consider the case of prediction in the presence of noise. Suppose that

$$x_t = \xi_t + \zeta_t, \qquad y_t = \xi_{t+\nu}$$

and $E(\xi_t\zeta_s) = 0$ (all s and t). Then

$$R_{yx}(u) = R_{\xi\xi}(u+\nu), \qquad f_{yx}(\lambda) = \mathrm{e}^{\mathrm{i}\nu\lambda} f_{\xi\xi}(\lambda),$$

where $R_{\xi\xi}(u)$ and $f_{\xi\xi}(\lambda)$, respectively, denote the autocovariance function and the power spectral density functions of ξ_t process. Hence

$$H(\lambda) = \frac{2\pi}{\sigma^2} A(\lambda)[f_{\xi\xi}(\lambda)\,\mathrm{e}^{\mathrm{i}\nu\lambda}\overline{A(\lambda)}]_+, \tag{3.4}$$

where $A(\lambda)$ and σ^2 are obtained by factorising the spectral density function,

$f_{xx}(\lambda)$, of x_t, $f_{xx}(\lambda)=f_{\zeta\zeta}(\lambda)+f_{\xi\xi}(\lambda)$ and $f_{\zeta\zeta}(\lambda)$ denotes the spectral density function of the ζ_t process.

Also

$$\Gamma(\lambda)=\{e^{i\nu\lambda}f_{\xi\xi}(\lambda)\}/\{f_{\xi\xi}(\lambda)+f_{\zeta\zeta}(\lambda)\}, \tag{3.5}$$

which reduces to the expression given by Whittle [24, p. 58], if $\nu=0$.

Third, consider the system (2.28), but now assume that the processes x_t and z_t are correlated and let $R_{zx}(u)=E(z_{t+u}x_t)$ be the cross-covariance function of $\{z_t, x_t\}$ and

$$f_{yx}(\lambda)=L(\lambda)f_{xx}(\lambda)+f_{zx}(\lambda),$$

be their cross-spectral density function. Then

$$H(\lambda)=L(\lambda)+\frac{2\pi}{\sigma^2}A(\lambda)[\overline{A(\lambda)}f_{zx}(\lambda)]_+,$$

$$\Gamma(\lambda)=L(\lambda)+\{f_{zx}(\lambda)/f_{xx}(\lambda)\}.$$

Thus, in this case, $H(\lambda)\neq L(\lambda)$ includes the contribution from the nonzero $f_{zx}(\lambda)$ and could be different from $\Gamma(\lambda)$. Related references are Akaike [1] and Priestley [21].

4. Estimation of the Wiener filter

So far, we have assumed that the spectra $f_{xx}(\lambda)$, $f_{yy}(\lambda)$ and $f_{yx}(\lambda)$ of the process $\{y_t, x_t\}$ are known exactly. In practice, these are invariably unknown a priori, and have to be estimated from data. Suppose that we are given T observations, $\{X_1,\ldots,X_T\}$, $\{Y_1,\ldots,Y_T\}$ from each series. We consider estimation of the filter coefficients from the 'window' estimates of $f_{yx}(\lambda)$ and $f_{xx}(\lambda)$.

Let

$$f_{yx}^{(T)}(\lambda)=\frac{2\pi}{T}\sum_{s=1}^{T-1}K_T\left(\lambda-\frac{2\pi s}{T}\right)I_{yx}^{(T)}\left(\frac{2\pi s}{T}\right), \tag{4.1a}$$

$$f_{xx}^{(T)}(\lambda)=\frac{2\pi}{T}\sum_{s=1}^{T-1}K_T\left(\lambda-\frac{2\pi s}{T}\right)I_{xx}^{(T)}\left(\frac{2\pi s}{T}\right), \tag{4.1b}$$

be the 'window' estimates of $f_{yx}(\lambda)$ and $f_{xx}(\lambda)$, respectively, considered by Brillinger [9]. Here

$$I_{yx}^{(T)}(\lambda)=(2\pi T)^{-1}\sum_{t=1}^{T}\sum_{s=1}^{T}Y_tX_s\exp\{-i\lambda(t-s)\}, \tag{4.2a}$$

$$I_{xx}^{(T)}(\lambda)=(2\pi T)^{-1}\left|\sum_{t=1}^{T}X_t\exp\{-i\lambda t\}\right|^2, \tag{4.2b}$$

are, respectively, the cross-periodogram and (auto) periodogram functions,

$$K_T(\alpha) = \sum_{j=-\infty}^{\infty} B_T^{-1} K\{B_T^{-1}(\alpha + 2\pi j)\} \quad (-\infty < \lambda < \infty), \tag{4.3}$$

$\{B_T\}$ $(T = 1, 2, \ldots)$ is a sequence of constants, such that $B_T \to 0$, $TB_T \to \infty$ as $T \to \infty$ and $K(\alpha)$ is a fixed weight function satisfying Assumption I stated below.

ASSUMPTION I. Let $K(\alpha)$, $-\infty < \alpha < \infty$, be a real-valued, even function of bounded variation and suppose that

$$\int_{-\infty}^{\infty} K(\alpha)\, d\alpha = 1, \qquad \int_{-\infty}^{\infty} |\alpha|^3 |K(\alpha)|\, d\alpha < \infty, \qquad \int_{-\infty}^{\infty} K^2(\alpha)\, d\alpha < \infty.$$

Note that the estimates (4.1) are obtained by assuming that the observed time series have been mean corrected. In practice, the means Ex_t and Ey_t of the processes $\{x_t\}$ and $\{y_t\}$ are unknown. Hence this is not an unrealistic assumption.

The estimates (4.1) are obtained by directly smoothing the cross-, and auto-, periodogram functions. A closely related class of 'window' estimates are of the form (see, e.g., Hannan [17])

$$f_{yx}^{*(T)}(\lambda) = \frac{1}{2\pi} \sum_{u=-T+1}^{T-1} k_T(u) R_{yx}^{(T)}(u) \exp(-iu\lambda), \tag{4.3a}$$

$$f_{xx}^{*(T)}(\lambda) = (2\pi)^{-1} \sum_{u=-T-n}^{T-1} k_T(u) R_{xx}^{(T)}(u) \exp(-iu\lambda), \tag{4.3b}$$

where

$$R_{yx}^{(T)}(u) = T^{-1} \sum_{t=1}^{T-|u|}{}' Y_{t+u} X_t, \qquad R_{xx}^{(T)}(\lambda) = T^{-1} \sum_{t-1}^{T-|u|}{}' X_{t+u} X_t, \tag{4.4}$$

$k_T(u) = k(B_T u)$ and $k(x)$ is a fixed weight function. If we define

$$k(x) = \int_{-\infty}^{\infty} K(\alpha) \exp(ix\alpha)\, d\alpha, \tag{4.5}$$

where $K(\alpha)$, as before, satisfies Assumption I, then a uniform bound on the difference between $f_{xx}^{(T)}(\lambda)$ and $f_{xx}^{*(T)}(\lambda)$ is given by Brillinger [9, p. 265].

The estimation of the coefficients $h(j)$ and the transfer functions $H(\lambda)$ and $A(\lambda)$ involves estimation of the Fourier integrals, (2.5), (2.16) and (2.20), and infinite Fourier series, (2.7) and (2.15). Since, with a finite record $f_{yx}(\lambda)$ and $f_{xx}(\lambda)$ can only be estimated at a finite number of points, we estimate $h(j)$ by approximating the Fourier integrals by the Trapezoidal rule, and the infinite Fourier series by a finite Fourier series.

Suppose that $f_{yx}^{(T)}(\lambda)$ and $f_{xx}^{(T)}(\lambda)$ are calculated at N_T+1 points, $\lambda_j = j\pi/N_T$ $(j=0,1,\ldots,N_T)$ equally spaced in $[0,\pi]$, and let $P_T = 2N_T$. Here $N_T \geq 1$ is a sequence of constants tending to infinity simultaneously with T, but at a slower rate than B_T approaches 0. The estimate of $h(u)$ is given by

$$h^{(T)}(u) = P_T^{-1} \sum_{j=-N_T}^{N_T-1} H^{(T)}(\lambda_j) \exp(\mathrm{i}u\lambda_j) \quad (u=0,1,\ldots,N_T), \tag{4.6}$$

where

$$H^{(T)}(\lambda_j) = \frac{2\pi}{\hat{\sigma}^2} A^{(T)}(\lambda_j) \left\{ \sum_{u=0}^{N_T-1} d^{(T)}(u) \exp(-\mathrm{i}u\lambda_j) \right\}, \tag{4.7}$$

$$d^{(T)}(u) = P_T^{-1} \sum_{j=-N_T}^{N_T-1} f_{yx}^{(T)}(\lambda_j) \overline{A^{(T)}(\lambda_j)} \exp(\mathrm{i}u\lambda_j), \tag{4.8}$$

$$A^{(T)}(\lambda_j) = \exp\left\{ \mathrm{i} \sum_{v=1}^{N_T-1} c^{(T)}(v) \sin v\lambda_j \right\} \left(\frac{\hat{\sigma}^2}{2\pi} \right)^{1/2} \{ f_{xx}^{(T)}(\lambda_j) \}^{-1/2}, \tag{4.9}$$

$$\hat{\sigma}^2 = 2\pi \exp\{c^{(T)}(0)\}, \tag{4.10}$$

$$c^{(T)}(v) = P_T^{-1} \sum_{j=-N_T}^{N_T-1} \log f_{xx}^{(T)}(\lambda_j) \exp(\mathrm{i}v\lambda_j). \tag{4.11}$$

Let $h_N(u)$ be the corresponding quantity obtained from (4.6) to (4.11) when $f_{xx}(\lambda_j)$ and $f_{yx}(\lambda_j)$ are known exactly. Then, for each finite N_T, $h^{(T)}(u)$ may be viewed as estimating $h_N(u)$, rather than the true parameter $h(u)$. A bound on the difference between $h_N(u)$ and $h(u)$ may be given by using the results of Davis and Rabinowitz [10, p. 109]. Suppose that, for an integer $l_0 \geq 1$, $R_{yx}(u)$ and $R_{xx}(u)$ satisfy an additional regularity condition of the form

$$\sum_{u=-\infty}^{\infty} |u|^{2l_0+1} |R_{yx}(u)| < \infty, \qquad \sum_{u=-\infty}^{\infty} |u|^{2l_0+1} |R_{xx}(u)| < \infty. \tag{4.12}$$

When this condition is satisfied, up to $2l_0+1$ derivatives of $f_{yx}(\lambda)$ and $f_{xx}(\lambda)$ with respect to λ exist. Then, as $N_T \to \infty$, for all u,

$$|h_N(u) - h(u)| = \mathrm{O}(N_T^{-2l_0+1}) \tag{4.13}$$

and, $h_N(u) \to h(u)$ as $N_T \to \infty$.

The estimation of the two-sided filter coefficients (2.25) and the transfer function (2.22) has also received considerable attention in the literature; see, e.g., Hannan [17, Chapter VII], Brillinger [9, Chapter 8] and Wahba [23], amongst others. The corresponding estimate of $\Gamma(\lambda)$ is given by

$$\Gamma^{(T)}(\lambda) = f_{yx}^{(T)}(\lambda) / f_{xx}^{(T)}(\lambda) \tag{4.14}$$

and that of $g(u)$ by

$$g^{(T)}(u) = P_T^{-1} \sum_{j=-N_T}^{N_T-1} \Gamma^{(T)}(\lambda_j) \exp(\mathrm{i}u\lambda_j)\,. \tag{4.15}$$

The above 'frequency-domain' approach to estimating the filter coefficients should be contrasted with its 'time-domain' alternatives. Notable amongst these is the 'transfer function' model suggested by Box and Jenkins [8], which expresses the filter coefficients $h(u)$ as functions of a finite number of parameters. Dhrymes [11] gives a survey of the 'econometric' techniques currently in use for estimating the filter coefficients.

5. Asymptotic properties of the estimated filter coefficients

Bhansali [5] studied the large sample behaviour of the estimates $c^{(T)}(v)$, $A^{(T)}(\lambda_j)$ and $\hat{\sigma}^2$. Karavellas [19] has studied the corresponding large sample behaviour of the statistics $d^{(T)}(u)$, $h^{(T)}(u)$ and $g^{(T)}(u)$. His proofs are too lengthy to be given here. Only a summary of his main results is given below.

Suppose that the process $\{y_t, x_t\}$ satisfies Assumption 2.6.1 stated by Brillinger [9, pp. 25–26], the spectral density function, $f_{xx}(\lambda)$, of the process x_t is bounded away from 0 and that (4.12) holds with $l_0 = 3$. Further, suppose that the estimates $f_{xx}^{(T)}(\lambda)$ and $f_{yx}^{(T)}(\lambda)$ of $f_{xx}(\lambda)$ and $f_{yx}(\lambda)$ are obtained in accordance with the formula (4.1), where $K(\alpha)$ satisfies Assumption I, and that $B_T = \mathrm{O}(T^{-b})$, $N_T = \mathrm{O}(T^c)$ where b and c lie in the shaded region shown in Fig. 1. Set $Q(T) = TB_TP_T\{2\pi \int_{-\infty}^{\infty} K^2(\alpha)\,\mathrm{d}\alpha\}^{-1}$. Then, for given u and v, the random vari-

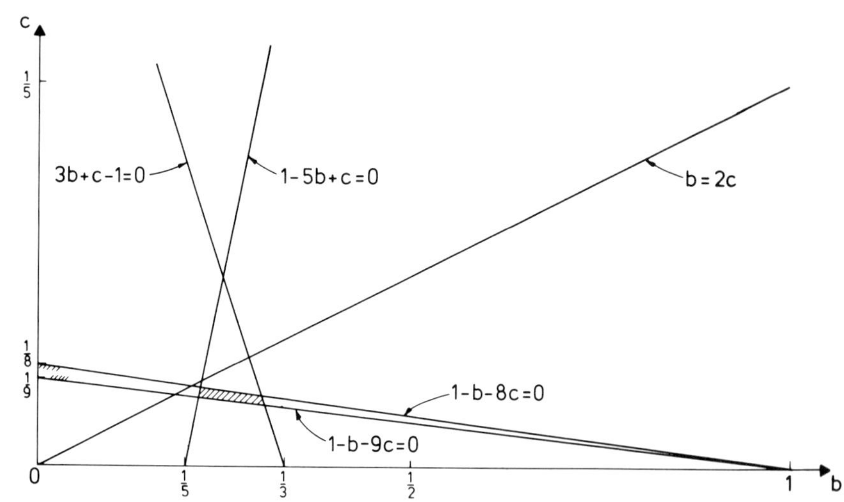

Fig. 1. Constraints on B_T and N_T when $B_T = \mathrm{O}(T^{-b})$, $N_T = \mathrm{O}(T^c)$. The shaded area shows the region where the constraints are satisfied.

ables $Q(T)^{1/2}\{d^{(T)}(u)-d(u)\}$ and $Q(T)^{1/2}\{d^{(T)}(v)-d(v)\}$ are, as $T\to\infty$, approximately normally distributed with zero means and covariance structure

$$\frac{\sigma^2}{4\pi^2}R_{yy}(u-v)+d(u)d(v)-\sum_{j=1}^{\infty}d(u+j)d(v+j)\,. \tag{5.1}$$

Also, for given u and v, the random variables $Q(T)^{1/2}\{h^{(T)}(u)-h(u)\}$ and $Q(T)^{1/2}\{h^{(T)}(v)-h(v)\}$ are, as $T\to\infty$, approximately normally distributed with zero means and covariance structure

$$\begin{aligned}&\frac{1}{\sigma^2}\sum_{l=0}^{\infty}\sum_{m=0}^{\infty}a(u-l)a(v-m)R_{yy}(l-m)\\&\quad-\frac{4\pi^2}{\sigma^4}\sum_{l=0}^{\infty}\sum_{m=0}^{\infty}\sum_{p=1}^{\infty}a(u-l)a(v-m)d(m+p)d(l+p)\\&\quad+\sum_{l=0}^{\infty}h(u-l)h(v-l)-\frac{2\pi}{\sigma^2}\sum_{l=0}^{\infty}\sum_{m=0}^{\infty}h(v-m)a(u-l)d(l-m)\\&\quad-\frac{2\pi}{\sigma^2}\sum_{l=0}^{\infty}\sum_{m=0}^{\infty}h(u-l)a(v-m)d(m-l)\,.\end{aligned} \tag{5.2}$$

Hannan [17, p. 480], Wahba [23] and Brillinger [9, p. 286] investigate the asymptotic distribution of the 'two-sided' estimates, $g^{(T)}(u)$, by assuming that the relationship between y_t and x_t is linear, time-invariant and open loop. If y_t and x_t are arbitrary stationary processes and the regularity conditions stated above hold, then, for given u and v, as $T\to\infty$, the random variables $Q(T)^{1/2}\{g^{(T)}(u)-g(u)\}$ and $Q(T)^{1/2}\{g^{(T)}(v)-g(v)\}$ are approximately normally distributed with zero means and covariance structure

$$\frac{1}{2\pi}\int_{-\pi}^{\pi}\{1-C_{yx}^2(\lambda)\}\frac{f_{yy}(\lambda)}{f_{xx}(\lambda)}\exp\{\mathrm{i}(u-v)\lambda\}\,\mathrm{d}\lambda\,, \tag{5.3}$$

which is consistent with the result of Brillinger [9, p. 318].

For comparing the behaviour of the $h^{(T)}(u)$ with that of $g^{(T)}(u)$, first suppose that $C_{yx}(\lambda)\equiv 0$, all λ, i.e. there is no linear relationship between y_t and x_t. Then, $h(u)\equiv 0$ ($u=0,1,\ldots$) and expressions (5.2) and (5.3) reduce to

$$\frac{1}{2\pi}\int_{-\pi}^{\pi}\frac{f_{yy}(\lambda)}{f_{xx}(\lambda)}\exp\{\mathrm{i}(u-v)\lambda\}\,\mathrm{d}\lambda\,, \tag{5.4}$$

so that the $h^{(T)}(u)$ and the $g^{(T)}(u)$ are asymptotically equivalent.

Next, suppose that the relationship between y_t and x_t is given by (2.28). Expression (5.2) then simplifies to

$$\frac{2\pi}{\sigma^2}\int_{-\pi}^{\pi}f_{zz}(\lambda)A_u^*(\lambda)\overline{A_v^*(\lambda)}\exp\{\mathrm{i}(u-v)\lambda\}\,\mathrm{d}\lambda\,,$$

where $A_u^*(\lambda) = \Sigma_{j=0}^{u} a(j)\exp(-ij\lambda)$ anf $f_{zz}(\lambda)$ is given by (2.30). In particular, if $u = v$, then the asymptotic variance of $h^{(T)}(u)$ is given by

$$\int_{-\pi}^{\pi} \frac{f_{zz}(\lambda)}{f_{xx}^*(\lambda)}\,d\lambda\,, \tag{5.5}$$

where $f_{xx}^*(\lambda) = \sigma^2(2\pi)^{-1}|A_u^*(\lambda)|^{-2}$, while that of $g^{(T)}(u)$ is given by

$$\int_{-\pi}^{\pi} \frac{f_{zz}(\lambda)}{f_{xx}(\lambda)}\,d\lambda\,. \tag{5.6}$$

Hence, the difference between their asymptotic variance is

$$\frac{2\pi}{\sigma^2}\int_{-\pi}^{\pi} f_{zz}(\lambda)\{|A_u^*(\lambda)|^2 - |A(\lambda)|^2\}\,d\lambda\,. \tag{5.7}$$

Note that if x_t is a finite autoregressive process of order m, then for $u \geqslant m$, (5.7) is zero and the two estimates are asymptotically equivalent. In other cases, the behaviour of these two estimates will depend upon that of $f_{zz}(\lambda)$. If $f_{zz}(\lambda) = P$, a constant, for all λ, then it is clear that (5.7) is negative for all λ, i.e. the asymptotic variance of $h^{(T)}(u)$ is smaller than that of $g^{(T)}(u)$.

On the other hand, suppose that

$$\frac{f_{zz}(\lambda)}{f_{xx}(\lambda)} = C^*\,, \text{ a constant, all } \lambda\,, \tag{5.8}$$

i.e. the 'signal-to-noise' ratio is constant (e.g. Fishman [13]). Then, by using a result of Parzen [20], (5.7) may be shown to be nonnegative for all u, and thus, for this case, the asymptotic variance of $h^{(T)}(u)$ is greater than that of $g^{(T)}(u)$. Note, however, that when (5.8) holds and $l(u) = 0$ for u sufficiently large, the $g^{(T)}(u)$ provide asymptotically efficient estimates of the (nonzero) $l(u)$, relative to the maximum likelihood in the Gaussian case, see Hannan [17, Chapter VII].

It is interesting to note that for prediction one step ahead, i.e. when $y_t = x_{t+1}$, expression (5.2) simplifies to

$$\sum_{r=0}^{\infty} a(u-r)a(v-r)\,, \tag{5.9}$$

which is consistent with the result of Bhansali [5], since, from (3.3), for this case, $h(j) = -a(j+1)$ $(j = 0, 1, \ldots)$.

6. A simulation study

The finite sample behaviour of the statistics $h^{(T)}(u)$ and $g^{(T)}(u)$ is extremely difficult to derive analytically. The usual practice is to use the asymptotic results obtained by letting $T \to \infty$, as approximations for the finite sample behaviour of these statistics. To examine the usefulness of this procedure, several different systems of known structure were generated on a computer and these estimates were computed. The detailed results are given by Karavellas [19]. To save space, in this paper, we only consider the following two systems, in which $\{u_t\}$ and $\{v_t\}$ are sequences of independent normal deviates with zero mean and variance 1:

System I

$$y_t = 0.5x_{t-1} + 0.15x_{t-2} + 0.05x_{t-3} + 0.5z_t,$$

$$x_t = 0.75x_{t-1} - 0.5x_{t-2} + u_t,$$

$$z_t = 0.75z_{t-1} - 0.5z_{t-2} + v_t;$$

System II

$$y_t = 0.5x_{t-1} + 0.5u_t,$$

$$x_t = 0.55x_{t-1} + 0.05x_{t-2} + v_t.$$

The $\{u_t\}$ and $\{v_t\}$ were generated using the two-sequence method, described by Bhansali [6]. A stretch of $T + 100$ observations on $\{y_t\}$ and $\{x_t\}$ was generated from these systems, but the first 100 observations were discarded to avoid the transients. The modified Daniell window, which takes a weighted average of $2m + 1$ periodogram and cross-periodogram ordinates, was used to estimate $f_{xx}(\lambda)$ and $f_{yx}(\lambda)$ at $N_T + 1$ equally spaced points in $[0, \pi]$. Several different values of T, and, for each T, several values of m and N_T were considered. However, to save space, only the results for $T = 960$, $m = 5$ and $N_T = 48$ are presented here. Note that the choice of N_T is in accordance with the results of Bhansali [4]. The estimates, $h^{(T)}(u)$ and $g^{(T)}(u)$, of the filter coefficients were computed by using formulae (4.6) and (4.15), respectively. These calculations were repeated 100 times, with a different set of observations each time, and the observed means and variances of the estimated filter coefficients were calculated. The computing was done on the 1906A/7600 computer of the University of Manchester.

The observed means of $h^{(T)}(u)$ and $g^{(T)}(u)$ $(u = 0, 1, \ldots, 5)$ are presented in Table 1 for System I, and in Table 2 for System II.

The $h^{(T)}(u)$ and the $g^{(T)}(u)$ are seen to provide biased estimates of the filter coefficients. Moreover, the bias in estimating the nonzero coefficients is greater than that for the zero coefficients. The actual magnitude of the bias is about the same for both these estimators, since their means differ from each other only in

Table 1
Means of $h^{(T)}(u)$ and $g^{(T)}(u)$ for System I

u	Mean of $h^{(T)}(u)$	Mean of $g^{(T)}(u)$	$h(u)$
0	0.00315	0.00263	0.0
1	0.47348	0.47303	0.5
2	0.14149	0.14106	0.15
3	0.04531	0.04527	0.05
4	0.00083	0.00043	0.0
5	0.00160	0.00215	0.0

Table 2
Means of $h^{(T)}(u)$ and $g^{(T)}(u)$ for System II

u	Mean of $h^{(T)}(u)$	Mean of $g^{(T)}(u)$	$h(u)$
0	0.00357	0.00298	0.0
1	0.47172	0.47167	0.5
2	−0.00067	−0.00087	0.0
3	−0.00181	−0.0016	0.0
4	0.00147	0.00115	0.0
5	0.00147	0.00222	0.0

the fourth decimal place. Both the systems generated here are linear, time-invariant, open loop and physically realizable. Hence the last finding is in accordance with the results of Section 2. The $h^{(T)}(u)$ and the $g^{(T)}(u)$ are based on the 'window' estimators $f_{yx}^{(T)}(\lambda)$ and $f_{xx}^{(T)}(\lambda)$. The latter are known to be biased for $f_{yx}(\lambda)$ and $f_{xx}(\lambda)$, and their bias may partly account for the bias of $h^{(T)}(u)$ and $g^{(T)}(u)$ in estimating the filter coefficients. A relevant reference is Hannan [17, p. 479].

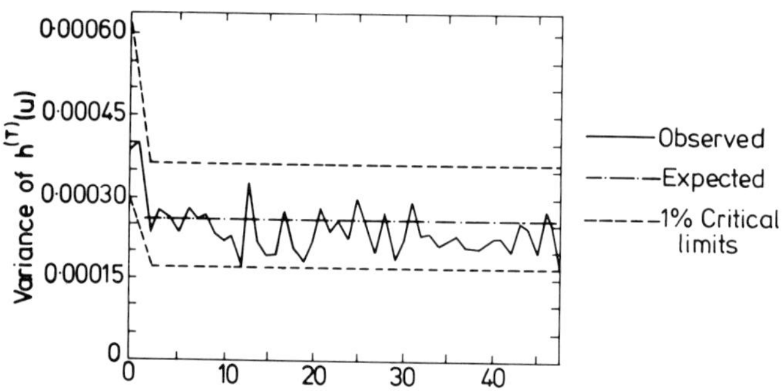

Fig. 2A. Plot of the variance of $h^{(T)}(u)$ for System I [$T = 960$, $m = 5$, $N_T = 48$].

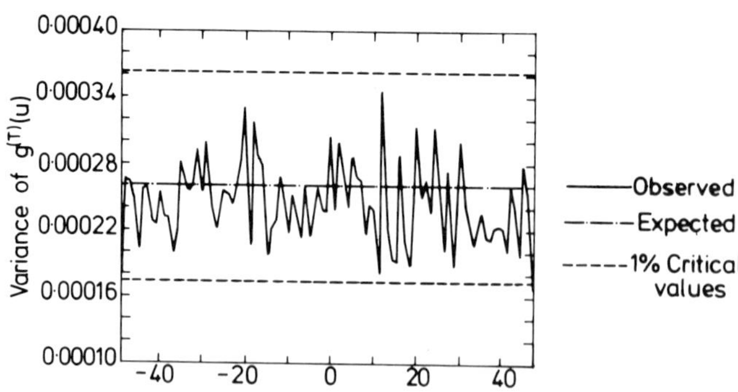

Fig. 2B. Plot of the variance of $g^{(T)}(u)$ for System I [$T = 960$, $m = 5$, $N_T = 48$].

The observed variances of $h^{(T)}(u)$ and $g^{(T)}(u)$ are plotted in Figs. 2A and 2B, for System I and Figs. 3A and 3B for System II, along with the corresponding 'expected' variances, which were calculated by using expressions (5.5) and (5.6), respectively. For checking whether the observed variances differ significantly from the 'expected', the 1% critical limits are also shown. These enable the χ^2 test for variances to be applied graphically.

For System I, the signal-to-noise ratio, (5.8), is constant for all λ, while, for System II, the residual process u_t is serially uncorrelated. Hence, as discussed earlier in Section 5, for System I, the $g^{(T)}(u)$ are expected to be asymptotically

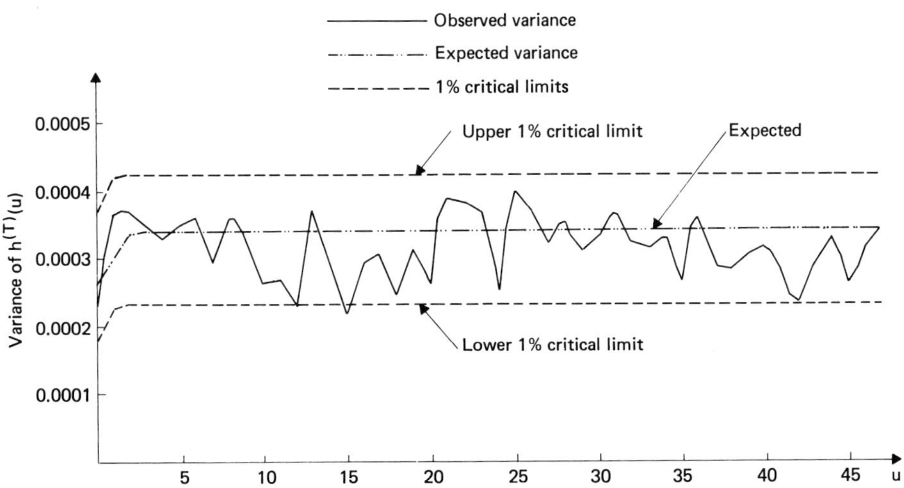

Fig. 3A. Plot of the variance of $h^{(T)}(u)$ for System II [$T = 960$, $m = 5$, $N_T = 48$].

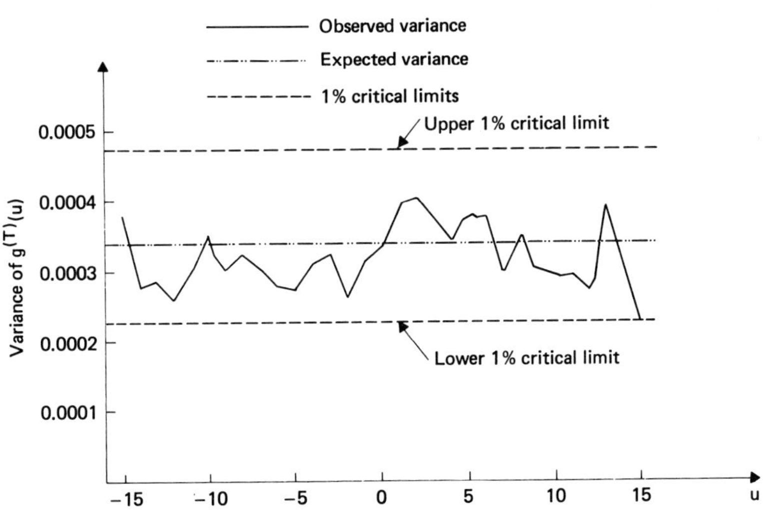

Fig. 3B. Plot of the variance of $g^{(T)}(u)$ for System I [$T = 960$, $M = 192$, $N = 48$].

more efficient than the $h^{(T)}(u)$, but conversely for System II. This asymptotic comparison is seen to hold also with a finite T, since the observed variances are closely approximated by the asymptotic variances.

Acknowledgements

The authors are grateful to Professor A. M. Walker for helpful comments.

References

[1] Akaike, H. (1967). Some problems in the application of the cross-spectral methods. In: B. Harris, ed., *Advanced Seminar on Spectral Analysis of Time Series.* Wiley, New York.

[2] Akaike, H. (1974). Markovian representation of stochastic processes and its application to the analysis of autoregressive moving average processes. *Ann. Inst. Statist. Math.* **26**, 363–387.

[3] Bhansali, R. J. (1973a). Estimation of the Wiener filter. *Bull. Int. Statist. Inst.* **45**, 159–165.

[4] Bhansali, R. J. (1973b). A simulation study of the Wiener–Kolmogorov predictor. *Sankhya* **A35**, 357–376.

[5] Bhansali, R. J. (1974). Asymptotic properties of the Wiener–Kolmogorov Predictor. I. *J. R. Statist. Soc.* **B36**, 61–73.

[6] Bhansali, R. J. (1977). Asymptotic properties of the Wiener–Kolmogorov Predictor. II. *J. R. Statist. Soc.* **B39**, 66–72.

[7] Bhansali, R. J. (1978). Linear prediction by autoregressive model fitting in the time domain. *Ann. Statist.* **6**, 224–231.

[8] Box, G. E. P. and Jenkins, G. M. (1970). *Time Series Analysis: Forecasting and Control.* Holden-Day, San Francisco.

[9] Brillinger, D. R. (1975). *Time Series: Data Analysis and Theory.* Holt, Rinehart and Winston, New York.

[10] Davis, P. J. and Rabinowitz, P. (1975). *Methods of Numerical Integration.* Academic Press, New York.

[11] Dhrymes, P. J. (1971). *Distributed Lags: Problems of Estimation and Formulation.* Holden-Day, San Francisco.

[12] Doob, J. L. (1953). *Stochastic Processes.* Wiley, New York.

[13] Fishman, A. S. (1969). *Spectral Methods in Econometrics.* Harvard University Press, Cambridge, Mass.

[14] Fuller, W. A. (1976). *Statistical Time Series.* Wiley, New York.

[15] Granger, C. W. J. and Hatanaka, M. (1964). *Spectral Analysis of Economic Time Series.* Princeton University Press, Princeton, N.J.

[16] Grenander, U. and Rosenblatt, M. (1957). *Statistical Analysis of Stationary Time Series.* Wiley, New York.

[17] Hannan, E. J. (1970). *Multiple Time Series.* Wiley, New York.

[18] Jenkins, G. M. and Watts, D. G. (1968). *Spectral Analysis and Its Applications.* Holden-Day, San Francisco.

[19] Karavellas, D. D. (1980). Asymptotic Properties of the Wiener Filter. Ph.D. thesis, Liverpool University.

[20] Parzen, E. (1974). Some recent advances in time series modelling. *IEEE Trans. on Auto. Control.* **AC-19**, 723–730.

[21] Priestley, M. B. (1969). Estimation of transfer functions in closed loop stochastic systems. *Automatica* **5**, 623–632.

[22] Priestley, M. B. (1971). Fitting relationships between time series. *Bull. Int. Statist. Inst.* **44**. 295–321.

[23] Wahba, G. (1969). Estimation of the coefficients in a multidimensional distributed lag model. *Econometrica* **37**, 398–407.

[24] Whittle, P. (1963). *Prediction and Regulation by Linear Least-Squares Methods*. English University Press, London.

[25] Wiener, N. (1949). *Extrapolation, Interpolation and Smoothing of Stationary Time Series*. Wiley, New York.

D. R. Brillinger and P. R. Krishnaiah, eds., *Handbook of Statistics, Vol. 3*
© Elsevier Science Publishers B.V. (1983) 21–37

2

The Finite Fourier Transform of a Stationary Process

*David R. Brillinger**

1. Introduction

The Fourier transform has proved of substantial use in most fields of science. It has proved of special use to statisticians concerned with stationary process data or concerned with the analysis of linear time-invariant systems. The intention of this paper is to survey some of the uses and properties of Fourier transforms of stochastic processes.

In the case of an observed function $X(t)$, $0 < t < T$, the *finite Fourier transform* is defined as

$$d^{\mathrm{T}}(\lambda) = \int_0^T X(t) \exp\{-\mathrm{i}\lambda t\}\, \mathrm{d}t \tag{1.1}$$

$-\infty < \lambda < \infty$. The computation of the quantity (1.1) was suggested, for example, by Stokes (1879) to test the observed function for the period $2\pi/\lambda$. In the case of discrete data $X(t)$, $t = 0, \ldots, T-1$, Schuster (1898) proposed the computation of

$$d^{\mathrm{T}}(\lambda) = \sum_{t=0}^{T-1} X(t) \exp\{-\mathrm{i}\lambda t\} \tag{1.2}$$

whose real and imaginary parts appear in the sample correlation of the values $X(t)$ with the values $\cos \lambda t$ and $\sin \lambda t$, respectively. Schuster further suggested the computation of the periodogram

$$I^{\mathrm{T}}(\lambda) = (2\pi T)^{-1} |d^{\mathrm{T}}(\lambda)|^2 \tag{1.3}$$

in a search for hidden periodicities in the series $X(\cdot)$.

In the case that the quantity (1.2) is computed for the particular frequencies $\lambda = 2\pi s/T$, $s = 0, \ldots, T-1$, the corresponding operation is referred to as the

*This work was supported by NSF Grant PFR-790642.

discrete Fourier transform. It turns out that for various values of T in this case, the transform may be computed much more rapidly than might have been expected. If such a computation is employed, one speaks of the *fast Fourier transform*.

The Fourier transform turns up in problems of functional approximation and interpolation. The particular value $T^{-1}d^{T}(0)$ corresponds to the sample average value, so often used as a summary statistic for a set of data. The value $T^{-1}d^{T}(\lambda)$ occurs as the maximum likelihood estimate of the parameter $\frac{1}{2}\rho\,\exp\{i\phi\}$ in the model

$$X(t)=\rho\cos(\lambda t+\phi)+\varepsilon(t)\,, \tag{1.4}$$

$t=0,\ldots,T-1$, with the $\varepsilon(t)$ a sample from a zero mean normal distribution and λ of the form $2\pi s/T$, s an integer. In the case that $X(\cdot)$ is a stationary time series with mean 0 and power spectral density $f(\lambda)$, the expected value of the periodogram, (1.3), is close to $f(\lambda)$ suggesting that estimates of $f(\lambda)$ be based on the values (1.2). In seismic engineering the Fourier transforms of observed strong motion records are taken as design inputs and corresponding responses of structures evaluated prior to construction (see, for example, Vanmarcke, 1976). On other occasions responses of systems to sinusoidal input, at frequency λ, are recorded and the Fourier transform (1.2), (or (1.1)), computed in system identification. For example, Regan (1977) proposes the examination of an individual's visual system by having him view a sinusoidally oscillating light as his EEG is recorded. The EEG is subjected to Fourier analysis at the frequency of oscillation (and some of its harmonics).

The transforms (1.1) and (1.2) refer to the cases of continuous and discrete equispaced time, respectively. The Fourier transform

$$d^{T}(\lambda)=\sum_{j=1}^{M}X(\sigma_j)\exp\{-i\lambda\sigma_j\}\,, \tag{1.5}$$

with the σ_j irregularly spaced, is also of important practical use (especially in the case that $X(\cdot)\equiv 1$, when one speaks of point process data). So too is the transform

$$\begin{aligned}d^{T}(\lambda_1,\ldots,\lambda_p)=\int_0^T\cdots\int_0^T &X(t_1,\ldots,t_p)\\ &\times\exp\{-i(\lambda_1t_1+\cdots+\lambda_pt_p)\}\,dt_1\cdots dt_p\end{aligned} \tag{1.6}$$

of spatial data. (The domain of $X(\cdot)$ may even be an abstract group.)

In another form of extension, the Fourier transform may be best viewed as a functional defined on a convenient function space. The entity, X, of concern may be described by a differential equation and the equation may have solutions only in a generalized function (Schwartz distribution) sense. In a related procedure, one sets

$$d^{\mathrm{T}}(\lambda) = \sum \phi^{\mathrm{T}}(t)X(t)\exp\{-\mathrm{i}\lambda t\} \tag{1.7}$$

with ϕ^{T} vanishing if $t<0$ or $t \geq T$. The function ϕ^{T} is usually called a data window or taper here. This form of extension makes the Fourier transform more useful and a more powerful tool. Details will be provided later in the paper.

There are various classes of functions that may be viewed as subject to a harmonic analysis. These include the functions belonging to some l_p space, that is satisfying

$$\sum_{t=-\infty}^{\infty} |X(t)|^p < \infty \tag{1.8}$$

from some $p \geq 1$. Examples include $X(t) = t\exp\{-\beta t\}$ and $X(t) = \Sigma_k \alpha_k \exp\{-\beta_k t\}\cos(\gamma_k t + \delta_k)$. Such functions provide models for transients. They have Fourier representations

$$X(t) \sim \int_{-\pi}^{\pi} \exp\{\mathrm{i}\lambda t\} z(\lambda)\,\mathrm{d}\lambda\,. \tag{1.9}$$

From (1.2), for such functions

$$\mathrm{d}^T(\lambda) \sim \int_{-\pi}^{\pi} \exp\{-\mathrm{i}\alpha(T-1)/2\} \frac{\sin \alpha T/2}{\sin \alpha/2} z(\lambda - \alpha)\,\mathrm{d}\alpha\,. \tag{1.10}$$

The function $D_T(\alpha) = (\sin \alpha T/2)/(\sin \alpha/2)$ is called the Dirichlèt kernel. It integrates to 1 and has most of its mass in the interval $(-2\pi/T, 2\pi/T)$. The finite Fourier transform might be expected to be near $z(\lambda)$ in this case. Classical Fourier analysis (see, for example, Timan (1963) or Lorentz (1966)) is concerned with just how near it is. It is further concerned with how the nearness may be increased by the insertion of convergence factors, ϕ^{T}, as in (1.7). In this case

$$d^{\mathrm{T}}(\lambda) \sim \int_{-\pi}^{\pi} \Phi^{\mathrm{T}}(\alpha) z(\lambda - \alpha)\,\mathrm{d}\alpha \tag{1.11}$$

with $\Phi^{\mathrm{T}}(\alpha) = \Sigma_t\, \phi^{\mathrm{T}}(t)\exp\{-\mathrm{i}\alpha t\}$. The particular data window employed is seen to affect the result directly.

Quite a different class of functions is provided by the realizations of stationary stochastic processes. Suppose that one has functions $X(t, \omega)$ indexed by the values of a random variable ω. If $E|X(t, \omega)|^2 < \infty$, and $\mathrm{cov}\{X(t+u, \omega), X(t, \omega)\}$ does not depend on t, then one has the spectral (or Cramér) representation

$$X(t, \omega) \sim \int_{-\pi}^{\pi} \exp\{\mathrm{i}t\lambda\} Z(\mathrm{d}\lambda, \omega) \tag{1.12}$$

with Z a stochastic measure satisfying

$$\operatorname{cov}\{Z(I, \omega), Z(J, \omega)\} = F(I \cap J) \tag{1.13}$$

for intervals I and J. F is a nonnegative measure on the interval $(-\pi, \pi]$. If this measure is absolutely continuous, its density $f(\lambda)$ is called the power spectrum of the process X. (The covariance in (1.13) is defined via $\operatorname{cov}\{U, V\} = E(U - EU)\overline{(V - EV)}$.)

Suppressing the dependence on ω, one can write

$$\operatorname{cov}\{Z(\mathrm{d}\lambda), Z(\mathrm{d}\mu)\} = \delta(\lambda - \mu) f(\lambda)\, \mathrm{d}\lambda\, \mathrm{d}\mu \tag{1.14}$$

in the absolutely continuous case with $\delta(\cdot)$ the Dirac delta function. In an important class of situations, all the moments of Z exist and are given by

$$\begin{aligned} &\operatorname{cum}\{Z(\mathrm{d}\lambda_1), \ldots, Z(\mathrm{d}\lambda_{k+1})\} \\ &\quad = \delta(\lambda_1 + \cdots + \lambda_{k+1}) f(\lambda_1, \ldots, \lambda_k)\, \mathrm{d}\lambda_1 \cdots \mathrm{d}\lambda_{k+1} \end{aligned} \tag{1.15}$$

for $k = 1, 2, \ldots$. (Here cum denotes the joint cumulant of the variates involved. It is defined and discussed in Brillinger (1975a,b), for example.) An effective way of ensuring that values of the process well separated in time are only weakly dependent (the process is mixing) is to require that the $f(\lambda_1, \ldots, \lambda_k)$ be absolutely integrable. For then, one has the representation

$$\begin{aligned} &\operatorname{cum}\{X(t + u_1), \ldots, X(t + u_k), X(t)\} \\ &\quad = \int \cdots \int \exp\{\mathrm{i}(u_1\lambda_1 + \cdots + u_k\lambda_k)\} f(\lambda_1, \ldots, \lambda_k)\, \mathrm{d}\lambda_1 \cdots \mathrm{d}\lambda_k \end{aligned} \tag{1.16}$$

and the cumulant is seen to tend to 0 as any $|u_j| \to \infty$, by the Riemann–Lebesgue lemma.

The spectral representation (1.12) is useful for indicating the result of linear filtering the series X. Specifically if $A(\lambda)$, the transfer function of the filter, satisfies $\int_{-\pi}^{\pi} |A(\alpha)|^2 F(\mathrm{d}\alpha) < \infty$, then the filtered series given by

$$\int_{-\pi}^{\pi} \exp\{\mathrm{i}t\lambda\} A(\lambda) Z(\mathrm{d}\lambda)\,. \tag{1.17}$$

Similarly, the finite Fourier transform (1.7) may be written

$$\int_{-\pi}^{\pi} \Phi^{\mathrm{T}}(\lambda - \alpha) Z(\mathrm{d}\alpha) \tag{1.18}$$

showing that, when Φ^{T} is a weight function with mass concentrated near the origin, the value of (1.7) is proportional to the value of Z for a neighborhood of λ. Further, from (1.14),

$$\operatorname{var} d^{\mathrm{T}}(\lambda) \sim \int |\Phi^{\mathrm{T}}(\lambda-\alpha)|^2 f(\alpha)\,\mathrm{d}\alpha$$

$$\sim f(\lambda) \int |\Phi^{\mathrm{T}}(\alpha)|^2\,\mathrm{d}\alpha \tag{1.19}$$

if f is continuous at λ. Similarly, from (1.15),

$$\begin{aligned} &\operatorname{cum}\{d^{\mathrm{T}}(\lambda_1), \ldots, d^{\mathrm{T}}(\lambda_{k+1})\} \\ &\quad \sim \int \cdots \int \Phi^{\mathrm{T}}(\lambda_1-\alpha_1)\cdots\Phi^{\mathrm{T}}(\lambda_k-\alpha_k)\Phi^{\mathrm{T}}(\lambda_{k+1}+\alpha_1+\cdots+\alpha_k) \\ &\quad \times f(\alpha_1, \ldots, \alpha_k)\,\mathrm{d}\alpha_1\cdots\mathrm{d}\alpha_k\,. \end{aligned} \tag{1.20}$$

The results (1.19) and (1.20) are useful in practice because the moments of a random quantity provide essential information concerning its statistical distribution.

The just-indicated results refer to the case of a univariate series and discrete time. In the case of an r vector-valued series, the spectral representation (1.12) becomes

$$\boldsymbol{X}(t) \sim \int_{-\pi}^{\pi} \exp\{\mathrm{i}t\lambda\}\boldsymbol{Z}(\mathrm{d}\lambda) \tag{1.21}$$

with $\boldsymbol{Z}$ r vector-valued and such that

$$\operatorname{cov}\{\boldsymbol{Z}(\mathrm{d}\lambda), \boldsymbol{Z}(\mathrm{d}\mu)\} = \delta(\lambda-\mu)\boldsymbol{F}(\mathrm{d}\lambda)\,\mathrm{d}\mu\,, \tag{1.22}$$

$\boldsymbol{F}$ being an $r \times r$ Hermitian matrix having nonnegative definite increments. In many cases $\boldsymbol{F}(\mathrm{d}\lambda)$ will be of the form $\boldsymbol{f}(\lambda)\,\mathrm{d}\lambda$. The matrix $\boldsymbol{f}$ is called the spectral density matrix of the series. In the case that time is continuous, the representation (1.12) becomes

$$X(t) \sim \int_{-\infty}^{\infty} \exp\{\mathrm{i}t\lambda\}Z(\mathrm{d}\lambda)\,. \tag{1.23}$$

In the case of a spatial process, $X(t_1, \ldots, t_p)$, with $-\infty < t_1, \ldots, t_p < \infty$, one has

$$X(t_1, \ldots, t_p) \sim \int_{-\infty}^{\infty} \cdots \int_{-\infty}^{\infty} \exp\{\mathrm{i}(t_1\lambda_1 + \cdots + t_p\lambda_p)\}Z\,(\mathrm{d}\lambda_1, \ldots, \mathrm{d}\lambda_p) \tag{1.24}$$

with

$$\operatorname{cov}\{Z(\mathrm{d}\lambda_1, \ldots, \mathrm{d}\lambda_p),\ Z(\mathrm{d}\mu_1, \ldots, \mathrm{d}\mu_p)\}$$
$$= \delta(\lambda_1 - \mu) \cdots \delta(\lambda_p - \mu_p) F(\mathrm{d}\lambda_1, \ldots, \mathrm{d}\lambda_p)\, \mathrm{d}\mu_1 \cdots \mathrm{d}\mu_p\,. \tag{1.25}$$

All that is changed is the domain of the functions involved.

As a final example to illustrate just how unifying the concept of the spectral representation is, consider the case of a nonstationary process with stationary increments. The spectral representation now takes the form

$$X(t) \sim \int_{-\infty}^{\infty} \frac{\exp\{\mathrm{i}t\lambda\} - 1}{\mathrm{i}\lambda} Z(\mathrm{d}\lambda) \tag{1.26}$$

with

$$\operatorname{cov}\{Z(\mathrm{d}\lambda), Z(\mathrm{d}\mu)\} = \delta(\lambda - \mu) F(\mathrm{d}\lambda)\, \mathrm{d}\mu \tag{1.27}$$

as before (see Yaglom, 1958 or Brillinger, 1972). However, suppose one defines the finite Fourier transform, including a data window, as

$$d^{\mathrm{T}}(\lambda) = \int \phi^{\mathrm{T}}(t) \exp\{-\mathrm{i}\lambda t\}\, \mathrm{d}X(t)\,, \tag{1.28}$$

then, using (1.26), one sees that

$$d^{\mathrm{T}}(\lambda) \sim \int \Phi^{\mathrm{T}}(\lambda - \alpha) Z(\mathrm{d}\alpha) \tag{1.29}$$

as in (1.18), with $\Phi^{\mathrm{T}}(\alpha) = \int \exp\{\mathrm{i}\alpha t\} \phi^{\mathrm{T}}(\alpha)\, \mathrm{d}t$. By considering frequency rather than time-domain statistics, one finds oneself working with expressions of identical form. This phenomenon holds as well for generalized processes (random distributions) defined only by the values of certain linear functionals based on them. The expression (1.29) continues to describe an appropriate statistic (see Brillinger, 1974, 1981).

This paper will consider, in particular: the large sample distribution of the finite Fourier transform d^{T} for a broad variety of stationary processes, the use of d^{T} in linear models, the use of d^{T} in estimating finite-dimensional parameters and finally, some interesting related results.

2. Central limit theorems

In the case that the time series $X(t)$, $-\infty < t < \infty$ is stationary with power spectrum $f(\lambda)$ and mixing (see Appendix), one has the following large sample results concerning the finite Fourier transform (1.1).

(i) For $\lambda \neq 0$, $d^{\mathrm{T}}(\lambda)$ is asymptotically complex normal with mean 0 and variance $2\pi T f(\lambda)$. (The complex normal is defined in the Appendix.)

(ii) For $0 < \lambda_1 < \cdots < \lambda_K$, $d^{\mathrm{T}}(\lambda_1), \ldots, d^{\mathrm{T}}(\lambda_K)$ are asymptotically independent.

(iii) For $\lambda_k^T = 2\pi s_k^T/T \to \lambda$, with the s_k^T distinct nonzero integers, $d^T(\lambda_1^T), \ldots, d^T(\lambda_K^T)$ are asymptotically independent complex normals with mean 0 and variance $2\pi T f(\lambda)$.

(iv) For $\lambda \neq 0$, $V = T/K$ and

$$d^T(\lambda, k) = \int_{(k-1)V}^{kV} X(t) \exp\{-i\lambda t\} \, dt, \tag{2.1}$$

$k = 1, 2, \ldots, K$, $d^T(\lambda, 1), \ldots, d^T(\lambda, K)$ are asymptotically independent complex normals with mean 0 and variance $2\pi V f(\lambda)$.

(v) For $\lambda \neq 0$, $\phi_k^T(t) = \phi_k(t/T)$, ϕ_k bounded and integrable, and

$$d^T(\lambda, k) = \int \phi_k^T(t) X(t) \exp\{-i\lambda t\} \, dt, \tag{2.2}$$

$\{d^T(\lambda, 1), \ldots, d^T(\lambda, K)\}$ is asymptotically $N_K^C(\mathbf{0}, \boldsymbol{\Sigma}^T f(\lambda))$ with the entry in row j and column k of $\boldsymbol{\Sigma}^T$ being $2\pi \int \phi_j^T(t)\phi_k^T(t) \, dt$. (The variate (2.1) is a particular case.)

(vi) For $\lambda_k^T \to \lambda$ with $T\lambda_k^T$, $T(\lambda_j^T - \lambda_k^T) \to \infty$, with $\Phi(\lambda)$ the Fourier transform of ϕ bounded by $L(1 + |\lambda|)^{-\alpha}$, $\alpha > 2$ and

$$d^T(\lambda) = \int \phi^T(t) X(t) \exp\{-i\lambda t\} \, dt, \tag{2.3}$$

$d^T(\lambda_1^T), \ldots, d^T(\lambda_K^T)$ are asymptotically independent complex normals with mean 0 and variance $2\pi \int \phi^T(t)^2 \, dt f(\lambda)$.

In the case that the mixing condition assumed is one based on joint cumulants of the process, these results are proved directly and simply by demonstrating that the standardized joint cumulants of order greater than 2 tend to 0, i.e. to the cumulants of a normal variate. Details may be found in Brillinger (1970, 1975a,b, 1981). References to central limit theorems for finite Fourier transforms, or equivalently for narrow band-pass filtered series include: Leonov and Shiryaev (1960), Picinbono (1960), Rosenblatt (1961), Hannan (1970) and Brillinger (1974).

The results (i) to (vi) suggest that in practice it may be reasonable to approximate the distribution of the Fourier transform of a long data stretch (or a series such that well-separated values are approximately independent) by a normal distribution. Further, Fourier transforms at distinct frequencies and based on nonintersecting data stretches may be approximated by independent normals. The variance of the approximating normal is proportional to the power spectrum of the series. This suggests how a direct estimate of the power spectrum may be constructed from the Fourier transform. (Details will be given in the next section.)

For result (i) to make sense, it is necessary that $f(\lambda) \neq 0$. In the case that $f(\lambda) = 0$, it is sometimes possible to demonstrate asymptotic normality, with the asymptotic variance of an order different than $O(T)$. Specifically, suppose that $f(\alpha) = (\alpha - \lambda)^\beta g(\alpha)$ with g continuous and nonzero at λ. Then the large sample variance of $d^T(\lambda)$ may be shown to be of order $T^{1-\beta}$, and provided that the large sample cumulants are of corresponding lower orders, asymptotic normality will follow.

In the case that the series is not mixing, asymptotic normality need not occur. Rosenblatt (1981) derives a non-Gaussian limit for the transform of a process with long-range dependence.

Results (i) to (vi) were set down for the case of a scalar-valued series. Corresponding results hold in the r vector-valued case. Suppose, for example, that $\boldsymbol{X}(t) = \{X_1(t), \ldots, X_r(t)\}$ and that

$$d_j^{\mathrm{T}}(\lambda) = \int \phi_j(t) X_j(t) \exp\{-\mathrm{i}\lambda t\}\, \mathrm{d}t, \tag{2.4}$$

then $\boldsymbol{d}^{\mathrm{T}}(\lambda) = \{d_1^{\mathrm{T}}(\lambda), \ldots, d_r^{\mathrm{T}}(\lambda)\}$ may be shown to be asymptotically $N_r^{\mathrm{C}}(\mathbf{0}, \boldsymbol{\Sigma}^{\mathrm{T}})$ with the entry in row j and column k of $\boldsymbol{\Sigma}^{\mathrm{T}}$ being $f_{jk}(\lambda) 2\pi \int \phi_j^{\mathrm{T}}(t) \phi_k^{\mathrm{T}}(t)\, \mathrm{d}t$. In the case that $\phi_j = \phi$ for all j, the covariance matrix of the large sample distribution is seen to be proportional to $\boldsymbol{f}(\lambda)$, the spectral density matrix of the series.

The above results continue to hold for other types of stationary processes and their corresponding finite Fourier transforms, such as (1.2), (1.5), (1.6) and (1.7). A distinct advantage of working with the Fourier transform is that the large sample results are the same for the frequency-domain statistics, whereas time-domain statistics have drastically differing appearances and properties.

Hannan and Thomson (1971) develop asymptotic normality under a different form of limit procedure. The hope is to obtain a better approximation to the joint distribution in a case like (iii) above when the values $f(\lambda_k^{\mathrm{T}})$, $k = 1, \ldots, K$ vary noticeably. The variates $d^{\mathrm{T}}(\lambda_k^{\mathrm{T}})$ are found to be asymptotically dependent with the limiting procedure adopted.

3. Direct estimation of second-order spectra

The results indicated in the previous section may be used to construct spectral estimates and to suggest approximate distributions for the estimates constructed. Specifically, result (iii) suggests taking

$$f^{\mathrm{T}}(\lambda) = K^{-1} \sum_s (2\pi T)^{-1} \left| d^{\mathrm{T}}\left(\frac{\pi s}{T}\right)\right|^2 \tag{3.1}$$

with the summation over K distinct integers with $2\pi s/T$ near λ, as an estimate of $f(\lambda)$. Further, it suggests approximating the distribution of $f^{\mathrm{T}}(\lambda)$ by that of $f(\lambda) K^{-1} \Sigma_s |z_s|^2$ where the z_s are independent complex normals having mean 0 and variance 1. (This distribution is the same as that of $f(\lambda)\chi_{2K}^2/2K$, see Brillinger, 1975.)

Results (iv) suggests the estimate

$$f^{\mathrm{T}}(\lambda) = K^{-1} \sum_{k=1}^{K} (2\pi V)^{-1} |d^{\mathrm{T}}(\lambda, k)|^2 \tag{3.2}$$

and the approximating distribution $f(\lambda)\chi^2_{2K}/2K$ once again.

The estimate (3.2) involves averaging periodograms based on disjoint stretches of data. Of course, periodograms based on overlapping stretches might be averaged to form an estimate (the shingled estimate). Result (v), taking $\phi_j^{\mathrm{T}}(t) = 1$ for the jth stretch and $=0$ otherwise, indicates that the large sample distribution of the estimate may be approximated by $f(\lambda)K^{-1}\sum_j |z_j|^2$ where the z_j are 0 mean, variance 1, complex normals as before; however, now the z_j are correlated in a manner depending on the overlapping employed.

Result (vi) suggests the estimate

$$f^{\mathrm{T}}(\lambda) = K^{-1} \sum_{k=1}^{K} \left(2\pi \int \phi^{\mathrm{T}}(t)^2\, \mathrm{d}t\right)^{-1} |d^{\mathrm{T}}(\lambda_k^{\mathrm{T}})|^2 \tag{3.3}$$

in the case that tapering has been employed, with the approximating distribution $f(\lambda)\chi^2_{2K}/2K$ if the λ_k^{T} are sufficiently far apart. Groves and Hannan (1968) discuss the above estimates in a comparative fashion.

The above estimates are for the scalar case. For a vector-valued process, the only change necessary is for the term $|d^{\mathrm{T}}|^2$ to be replaced by the matrix $\boldsymbol{d}^{\mathrm{T}}(\lambda)^{\tau}\overline{\boldsymbol{d}^{\mathrm{T}}(\lambda)}$, with $\boldsymbol{d}^{\mathrm{T}}$ the (row) vector of finite Fourier transforms of the component processes. The large sample approximating distributions will now be complex Wisharts rather than chi-squares (see Brillinger, 1975).

Direct estimates of higher-order spectra may also be formed from the finite Fourier transform. Such estimates are considered in Brillinger and Rosenblatt (1967) and Rosenblatt (1983, this volume) for example.

4. Linear models

The finite Fourier transform is of substantial use in the analysis of random process data assumed to satisfy a linear (time-invariant) model. Suppose that the data $\{X(t), Y(t)\}$, $0 < t < T$ is available and satisfies the model

$$Y(t) = \mu + \int a(t-u)X(u)\, \mathrm{d}u + \varepsilon(t), \tag{4.1}$$

where μ and $a(\cdot)$ are unknown parameters, ε is a zero mean stationary mixing process with power spectrum $f_{\varepsilon\varepsilon}(\lambda)$ and X is fixed. Set $A(\lambda) = \int a(u) \exp\{-\mathrm{i}\lambda u\}\, \mathrm{d}u$. Taking Fourier transforms of the relationship (4.1) leads to

$$d_Y^{\mathrm{T}}\left(\frac{2\pi s}{T}\right) \doteq A(\lambda) d_X^{\mathrm{T}}\left(\frac{2\pi s}{T}\right) + d_\varepsilon^{\mathrm{T}}\left(\frac{2\pi s}{T}\right) \tag{4.2}$$

for $2\pi s/T$ near λ. From the results in Section 2, in many situations it is reasonable to approximate the distribution of several $d_\varepsilon^T(2\pi s/T)$ with $2\pi s/T$ near λ by independent complex normals with mean 0 and variance $2\pi T f_{\varepsilon\varepsilon}(\lambda)$. Rewriting expression (4.2) as

$$y_k \doteq a x_k + \varepsilon_k \tag{4.3}$$

with k indexing K distinct frequencies near λ, shows (4.2) to be (approximately) the standard linear model. The estimate

$$\hat{a} = \sum_k y_k \bar{x}_k \Big/ \sum_k |x_k|^2 \tag{4.4}$$

of $a = A(\lambda)$ is the Gauss–Markov estimate. Its distribution may be approximated by a complex normal with mean $A(\lambda)$ and variance $2\pi T f_{\varepsilon\varepsilon}(\lambda)/\Sigma_k |x_k|^2$. The error spectrum may be estimated by the residual sum of squares

$$\sum_k |y_k - \hat{a} x_k|^2/(2\pi T(K-1)) \tag{4.5}$$

and the strength of the linear relationship may be estimated by (the coherence)

$$\left|\sum_k y_k \bar{x}_k\right|^2 \Big/ \left(\sum_k |x_k|^2\right)\left(\sum_k |y_k|^2\right). \tag{4.6}$$

These results are developed in detail in Brillinger (1975) for discrete time and for both the scalar and vector cases. Asymptotic distributions are derived and approximate confidence regions are constructed. The results for the (continuous time) model (4.1) are the same.

The approximate relationship (4.2) also occurs for other sorts of processes. Suppose that $\{X(t), Y(t)\}$ denotes a bivariate point process with $X(t)$ counting the number of points of one type in the interval $(0, t]$ and $Y(t)$ counting the number of points of a second type. Then the relationship

$$\mathrm{Prob}\{\mathrm{d}Y(t) = 1 \mid X\} = \left[\mu + \int a(t-u)\,\mathrm{d}X(u)\right]\mathrm{d}t \tag{4.7}$$

may be shown to yield (4.2) in the case that the process is stationary and mixing. The models (4.7) and (4.1) look very different in the time domain; however, in the frequency domain they have similar forms and analyses.

The extension of these results to the case of vector $\boldsymbol{X}$ is immediate and analogous to multiple regression. The extension to vector $\boldsymbol{Y}$ is also immediate. Details may be found in Brillinger (1980) where various extensions and tests of hypotheses are also given.

The finite Fourier transform is also of use in examining the traditional model of multiple regression, but with the errors stationary rather than uncorrelated. Specifically, consider the model

$$Y(t) = \boldsymbol{\theta}\boldsymbol{X}(t) + \varepsilon(t), \tag{4.8}$$

$t = 0, \ldots, T-1$, with $\boldsymbol{\theta}$ an r (row) vector, with $\boldsymbol{X}(t)$ an r (column) vector and with ε a stationary series having power spectrum $f_{\varepsilon\varepsilon}(\lambda)$. Taking the finite Fourier transform leads to

$$d_Y^T\left(\frac{2\pi s}{T}\right) = \boldsymbol{\theta}\boldsymbol{d}_X^T\left(\frac{2\pi s}{T}\right) + d_\varepsilon^T\left(\frac{2\pi s}{T}\right) \tag{4.9}$$

$s = 0, \ldots, T-1$. Treating the $d_\varepsilon^T(2\pi s/T)$ as uncorrelated zero mean, variance $2\pi T f_{\varepsilon\varepsilon}(2\pi s/T)$ normal variates leads to

$$\begin{aligned}\hat{\boldsymbol{\theta}} = &\left(\sum_s d_Y^T\left(\frac{2\pi s}{T}\right)\overline{\boldsymbol{d}_X^T\left(\frac{2\pi s}{T}\right)}^\tau w\left(\frac{2\pi s}{T}\right)\right) \\ &\times\left(\sum_s \boldsymbol{d}_X^T\left(\frac{2\pi s}{T}\right)\overline{\boldsymbol{d}_X^T\left(\frac{2\pi s}{T}\right)}^\tau w\left(\frac{2\pi s}{T}\right)\right)^{-1}\end{aligned} \tag{4.10}$$

with $w(\lambda) = f_{\varepsilon\varepsilon}(\lambda)^{-1}$ as the best linear unbiased estimate of $\boldsymbol{\theta}$. Further, the distribution of (4.10) may be approximated by a normal with mean $\boldsymbol{\theta}$ and covariance matrix

$$\begin{aligned}&2\pi T^{-1}\left(\int w(\lambda)\,\mathrm{d}\boldsymbol{F}_{XX}(\lambda)\right)^{-1}\left(\int w(\lambda)^2 f_{\varepsilon\varepsilon}(\lambda)\,\mathrm{d}\boldsymbol{F}_{XX}(\lambda)\right) \\ &\quad\times\left(\int w(\lambda)\,\mathrm{d}\boldsymbol{F}_{XX}(\lambda)\right)^{-1}\end{aligned} \tag{4.11}$$

assuming that the sequence $\boldsymbol{X}$ is subject to a generalized harmonic analysis and has spectral measure $\boldsymbol{F}_{XX}$. Specific assumptions leading to this approximation as the asymptotic distribution of $\hat{\boldsymbol{\theta}}$ may be found in Hannan (1973) and Brillinger (1975). The minimum of (4.11) occurs for $w(\lambda) = f_{\varepsilon\varepsilon}(\lambda)^{-1}$ and is

$$2\pi T^{-1}\left(\int f_{\varepsilon\varepsilon}(\lambda)^{-1}\,\mathrm{d}\boldsymbol{F}_{XX}(\lambda)\right)^{-1} \tag{4.12}$$

This last expression is of use in questions of experimental design, i.e. choice of the regressor series $\boldsymbol{X}$. It shows that it is advantageous to concentrate the power of the components of $\boldsymbol{X}$ at frequencies at which the noise spectrum is smallest. It will be further advantageous to take the components of $\boldsymbol{X}$ orthogonal to each other.

5. Parametric models

The linear model (4.8) is a particular case of the following model of considerable practical importance,

$$Y(t) = S(t, \theta) + \varepsilon(t) \tag{5.1}$$

with θ a finite-dimensional parameter, with S a function of known form and with ε a stationary series having power spectrum $f_{\varepsilon\varepsilon}(\lambda)$ as before. The problem is to estimate θ given the data $Y(t)$, $t = 0, \ldots, T-1$ say. For example, Whittle (1952) considered the case of

$$S(t, \theta) = \sum_j \alpha_j \cos(\gamma_j t + \delta_j) \tag{5.2}$$

with $\theta = (\alpha_1, \gamma_1, \delta_1, \ldots, \alpha_J, \gamma_J, \delta_J)$ while Bolt and Brillinger (1979) considered the case

$$S(t, \theta) = \sum_j \alpha_j \exp\{-\beta_j t\} \cos(\gamma_j t + \delta_j) \tag{5.3}$$

with $\theta = (\alpha_1, \beta_1, \gamma_1, \delta_1, \ldots, \alpha_J, \beta_J, \gamma_J, \delta_J)$. The problem is that of nonlinear time series regression. In many cases it is convenient to address the problem by means of finite Fourier transforms.

Taking the finite Fourier transform of the relationship (5.1) leads to

$$d_Y^T\left(\frac{2\pi s}{T}\right) = d_S^T\left(\frac{2\pi s}{T}, \theta\right) + d_\varepsilon^T\left(\frac{2\pi s}{T}\right) \tag{5.4}$$

$s = 0, \ldots, T-1$. Taking the $d_\varepsilon^T(2\pi s/T)$ to be independent zero mean, variance $2\pi T f_{\varepsilon\varepsilon}(2\pi s/T)$ normal variates gives (5.4) the form of the usual nonlinear regression model, considered for example in Jennrich (1969). The least-squares estimate of θ is the value minimizing

$$\sum_{s=0}^{T-1} |Y(t) - S(t, \theta)|^2 = \sum_{s=0}^{T-1} \left| d_Y^T\left(\frac{2\pi s}{T}\right) - d_S^T\left(\frac{2\pi s}{T}, \theta\right)\right|^2 . \tag{5.5}$$

It is also convenient to consider the weighted least-squares estimate minimizing

$$\sum_{s=0}^{T-1} \left|d_Y^T\left(\frac{2\pi s}{T}\right) - d_S^T\left(\frac{2\pi s}{T}, \theta\right)\right|^2 w\left(\frac{2\pi s}{T}\right) \tag{5.6}$$

with $w(\lambda) = f_{\varepsilon\varepsilon}(\lambda)^{-1}$ for example. The asymptotic properties of this estimate may be derived and, for example, approximate confidence regions constructed for θ, by linearization. That is by reducing the model (5.4) to the model (4.9) by making a Taylor series expansion of d_S^T as a function of θ in the neighborhood of its true value θ_0. Details for the cases (5.2) and (5.3) may be found in Whittle (1952), Bolt and Brillinger (1979) and Hasan (1983, this volume). The general case is discussed in Hannan (1971) and Robinson (1972) for example. In the case of models (5.2) and (5.3), it is convenient to minimize separately the terms in the sum (5.5) that are believed to be in the neighborhood of an individual γ_j. This reduces the computations involved and allows one to treat the weights $w(2\pi s/T)$ of (5.6) as constant. One can alternatively consider a stepwise procedure involving the estimation of $f_{\varepsilon\varepsilon}$ using the estimate of θ at the previous step and then minimizing (5.6) with $w = \hat{f}_{\varepsilon\varepsilon}^{-1}$.

The asymptotic properties of the finite Fourier transform, indicated in Section 2, suggest a means of estimating the value of an unknown finite-dimensional parameter in a circumstance of quite different form. Suppose that X is a stationary process with power spectrum $f(\lambda, \theta)$ of the known function form, but with the value of θ needing to be estimated. Were the values $d^T(2\pi s/T)$, $s = 1, 2, \ldots, (T-1)/2$ independent complex normals with mean 0 and variance $2\pi Tf(2\pi s/T, \theta)$, one could set down the likelihood function

$$\prod_s f\left(\frac{2\pi s}{T}, \theta\right)^{-1} \exp\left\{-\left|d^T\left(\frac{2\pi s}{T}\right)\right|^2 \Big/ \left(2\pi Tf\left(\frac{2\pi s}{T}, \theta\right)\right)\right\} \tag{5.7}$$

and consider as an estimate of θ the value maximizing (5.7). Once the expression (5.7) has been set down, one can consider the properties of the value maximizing it, quite separately from whatever motivated one to set the expression down. This has been done. See, for example, Whittle (1954, 1961), Hannan (1970) and Dzhaparidze and Yaglom (1974). It turns out that this estimate is consistent and asymptotically normal, under regularity conditions. It proves of special use in fitting ARMA and ARMAX models (see Hannan, 1976) and in dealing with data that has been modeled in continuous time, but observed in discrete time (see Brillinger, 1973). Asymptotic properties of the estimate are discussed for the case of point process data in Brillinger (1975b).

The results of this section provide another example of situations that have substantially different appearances in the time domain, yet essentially the same form in the frequency domain.

6. Other topics

This section presents an indication of some other results that have been derived concerning finite Fourier transforms.

Results (i) to (vi) of Section 2 all relate to finite collections of Fourier transform values. There are situations in which one is interested in a collection whose number goes to ∞ with the sample size, for example, the collection $d^T(2\pi s/T)$, $s = 0, \ldots, T-1$. Freedman and Lane (1980) demonstrate that the empirical distribution of these values tends to the complex normal distribution function, in the case that $X(t)$, $t = 0, \ldots$ is a sequence of independent identically distributed random variables with finite variance. In related work, Chen and Hannan (1980) prove that the empirical distribution of the standardized values $|d^T(2\pi s/T)|^2/(2\pi T f(2\pi s/T))$, $s = 1, \ldots, (T-1)/2$ tends to the distribution of $\chi_2^2/2$ (i.e. the exponential). There are situations in which one is interested in

$$\sup_\lambda |d^T(\lambda)| \quad \text{or} \quad \sup_s \left| d^T\left(\frac{2\pi s}{T}\right)\right| . \tag{6.1}$$

In probability and almost sure bounds are given in Whittle (1959) and Brillinger (1975a) for example. The asymptotic distribution of the second statistic of (6.1) is considered in Fisher (1929) and Whittle (1954).

The results of Section 2 lead to approximating the distribution of $|d^T|^2$ by a multiple of χ_2^2. Wittwer (1978) derives an improved approximation in the case that X is Gaussian.

Physical models involving echoes have led to the computation of log $d^T(\lambda)$ in quite a number of situations (see Childers et al., 1977). This statistic is known as the complex cepstrum or kepstrum. There are, further, quite a large number of situations in which essential information is provided by the computation of the finite Fourier transform for (possibly overlapping) segments of the series and displaying it as a function of frequency and time. See, for example, Levshin et al. (1972). Complex demodulation is an effective means of carrying through these computations (see Bingham et al., 1967 and Bolt and Brillinger, 1979).

The Fourier analysis considered in this paper has been that of sine and cosine transformations. There are situations in which the symmetries of the problem are such that other transformations are relevant. Hannan (1969) indicates a number of these. Morettin (1974) and Kohn (1980) consider the case of the Walsh transform.

The computation of the Fourier transform of a data stretch is essential to its use in statistics. One general reference to problems of computation is Digital Signal Processing (1972). Computer programs became available in the 1960s allowing the computation of the discrete Fourier transform of T data points with number of multiplications proportional to $T \log T$. The Winograd–Fourier transform algorithm (see Winograd, 1978) reduces this to a number proportional to T.

In summary, the Fourier transform proves an effective tool mathematically, statistically and computationally. It is of great use in mathematics because convolution occurs so often and is greatly simplified by the Fourier transform. It is of use in statistics, in part, because its (large sample) properties are much simpler than those of corresponding time-domain quantities. It is of use in computations because fast Fourier algorithms allow the evaluation of quantities of interest more rapidly and with smaller round-off error, than proceeding by direct evaluation.

Appendix

1. The complex normal distribution

An r vector-valued variate $\boldsymbol{U}$, with complex components, is said to have the complex normal distribution with mean $\boldsymbol{\mu}$ and covariance matrix $\boldsymbol{\Sigma}$, (denoted $N_r^C(\boldsymbol{\mu}, \boldsymbol{\Sigma})$), if the variate

$$\begin{bmatrix} \operatorname{Re} \boldsymbol{U} \\ \operatorname{Im} \boldsymbol{U} \end{bmatrix}$$

is distributed as

$$N_{2r}\left(\begin{bmatrix} \operatorname{Re} \boldsymbol{\mu} \\ \operatorname{Im} \boldsymbol{\mu} \end{bmatrix}, \begin{bmatrix} \operatorname{Re} \boldsymbol{\Sigma} & -\operatorname{Im} \boldsymbol{\Sigma} \\ \operatorname{Im} \boldsymbol{\Sigma} & \operatorname{Re} \boldsymbol{\Sigma} \end{bmatrix}\right)$$

N_{2r} denoting the usual multivariate normal. In the case that $\boldsymbol{\mu} = \mathbf{0}$ and that $\boldsymbol{\Sigma}$ is nonsingular, the probability element of $\boldsymbol{U}$ is

$$\pi^{-r}(\operatorname{Det} \boldsymbol{\Sigma})^{-1} \exp\{\bar{\boldsymbol{U}}^{\tau}\boldsymbol{\Sigma}^{-1}\boldsymbol{U}\} \prod_{j=1}^{r} (\mathrm{d} \operatorname{Re} U_j)(\mathrm{d} \operatorname{Im} U_j)\,.$$

2. Mixing

A random process is said to be mixing if well-separated (in time) values are only weakly dependent (statistically). The property has been formalized in a number of ways. In the case of a continuous time series these include:

(a) With F_s^t denoting the σ algebra of events generated by the random variables $X(u)$, $s \leqslant u \leqslant t$

$$\sup_{A \in F_{-\infty}^{t},\, B \in F_{t+u}^{\infty}} |\operatorname{Prob}\{A \cap B\} - \operatorname{Prob}\{A\} \operatorname{Prob}\{B\}| \leqslant \alpha(u)$$

with $\alpha(u) \downarrow 0$ as $u \to \infty$ (see Rosenblatt, 1956).

(b) With $c(u_1, \ldots, u_k) = \operatorname{cum}\{X(t+u_1), \ldots, X(t+u_k), X(t)\}$

$$\int \cdots \int |c(u_1, \ldots, u_k)| \, \mathrm{d}u_1 \cdots \mathrm{d}u_k < \infty$$

for $k = 1, 2, \ldots$ (see Leonov, 1960 and Brillinger, 1970).

(c) With $f(\lambda_1, \ldots, \lambda_k)$ the cumulant spectrum defined by (1.15)

$$\operatorname{vrai\,sup} |f(\lambda_1, \ldots, \lambda_k)| < \infty$$

for $k = 1, 2, \ldots$ (see Shiryaev, 1960 and Brillinger, 1981).

References

Bingham, C., Godfrey, M. D. and Tukey, J. W. (1967). Modern techniques of power spectrum estimation. *IEEE Trans. Audio Electro.* **AU-15**, 56–66.

Bolt, B. A. and Brillinger, D. R. (1979). Estimation of uncertainties in eigenspectral estimates from decaying geophysical time series. *Geophys. J. R. Astr. Soc.* **59**, 593–603.

Brillinger, D. R. (1970). The frequency analysis of relations between stationary spatial series. In: R. Pyke ed., *Proc. Twelfth Bien. Sem. Canadian Math. Congr.*, 39–81. Canadian Math. Congress, Montreal.

Brillinger, D. R. (1972). The spectral analysis of stationary interval functions. In: L. M. LeCam, J. Neyman and E. L. Scott, eds., *Proc. Sixth Berk. Symp. Math. Statist. Prob.*, 483–513. University of California Press, Berkeley.

Brillinger, D. R. (1973). An empirical investigation of the Chandler wobble and two proposed excitation processes. *Bull. Internat. Statist. Inst.* **45**, 413–433.

Brillinger, D. R. (1974). Fourier analysis of stationary processes. *Proc. IEEE* **62**, 1628–1643.

Brillinger, D. R. (1975a). *Time Series: Data Analysis and Theory*. Holt, Rinehart and Winston, New York.

Brillinger, D. R. (1975b). Statistical inference for stationary point processes. In: M. L. Puri, ed., *Stochastic Processes and Related Topics*, Vol. 1, 55–99. Academic Press, New York.

Brillinger, D.R. (1980). Analysis of variance and problems under time series models. In: P. R. Krishnaiah, ed., *Handbook of Statistics*, Vol. 1, 237–278. North-Holland, Amsterdam.

Brillinger, D. R. (1981). Asymptotic normality of finite Fourier transforms of stationary generalized processes. *J. Mult. Anal.* **11**, 64–71.

Brillinger, D. R. and Rosenblatt, M. (1967). Asymptotic theory of kth order spectra. In: B. Harris, ed., *Spectral Analysis of Time Series*, 153–188. Wiley, New York.

Chen, C.-G. and Hannan, E. J. (1980). The distribution of periodogram ordinates. *J. Time Series Analysis* **1**, 73–82.

Childers, D. G., Skinner, D. P. and Kemerait, R. C. (1977). The cepstrum: a guide to processing. *Proc. IEEE* **65**, 1428–1443.

Dzhaparidze, K. O. and Yaglom, A. M. (1974). Asymptotically efficient estimation of the spectrum parameters of stationary stochastic processes. In: *Proc. Prague Symp. Asymp. Statist*, Vol. 1, 55–105. Charles University Press, Prague.

Fisher, R. A. (1929). Tests of significance in harmonic analysis. *Proc. R. Soc.*, **Ser A 125**, 54–59.

Freedman, D. and Lane, D. (1980). The empirical distribution of Fourier coefficients. *Ann. Statistics* **8**, 1244–1251.

Groves, G. W. and Hannan, E. J. (1968). Time series regression of sea level on weather. *Rev. Geophysics* **6**, 129–160.

Hannan, E. J. (1969). Fourier methods and random processes. *Bull. Internat. Statist. Inst.* **42**, 475–469.

Hannan, E. J. (1970). *Multiple Time Series*. Wiley, New York.

Hannan, E. J. (1971). Non-linear time series regression. *J. Appl. Prob.* **8**, 767–780.

Hannan, E. J. (1973). Central limit theorems for time series regression. *Z. Wahrschein. verw. Geb.* **26**, 157–170.

Hannan, E. J. (1976). The identification and parameterization of ARMAX and state space forms. *Econometrica* **44**, 713–723.

Hannan, E. J. and Thomson, P. J. (1971). Spectral inference over narrow bands. *J. Appl. Prob.* **8**, 157–169.

Hasan, T. (1983). Complex demodulation: Some theory and applications. This volume (Chapter 7).

Jennrich, R. I. (1969). Asymptotic properties of non-linear least squares estimators. *Ann. Math. Statist.* **40**, 633–643.

Kohn, R. (1980). On the spectral decomposition of stationary time series using Walsh functions I. *Adv. Appl. Prob.* **12**, 183–199.

Leonov, V. P. (1960). The use of the characteristic functional and semi-invariants in the ergodic theory of stationary processes. *Soviet Math.* **1**, 878–881.

Leonov, V. P. and Shiryaev, A. N. (1960). Some problems in the spectral theory of higher-order moments, II. *Theory. Prob. Appl.* **4**, 319–329.

Levshin, A. L., Pisarenko, V. F. and Pogrebinsky, G. A. (1972). On a frequency-time analysis of oscillations. *Ann. Geophys.* **28**, 211–218.

Lorentz, G. G. (1966). *Approximation of Functions.* Holt, Rinehart and Winston, New York.

Morettin, P. A. (1974). Walsh-function analysis of a certain class of time series. *Stoch. Proc. Appl.* **2**, 183–193.

Picinbono, B. (1960). Tendance ver le caractere gaussien par filtrage selectif. *C. R. Acad. Sci. Paris* **250**, 1174–1176.

Rabiner, L. R. and Rader, C. M. (eds.) (1972). *Digital Signal Processing.* IEEE Press, New York.

Regan, D. (1977). Fourier analysis of evoked potentials; some methods based on Fourier analysis. In: J. E. Desmedt, ed., *Visual Evoked Potentials in Man*, 110–117. Oxford Univ. Press, Oxford.

Robinson, P. M. (1972). Non-linear regression for multiple time-series. *J. Appl. Prob.* **9**, 758–768.

Rosenblatt, M. (1961). Some comments on narrow band-pass filters. *Quart. Applied Math.* **18**, 387–394.

Rosenblatt, M. (1981). Limit theorems for Fourier transforms of functionals of Gaussian sequences. *Z. Wahrschein. verw. Geb.* **55**, 123–132.

Rosenblatt, M. (1983). Cumulants and cumulant spectra. This volume (Chapter 17).

Schuster, A. (1898). On the investigation of hidden periodicities with application to a supposed 26 day period in meteorological phenomena. *Terr. Magn.* **3**, 13–41.

Shiryaev, A. N. (1960). Some problems in the spectral theory of higher-order moments. *Theory Prob. Appl.* **5**, 265–284.

Stokes, G. G. (1879). *Proc. Roy. Soc.* **122**, 303.

Timan, A. F. (1963). *Theory of Approximation of Functions of a Real Variable.* Macmillan, New York.

Vanmarche, E. H. (1976). Structural response to earthquakes. In: C. Lomnitz and E. Rosenblueth, eds., *Seismic Risk and Engineering Decisions*, 287–337. Elsevier, Amsterdam.

Whittle, P. (1952). The simultaneous estimation of a time series harmonic and covariance structure. *Trab. Estad.* **3**, 43–57.

Whittle, P. (1954). Some recent contributions to the theory of stationary processes. In: H. Wold, ed., *A study in the Analysis of Stationary Time Series*, 169–228. Almqvist, Uppsala.

Whittle, P. (1959). Sur la distribution du maximum d'un polnome trigonometrique a coefficients aleatoires. *Colloq. Internat. Centre Nat. Rech. Sci.* **87**, 173–184.

Whittle, P. (1961). Gaussian estimation in stationary time series. *Bull. Internat. Statist. Inst.* **39**, 105–130.

Winograd, S. (1978). On computing the discrete Fourier transform. *Math. Comput.* **32**, 175–199.

Wittwer, G. (1978). Uber die asymptotische Verteilung des Periodogramms stationarer Gausscher zufalliger Folgen. *Math. Operationsforsch. Statist.*, **Ser. Stat. 9**, 357–368.

Yaglom, A. M. (1958). Correlation theory of processes with stationary *n*th order increments. *Amer. Math. Soc. Trans.* **8** Providence, 87–142.

D. R. Brillinger and P. R. Krishnaiah, eds., *Handbook of Statistics, Vol. 3*
© Elsevier Science Publishers B.V. (1983) 39–72

3

Seasonal and Calendar Adjustment

William S. Cleveland

1. Introduction

1.1. Seasonal and calendar adjustment

The results of seasonal and calendar adjustment procedures are directly consumed by millions of people via radio, television, magazines, newspapers, government reports, company business reports, and scientific journals. For example, each month thousands of economic and business series are adjusted, including the important and widely reported national economic series such as unemployment, housing starts, industrial production, money supply, and price indices. Surely this makes seasonal and calendar adjustment one of our most important statistical tools.

What does adjustment mean? Many time series, particularly economic and business time series, contain variation due to the time of the year or the arrangement of the calendar. Calendar and seasonal adjustment refers to the removal of this variation so that other variation can be revealed. (Calendar adjustment frequently is called 'trading-day adjustment'.)

As an example of a series with seasonal and calendar effects, consider the number of first-class postage stamps sold at a post office window. The series might vary, in part, due to the following factors:

Month-length effects: On the average more stamps will be sold in March than in February since March has more days.

Day-of-the-week and holiday calendar effects: Sales for Sunday and national holidays are zero, and sales for other days will depend on the day of the week. The changing proportion of holidays and Sundays in a month and the changing proportion of each day of the week in a month cause calendar variation in the series.

Seasonal effects: Each month will tend to have a different value due to the time of the year. For example, more stamps are sold in December than in other months due to Christmas mailing.

If the goal in compiling the postage series is to assess changes in the demand for stamps due to business factors such as quality of service, price, the general level of the economy, and availability of competitive services then we would want to remove the variation due to month-length, calendar, and seasonal effects to enable us to better assess the business factors variation. As we shall see, adjusting for month length is easy; we just divide an aggregated monthly series (a flow) by month length. But describing and removing seasonal and calendar variation requires much more methodology.

1.2. Seasonal variation

Many time series contain seasonal variation: a periodic or nearly periodic component. One example, shown in the top panel of Fig. 1, is monthly concentrations of atmospheric CO_2 in Mauna Loa, Hawaii from January 1960 to December 1978. The acquisition, calibration, and selected properties of the data are described by Keeling et al. (1976a,b). The yearly seasonal component is caused by vegetation in the Northern Hemisphere; plants take in CO_2 during the growing season and then release it in the late summer and fall.

A second example of a seasonal series, shown in the top panel of Fig. 2, is the number of telephones installed by the Bell System each month from January 1962 to December 1978. The seasonal variation in the installations series is in large part a result of the seasonal variation in household moves. The values plotted are the logarithms of the aggregated series corrected for month length:

$$\text{Value for month } m = \log_e\left(30.4375 \frac{\text{aggregated series for month } m}{\text{number of days in month } m}\right),$$

where $30.4375 = 365.25/12$ is the average month length.

The term *seasonal* is used since in most applications of seasonal adjustment the fundamental frequency of the component is one cycle/year, but the methodology does not in principle depend on the frequency; thus we could equally well deal with a daily series with a weekly periodic component. But we will suppose in this account that the series is monthly and has a yearly seasonal component, partly because it will make terminology simpler and partly because many computer implementations of seasonal adjustment procedures are principally for this case.

1.3. Describing or removing seasonal variation: Decomposition into trend, seasonal, and irregular

With a seasonal series there is often one of two goals:

- Describe the seasonal component in order to understand its behavior
- Remove the seasonal component since its variation obscures other important variation.

(Of course, in some applications we want to do both; that is, describe the seasonal component for understanding and then remove it to see the behavior of other components.) For example, we want to study the seasonal component

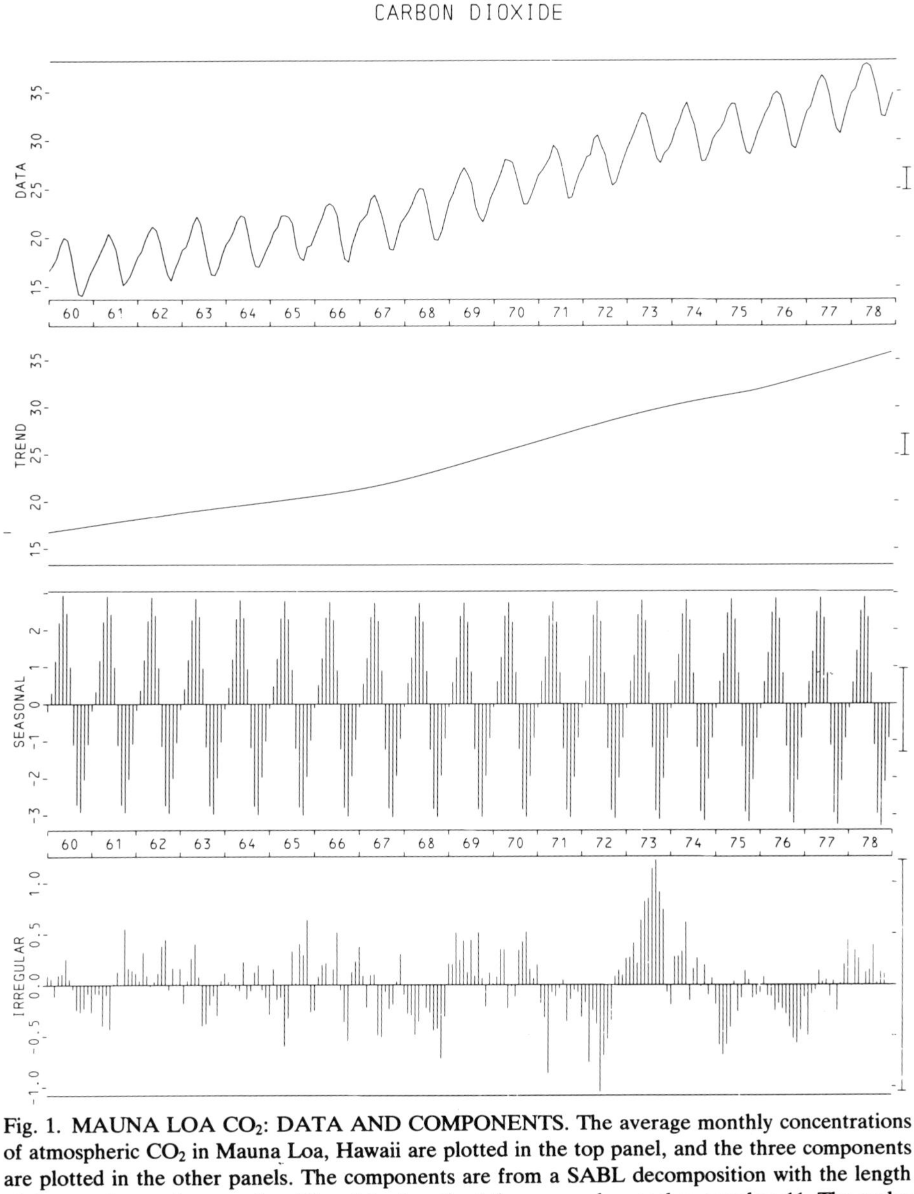

Fig. 1. MAUNA LOA CO_2: DATA AND COMPONENTS. The average monthly concentrations of atmospheric CO_2 in Mauna Loa, Hawaii are plotted in the top panel, and the three components are plotted in the other panels. The components are from a SABL decomposition with the length of the trend smoother equal to 75 and the length of the seasonal smoother equal to 11. The scales of the panels are not the same. The bars at the right portray the relative scaling by representing the same amount of change in each panel. The upward trend in the concentrations is the result of emissions from the burning of fossil fuels. The seasonal component is the result of photosynthetic activity of the Earth's vegetation.

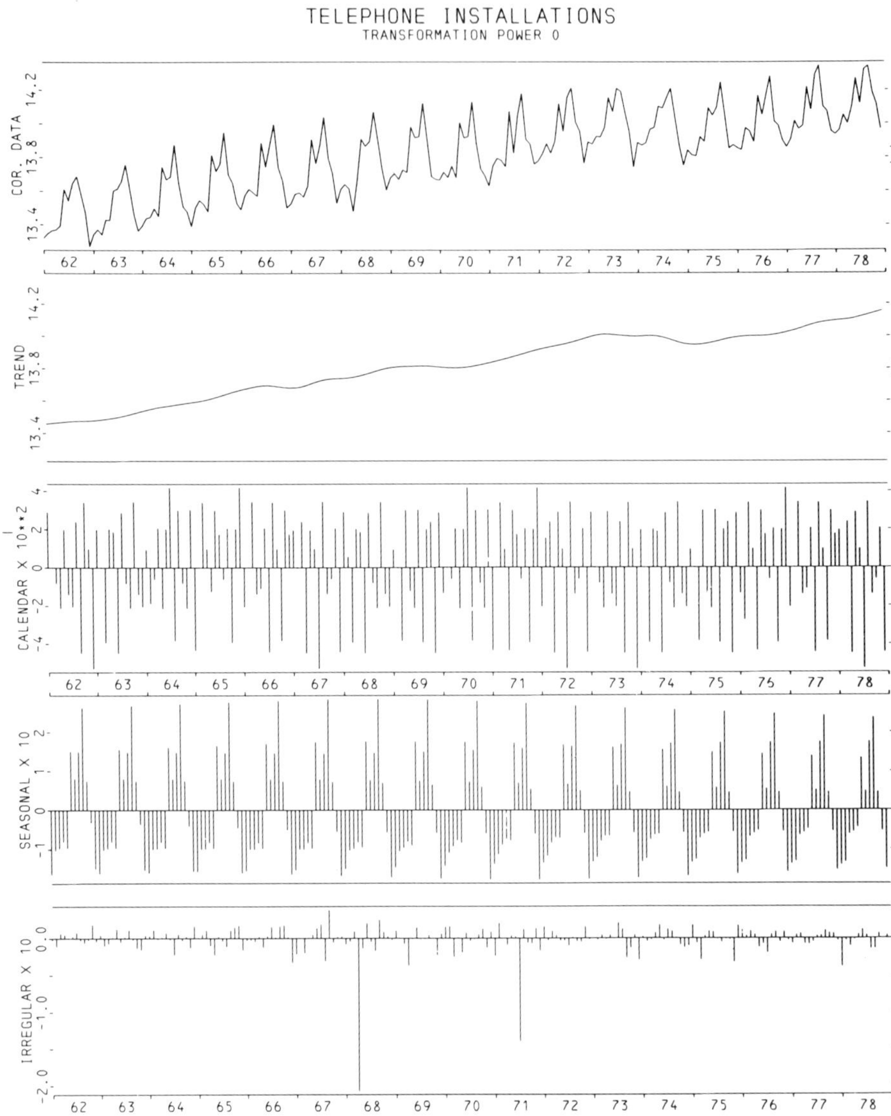

Fig. 2. TELEPHONE INSTALLATIONS: DATA AND COMPONENTS. The natural logarithms of month-length corrected Bell System telephone installations are plotted in the top panel, and the four components are plotted in the other panels. For this example the SABL decomposition was run with the length of the trend smoother equal to 15 and the length of the seasonal smoother equal to 11. The seasonal and calendar components account for a substantial amount of variation in the series. The irregular component reveals two outliers, one in 1968 and one in 1971, both of which are the result of strikes.

of CO_2 in order to make inferences about the mechanism causing the seasonality. However, for the telephone installations series the main goal is to remove the seasonal variation in order to get a sense of the overall growth in installations.

The second of the two goals, removing the seasonal component, is referred to as *seasonal adjustment*. But the methodology for removing the seasonal component involves first describing it, so the methodologies for the two goals are identical. In both cases the series is decomposed into components:

- Trend component—describes the long-term change in the level of the series
- Seasonal component—describes the variation in the data that repeats itself or nearly repeats itself every 12 months; this pattern can be exactly periodic, that is, exactly repeat itself, or it can slowly evolve through time
- Irregular component—describes the remaining variation.

As we shall see shortly, a calendar component is added to the decomposition when calendar effects are present.

1.4. Decomposition of the CO_2 series

The lower three panels of Fig. 1 show a decomposition of the CO_2 series into trend, seasonal, and irregular that is additive. If $x(m)$ is the CO_2 concentration for month m, then

$$x(m) = t(m) + s(m) + i(m) . \tag{1.1}$$

The procedure used to carry out the decomposition is SABL (W. S. Cleveland, Devlin and Terpenning, 1982. N.B. There are references to W. S. Cleveland and W. P. Cleveland; initials will be used in all cases in order to differentiate). SABL will be described, together with other procedures, in Section 2. The increase in the level of the CO_2 series, which is described by the trend component, has received great publicity because of predictions that further increases in CO_2 may have the potential to produce changes in global climate (Hansen et al., 1981; Kukla and Gavin, 1981).

A careful look also shows that the seasonal component of the CO_2 series reveals a consistent pattern through time (W. S. Cleveland, Freeny and Graedel, 1982). Fig. 3 is a monthly subseries plot of the seasonal component. First the January values of the seasonal are plotted for successive years, then the February values, and so forth. For each monthly subseries the midmean of the values (the average of all values between the quartiles) is portrayed by a horizontal line. The individual values of the subseries are portrayed by vertical lines emanating from the midmean line. The seasonal subseries plot allows an assessment of the overall pattern of the seasonal component (as portrayed by the horizontal midmean lines) and also of the behavior of each monthly subseries. Since all of the values are on the plot we can see whether the change in any subseries is large or small compared with the overall pattern of the seasonal component.

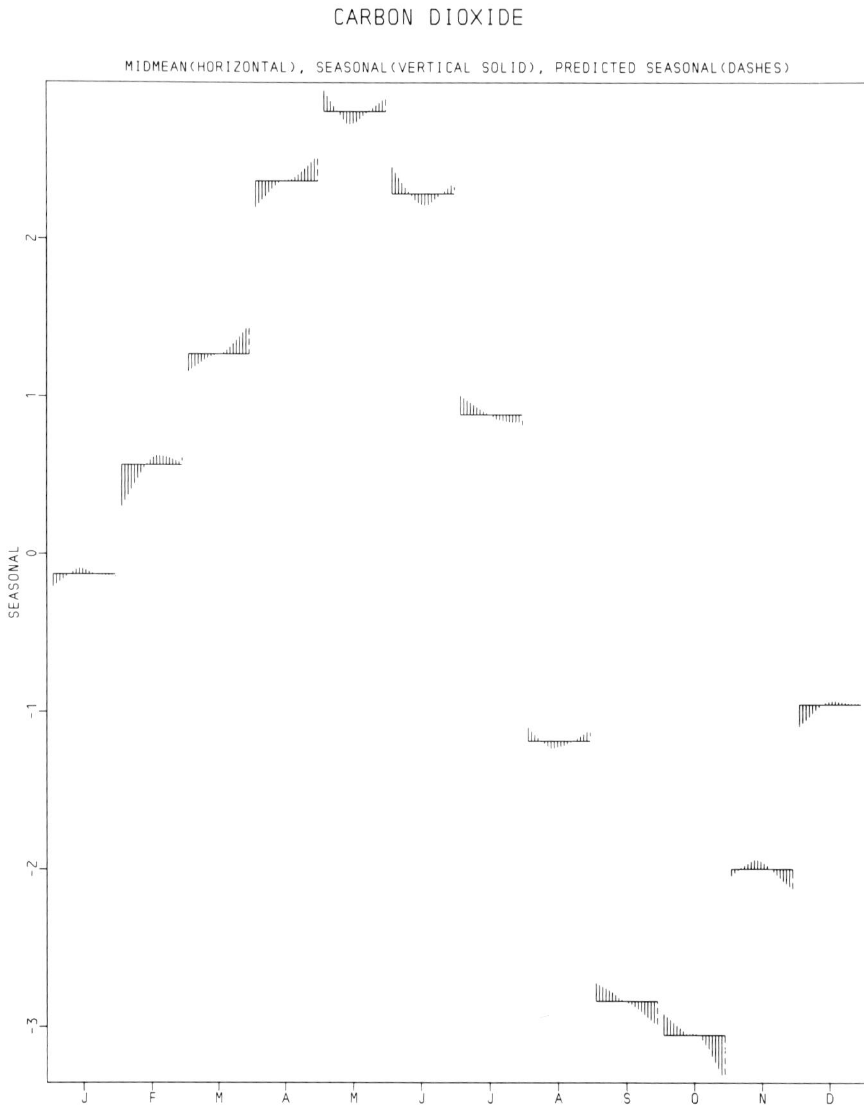

Fig. 3. MAUNA LOA CO_2: SEASONAL SUBSERIES. The seasonal component for Mauna Loa from the third panel of Fig. 1 is plotted. For each monthly subseries of the seasonal component—for example, the January values—the midmean of the values is portrayed by a horizontal line; the values of the subseries are portrayed by the vertical lines emanating from the midmean line. The overall seasonal pattern, with a May maximum and an October minimum, is attributed mainly to photosynthesis by the Earth's vegetation. The consistent changes in the seasonal subseries might be due to increasing global photosynthetic activity.

The seasonal subseries plot for Mauna Loa contains two features of substantial interest. The first is the overall seasonal pattern, with a May maximum and an October minimum. This pattern has long been recognized (e.g. Keeling et al., 1976a), and is attributed mainly to photosynthesis by the Earth's vegetation, particularly the forests (e.g. Lieth, 1963; Woodwell, 1978). However, ocean and altitude effects are also present (Fraser, Hyson and Pearman, 1981). The second feature is the patterns in the monthly subseries. The values for the months November to May—the time of year when the CO_2 yearly cycle is rising—are stable or increasing from year to year. The biggest year-to-year decreases occur during the months September and October.

A seasonal subseries plot for CO_2 measurements at the South Pole shows a very similar pattern except that, since the South Pole is within the Southern Hemisphere, the effects are shifted by about six months. The changes in the monthly series through time at Mauna Loa and the South Pole are most likely due to changes in global photosynthetic activity due to the increasing level of CO_2 concentrations, but changes in the pattern of fossil fuel use or changes in ocean temperatures cannot be ruled out, at this time, as causes (W. S. Cleveland, Freeny and Graedel, 1982).

1.5. Calendar variation

Many monthly time series that represent a total of some variable for each month contain variation as a result of a *weekly* cycle in the daily data. One example is the installations series. The number of installations on a particular day depends on the day of the week and whether the day is a holiday; on Sundays and many holidays, for example, there are no installations. Thus the monthly installations series will have variation due to the changing fraction of each day of the week in the months and variation due to the changing fraction of holidays. This is referred to as *calendar variation.*

As with seasonal variation we are often not interested in the calendar variation since it obscures the important movement in the series; for example, for the installations series we would not want to misinterpret a decrease due to a larger than average number of Sundays and holidays in a particular month as a decrease in the demand for telephones. Removal of the calendar variation from the series is referred to as *calendar adjustment.* And as with seasonal adjustment, calendar adjustment is carried out by first describing the calendar component and then subtracting it from the series.

1.6. Decomposition of the telephone installations series

Fig. 2 shows a decomposition of the telephone installations series into trend, seasonal, calendar, and irregular components, and Fig. 4 is a monthly subseries plot of the seasonal component. If $x(m)$ is the natural logarithm of the monthly installations divided by month length and multiplied by 30.4375, then

$$x(m) = t(m) + s(m) + c(m) + i(m) .$$

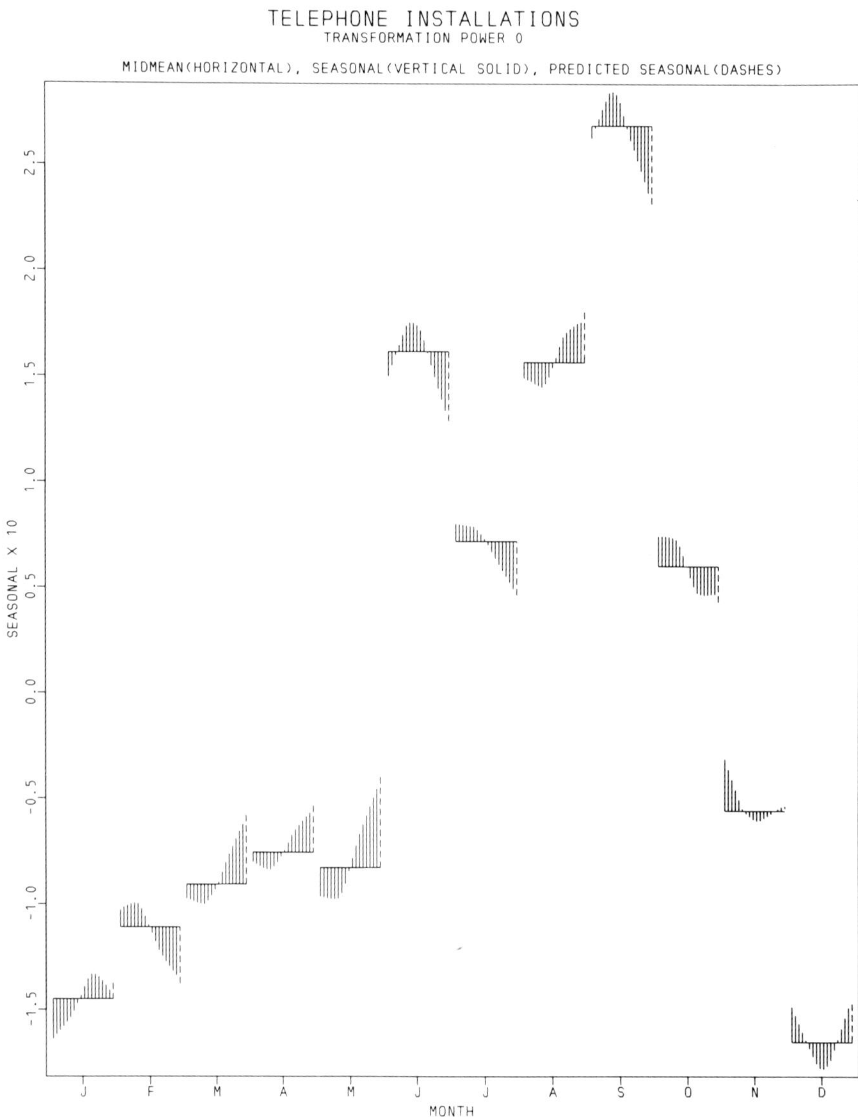

Fig. 4. TELEPHONE INSTALLATIONS: SEASONAL SUBSERIES. The seasonal component for telephone installations from the fourth panel of Fig. 2 is plotted. The dashed vertical lines are the predicted values for one year beyond the end of the series. The overall pattern of the seasonal is one in which September is the highest month; this is caused by the large number of household moves that occur in September. The monthly subseries are nearly stable in the sense that the changes in the subseries are small compared with the changes from month to month in the midmeans.

The decomposition procedure in this case is SABL (W. S. Cleveland, Devlin and Terpenning, 1982), which will be described in the next section. It is clear that the seasonal and calendar components account for a large fraction of the variation in the series.

1.7. Decomposition magnifies subtle effects

If we look very carefully at the top panel of the telephone installations series we can see two somewhat unusual points. One, toward the beginning of 1968, is a somewhat unusually low value compared with installations at the same time of year in other years. A second, in the middle of 1971, is a trough between the double peaks, lower than in other years. Both of these effects do not stand out in an obtrusive way because their magnitude is not large compared with the total variation in the series. But in the irregular component, the variation after the trend and seasonal variation are removed, the two effects appear as two very prominently low values; they were both caused by strikes, which substantially reduced the number of installations during the months in which they occurred.

For the CO_2 series, after decomposing and plotting we found a subtle effect in the seasonal—an increasing amplitude. The plot of the data in the top panel of Fig. 1 does not provide any suggestion of this effect. In fact, even knowing that the effect occurs, it would be quite difficult to measure it just from the data itself without a decomposition. Again, the reason why the decomposition has helped is that the effect is subtle compared with the overall variation in the data.

2. Methods for seasonal adjustment

2.1. X-11

The traditional seasonal adjustment package is X-11, which is widely used throughout the world. The initial development of the X-11 procedures began in the 1950s (Shiskin, 1955); they evolved through time and culminated in the current version (Shiskin, Young and Musgrave, 1967). The computer implementation of the first versions of X-11 was one of the first uses of the electronic computer to carry out complex statistical procedures. While it will be argued shortly that substantial improvements in X-11 methodology are not possible, one cannot argue that X-11 has not been a reliable performer over the years. Criticisms of X-11 have come (Nerlove, 1965) and gone (Grether and Nerlove, 1970).

Part of the X-11 reliability undoubtedly stems from the fact that the designers started with the data, knew the intricacies of the data, and built the procedures to realistically face the data. Robust estimation is one example. The X-11 designers were aware that real time series in need of adjustment contain outliers and that, unless special procedures are used to deal with these outliers, they will frequently distort seasonal adjustments. Robust estimation is now a popular topic after nearly two decades of research (e.g. Huber, 1964; Andrews et al., 1972; Hampel, 1974; Mosteller and Tukey, 1977; Mallows, 1980). Today we can substantially improve on the X-11 robustness procedures, but we should remember that the X-11 designers developed and implemented automated methods of robust estimation long before the subject became popular.

X-11 carries out the decomposition into trend, seasonal, and irregular by applying a series of weighted moving averages, which we shall refer to as smoothers or filters. The smoothers are designed to accommodate series that have persistent trends such as those in the CO_2 and installations series. The characteristics of the smoothers can be varied somewhat in order to vary the smoothness of the trend and seasonal components. Young (1968) describes linear approximations to these filters.

X-11 has an additive version and a multiplicative version. In the first, the decomposition is additive as in (1.1) and the seasonal and irregular components both vary around 0. In the second, the decomposition is

$$x(m) = t(m)s(m)i(m)$$

and the seasonal and irregular components vary around 1 (or 100 depending on how the output is scaled).

The basic design of the X-11 procedures in the multiplicative version goes back to Macaulay (1931, Appendix I). The procedure suggested by Macaulay for a monthly series of call money rates is the following:

(1) Apply a 43-term weighted moving average to the data to get an estimate of the trend.

(2) Divide the result of (1) into the data.

(3) Smooth each monthly subseries of (2) by a moving trimmed mean of length 9; each trimmed mean is computed by dropping the smallest and largest values and averaging the remaining.

It is quite interesting that the result of (3), which is the seasonal component, utilizes a robust smoother—a moving trimmed mean. But it was not until the book of Tukey (1977), where moving medians are used, that robust smoothing became a widely used tool for data analysis.

The X-11 procedures achieve a degree of robustness by iteration:

(1) Initial trend and seasonal components are computed.

(2) An irregular component is computed by subtracting (in the additive version) the trend and seasonal components from the data.

(3) The data are altered on the basis of the irregular; if the irregular is very large at a particular time point, the observation at that time is replaced by a linear combination of other observations.

(4) Step (1) is repeated with the altered data.

(5) Steps (2) to (4) are repeated several times.

2.2. DeForest extension

Filtering methods for seasonal adjustment, such as X-11, apply a series of symmetric weighted moving averages, or symmetric filters. If the filter length is $2k+1$ and the values being smoothed are $v(1), \ldots, v(n)$, then the smoothed values for $m = k+1, \ldots, n-k$ are

$$\sum_{j=-k}^{k} \alpha_j v(m+j)$$

where the filter coefficients are symmetric: $\alpha_j = \alpha_{-j}$. The problem is to extend the smoothed values to the ends, that is, to compute smoothed values for $m = 1, \ldots, k$ and $m = n-k+1, \ldots, n$.

DeForest (1877) suggested a method for extending a symmetric filter to the ends of a series: forecast the series both forward and backward and apply the filter to a new series consisting of the old series together with the forecasts appended fore and aft. The forecasts need to extend k time units beyond the ends of a filter of length $2k+1$. Greville (1979) gives a thorough discussion of DeForest's method and other methods for taking a filter to the ends. For seasonal adjustment this idea has been utilized by Dagum (1978) and by Kenney and Durbin (1982). In the first reference, forecasting is carried out by fitting ARIMA (autoregressive integrated moving average) time-series models to the data; in the second reference, autoregressive models are fit with the lags selected by stepwise autoregression. The X-11 trend and seasonal filters are symmetric at time points sufficiently far from the ends and become more and more asymmetric near the ends (as they must, of course). The reason for appending forecasts and then applying X-11 is to improve the performance of the seasonal adjustments at the ends of the series. This idea is discussed further in Section 2.5.

2.3. ARIMA modeling of data and seasonal components

One approach to decomposition that recently has received much attention is to model the data and each of the three components by time-series models and then use signal extraction techniques to estimate the trend and the seasonal components. The first papers using this approach appear to have been (Couts, Grether and Nerlove, 1966) and (Grether and Nerlove, 1970), in which ARIMA models are used; this work culminated in a book (Nerlove, Grether and Carvalho, 1979), several chapters of which are devoted to such modeling. The ARIMA approach has been investigated and further developed in a more recent series of papers: (Brewer, Kagan and Perazzeli, 1975), (Box, Hillmer and Tiao, 1978), (Pierce, 1978), (Burman, 1980), (Hillmer and Tiao, 1982), and (Hillmer, Bell and Tiao, 1982). A similar approach has been taken by Abrahams and Dempster (1979) who fit FRIMA (fractional integrated moving average) models to the data.

To illustrate the methodology we will consider the work of Hillmer and Tiao

(1982). Here the data $x(m)$ are assumed to be an additive decomposition into trend, seasonal, and irregular:

$$x(m) = t(m) + s(m) + i(m)$$

where the data and all components follow Gaussian ARIMA models, but with certain constraints. For example, the seasonal component is assumed to obey an ARIMA model of the form

$$(1 + B + \cdots + B^{11})s(m) = \theta(B)b(m)$$

where B is the backward shift operator defined by $Bs(m) = s(m-1)$, $\theta(B)$ is a polynomial of degree no more than 11, and $b(m)$ is white noise. These constraints on $s(m)$ imply certain constraints on $x(m)$; if the identified model for $x(m)$ did not satisfy these constraints, the seasonal component could not then be specified in this way, but in fact the constraints on $x(m)$ are very reasonable for the data to which seasonal adjustment methods are usually applied. Having identified a model for $x(m)$ that is consistent with the seasonal model, $\theta(B)$ is derived, and $s(m)$ is estimated by using signal extraction techniques. Thus these models lead to filters for the trend and seasonal components and to sampling distributions for the components, conditional on the model being correct, of course.

2.4. *SABL: Another filtering approach*

The starting point in the development of the SABL seasonal adjustment procedures (W. S. Cleveland, Devlin and Terpenning, 1982) was methodology that already existed in statistics—methodology that was a part of the seasonal adjustment domain and methodology that was brought to this domain from other areas of statistics. The chief sources were the following:

- The X-11 seasonal adjustment procedures
- Methods of robust estimation developed during the past 15 years for domains other than seasonal adjustment
- The work of John W. Tukey on nonadditivity in two-way tables, power transformations, and smoothing
- New developments in statistical graphics.

The SABL decomposition procedure has employed many of the smoothing ideas in X-11:

- Forming successive estimates of trend and seasonal components by estimating one, subtracting it out, and then estimating the other
- Smoothing each monthly subseries separately to estimate the seasonal component
- Applying a trend component smoother to the seasonal component and subtracting the result to center the seasonal component at zero.

But the actual smoothers used in SABL are quite different from those in X-11. For example, the SABL smoothers reproduce linear effects, either in the trend component or in a monthly subseries of the seasonal component. X-11, on the other hand, underestimates the magnitude of the slope of a linear effect at the beginning and end.

SABL takes advantage of the recent revolution in computer graphics to produce eight graphical displays that are powerful tools for judging the adequacy of the adjustment of a series and for understanding the behavior of the trend, seasonal, calendar, and irregular components (W. S. Cleveland and Devlin, 1980; W. S. Cleveland and Terpenning, 1982). The importance of graphics for seasonal and calendar adjustment will be discussed in more detail in Section 4. All of the displays in this chapter were generated by the SABL package.

In X-11 there is a set of procedures that carries out a purely additive decomposition and another set of procedures that carries out a purely multiplicative decomposition. In SABL these possibilities are expanded by using power transformations just as they are used in other areas of statistics (Box and Cox, 1974; Tukey, 1957). The class of power transformations is defined by

$$x^{(p)} = \begin{cases} x^p & \text{if } p > 0\,, \\ \log_e x & \text{if } p = 0\,, \\ -x^p & \text{if } p < 0\,. \end{cases}$$

In SABL a value of p is selected and $x^{(p)}(m)$ is additively decomposed. If $p = 1$ the decomposition of the original series, $x(m)$, is purely additive and if $p = 0$ the decomposition of $x(m)$ is purely multiplicative. In effect, the use of a power transformation makes the decomposition on the transformed scale additive and thereby removes the interaction between the trend and seasonal components. The motivation for using a power transformation to remove a trend-seasonal interaction came from work on removable nonadditivity in two-way tables (Tukey, 1949).

The idea that outliers can have disastrous effects on statistical procedures unless care is taken seems to have been clearly articulated throughout the history of the development of statistical methodology. One striking statement of this for economics can be found in Kuznets (1933). Robust methods of estimation have undergone substantial development during the past 15 years through the use of statistical theory (e.g. Huber, 1964; Hampel, 1974), through extensive Monte Carlo experimentation (e.g. Andrews et al., 1972), and through applications (e.g. Mallows, 1980) to attain procedures which are nearly efficient over a wide range of conditions and which are not distorted by outliers. One high performance procedure is M-estimation (Huber, 1964). In SABL, M-estimation is tailored to the trend-seasonal-irregular decomposition and to the power transformation selection. In the decomposition, preliminary estimates are computed and then iterations are carried out in which the irregular is used to determine weights for the next step. This is analogous to the use of iterated weighted least squares (Andrews, 1971) in location estimation and regression.

2.5. *Filtering with DeForest extension and ARIMA methods are much more similar than it might seem*

ARIMA and filtering methods, provided the filter is applied to the series extended fore and aft by forecasts based on a model for the series, are not substantially different.

To see why this is so we first need to describe a fact about optimal forecasting and signal extraction, which we shall refer to collectively as projection. Suppose that

$$x(m) = t(m) + s(m) + i(m),$$

where the probabilistic mechanism for the stochastic behavior of the components is known, and that all series are Gaussian with finite variances so that mean square error projection is optimal projection. Let $x(1), \ldots, x(n)$ be a finite stretch of the data. Let $x^*(m)$, $-\infty < m < \infty$, be a series that is $x(m)$ for $m = 1, \ldots, n$, and for m not in this set, is equal to the optimal projection of $x(m)$ on $x(1), \ldots, x(n)$. Now suppose the optimal projection (extraction) of $s(v)$ on all $x(m)$, $-\infty < m < \infty$, is

$$\hat{s}_{-\infty,\infty}(v) = \sum_{j=-\infty}^{\infty} \alpha_j x(v+j)$$

then the optimal extraction of $s(v)$ from $x(1), \ldots, x(n)$ is

$$\hat{s}_{1,n}(v) = \sum_{j=-\infty}^{\infty} \alpha_j x^*(v+j). \tag{2.1}$$

Equation (2.1) is often attributed to W. P. Cleveland (1972) but it is actually a special case of a simple fact about geometry. If $\boldsymbol{S}_1$ and $\boldsymbol{S}_2$ are two subspaces with $\boldsymbol{S}_1 \subset \boldsymbol{S}_2$ and P_1 and P_2 are projection operators onto $\boldsymbol{S}_1$ and $\boldsymbol{S}_2$ then

$$P_1 y = P_1 P_2 y.$$

That is, to project on $\boldsymbol{S}_1$ we can project onto $\boldsymbol{S}_2$ first and then project the result onto $\boldsymbol{S}_1$. To see this note that

$$y = P_2 y + o_2$$

where o_2 is in the orthogonal complement of $\boldsymbol{S}_2$ and therefore in the orthogonal complement of $\boldsymbol{S}_1$. Then

$$P_1 P_2 y = P_1(y - o_2) = P_1 y - P_1 o_2 = P_1 y.$$

Equation (2.1) is now clearly true if we take $\boldsymbol{S}_1$ to be the space spanned by $x(1), \ldots, x(n)$ and $\boldsymbol{S}_2$ to be the space spanned by $x(m)$, $-\infty < m < \infty$.

This fact about signal extraction and forecasting is helpful for several reasons. First, it gives a convenient method for computing extractions based on a finite amount of data since formulas for forward and backward prediction and extraction from the infinite $x(m)$ series are typically simple. (We cannot, of course, actually use all of the α_j, but rather just a very large finite number.) Second, we can easily see that in principle (practical considerations aside) seasonal adjustment by ARIMA modeling and signal extraction is the same as seasonal adjustment by filtering and DeForest extension. (For simplicity we shall assume $x(m)$ is stationary, but the argument does not depend on this.) The seasonal filters of X-11 and SABL for values of the series not too close to the ends are symmetric in $x(m)$; let us write any such seasonal filter as

$$\sum_{j=-v}^{v} \beta_j x(m+j) \tag{2.2}$$

where $\beta_j = \beta_{-j}$. Now suppose the spectrum of $x(m)$ is $S_{xx}(f)$. Then the spectrum of the estimated seasonal is

$$\left| \sum_{j=-v}^{v} \beta_j \, \mathrm{e}^{2\pi ijf} \right|^2 S_{xx}(f) .$$

The spectrum of an optimally extracted $s(m)$ (based on $x(m)$ for $-\infty < m < \infty$) is

$$\frac{S_{ss}^2(f)}{S_{xx}(f)} ,$$

where $S_{ss}(f)$ is the spectrum of $s(m)$. Thus if seasonal adjustment is carried out using DeForest extension and applying the seasonal filter (2.2), the procedure can be thought of as optimal extraction when the seasonal is taken to be a process whose spectrum is of the form

$$\left| \sum_{j=-v}^{v} \beta_j \, \mathrm{e}^{2\pi ijf} \right| S_{xx}(f) .$$

All of this provides a powerful argument for adding DeForest extension to current filtering methods such as X-11 (Dagum, 1978; Kenney and Durbin, 1982) and SABL. It also shows that choosing between filtering and component ARIMA modeling methods is largely one of practicality. Is it easier to specify the extraction filter directly and thereby specify a model for the component, or is it easier to specify the component and derive the filter? This is likely to be a personal choice based on the user's background and experience. It should be appreciated that *in neither case will one be led to a unique answer.* Without more criteria, the choice of the seasonal must in part be subjective; and an arbitrary fixed choice does not remove the subjectivity. In Section 4 we shall discuss graphs and a soft criterion that give some assistance in judging the adequacy of the estimated seasonal component.

The author's preference is to choose the filter directly and thereby not go through the trouble of specifying a model for the seasonal and deriving the filter. This allows direct control of the amount of smoothness in the monthly subseries of the seasonal component, which can then be judged by the graphical displays described in Section 4. No implementation of the ARIMA component modeling approach yet allows such a direct control over smoothness, although that does not mean such an approach does not exist.

2.6. *In the footsteps of Whittaker and Henderson: Seasonal adjustment by criterion optimization*

In the 1920s Whittaker (1923) and Henderson (1924) independently suggested a method for smoothing a time series (or any sequence of numbers), $x(m)$. The method is to take the smoothed values to be the numbers, $t(m)$, that minimize the expression

$$\alpha \sum_m (\Delta^3 t(m))^2 + \sum_m (x(m) - t(m))^2 \tag{2.3}$$

where α is a positive parameter and Δ is the difference operator

$$\Delta t(m) = t(m) - t(m-1) .$$

The sum of squares of the third differences of $t(m)$ in (2.3) is a measure of the smoothness of $t(m)$; a smaller sum means a smoother function. The sum of squares of deviations of $t(m)$ from $x(m)$ measures how well the smoothed values fit the data. As α increases, $t(m)$ becomes smoother, so α serves, in effect, as a smoothness parameter.

This idea was first generalized to trend-seasonal-irregular decompositions by Leser (1963), simply by adding a stable seasonal component in the minimization. That is, we now find the $t(1), \ldots, t(n)$ and the $s(1), \ldots, s(n)$ that minimize

$$\alpha \sum_m (\Delta^2 t(m))^2 + \sum_m (x(m) - t(m) - s(m))^2 ,$$

where $s(m)$ is exactly periodic with period 12. Note that second differences are used rather than third differences to measure smoothness.

A method by which the Whittaker–Henderson criterion can be generalized to trend-seasonal-irregular decomposition and allow for an evolving seasonal is the following: choose $s(m)$ and $t(m)$ to minimize

$$\begin{aligned} \alpha \sum_m (\Delta^2 t(m))^2 &+ \beta \sum_m (s(m) - s(m-12))^2 \\ &+ \gamma \sum_m (s(m) + s(m-1) + \cdots + s(m-11))^2 \\ &+ \sum_m (x(m) - t(m) - s(m))^2 \end{aligned} \tag{2.4}$$

where α, β and γ are positive parameters. The first term, as before, is a measure of smoothness of the trend. The second term is a measure of smoothness of *each monthly subseries of the seasonal* and is therefore a measure of stability of the seasonal; as the measure decreases the subseries becomes smoother so that the seasonal becomes more nearly stable. The third term, which tends to keep the overall level of the seasonal from wandering too far from zero, is the sum of squares of a moving average of $s(m)$ of length 12.

Interestingly, the generalization in (2.4) of the Whittaker–Henderson method appears to have arisen independently in both Akaike (1980) and Schlicht (1981). The minimization in (2.4) can in fact be carried out relatively quickly, since it involves inverting matrices with simple structure. Akaike adds to this approach an assumption of Gaussian processes and, using what he calls a Bayesian information criterion, is able to specify values for α, β and γ. The Akaike procedure has been further developed by Kitagawa and Gersch (1982).

2.7. Other methods

Quite a few other methods of seasonal adjustment have been suggested, investigated, and put into use in certain areas (Joy and Thomas, 1928; Kuznets, 1932; Wald, 1936, Menderhausen, 1937; Lovell, 1963; Burman, 1965; Nullau et al., 1969; Hannan, Terrell and Tuckwell, 1970; Stephenson and Farr, 1972; Haan, 1974; Durbin and Murphy, 1975; Havenner and Swamy, 1981; Raveh, 1982). This is not the place for a full discussion of all of them but the interested reader can consult Kuiper (1978) for a discussion and comparison of methods in use by statistical agencies in countries in North America and Europe.

3. Calendar adjustment

Recall that calendar adjustment is appropriate for aggregated *monthly* data when there is a weekly cycle in the aggregated *daily* data, as there often is. Let $X(D)$ be the aggregated daily series for the Dth day. Then the aggregated monthly series for the mth month is

$$\sum_{m} X(D),$$

where Σ_m means the sum over all days in month m. In this section we shall hypothesize a model for $X(D)$ and then derive the properties of the aggregated monthly series. The approach is that of W. S. Cleveland and Devlin (1982).

2.1. The daily model

We shall suppose that the aggregated daily data has four additive components,

$$X(D) = T(D) + S(D) + C(D) + I(D)\,. \tag{3.1}$$

$T(D)$ is the trend component in the daily series; $S(D)$ is the seasonal component with a period of one year; $I(D)$ is the irregular component; and $C(D)$ accounts for a day-of-the-week effect (i.e. a weekly cycle) in the series:

$$C(D) = \alpha_k\,,$$

if D is the kth day of the week, where

$$\sum_{k=1}^{7} \alpha_k = 0\,.$$

Modifications of this model to account for the effect of holidays will be given later.

3.2. The monthly model

A lower case letter will be used to denote a month-length corrected value of an aggregated monthly series. Thus

$$x(m) = \frac{30.4375\, \Sigma_m\, X(D)}{\text{number of days in month } m}\,;$$

and $t(m)$, $s(m)$, $c(m)$, and $i(m)$ are similarly defined. Finally we will let $d_k(m)$ be the fraction of times the kth day of the week occurs in month m multiplied by 30.4375.

Summing both sides of (3.1), dividing each side by the number of days in month m, and multiplying by 30.4375 gives

$$x(m) = t(m) + s(m) + c(m) + i(m)\,,$$

where

$$c(m) = \sum_{k=1}^{7} \alpha_k d_k(m)\,. \tag{3.2}$$

3.3. Holidays

To take account of holidays we need to modify the model in (3.1) in the following way:

$$X(D) = T(D) + S(D) + C(D) + I(D) + H(D) \tag{3.3}$$

where $C(D)$ is now modified to be 0 if D is a holiday, where $H(D)$ is β_j when D is the day of the jth holiday of the year, and where $H(D)$ is 0 otherwise.

Model (3.3) assumes that the holiday effects are additive and that the effect of each holiday is the same from year to year and does not change, for example, if there is a changing day of the week on which the holiday occurs.

Suppose the jth holiday always occurs within the same month, then it is clear that the holiday's effect is a purely seasonal one and can therefore be included as part of the seasonal component. In the U.S. the only holiday of any consequence that changes months is Easter. Thus we will redefine the $S(D)$ and $H(D)$ components so that $S(D)$ includes all holiday effects except Easter and $H(D)$ describes Easter.

Suppose the jth holiday always occurs on the same day of the week and within the same month. Let D be a day on which the jth holiday occurs and suppose it is the kth day of the week. Then we can alter $C(D)$ by changing it from 0 to α_k, and we can alter $S(D)$ by changing it from $S(D)$ to $S(D)-\alpha_k$; eq. (3.3) still holds and our new $S(D)$ is still legitimately a seasonal component since the value of α_k subtracted is the same each year. Of course, this same change cannot be done for a holiday that occurs on different days of the week from one year to the next (such as Christmas and January 1), since then the α_k subtracted would be different from one year to the next.

When we aggregate (3.3) over months and divide by month-length we now get

$$x(m)=t(m)+s(m)+c(m)+i(m)+h(m) \tag{3.4}$$

where $c(m)$ is defined as in (3.2), but $d_k(m)/30.4375$ now equals the fraction of times the kth day of the week occurs in the mth month minus the fraction of days on which the kth day of the week is a holiday that does not occur within the same month and on the same day of the week. Let γ be the Easter effect, then $h(m)$ is γ if Easter occurs in the mth month and is 0 otherwise; thus if Easter is likely to have an effect very different from that of other Sundays, its effect can be included in the calendar model as a dummy variable with γ as its coefficient, and Easter should not be counted in $d_k(m)$ for Sunday.

In the remainder of the chapter we shall suppose that the occurrence of Easter has a negligible effect on the series and will not consider it in our modeling. The moving holidays used in computing $d_k(m)$ will be New Year's Day, Memorial Day, July 4, and Christmas.

3.4. Calendar adjustment with SABL

The model just derived for the calendar component can be fit to the data by regression techniques and trend-seasonal-irregular decomposition methodology. In this section we will use the SABL decomposition procedures and illustrate the fitting with the installations series, which has a substantial calendar effect. First, we have chosen to work with the logarithms of the month-length corrected aggregated series since the logs stabilize the seasonal oscillations; it turns out that if $x(m)$ has been transformed by a power transformation, then it is still appropriate to use the model in (3.4) for the calendar effects (W. S. Cleveland and Devlin, 1982).

In SABL the first step in fitting the calendar effects is to decompose the transformed series into trend plus seasonal plus irregular. The calendar variation is in the irregular, so the α_k can be estimated by robustly regressing the irregular on the seven calendar explanatory variables, $d_k(m)$, but not before the calendar variables are altered somewhat by a procedure called matched processing that takes account of the fact that it is processed data and not $x(m)$ that is used in the regression.

When this procedure is applied to the telephone installations data the estimates of the coefficients are

Mon	$\hat{\alpha}_1 = 0.04,$
Tue	$\hat{\alpha}_2 = 0.14,$
Wed	$\hat{\alpha}_3 = 0.13,$
Thur	$\hat{\alpha}_4 = 0.04,$
Fri	$\hat{\alpha}_5 = 0.04,$
Sat	$\hat{\alpha}_6 = -0.17,$
Sun	$\hat{\alpha}_7 = -0.22.$

The coefficients are low on Saturdays and Sundays, reflecting a small number of installations on these days; in addition, the numbers are greater on the midweek days Tuesday and Wednesday than on the other weekdays. When the calendar coefficients have been estimated, the calendar component can be estimated by

$$c(m) = \sum_{k=1}^{7} \hat{\alpha}_k d_k(m) .$$

Now the decomposition procedure is again run, but this time on $x(m) - c(m)$, to yield final trend, seasonal, and irregular components.

Fig. 2 shows the decomposition of the installations series. Note that the $x(m)$ in the top panel, which is the series that is decomposed, is the logarithms of the month-length corrected data:

$$x(m) = \log_e \left[\frac{30.4375 \text{ (installations for month } m)}{\text{number of days in month } m} \right] .$$

3.5. *Calendar adjustment with* X-11

The X-11 calendar adjustment procedure is similar to the procedure just described for SABL. The irregular from an initial decomposition is regressed on seven calendar variables similar to those in Section 3.4; in X-11 there is no division by month length in the additive version, no correction of the calendar variables for holidays, and no matched processing. The methodology in X-11 is based on the work of Young (1965).

3.6. Calendar adjustment with ARIMA modeling

Calendar adjustment can be introduced into ARIMA modeling procedures by taking $x(m) - \Sigma_{k=1}^{7} \hat{\alpha}_k d_k(m)$ to be an ARIMA model and using the techniques of Section 2.3. This approach has been followed by Bell and Hillmer (1982); Hillmer, Bell and Tiao (1982); and W. P. Cleveland and Grupe (1982).

As an example let us consider the analysis of U.S. hardware sales data from January 1967 to November 1979 carried out by Hillmer, Bell and Tiao (1982). The log series, $x(m)$, is modeled by

$$(1 - B)(1 - B^{12})(x(m) - c(m)) = (1 - \theta_1 B)(1 - \theta_{12} B^{12}) a_t \,.$$

The definition of $c(m)$ is similar to that in (3.2) but there is no correction for holidays and no division by month length, just as in X-11. The estimates of the model parameters are

	$\hat{\theta}_1 = 0.22,$
	$\hat{\theta}_{12} = 0.75,$
Mon	$\hat{\alpha}_1 = 0.001,$
Tue	$\hat{\alpha}_2 = 0.013,$
Wed	$\hat{\alpha}_3 = 0.004,$
Thur	$\hat{\alpha}_4 = 0.011,$
Fri	$\hat{\alpha}_5 = 0.001,$
Sat	$\hat{\alpha}_6 = -0.015,$
Sun	$\hat{\alpha}_7 = -0.015.$

Then the ARIMA decomposition procedure of Section 2.3 is applied to the data minus the estimated calendar component. The data and the four components from this decomposition are shown in Fig. 5.

3.7. Calendar adjustment with criterion optimization

A calendar component can be added in a straightforward way to the generalization of the Whittaker–Henderson method described in Section 2.6. Thus the changed optimization would be to find $t(m)$, $s(m)$, and $\alpha_1, \ldots, \alpha_7$ that minimize

$$\begin{aligned} &\alpha \sum_m (\Delta^2 t(m))^2 + \beta \sum_m (s(m) - s(m-12))^2 \\ &+ \gamma \sum_m (s(m) + s(m-1) + \cdots + s(m-11))^2 \\ &+ \sum_m (x(m) - t(m) - s(m) - c(m))^2 \end{aligned}$$

where $c(m)$ is defined as in (3.2). A similar approach has been investigated by Ishiguro and Akaike (1981).

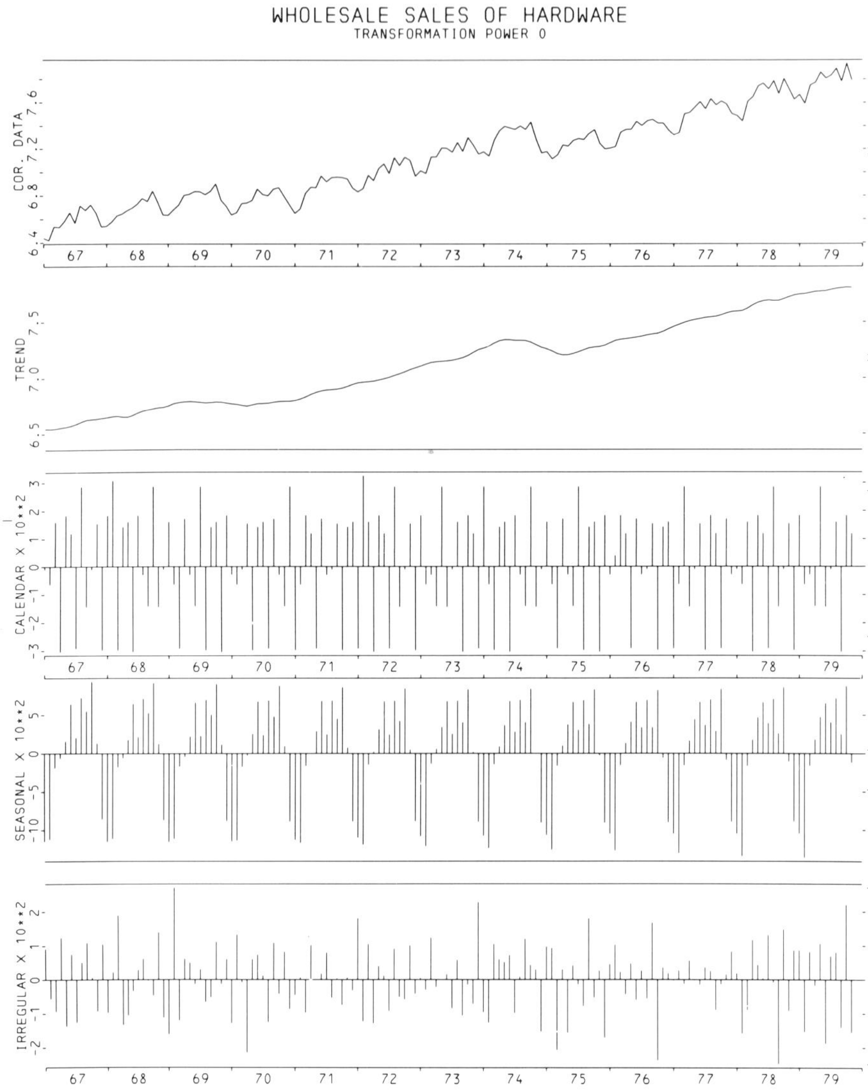

Fig. 5. HARDWARE SALES: DATA AND COMPONENTS. The top panel is the natural logarithms of monthly U.S. hardware sales (with no month-length correction). The other panels show the four components resulting from the Hillmer–Bell–Tiao ARIMA modeling.

4. Graphics for seasonal and calendar adjustment

Whatever method of decomposition and calendar estimation is used we need methods for assessing the performance of the procedures. Graphs can provide powerful tools for doing this. A full discussion of graphical methodology for calendar and seasonal adjustment is given by W. S. Cleveland and Devlin

(1980) and W. S. Cleveland and Terpenning (1982). Here we will be content to show a few examples.

We have already seen two kinds of displays: the data and components plot (Figs. 1, 2 and 5) and the seasonal subseries plot (Figs. 3 and 4). The first provides a useful first look at the decomposition. The panels have been arranged in a vertical array so that time is a common horizontal axis, which allows the viewer to easily study the variation of all components over the same intervals of time. The seasonal subseries plot allows an assessment of the overall pattern of the seasonal as portrayed by the horizontal midmean lines and also of the behavior of each monthly subseries.

Another type of graph, the seasonal-irregular plot, provides an assessment of the adequacy of the seasonal component. It is important to keep clearly in mind what variation in the series is to be described by the trend component and by the seasonal component. The trend component is a portrayal of the long-term variation in the series. Thus it should appear like a smooth curve drawn through the entire series. The seasonal component portrays the periodic variation with a period of 12 time units. Each monthly subseries of $s(m)$—for example, the January values—should describe the long-term variation in the corresponding monthly subseries of

$$x(m) - t(m) - c(m) = s(m) + i(m)$$

if there is a calendar component, or of

$$x(m) - t(m) = s(m) + i(m)$$

if there is not a calendar component. Thus each monthly subseries of $s(m)$ should appear like a smooth curve drawn through the corresponding monthly subseries of $s(m) + i(m)$.

The seasonal smoothers in SABL and X-11 have window lengths; increasing the window length of a seasonal smoother increases the smoothness of each monthly subseries of the seasonal. For decompositions using ARIMA modeling the smoothness is controlled by the form of the model chosen for the seasonal component. For decompositions using the Whittaker–Henderson generalization the smoothness is controlled by the choices of α, β and γ or by the form of prior distributions on the parameters. The critical point to appreciate in all of these methods is that the *amount of smoothness is either explicitly chosen by the user.* We need methods for assessing the adequacy of seasonal smoothing; the seasonal-irregular plot can help us do this.

The seasonal subseries should be as smooth as possible subject to the constraint of reproducing the overall long-term pattern in the seasonal-plus-irregular component. This is the soft criterion referred to in Section 2.5. One way to judge the adequacy of the smoothing is to plot, for each month, the monthly subseries of the seasonal component and of the seasonal-plus-irregular component. This has been done in Figs. 6 to 8. The seasonal-plus-irregular component is plotted using the symbol 'o' at the plotting locations; the seasonal

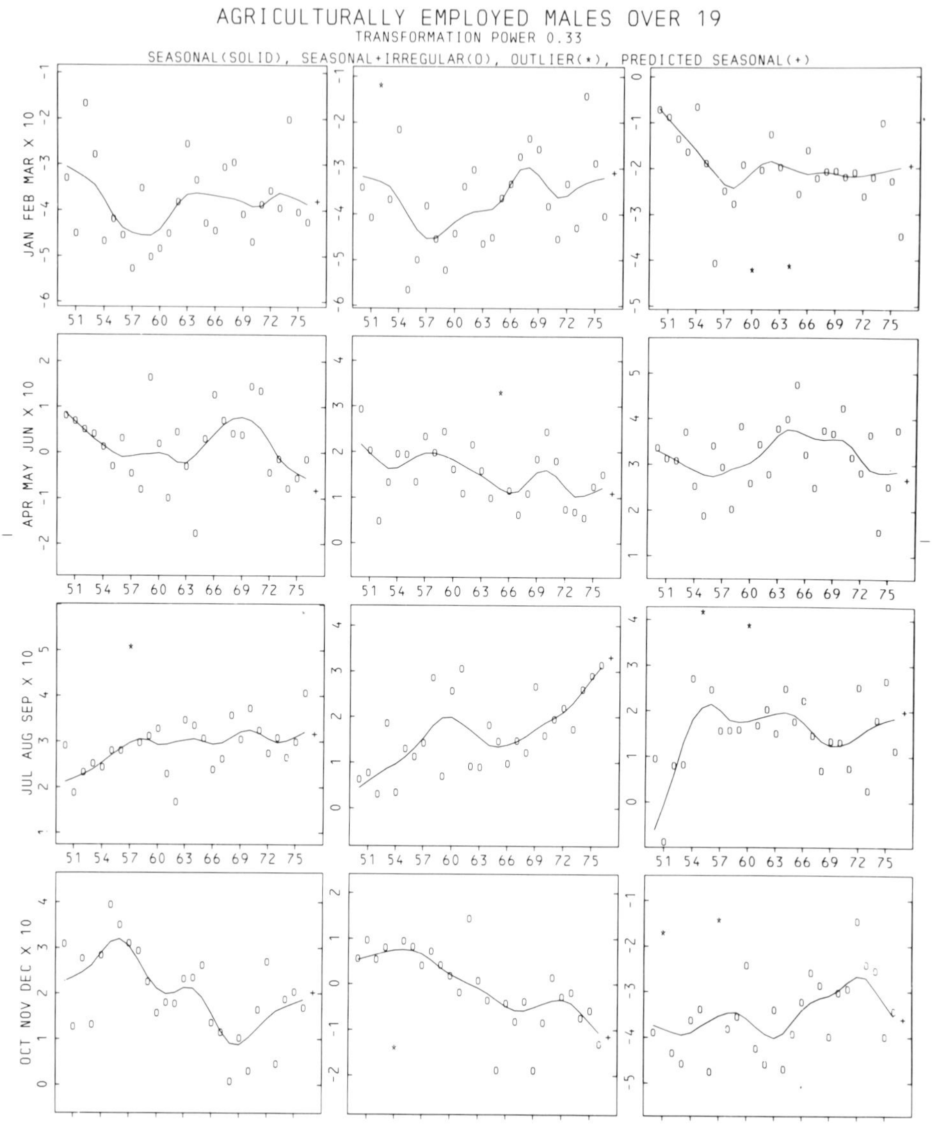

Fig. 6. U.S. AGRICULTURALLY EMPLOYED MALES OVER 19: SEASONAL-IRREGULAR. The monthly subseries of the seasonal component and the seasonal component plus the irregular component for the cube roots of the number of agriculturally employed males over 19 in the U.S. are plotted. The scales on all panels are the same. The cube roots were decomposed using SABL with the length of the trend smoother equal to 15 and the length of the seasonal smoother equal to 7. Each monthly subseries of the seasonal should represent as smoothly as possible the long-term change in the monthly subseries of the seasonal plus-irregular component. But in this example the seasonal values follow too closely the values of the seasonal-plus-irregular and thus are not sufficiently smooth.

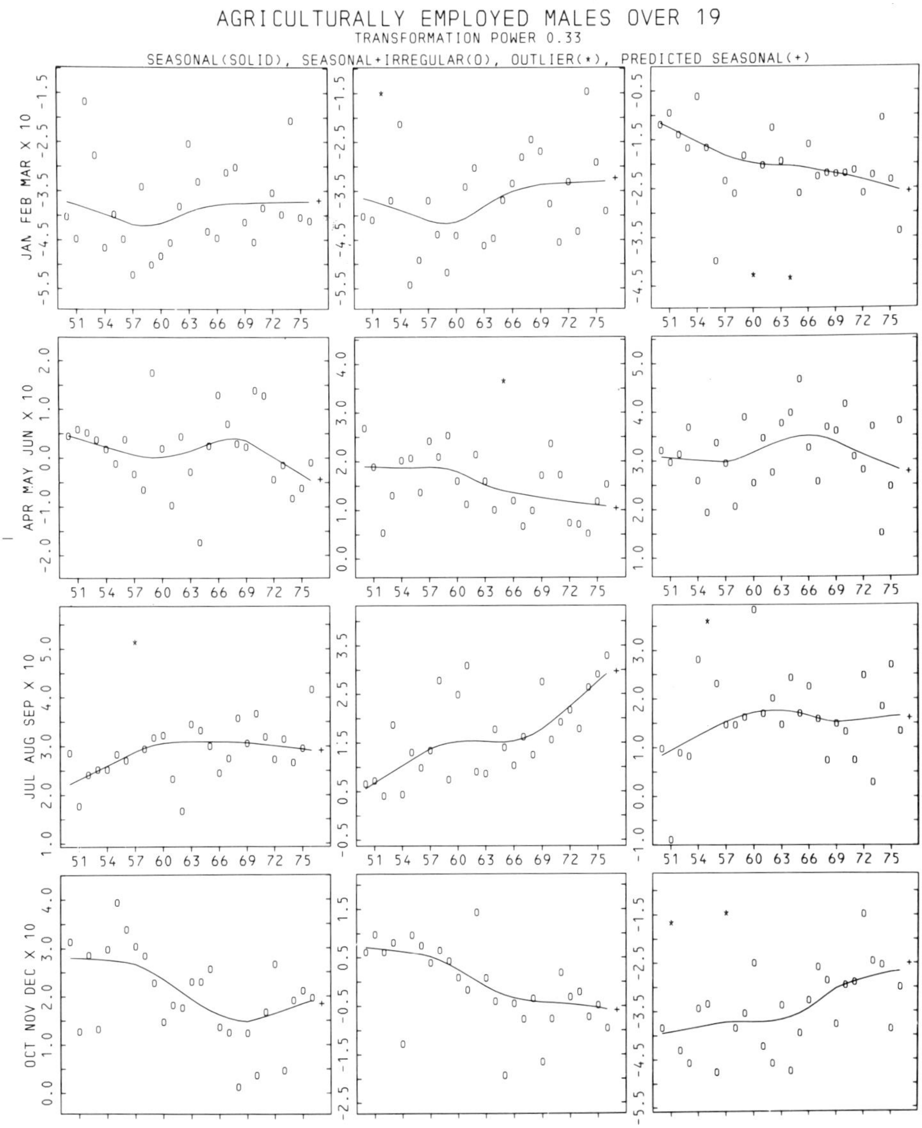

Fig. 7. U.S. AGRICULTURALLY EMPLOYED MALES OVER 19: SEASONAL-IRREGULAR. The details of this figure are the same as those for Fig. 6 except that in the decomposition the length of the seasonal smoother has been increased to 15. Now the unwanted variation in the seasonal subseries has been removed.

component is plotted using a connected plot in which successive plotting locations are connected by straight lines; the values of the seasonal component predicted 12 time units beyond the end of the data are plotted by the symbol '+'. Sometimes, as in the July panel of Fig. 8, an '*' is used as a plotting character instead of 'o'; we will explain this later.

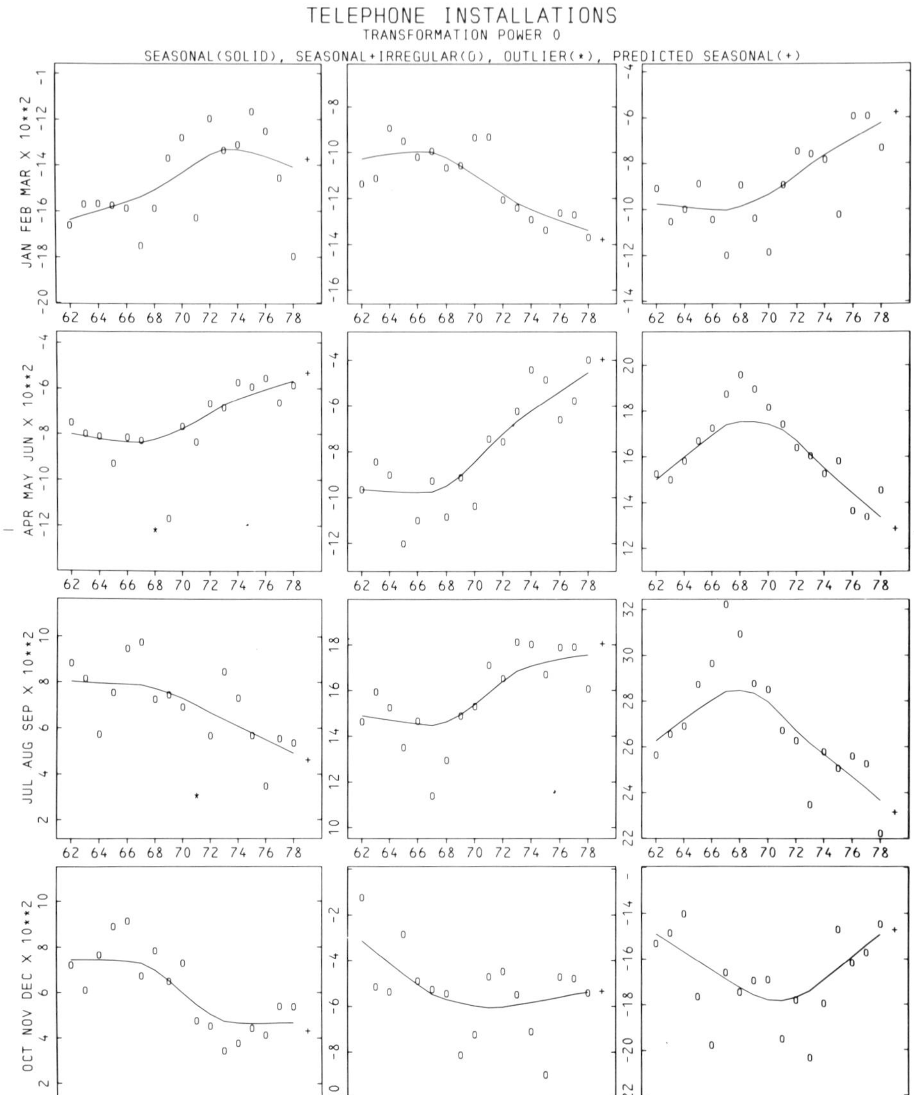

Fig. 8. TELEPHONE INSTALLATIONS: SEASONAL-IRREGULAR. The monthly subseries of the seasonal component and the seasonal component plus the irregular component for the natural logarithms of telephone installations are plotted. The components are from the SABL decomposition shown in Fig. 2. For June and September the seasonal does not adequately describe the effect in the seasonal-plus-irregular component.

To assess the appropriateness of the smoothing we do not need any information about the overall level of the values in each panel of a seasonal-irregular plot. Thus the maxima of the vertical scales are not the same for all panels nor are the minima. However, we do need to compare the variation of the values in one panel with the variation in another if we are to compare the smoothness of the seasonal component for different months. To allow this comparison the scales on the panels are the same in the sense that one cm represents the same number of units on each panel.

The values plotted in Fig. 6 are from the decomposition of the cube roots of agriculturally employed males over 19 in the U.S. from January 1950 to December 1976. The SABL procedure has been used with the length, NSEASONAL, of the seasonal smoother equal to 7. The overall impression from this figure is that the seasonal is not sufficiently smooth. The seasonal component values for many of the months follow too closely the values of the seasonal-plus-irregular component. To increase the smoothness of the seasonal component the decomposition was run again with NSEASONAL = 15. The results are shown in Fig. 7; the unwanted variation in the seasonal component has now been removed without distorting what appear to be important patterns in the seasonal-plus-irregular values.

Such distortion of patterns does occur in Fig. 8 for the telephone installations series. For example, for the month of June the peak in the seasonal-plus-irregular values is not described well by the seasonal component. One solution might be to reduce NSEASONAL, which is 11, but a better solution is to carry out the decomposition for just the last 11 years of data, 1968–1978, rather than the entire 17 years of data. This assumes, of course, that seasonally adjusted values are not needed for the years prior to 1968.

5. Spectrum analysis

5.1. Seasonal component

The spectrum of the seasonal component is, of course, concentrated at and near the seasonal frequencies

$$\frac{1}{12}, \frac{1}{6}, \frac{1}{4}, \frac{1}{3}, \frac{5}{12}, \frac{1}{2} \text{ cycles/month} .$$

If the seasonal component were perfectly stable (i.e. periodic) the spectrum would have point masses (i.e. lines) at these frequencies. But such lines are asking too much of real-world phenomena, which cannot be expected to behave like perfect clocks; this is particularly true of the kinds of series—many of them economic series—to which seasonal adjustment techniques are applied.

A realistic expectation for the seasonal component is that its spectrum will be concentrated in narrow bands around the seasonal frequencies. The more nearly stable the seasonal component is, the narrower the bands will be.

An estimate of the spectrum of the irregular provides another diagnostic tool for assessing the adequacy of the seasonal component. Since the seasonal behavior should be in the seasonal component, there should not be peaks in an estimate of the spectrum of the irregular at the seasonal frequencies; if there are peaks, seasonal behavior has leaked into the irregular component. (A check of the spectrum of the trend might also be reasonable.)

We can, however, expect troughs in the spectrum of the irregular. This has, unfortunately, led to some confusion. Nerlove (1965) criticized X-11 adjustment procedures on the basis of 'overadjustment' since there were troughs in the estimated spectra of seasonally adjusted values; but later, Grether and Nerlove (1970) pointed out that troughs were a natural consequence of any sensible adjustment procedure. All of this was reenacted in (Zellner, 1978); Granger (1978) states that the spectrum of the adjusted series should 'not have dips at seasonal frequencies' and both Sims and Tukey, who discuss Granger's paper, rebut.

One way to see why troughs are inevitable is to think about signal extraction. Suppose $x(m)$ is made up of stationary Gaussian seasonal and irregular components (for simplicity we will ignore the trend component),

$$x(m) = s(m) + i(m),$$

where s and i are independent series. Let $S_{ss}(f)$ and $S_{ii}(f)$ be the spectrum densities of the seasonal and irregular components. Consider the estimation (extraction) of $s(m)$ from $x(v)$ for $-\infty < v < \infty$. $x(m) - \hat{s}(m)$, the error of extraction, which is also equal to $\hat{i}(m)$, the estimated irregular, has spectrum density

$$\frac{S_{ii}^2(f)}{S_{ii}(f) + S_{ss}(f)}.$$

Clearly, the optimally extracted irregular will tend to have troughs in its spectrum since $S_{ss}(f)$ has peaks in its spectrum.

5.2. Calendar component

The situation for the calendar component is not quite so simple as for the seasonal component. To make any headway we shall adopt the unrealistic assumption that the α_k in (3.2) are perfectly constant through time, which will lead us to the unrealistic result that the spectrum of $c(m)$ has lines at certain calendar frequencies; but if we keep in mind that there are likely to be changes in α_k through time, perhaps even small ones, which will lead to power in bands about the calendar frequencies, we will still be well served by the following derivations, which are from (W. S. Cleveland and Devlin, 1980), despite the unrealistic assumption.

We shall think of the unaggregated series as a continuous parameter time series $X(T)$, where the units of the parameter T are days. Let T_0 be the beginning of the first month and let T_m be the time at the end of the mth month. Then the aggregated monthly series is

$$ax(m) = \int_{T_{m-1}}^{T_m} X(T)\,\mathrm{d}T,$$

for $m = 1, 2, \ldots$. Let $C(T)$ be a weekly periodicity in $X(T)$ (i.e. $C(T+7) = C(T)$) whose integral over a period of 7 days is 0, and suppose $X(T) = C(T) + R(T)$. Then $ax(m) = ac(m) + ar(m)$ where ac and ar are the aggregates of C and R, respectively. (Note that $ac(m)$ differs from $c(m)$ in (3.2) in that $ac(m)$ does not take account of holidays and is not divided by month length.)

Since $C(T)$ is a periodic function with period equal to 7 days, we can write

$$C(T) = \sum_{k=1}^{\infty} \gamma_k \cos\left(2\pi \frac{kT}{7} + \phi_k\right),$$

where γ_k is the amplitude of the cosine at frequency $k/7$ cycles/day and ϕ_k is the phase. Thus for the aggregated calendar effects

$$ac(m) = \sum_{k=1}^{\infty} \gamma_k h_k(m) \tag{5.1}$$

where

$$h_k(m) = \int_{T_{m-1}}^{T_m} \cos\left(2\pi \frac{kT}{7} + \phi_k\right) \mathrm{d}T. \tag{5.2}$$

For two reasons the contributions in (5.1) for small k are the most important ones and those for larger k have a negligible effect. The first is that the spectrum of $h_k(m)$ becomes small for large k. Table 1 shows the important calendar frequencies for $ac(m)$, which are defined to be frequencies at which the spectrum of some $h_k(m)$ is greater than 0.1. Only values of k equal to 1 or 2 appear. The second reason is an empirical result; for most weekly patterns, γ_k, which depends on the shape of the weekly pattern, will tend to be small except for small values of k.

A heuristic explanation can be given for the importance of the calendar frequency 0.348 cycles/month, which has the largest spectrum value in Table 1. Suppose the lengths of all months were equal to the average,

$$\frac{365.25}{12} \text{ days} = 30.4375 \text{ days}.$$

Table 1
Values of spectrum of $h_k(m)$ greater than 0.1

Frequency	Spectrum	k
0.220	0.151	2
0.304	0.157	2
0.348	2.649	1
0.402	0.119	1
0.432	0.473	1

Suppose a cosine with a period of 7 days is sampled every month. Then the sampled series has a frequency of

$$\frac{\text{cycles}}{7\text{ days}} = \frac{30.4375/7\text{ cycles}}{\text{month}} = 4.348\text{ cycles/month}$$

and the alias of this frequency is 0.348 cycles/month.

In practice the two important calendar frequencies are 0.348 cycles/month and 0.432 cycles/month. These are the frequencies with the two largest values in Table 1.

There are two stages in the overall analysis of calendar effects in which spectrum estimates can be used. The first stage is one in which the techniques would be used to decide if calendar effects are present and are sufficiently important to warrant modeling and adjustment. The second stage occurs after an adjustment has been carried out. Here the techniques can be applied to check the adequacy of the adjustment by checking for the presence of remaining calendar effects in the adjusted series. Thus the use of the detection procedures is analogous to the use of the usual summed, lagged product

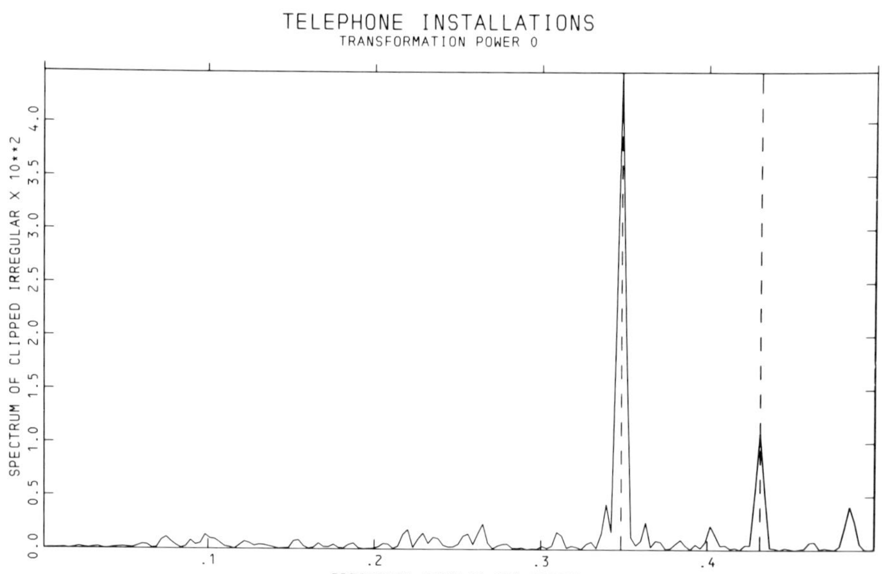

Fig. 9. TELEPHONE INSTALLATIONS: SPECTRUM OF THE CLIPPED IRREGULAR. An estimate of the spectrum of the irregular of telephone installations is plotted against frequency. The irregular component is from a SABL decomposition the same as that in Fig. 2, except that no calendar component was included in the decomposition. The dashed vertical lines show the two important calendar frequencies, 0.348 cycles/month and 0.432 cycles/month. Outliers in the irregular were first clipped to prevent the spectrum estimate from being distorted. The large peaks at the calendar frequencies show that there is a substantial calendar effect in the installations series.

estimates of autocorrelation in time-series modeling (Box and Jenkins, 1970). The autocorrelation function, which is relatively simple to estimate, is used initially to determine if autocorrelation is present. If so, a model is used to account for it, an adjusted series (residuals) is computed, and the autocorrelation function of the adjusted series is studied to determine if there is any residual autocorrelation.

The place to look for calendar effects is in the irregular component. This can be seen from Table 1; variation at the main calendar frequencies is not likely to be captured by reasonable estimates of the trend and seasonal. Fig. 9 shows an estimate of the spectrum of the irregular component (with outliers clipped) of telephone installations with no calendar component included in the decomposition; it is quite clear that a substantial calendar component is present. Fig. 10 shows an estimate of the spectrum of the irregular after a calendar component has been estimated and removed. The spectrum has been substantially reduced; the maximum value of the spectrum in Fig. 9 is about 0.04 and in Fig. 10 is about 0.001. Furthermore, it is clear that the calendar variation has been satisfactorily removed since no peaks remain at the calendar frequencies.

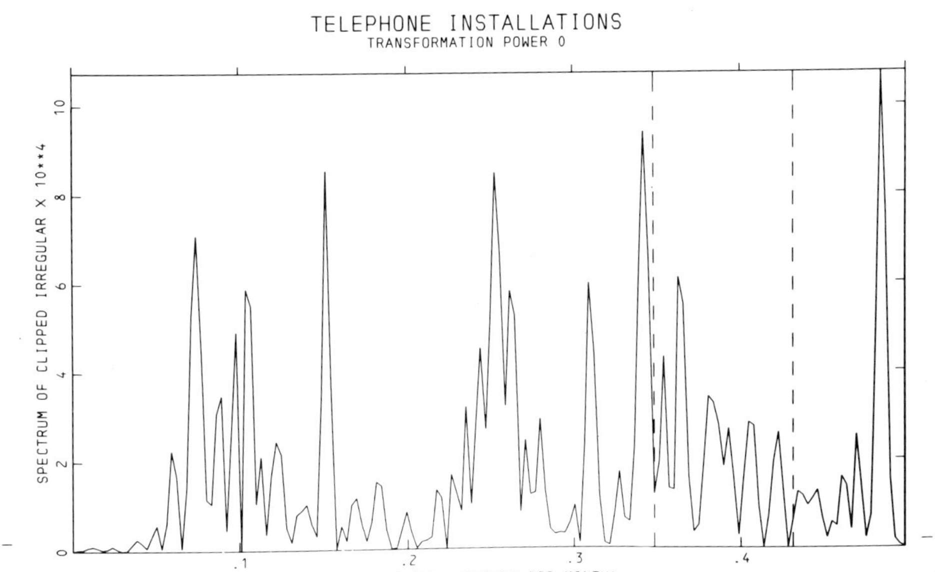

Fig. 10. TELEPHONE INSTALLATIONS: SPECTRUM OF THE CLIPPED IRREGULAR. The details of this figure are the same as those for Fig. 9 except that a calendar component was included in the decomposition. The values of the spectrum are now much smaller and the peaks at the calendar frequencies are gone, which means the estimated calendar component has adequately accounted for the calendar effects.

Acknowledgments

I am greatly indebted to David Findlay, Irma Terpenning, Allan Young, Marc Nerlov, Bill Cleveland and Henry Pollak for many useful comments on an earlier version of this paper.

References

Abrahams, D. M. and Dempster, A. P. (1979). Research on seasonal analysis. ASA/Census/NSF project report.

Akaike, H. (1980). Seasonal adjustment by Bayesian modeling. *Journal of Time Series Analysis* **1**, 1–13.

Andrews, D. F. (1971). A robust method for multiple linear regression. *Technometrics* **16**, 523–531.

Andrews, D. F., Bickel, P. J., Hampel, F. R., Huber, P. J., Rogers, W. H. and Tukey, J. W. (1972). *Robust Estimates of Location—Survey and Advances.* Princeton University Press, Princeton, N.J.

Bell, W. R. and Hillmer, S. C. (1982). Modeling time series with calendar variation. *Journal of the American Statistical Association*, to appear.

Box, G. E. P. and Cox, D. R. (1974). An analysis of transformations. *Journal of the Royal Statistical Society* **B26**, 211–243.

Box, G. E. P., Hillmer, S. C. and Tiao, G. C. (1978). Analysis and modelling of seasonal time series. In: A. Zellner, ed., *Seasonal Analysis of Time Series*, 309–334. U.S. Census Bureau, Washington, D.C.

Box, G. E. P. and Jenkins, G. M. (1970). *Time Series Analysis: Forecasting and Control.* Holden-Day, Inc., San Francisco.

Brewer, K. R. W., Kagan, P. J. and Perazzeli, P. (1975). Seasonal adjustment using Box–Jenkins models. *Proceedings of the 40th Session of the International Statistical Institute*, 133–139.

Burman, J. P. (1965). Moving seasonal adjustment of economic time series. *Journal of the Royal Statistical Society* **A128**, 534–558.

Burman, J. P. (1980). Seasonal adjustment by signal extraction. *Journal of the Royal Statistical Society* **A143**, 321–337.

Cleveland, W. P. (1972). Analysis and Forecasting of Seasonal Time Series. Ph.D. dissertation, Dept. of Statistics, University of Wisconsin.

Cleveland, W. P. and Grupe, M. R. (1982). Modeling time series when calendar effects are present. In: A. Zellner, ed., *Applied Time Series Analysis of Economic Data*, to appear. U.S. Census Bureau, Washington, D.C.

Cleveland, W. S. and Devlin, S. J. (1980). Calendar effects in monthly time series: detection by spectrum analysis and graphical methods. *Journal of the American Statistical Association* **75**, 487–496.

Cleveland, W. S. and Devlin, S. J. (1982). Calendar effects in monthly time series: modeling and adjustment. *Journal of the American Statistical Association* **77**, 520–528.

Cleveland, W. S., Devlin, S. J. and Terpenning, S. J. (1982). The SABL seasonal and calendar adjustment procedures. In: O. D. Anderson, ed., *Time Series Analysis: Theory and Practice 1*, 539–564. North-Holland, New York.

Cleveland, W. S., Freeny, A. E. and Graedel, T. E. (1982). Atmospheric CO_2 concentrations: properties and implications of the seasonal component. Submitted for publication.

Cleveland, William S. and Terpenning, Irma J. (1982). Graphical methods for seasonal adjustment. *Journal of the American Statistical Association* **77**, 52–62.

Couts, D., Grether, D.M. and Nerlove, M. (1966). Forecasting non-stationary economic time series. *Management Science* **13**, 1–21.

Dagum, E. B. (1978). Modelling, forecasting, and seasonally adjusting economic time series with the X-11 ARIMA method. *The Statistician* **27**, 203–216.

DeForest, E. L. (1877). On adjustment formulas. *The Analyst* **4**, 79–86 and 107–113.

Durbin, J. and Murphy, M. J. (1975). Seasonal adjustment based on a mixed additive-multiplicative model. *Journal of the Royal Statistical Society* **A138**, 385–410.

Fraser, P. J., Hyson, P. and Pearman, G. I. (1981). Some considerations of the global measurements of background atmospheric carbon dioxide. *WMO/UNEP/ICSU Conference on Analysis and Interpretation of* CO_2 *Data*, 179–186.

Granger, C. W. J. (1978). Seasonality: causation, interpretation and implications. In: A. Zellner, ed., *Seasonal Analysis of Economic Time Series*, 33–46. U.S. Census Bureau, Washington, D.C.

Grether, D. M. and Nerlove, M. (1970). Some properties of "optimal" seasonal adjustment. *Econometrica* **38**, 682–703.

Greville, T. N. E. (1979). Moving-weighted-average smoothing extended to the extremities of the data. *Technical Summary Report #2025*, Mathematics Research Center, University of Wisconsin, Madison, Wisconsin.

Haan, R. J. A. den (1974). *A Mechanized Method of Seasonal Adjustment.* Central Planning Bureau, The Hague.

Hampel, F. R. (1974). The influence curve and its role in robust estimation. *Journal of the American Statistical Association* **69**, 383–393.

Hannan, E. J., Terrell, R. D. and Tuckwell, N. E. (1970). The seasonal adjustment of economic time series. *International Economic Review* **11**, 24–52.

Hansen, J., Johnson, D., Lacis, A., Lebedeff, S., Lee, P., Rind, D. and Russel, G. (1981). Climate impact of increasing atmospheric carbon dioxide. *Science* **213**, 957–966.

Havenner, A. and Swamy, P. A. V. B. (1981). A random coefficient approach to seasonal adjustment of economic time series. *Journal of Econometrics* **15**, 177–209.

Henderson, R. (1924). A new method of graduation. *Transactions of the Acturial Society of America* **25**, 29–40.

Hillmer, S. C. and Tiao, G. C. (1982). An ARIMA model-based approach to seasonal adjustment. *Journal of the American Statistical Association* **77**, 63–70.

Hillmer, S. C., Bell, W. R. and Tiao, G. C. (1982). Modeling considerations in the seasonal adjustment of economic time series. In: A. Zellner, ed., *Applied Time Series Analysis of Economic Data*, to appear. U.S. Census Bureau, Washington, D.C.

Huber, P. J. (1964). Robust estimation of a location parameter. *Annals of Mathematical Statistics* **35**, 73–101.

Ishiguro, M. and Akaike, H. (1981). A bayesian approach to trading day adjustment of monthly data. In: Perryman and Anderson, eds., *Time Series Analysis*, 213–226. North-Holland, New York.

Joy, A. and Thomas, W. (1928). The use of moving averages in the measurement of seasonal variations. *Journal of the American Statistical Association* **23**, 241–252.

Keeling, C. D., Bacastow, D. B., Bainbridge, A. E., Ekdahl, Jr., C. A., Guenther, P. R., Waterman, L. S. and Chin, J. F. S. (1976a). Atmospheric carbon dioxide variations at Mauna Loa Observatory. *Tellus* **28**, 538–551.

Keeling, C. D., Adams, Jr., J. A., Ekdahl, Jr., C. A. and Guenther, P. R. (1976b). Atmospheric carbon dioxide variation at the South Pole. *Tellus* **28**, 552–564.

Kenney, P. B. and Durbin, J. (1982). Local trend estimation and seasonal adjustment of economic and social time series. *Journal of the Royal Statistical Society* **A145**, 1–41.

Kitagawa, G. and Gersch, W. (1982). A smoothing problem approach to the modeling of time series with trend and seasonality. Submitted for publication.

Kuiper, J. (1978). A survey and comparative analysis of various methods of seasonal adjustment. In: A. Zellner, ed., *Seasonal Analysis of Economic Time Series*, 59–76. U.S. Census Bureau, Washington, D.C.

Kukla, G. and Gavin, J. (1981). Summer ice and carbon dioxide. *Science* **214**, 497–503.

Kuznets, S. (1932). Seasonal pattern and seasonal amplitude: measurement of their short-time variation. *Journal of the American Statistical Association* **27**, 9–20.

Kuznets, S. (1933). On moving correlation of time series. *Journal of the American Statistical Association* **28**, 179–184.

Leser, C. E. V. (1963). Estimation of quasi-linear trend and seasonal variation. *Journal of the American Statistical Association* **58**, 1033–1043.

Lieth, H. (1963). The role of vegetation in the carbon dioxide content of the atmosphere. *Journal of Geophysical Research* **68**, 3887–3898.

Lovell, M. C. (1963). Seasonal adjustment of economic time series and multiple regression analysis. *Journal of the American Statistical Association* **58**, 993–1010.

Macaulay, F. R. (1931). *The Smoothing of Time Series.* National Bureau of Economic Research, New York.

Mallows, C. L. (1980). Robust methods—some examples of their use. *The American Statistician* **33**, 179–184.

Menderhausen, H. (1937). Methods of computing and eliminating changing seasonal fluctuations. *Econometrica* **5**, 234–262.

Mosteller, F. and Tukey, J. W. (1977). *Data Analysis and Regression.* Addison-Wesley, Reading, Mass.

Nerlove, M. (1965). A comparison of a modified 'Hannan' and the BLS seasonal adjustment filters. *Journal of the American Statistical Association* **60**, 442–491.

Nerlove, M., Grether, D. M. and Carvalho, J. L. (1979). *Analysis of Economic Time Series.* Academic Press, New York.

Nullau, B., Heiler, S., Wasch, P., Meisner, B. and Filip, D. (1969). *Das Berliner Verfahren, Ein Beitrag zur Zeitreihenanalyse.* Deutsches Institut fur Wirtschaftsforschung, Berlin.

Pierce, D. A. (1978). Seasonal adjustment when both deterministic and stochastic seasonality are present. In: A. Zellner, ed., *Seasonal Analysis of Economic Time Series*, 242–269. U.S. Census-Bureau, Washington, D.C.

Raveh, A. (1982). A nonmetric technique for deseasonalized time series integrated with Box–Jenkins approach for analysis and forecasting. Technical report, Department of Statistics, Hebrew University, Jerusalem.

Schlicht, E. (1981). A seasonal adjustment principle and a seasonal adjustment method derived from this principle. *Journal of the American Statistical Association* **76**, 374–378.

Shiskin, J. (1955). Seasonal Computations of Univac. *The American Statistician* **9**, 19–23.

Shiskin, J., Young, A. H. and Musgrave, J. C. (1967). The X-11 variant of the Census method II seasonal adjustment program. Bureau of the Census Technical Paper. U.S. Department of Commerce, Washington, D.C.

Stephenson, J. A. and Farr, H. T. (1972). Seasonal adjustment of economic data by application of the general linear statistical model. *Journal of the American Statistical Association* **67**, 37–45.

Tukey, J. W. (1949). One degree of freedom for nonadditivity. *Biometrics* **5**, 232–242.

Tukey, J. W. (1957). On the comparative anatomy of transformations. *Annals of Mathematical Statistics* **28**, 602–632.

Tukey, J. W. (1977). *Exploratory Data Analysis.* Addison-Wesley, Reading, Mass.

Wald, A. (1936). Berechnung und ausschaltung von saisonschwankungen. *Beitrage zur Konjunkturforschung.* Vienna, Austria.

Whittaker, E. T. (1923). On a new method of graduation. *Proceedings of the Edinburgh Mathematical Society* **41**, 63–75.

Woodwell, G. M. (1978). The carbon dioxide question. *Scientific American* **238**, 34–43.

Young, A. (1965). Estimating trading-day variation in monthly economic time series. Technical Paper No. 12. U.S. Census Bureau, Washington, D.C.

Young, A. H. (1968). Linear approximation to the Census and BLS seasonal adjustment methods. *Journal of the American Statistical Association* **63**, 445–471.

Zellner, A. (editor) (1978). *Seasonal Analysis of Economic Time Series.* Bureau of the Census, Washington, D.C.

D. R. Brillinger and P. R. Krishnaiah, eds., *Handbook of Statistics, Vol. 3*
© Elsevier Science Publishers B.V. (1983) 73–92

4

Optimal Inference in the Frequency Domain

Robert B. Davies

1. Introduction

We observe n consecutive observations $x_0, \ldots, x_{n-1}$ from a stationary r-dimensional Gaussian time-series. Suppose we have a model for the covariance structure of the process depending on a finite number of parameters $\theta_1, \ldots, \theta_s$, denoted collectively by θ. For example, our series might be just a one-dimensional, first-order autoregressive process with unknown autoregressive parameter θ_1 and unknown residual variance θ_2. Or we might have a model especially developed for a time-series we are investigating, involving quite a number of unknown parameters. We might want to estimate θ or we might wish to carry out test to see, for example, if some of its components could be zero.

For any given values of θ the 'likelihood' of our observations can be approximately expressed in terms of the periodogram (or Fourier transform) of the observations together with the theoretical spectrum corresponding to θ. Tests and estimates can be based on this approximate likelihood. Such tests and estimates have the advantage of often requiring less computation than those based directly on the exact likelihood. In addition, they can be based on only part of the periodogram to reduce the influence of, for example, seasonal cycles or slow fluctuations that are not catered for in the model we are trying to fit. On the other hand, the periodgram/spectrum representation gives only an approximation to the likelihood and so is applicable only when n is quite large.

The purpose of this article is to review the asymptotic optimality of various tests and estimators which are based on the approximate likelihood. In Section 2 we summarize some general theory on asymptotic inference in order to put the various techniques into context. In particular, we look at some work of LeCam and see how this fits in with the more traditional likelihood theory. In Section 3 we apply this theory to the time-series problem. This section is based largely on the work of Whittle (1953), Davies (1973), and Dunsmuir and Hannan (1976). Finally, in Section 4 we give a brief summary and example.

Related topics not considered in this paper include regression analysis with the residual error being from a stationary time-series—see Brillinger (1980) and Hannan, Dunsmuir and Deistler (1980), and the handling of missing observations—see Dunsmuir and Robinson (1981).

We reiterate that we will be assuming that our observations come from a Gaussian (normal) process. Most studies in this area show that the derivation of the asymptotic distributions of the estimators does not require the full Gaussian condition. However, in the non-Gaussian case there may be more information in the higher-order cumulants or periodograms and so the methods described here may not be optimal. At the present time, discussion of optimal inference must be limited to the Gaussian case.

2. Theory of asymptotic inference

In this section we go over some of the standard theory of asymptotic inference theory to put into context the various procedures discussed in the next section. In particular, we describe some of traditional theory usually applied to the independent, identically distributed (i.i.d.), random sample situation and indicate how this relates to the more general work of LeCam.

2.1. Traditional likelihood theory

We suppose in the tradition of asymptotic theory that we have a sequence of hypothetical inference problems, the nth one, for example, corresponding to the situation where we observe a sequence of n random variables which we represent by the symbol X_n. That is, X_n represents the complete sequence of n random variables. In each of these hypothetical problems we suppose that there is the same set of unknown parameters $\theta = (\theta_1, \ldots, \theta_s)$, which determines the distribution of the X_n.

Asymptotic theory, in the sense used in statistics, is concerned with the limits of distributions of functions of the X_n as n tends to infinity. One hopes that the results obtained will hold reasonably accurately for finite values of n corresponding to real practical experiments. Ideally, such 'hopes' should be confirmed by simulation or by a more precise theory. In fact, experience suggests that in many cases asymptotic theory is sufficiently accurate to be relevant to real situations with moderate sample sizes.

Let $L_n(\theta, X_n)$ denote the likelihood function (or, more precisely, a representation of it) and $l_n(\theta, X_n)$ the log-likelihood:

$$l_n(\theta, X_n) = \log L_n(\theta, X_n) .$$

The maximum likelihood estimator of θ is the value $\hat{\theta}_n(X_n)$ that maximizes $L_n(\hat{\theta}_n, X_n)$ or equivalently maximizes the log-likelihood (assuming there exists such a value). Various results have been proved for the maximum likelihood estimator when X_n represents a sequence of independent identically distributed random variables and where certain regularity conditions have been satisfied. These results have been extended to a variety of other situations, for example, Markov chains and processes (Billingsley, 1961).

We denote the probability measure defining the distribution of X_n when θ is the 'true' value of the unknown parameter by P_θ and the expectations and variances under P_θ by E_θ and Var_θ. Distributions are denoted by $\mathscr{L}_\theta$.

Letting l_n denote the log-likelihood and regarding $\partial l_n/\partial\theta$ as a column vector when θ is a vector and $\partial^2 l_n/\partial\theta^2$ as the matrix of second derivatives, again under regularity conditions we have

$$E_\theta(\partial l_n/\partial\theta) = 0, \tag{2.1}$$

$$E_\theta\{\partial l_n/\partial\theta \cdot (\partial l_n/\partial\theta)^*\}/n = -E_\theta(\partial^2 l_n/\partial\theta^2)/n \tag{2.2}$$

$$= \Gamma_n(\theta), \tag{2.3}$$

say, where $*$ denotes transpose of a vector.

Suppose for all θ

$$\lim_{n\to\infty} \Gamma_n(\theta) = \Gamma(\theta)\,.$$

(In the i.i.d. case $\Gamma_n(\theta) = \Gamma(\theta)$ for all n.) Then the standard property of the maximum likelihood estimator $\hat{\theta}_n$ in 'regular' situations is that the distribution of

$$n^{1/2}(\hat{\theta}_n - \theta)$$

tends, as n tends to infinity, to the multivariate normal distribution with mean 0 and variance/covariance matrix $\Gamma^{-1}(\theta)$. That is, for large n, $\hat{\theta}_n$ is approximately normally distributed with mean θ and variance/covariance matrix $\Gamma_n^{-1}(\theta)/n$.

An indication that the estimate is, in some sense, asymptotically optimal, can be obtained by noting that the Cramer–Rao inequality gives $\Gamma_n^{-1}(\theta)/n$ as a lower bound on the variance/covariance matrix for unbiased estimators. That is,

$$V_n(\theta) - \Gamma_n^{-1}(\theta)/n$$

is positive semidefinite if $V_n(\theta)$ is the variance/covariance matrix of an unbiased estimator of θ for the nth hypothetical problem. A more precise statement for the i.i.d. case is given by Bahadur (1960).

Now consider the problem of actually calculating $\hat{\theta}_n$. One approach would be to solve the equations

$$\partial l_n/\partial\theta = 0 \tag{2.4}$$

for θ. In the situations we will encounter there will be no exact analytic solution for (2.4) and numerical methods will have to be used. Suppose $\bar{\theta}_n$ is an approximate solution for (2.4). Then one iteration of Newton–Raphson would give the improved solution

$$\bar{\bar{\theta}}_n = \bar{\theta}_n - (\partial^2 l_n / \partial\theta^2)^{-1} \partial l_n / \partial\theta \tag{2.5}$$

with the derivatives being evaluated at $\theta = \bar{\theta}_n$. In view of (2.2) one might replace the matrix of second derivatives in (2.5) by $n\Gamma_n(\theta)$ or $n\Gamma(\theta)$. To solve (2.4), of course, one might keep iterating (2.5), possibly with the second derivative matrix replaced by $n\Gamma_n(\theta)$ or $n\Gamma(\theta)$, until each iteration made little difference to $\bar{\bar{\theta}}_n$. In practice, it seems necessary to be a little cleverer than this and use, for example, methods incorporating line searches (see Fletcher, 1980), in order to be reasonably sure that the method will converge. However, the result we want is that if $\bar{\theta}_n$ satisfies

$$\lim_{c\to\infty} \limsup_{n\to\infty} P_\theta\{|n^{1/2}(\bar{\theta}_n - \theta)| > c\} = 0\,, \tag{2.6}$$

that is, the error in $\bar{\theta}_n$ is of order $n^{-1/2}$, for example, $\bar{\theta}_n$ might be a method of moments estimator, then $\bar{\bar{\theta}}_n$ has the same asymptotic distribution as the maximum likelihood estimator. Thus, asymptotically, one iteration of (2.5) is enough. In the i.i.d. situation, this is Fisher's optimum scoring method (see Rao, 1965, p. 302). In practice, it seems more satisfying to continue iterating until (hopefully) the maximum likelihood estimator is obtained but we will see that there are good reasons for studying $\bar{\bar{\theta}}_n$.

The preceding theory was originally worked out and made rigorous for the independently identically distributed random sample case and various parts have been proved for other situations. One would like to show that at least some of these results hold for our time-series situation and this was the approach of, for example, Dunsmuir and Hannan (1976). On the other hand, LeCam (1960, 1969, 1974) has derived a set of conditions under which results similar to the preceding ones can be derived. A closely related method has also been developed by Hájek (1972). LeCam's results, in particular, give a rather more satisfactory statement of the optimality of the techniques than was traditionally available. To apply them to the time-series problem, one need show only that LeCam's conditions are satisfied. This was the approach of Davies (1973).

2.2. *LeCam's asymptotic theory*

We now very briefly summarize some of LeCam's work, particularly that in his (1969) lecture notes (pp. 57–87), but in a slightly more restricted form. Our notation is as before. The conditions are:

(A0) Θ, the set of possible values of θ is an open set in $\mathcal{R}^s$.

(A1) The sequence of probability measures defined by X_n under θ is contiguous to the sequence defined by X_n under $\theta + n^{-1/2}t$ for each $\theta \in \Theta$ and

s-dimensional vector t. See LeCam (1960, 1969) or Davies (1973) for definitions of contiguity.

(A2) For each $\theta \in \Theta$ there exists a sequence of s-dimensional random vectors $\Delta_n(\theta)$ and an $s \times s$ matrix $\Gamma(\theta)$ such that

$$l_n(\theta + n^{1/2}t) - l_n(\theta) - t^*\Delta_n(\theta) + \tfrac{1}{2}t^*\Gamma(\theta)t \to 0$$

in P_θ probability for each s-dimensional vector t.

(A3) $$l_n(\theta + n^{-1/2}t_n) - l_n(\theta + n^{-1/2}t) \to 0$$

in P_θ probability when $t_n \to t$.

(A4) $\Gamma(\theta)$ is nonsingular for each $\theta \in \Theta$.

(A5) There exists a 'preliminary' estimator $\bar{\theta}_n$, such that for each $\theta \in \Theta$ (2.6) is satisfied. We will suppose that $\bar{\theta}_n$ is chosen to take values only on a lattice of points with spacing $n^{-1/2}$.

LeCam's estimator is

$$T_n = \bar{\theta}_n + n^{-1/2}\Gamma^{-1}(\bar{\theta}_n)\Delta_n(\bar{\theta}_n)\,. \tag{2.7}$$

If Δ_n is chosen to be $n^{-1/2}$ times the derivative of the log-likelihood, then this is essentially Fisher's scoring estimator.

LeCam shows when these conditions are satisfied that T_n is asymptotically normally distributed with mean θ and variance $\Gamma^{-1}(\theta)/n$. More precisely

$$\mathcal{L}_\theta\{n^{1/2}(T_n - \theta)\} \to \mathcal{N}\{0, \Gamma^{-1}(\theta)\}\,. \tag{2.8}$$

Various optimality properties can be proved for T_n. LeCam shows that T_n is 'asymptotically sufficient', that is, for n large enough, T_n contains most of the information in X_n concerning the value of θ. One quite simple result that can be deduced from the results of LeCam (1969) is the following:

THEOREM. *Suppose the conditions* (A0)–(A5) *are satisfied and* $S_n(X_n)$ *is such that*

$$\lim_{c\to\infty} \limsup_{n\to\infty} P_\theta\{|S_n(X_n)| > c\} = 0\,,$$

that is, $S_n(X_n) = O_P(1)$ *under* P_θ. *Suppose also* T *is an* s*-dimensional normal random variable with expected value* t *and variance/covariance matrix* $\Gamma^{-1}(\theta)$. *Then there is a subsequence and a possibly randomized function,* $S_\theta(T)$, *of* T *such that for each* K

$$\mathcal{L}_{\theta+n^{-1/2}t}\{S_n(X_n)\} \to \mathcal{L}_t\{S_\theta(T)\}$$

along the subsequence, uniformly for $\|t\| < K$.

Conversely, if $S(T)$ is a function of T there exists a sequence of random variables $S_{n,\theta}(X_n)$ such that

$$\mathscr{L}_{\theta+n^{-1/2}t}\{S_{n,\theta}(X_n)\} \to \mathscr{L}_t\{S(T)\}\,.$$

If S is continuous almost everywhere we can take

$$S_{n,\theta}(X_n) = S\{n^{1/2}(T_n - \theta)\}\,.$$

In effect, this means that for each θ_0, making inferences about values of θ in a neighbourhood of θ_0 of size $O(n^{-1/2})$ is asymptotically equivalent to making inferences about the expected values of a multivariate normal distribution with known variance/covariance matrix $\Gamma^{-1}(\theta_0)$, given one observation. The first part of the theorem shows that a function of X_n, after suitable normalization, can be mapped to the multivariate normal situation and the converse shows how to transfer a technique appropriate for the normal situation back to the X_n.

For example, suppose an estimator $\hat{\theta}_n$ satisfies $(\hat{\theta}_n - \theta) = O_P(n^{-1/2})$ under P_θ. Let

$$S_n = n^{1/2}(\hat{\theta}_n - \theta)$$

and work along any subsequence along which $\mathscr{L}_\theta(S_n)$ converges. Suppose $\hat{\theta}_n$ is asymptotically unbiased in the sense that

$$\lim_{c\to\infty} \lim_{n\to\infty} E_{\theta+n^{-1/2}t} \mathscr{T}_c\{n^{1/2}(\hat{\theta}_n - \theta)\} = t \tag{2.9}$$

for all t, where $\mathscr{T}_c(x) = x$ if $|x| < c$, 0 if $|x| \geqslant c$, that is $\mathscr{T}_c$ truncates its argument at $\pm c$. This truncation function is necessary to avoid dealing with L_1 convergence. Then, according to the theorem, there exists $S(T)$ with S_n tending in distribution to $S(T)$. Hence, from (2.9)

$$E_t S(T) = t$$

all t and also

$$\lim_{c\to\infty} \lim_{n\to\infty} [\operatorname{Var}_\theta \mathscr{T}_c\{n^{1/2}(\hat{\theta}_n - \theta)\}] = \operatorname{Var}_\theta S(T)\,. \tag{2.10}$$

From unbiased estimator theory we have

$$\operatorname{Var}_\theta S(T) \geqslant \Gamma^{-1}(\theta)$$

with equality if $S(T) = T$. In view of the second part of the theorem, putting $S(T) = T$, or using (2.8) directly we can say that the estimator, T_n, minimizes the asymptotic variance (2.10) amongst estimators which are asymptotically unbiased, that is, satisfy (2.9).

See Hájek (1972) for other optimality properties. However, perhaps a better approach is to say that if one is happy to use T to estimate t when T has a $\mathcal{N}(t, \Gamma^{-1}(\theta))$ distribution, then one should be satisfied with T_n for estimating θ (at least when n is large enough). On the other hand, if one believes one should use, for example, James–Stein estimators, then the preceding theory would enable one at least to begin to set up the corresponding asymptotic estimators.

The preceding results, of course, apply to LeCam's estimator, T_n, defined by (2.7). In fact we would like to avoid discretizing the preliminary estimator and in fact one can show that this is unnecessary if

$$\sup_{\|\xi\|<\delta,\, \|\xi-\eta\|<\varepsilon} \|\Delta_n(\theta + n^{-1/2}\xi) - \Delta_n(\theta + n^{-1/2}\eta)\| \to 0 \tag{2.11}$$

in P_θ probability, uniformly in n as $\varepsilon \to 0$ for each $\theta \in \Theta$ and $\delta > 0$.

LeCam (1969) does not deal with the maximum likelihood estimator. It is probably not possible to find reasonable general conditions for it to satisfy (2.8) since it is possible, even in apparently regular situations for the likelihood function to have 'spurious' maxima which bear no relation to the true value of θ. However, we can say that if $\bar{\theta}_n$ is the maximum likelihood estimator or alternatively a root of the likelihood equation (2.4) which has been appropriately selected and if $\bar{\theta}_n$ satisfies (2.6), $\Delta_n(\theta) = n^{-1/2}\partial l_n/\partial\theta$, (A0)–(A4) and (2.11) are satisfied, then $\bar{\theta}_n = T_n$ and so $\bar{\theta}_n$ does have the asymptotic optimality properties we have discussed above.

2.3. *Hypothesis testing*

Now consider the hypothesis testing problem. Suppose the vector of unknown parameters is partitioned into two components, say $\theta = (\theta^{(1)}, \theta^{(2)})$ where $\theta^{(1)} = (\theta_1, \ldots, \theta_p)$, $\theta^{(2)} = (\theta_{p+1}, \ldots, \theta_s)$ and we are writing column vectors as rows to simplify printing. We might want to test the hypothesis

$$\theta^{(1)} = 0 \tag{2.12}$$

against the alternative that at least one component of $\theta^{(1)}$ is nonzero. The generalized likelihood ratio test is a commonly used asymptotic method for this problem. However, a simpler approach is the $C(\alpha)$ test developed by Neyman (1959), see also Bühler and Puri (1966), Bartlett (1955) and Moran (1970), which has similar asymptotic properties for $\theta^{(1)} = O(n^{-1/2})$. Let

$$\bar{\theta}_n = (0, \bar{\theta}_n^{(2)})$$

be an estimator of θ satisfying (2.6) when the hypothesis (2.12) holds; for example, $\bar{\theta}_n$ might be the maximum likelihood estimator under the hypothesis.

Then calculate the first component of $\bar{\bar{\theta}}_n$, that is $\bar{\bar{\theta}}_n^{(1)}$ according to (2.5) using the expected second derivatives

$$\bar{\bar{\theta}}_n^{(1)} = [\Gamma_n^{-1}(\bar{\theta}_n)\partial l_n/\partial\theta]^{(1)}/n\,, \tag{2.13}$$

with the derivative being evaluated at $\bar{\theta}_n$. Let V_n be the $p\times p$ principal submatrix of $\Gamma_n^{-1}(\bar{\theta}_n)$. Then in the i.i.d. situation, under the hypothesis $\theta^{(1)}=0$

$$n\{\bar{\bar{\theta}}_n^{(1)}\}^* V_n^{-1}\bar{\bar{\theta}}_n^{(1)} \tag{2.14}$$

is asymptotically chi-squared distributed with p degrees of freedom and will tend to be large when $\theta^{(1)}\neq 0$. When $p=1$,

$$n^{1/2}V_n^{-1/2}\bar{\bar{\theta}}_n^{(1)} \tag{2.15}$$

is asymptotically $\mathcal{N}(0,1)$ under the hypothesis and can be used for one-sided tests.

Application of the formulae for the inverses of partitioned matrices leads to the formulae given by Neyman (1959) and Bühler and Puri (1966). The particular advantage of tests based on (2.8) and (2.9) is that once $\bar{\theta}_n$ has been calculated, they are noniterative and the calculation of $\bar{\theta}_n$ and the expected values required for $\Gamma_n(\bar{\theta}_n)$ are worked out under the hypothesis, often leading to quite simple formulae. When $\bar{\theta}_n$ is the maximum likelihood estimator (under the hypothesis) $[\partial l_n/\partial\theta]^{(2)}$ vanishes when θ is replaced by $\bar{\theta}_n$ and the $C(\alpha)$ test reduces to the Lagrange-multiplier test of Aitchison and Silvey (1958—see Hosking, 1980, for additional references).

Of course, $C(\alpha)$ tests were developed for the i.i.d. random sample case. To apply LeCam's work, first note that under conditions (A1)–(A5) the testing problem is asymptotically equivalent to testing the hypothesis $t^{(1)}=0$ given a multivariate normal random variable T with variance/covariance matrix $\Gamma^{-1}(\theta)$ and $E(T)=(t^{(1)},t^{(2)})$.

The obvious test statistics for the normal situation are $[T^{(1)}]^*V^{-1}T^{(1)}$ and $V^{-1/2}T^{(1)}$ for multivariate and univariate $t^{(1)}$, where V is the $p\times p$ principal submatrix of $\Gamma^{-1}(\theta)$. Applying the converse part of the theorem of Section 2.2, noting that $T_n^{(1)}$ may be replaced by $\bar{\bar{\theta}}_n^{(1)}$ and V by V_n leads to (2.14) and (2.15).

More precisely, the univariate case, one uses an argument analogous to that used in Section 2.2 and the theory of similar tests to show that, with an appropriate critical point, the test based on (2.15) maximizes the asymptotic power

$$\liminf_{n\to\infty} P_{\theta+n^{-1/2}t}(\text{reject hypothesis})$$

when $\theta_1=0$ and $t_1>0$ amongst tests which satisfy

$$\lim_{n\to\infty} P_{\theta+n^{-1/2}t}(\text{reject hypothesis}) = \alpha$$

for all $t^{(2)}$, when $\theta_1 = 0$ and $t_1 = 0$.

Similarly, tests based on (2.14) are asymptotically most stringent when $\theta^{(1)}$ is multivariate.

2.4. *Inference using approximations to the likelihood*

In Section 3 of this paper we will want to base our tests and estimators on a function $w_n = w_n(\theta, X_n)$ that only approximates the log-likelihood. Naturally, if conditions (A0)–(A5) are satisfied when

$$\Delta_n(\theta) = n^{-1/2}\partial w_n/\partial\theta\,,$$

then one can base estimators similar to Fisher's scoring estimator and tests similar to $C(\alpha)$ tests on w_n rather than on l_n. Similarly, if condition (2.11) is satisfied and the estimator obtained by maximizing w_n satisfies (2.6), then it too is a version of T_n and so has the asymptotic optimality properties we have considered.

2.5. *Inference using only part of the data*

It will sometimes be convenient to base one's estimates on only part of the data, for example, only the high-frequency part of a periodogram when there are low-frequency trends in the data that are not of interest. Suppose $\tilde{X}_n$ represents the part of X_n on which we do want to base our estimates. An obvious question is, if the conditions (A0)–(A4) are satisfied for X_n, are they also satisfied for $\tilde{X}_n$? In fact, one can show that if $\tilde{\Delta}_n(\theta)$ is a function of $\tilde{X}_n$, $\tilde{\Gamma}(\theta)$ is a nonrandom nonsingular matrix and

$$\mathcal{L}_0\{\Delta_n(\theta)|\tilde{X}_n\} \to \mathcal{N}\{\tilde{\Delta}_n(\theta), \Gamma(\theta) - \tilde{\Gamma}(\theta)\}$$

in the sense of convergence of c.d.f.s. in $\tilde{P}_\theta$ probability, then (A0)–(A4) are satisfied for the probabilities generated by $\tilde{X}_n$ if $\Delta_n(\theta)$ and $\Gamma(\theta)$ are replaced by $\tilde{\Delta}_n(\theta)$ and $\tilde{\Gamma}(\theta)$.

Tests and estimators which are asymptotically optimal amongst those that depend only on $\tilde{X}_n$ can then be found, provided that (A5) is also satisfied. If

$$\tilde{\Delta}_n(\theta) = n^{-1/2}\partial\tilde{w}_n(\theta)/\partial\theta$$

where $\tilde{w}_n(\theta)$ is a function of $\tilde{X}_n$, then one may be able to define an estimator by maximizing $\tilde{w}_n(\theta)$. Provided that (2.6) and the analogue of (2.11) were satisfied, this would provide an asymptotically optimal estimator.

3. Inference in the frequency domain

This section is based primarily on the papers of Davies (1973) and Dunsmuir and Hannan (1976). However, many of the main ideas have their basis in the pioneering work of Whittle (1953). Other relevant early references are Whittle (1962) and Walker (1964).

3.1. *Specification of the problem*

Returning to the time-series problem: we observe $X_n = (x_0, \dots, x_{n-1})$, a series of n r-dimensional observations from a stationary normal time-series. We suppose that the covariance structure is determined by the set of unknown parameters $\theta = (\theta_1, \dots, \theta_s)$. We also suppose that the expectation of the process does not depend on θ and it is convenient to suppose that it is zero. In fact all the asymptotic results continue to hold when each x_k is replaced by $x_k - \bar{x}$, where $\bar{x}$ is the sample average so this is no real restriction. Regarding the x_k as r-dimensional column vectors, and X_n as an nr-dimensional column vector, and letting A^* denote the (conjugate) transpose of a (complex) matrix or vector A, define

$$c_m(\theta) = \operatorname{cov}_\theta(x_k, x_{k+m}) = E_\theta(x_k \cdot x_{k+m}^*) \tag{3.1}$$

since we are supposing $E(x_k) = 0$, and

$$C_n(\theta) = \operatorname{cov}_\theta(X_n, X_n) = E_\theta(X_n \cdot X_n^*)$$
$$= \begin{pmatrix} c_0, & c_1, & \dots, & c_{n-1} \\ c_{-1}, & c_0, & \dots, & c_{n-2} \\ \dots & \dots & \dots & \dots \\ c_{-n+1}, & c_{-n+2}, & \dots, & c_0 \end{pmatrix} \tag{3.2}$$

The log-likelihood is given (apart from an additive constant) by

$$l_n(\theta) = -\tfrac{1}{2}\{\log \det C_n(\theta) + X_n^* C_n^{-1}(\theta) X_n\}. \tag{3.3}$$

Our parametrization is a little different from that used by some others, for example, Dunsmuir and Hannan (1976). They use the moving average representation of the process

$$x_k = \varepsilon_k + \sum_1^\infty a_j(\theta)\varepsilon_{k-j}, \tag{3.4}$$

where $\{\varepsilon_k;\ k = 0, \pm 1, \pm 2, \dots\}$ is a sequence of independent Gaussian (for the Gaussian case) r-dimensional random variables with the ε_k having the same variance/covariance matrix, $\sigma(\theta)$, and $\{a_j(\theta)\}$ is a sequence of $r \times r$ matrices. In this case,

$$c_m(\theta)=\sum_{j=0}^{\infty} a_j(\theta)\sigma(\theta)a^*_{j+m}(\theta)\,, \tag{3.5}$$

where $a_0(\theta)=I$, the identity matrix. When $\{a_j(\theta)\}$ and $\sigma(\theta)$ depend on disjoint subsets of $(\theta_1,\ldots,\theta_s)$, the particular advantage of this parametrization is that the asymptotic distribution of the maximum likelihood and related estimators of the components of θ on which only the a_j depend does not depend on the distribution of the ε_k. That is, they need not be Gaussian although independence or the weaker condition of Dunsmuir and Hannan (1976) is still required. However, the representation (3.4) can be unnatural and difficult to find, particularly in the multivariate situation, and the independence assumption very difficult to verify. Since this paper is primarily concerned with the Gaussian case, we do not use (3.4).

Following from (3.3), we have

$$\frac{\partial l_n}{\partial\theta_u}=\operatorname{tr}\left[C_n^{-1}(\theta)\frac{\partial}{\partial\theta_u}C_n(\theta)C_n^{-1}(\theta)\{X_nX_n^*-C_n(\theta)\}\right]. \tag{3.6}$$

In fact, it might be possible to develop numerical techniques to handle (3.6) for n up to a few hundred using Toeplitz matrix techniques (see Cybenko, 1980, for references) and one would expect this to be a good approach for n less than, say, 100. For autoregressive/moving average processes, various exact and approximate formulae have been developed for the likelihood and so when one does want to fit such processes they are the appropriate formulae to use. See, for example, Gardener, Harvey and Phillips (1980). However, for larger values of n, computations with (3.3) and (3.6) become impossible and frequency-domain methods are appropriate. We should note though that recent work by Brent (1979) shows that it is possible to evaluate expressions such as (3.6) with $O(n\log^2 n)$ operations and so the computational reasons for using frequency-domain methods may disappear.

3.2. Frequency-domain approximation

Define the spectrum of the process

$$f(\lambda,\theta)=\sum_{-\infty}^{\infty} c_m(\theta)\,\mathrm{e}^{2\pi i m\lambda} \tag{3.7}$$

and

$$F_n(\theta)=\operatorname{diag}\{f(0,\theta),f(1/n,\theta),\ldots,f((n-1)/N,\theta)\}\,. \tag{3.8}$$

Let Ω_n be an $nr\times nr$ unitary matrix composed of $n\times n$ blocks of $r\times r$ submatrices; the (j,k)th block $(0\le j,k\le n-1)$ being the unit matrix multiplied by

$$n^{-1/2}\, e^{2\pi ijk/n}\,.$$

In effect, multiplication by Ω_n is the taking of a discrete Fourier transform. If

$$Z_n = \begin{pmatrix} z_{n,0} \\ z_{n,1} \\ \vdots \\ z_{n,n-1} \end{pmatrix} = \Omega_n X_n\,, \tag{3.9}$$

then

$$z_{n,k} = n^{-1/2} \sum_{j=0}^{n-1} x_j\, e^{2\pi ijk/n}\,. \tag{3.10}$$

Hence multiplication by Ω_n can be carried out very efficiently using a fast Fourier transform program (the number of operations required is of order $n \log n$ as opposed to n^3 for ordinary matrix multiplication). Davies (1973) shows that

$$G_n(\theta) = \Omega_n C_n(\theta) \Omega_n^* - F_n(\theta)$$

is, in a certain sense, small for large n. Thus the Ω_n transformation, approximately, simultaneously transforms the $C_n(\theta)$ into block diagonal matrices and the X_n into n approximately uncorrelated complex r-dimensional vectors. This suggests replacing the log-likelihood (3.3) by

$$\begin{aligned} w_n(\theta) &= -\tfrac{1}{2}\{\log \det F_n(\theta) + Z_n^* F_n^{-1}(\theta) Z_n\} \\ &= -\tfrac{1}{2} \sum_{0}^{n-1} [\log \det f(j/n, \theta) + z_{n,j}^* \{f(j/n, \theta)\}^{-1} z_{n,j}] \end{aligned} \tag{3.11}$$

and its derivatives (3.6) by

$$\begin{aligned} \frac{\partial w_n}{\partial \theta_u} &= \operatorname{tr}\left[F_n^{-1}(\theta) \frac{\partial}{\partial \theta_u} F(\theta) F_n^{-1}(\theta) \{Z_n Z_n^* - F_n(\theta)\} \right] \\ &= \sum_{0}^{n-1} \operatorname{tr}\left[\{f(j/n, \theta)\}^{-1} \frac{\partial}{\partial \theta_u} f(j/n, \theta) \{f(j/n, \theta)\}^{-1} \{z_{n,j} z_{n,j}^* - f(j/n, \theta)\} \right]. \end{aligned} \tag{3.12}$$

Note that only half the terms in (3.11) and (3.12) need be calculated in practice since the $(n-j)$th terms of $f(j/n, \theta)$ and $z_{n,j}$ are just the complex conjugates of the jth terms. These expressions are vastly more workable than (3.3) and (3.6) being usable for sample sizes of many thousand. Several variants of (3.11) and (3.12) have been proposed. Whittle (1953) used integrals in the place of the sums in (3.11) and (3.12); Dunsmuir and Hannan (1976) introduce a modification of (3.11) where the Fourier transform of the data and the spectrum are sampled at intervals smaller than the $1/n$ used by (3.11). Whittle's version was suggested before fast Fourier transform programs became available and is now mainly of theoretical interest. Dunsmuir and Hannan's version is particularly convenient when the sample size, n, is not a suitable number for the fast Fourier transform program. Davies (1973) suggests replacing $f(\lambda, \theta)$ by

$$\sum_{-n+1}^{n-1} (1-|m|/n)c_m(\theta)\, e^{2\pi i m\lambda} \tag{3.13}$$

to reduce the bias for finite n.

3.3. Conditions for LeCam's results to hold

The conditions given by Davies (1973) for (A0)–(A4) to hold with

$$\Delta_n(\theta) = n^{-1/2}\partial l_n(\theta)/\partial\theta \tag{3.14}$$

and $\Gamma(\theta)$ defined by

$$[\Gamma(\theta)]_{u,v} = \tfrac{1}{2}\int_0^1 \operatorname{tr}\left\{f^{-1}(\lambda,\theta)\frac{\partial}{\partial\theta_u}f(\lambda,\theta)f^{-1}(\lambda,\theta)\frac{\partial}{\partial\theta_v}f(\lambda,\theta)\right\} d\lambda \tag{3.15}$$

are:

(B0) The set of possible values of θ is an open set Θ in s-dimensional Euclidean space.
(B1.1) The $c_m(\theta)$ are differentiable functions of θ.

(B1.2) $$\sum_{m=-\infty}^{\infty} \|c_m(\theta)\| < \infty,$$

$$\lim_{\varepsilon\to 0}\sum_{m=-\infty}^{\infty} \|c_m(\theta+\varepsilon)-c_m(\theta)\| = 0 \quad \text{for all } \theta\in\Theta.$$

(B1.3) $$\sum_{m=-\infty}^{\infty}\left\|\frac{\partial}{\partial\theta_k}c_m(\theta)\right\|^2 < \infty,$$

$$\lim_{\varepsilon\to 0}\sum_{m=-\infty}^{\infty}\left\|\frac{\partial}{\partial\theta_k}c_m(\theta+\varepsilon)-\frac{\partial}{\partial\theta_k}c_m(\theta)\right\|^2 = 0 \quad \text{for all } k;\ 1\le k\le s \text{ and } \theta\in\Theta.$$

(B1.4) $\det f(\lambda,\theta) > 0$ for all λ; $0\le\lambda\le 1$ and $\theta\in\Theta$.

(B2) $$\sum_{k=1}^{s} t_k\frac{\partial}{\partial\theta_k}f(\lambda,\theta)\neq 0$$

for some nonnull set of λ for each r-dimensional vector $t\neq 0$ and $\theta\in\Theta$.

In order to use the version of $\Delta_n(\theta)$ based on (3.12),

$$\Delta_n(\theta) = n^{-1/2} \partial w_n(\theta)/\partial\theta , \tag{3.16}$$

one requires in addition

(B3.1) $\sum_1^\infty m^{1/2} \|c_m(\theta)\| < \infty .$

(B3.2) $\sum_1^\infty \left\| \frac{\partial}{\partial\theta_k} c_m(\theta) \right\| < \infty .$

Davies (1973) also shows that the discretization of the preliminary estimate can be avoided if (B3.2) and

(B3.3) $\sup_{\substack{\|\xi\|<\delta \\ \|\eta\|<\delta \\ \xi\neq\eta}} \sum_{-\infty}^{\infty} \left\| \frac{\partial}{\partial\theta_k} c_m(\theta+\xi) - \frac{\partial}{\partial\theta_k} c_m(\theta+\eta) \right\| \Big/ \|\xi-\eta\| < \infty$ for some δ and each k and θ

are satisfied.

If $f(\lambda, \theta)$ is defined by (3.13) in the expression $w_n(\theta)$, condition (B3.1) is not required. On the other hand, to ensure reasonable rates of convergence, Davies (1973) suggests

(B4.1) $\sum_1^\infty m^{1/2} \left\| \frac{\partial}{\partial\theta_k} c_m(\theta) \right\|^2 < \infty$

should also be satisfied for estimators based on $l_n(\theta)$;

(B4.2) $\sum_1^\infty m^{1/2} \left\| \frac{\partial}{\partial\theta_k} c_m(\theta) \right\| < \infty$

should be satisfied for estimators based on $w_n(\theta)$ with $f(\lambda, \theta)$ as in (3.13) and in addition

(B4.3) $\sum_1^\infty m \|c_m(\theta)\| < \infty$

should be satisfied when the usual version of $w_n(\theta)$ is used.

Similar results to those of Davies (1973) have also been obtained by Dzhaparidze (1977). These results are for the one-dimensional case with the integral version of (3.11). However, they have been extended to include the situation where the spectrum is of the form

$$f(\lambda, \theta) = \prod_1^q |(e^{2\pi i\lambda} - e^{2\pi i\lambda_j})|^2 f_0(\lambda, \theta) ,$$

where $\lambda_1, \ldots, \lambda_q$ are real constants which do not depend on θ. Our condition (B2) is then weakened to requiring only that $f_0(\lambda, \theta)$ has no zeros. This is the situation that arises when a time-series has been filtered to remove specific discrete known frequencies, either by the process generating the data or analytically after the raw data has been collected. Note that the problem becomes rather more complicated if the λ_j are allowed to depend on θ and will probably not fall within the present framework.

The preceding conditions imply that the estimators of the Fisher scoring type (2.7) based on either of the various versions of $\Delta_n(\theta)$ we have considered have the optimality and distributional properties discussed in Section 2, provided that a preliminary estimator $\bar{\theta}_n$ is available. In particular,

$$\mathscr{L}\{n^{1/2}(T_n - \theta)\} \to \mathcal{N}(0, \Gamma^{-1}(\theta)), \tag{3.17}$$

where $\Gamma(\theta)$ is given by (3.15) and $\Gamma^{-1}(\theta)$ is the 'smallest' possible variance/covariance matrix for estimators obeying (2.9).

Note that estimators of the Fisher scoring type have also been considered by Hannan (1970), Parzen (1971) and Nicholls (1977).

Our results also establish the asymptotic optimality of $C(\alpha)$ tests based on either the likelihood function or its frequency-domain approximation. In particular, the optimality of Hosking's (1980, 1981) Lagrange-multiplier tests is established and one can see how to find corresponding tests based on the periodogram.

3.4. Maximum likelihood estimation

To extend our results to include the maximum likelihood estimator or the estimator based on maximizing the approximate likelihood (3.11) one must show that these estimators are 'root n consistent', that is, (2.6) must be satisfied. As we have already indicated, we would not expect to be able to find usable general conditions for this to be so. However, if Θ is compact (i.e. closed and bounded) with the true value θ belonging to the interior of θ, then the situation is much more satisfactory. The relevant results are given by Dunsmuir and Hannan (1976), Deistler, Dunsmuir and Hannan (1978). They assume the representation (3.4) and a restricted version of their conditions is as follows.

(D0) Θ is a compact subset of $\mathscr{R}^s$ and θ belongs to the interior of Θ.

(D1) $$\sum_{j=0}^{\infty} j\|a_j(\theta)\|^2 < \infty$$

(D2) If $\theta_1 \neq \theta_2$, then $c_m(\theta_1) \neq c_m(\theta_2)$ for some m.

(D3) The elements of

$$k(z, \theta) = \sum_{0}^{\infty} a_j(\theta) z^j$$

are analytic within the unit circle, are continuous functions of θ and det $k(z, \theta)$ has no zeros within or on the unit circle.
(D4) $k(e^{i\lambda}, \theta)$ and $\sigma(\theta)$ have second derivatives with respect to θ which are continuous in θ and λ.
(D5) Condition (B2) holds.

Dunsmuir and Hannan (1976) allow these conditions to be weakened on the boundary of Θ so that, for example, their results can be applied to autoregressive/moving average models with Θ as the natural parameter space. They show that the estimators obtained by maximizing (3.3), (3.11) or its integral version are root n consistent and (in the case of the components of θ on which only the a_j depend), in fact, have the limiting distribution given by (3.17). We do not try to find the relation between conditions (B1)–(B3) and (D1)–(D5) but in practice one would not expect to find many time-series that fulfil one set and not the other. The results of Dunsmuir and Hannan (1976) establish the consistency and limiting distributions of the various estimators that rely on maximizing the exact or approximate likelihood, the results of Davies establish their asymptotic optimality and sufficiency and also that of the related Fisher scoring type of estimator.

The condition (D0), in reality, is not a serious problem since one will usually have some idea of what values of θ are appropriate and so, in effect, the range of values of θ to be considered is limited to a bounded subset of those for which the model is defined.

3.5. Inference based on only part of the periodogram

We now consider frequency-domain inference based on only part of the periodogram. This is appropriate when one is fitting a model which is intended to model short-term effects, but where there are also long-term fluctuations which one wishes to ignore or where there is a varying seasonal effect that affects a band of frequencies which one similarly wishes to ignore.

Suppose Λ represents the union of a finite set of disjoint intervals from [0, 1] such that $\lambda \in \Lambda \Leftrightarrow 1-\lambda \in \Lambda$, and we wish to base our estimator on $\{z_{j,n}: j/n \in \Lambda\}$. Using the kind of arguments in Davies (1973), if conditions (B0)–(B3) hold and

$$[\tilde{\Delta}_n(\theta)]_u = n^{-1/2} \sum \operatorname{tr}\Big[\{f(j/n, \theta)\}^{-1} \frac{\partial}{\partial \theta_u} f(j/n, \theta)\{f(j/n, \theta)\}^{-1} \times \{z_{n,j} z^*_{n,j} - f(j/n, \theta)\}\Big] \tag{3.18}$$

where the sum is over values of j which satisfy

$$j/n \in \Lambda, \quad 0 \leq j \leq n,$$

one can show

$$\mathscr{L}\{\Delta_n(\theta)-\tilde{\Delta}_n(\theta)\mid z_{j,n}: j/n\in\Lambda\}\to\mathcal{N}\{0,\Gamma(\theta)-\tilde{\Gamma}(\theta)\}$$

where

$$\tilde{\Gamma}(\theta)_{u,v}=\int_{\lambda\in\Lambda}\operatorname{tr}\Big\{f^{-1}(\lambda,\theta)\frac{\partial}{\partial\theta_u}f(\lambda,\theta)f^{-1}(\lambda,\theta)\Big\}\,\mathrm{d}\lambda\,.$$

In view of the discussion in Section 2.5, tests and estimators which are optimal amongst those based only on $\{z_{n,j}: j/n\in\Lambda\}$ can be derived from $\tilde{\Delta}_n(\theta)$ and $\tilde{\Gamma}(\theta)$ provided a preliminary estimator is available. Formula (3.18) can be obtained by differentiating

$$-\tfrac{1}{2}\sum\,[\log\det f(j/n,\theta)+z^*_{n,j}\{f(j/n,\theta)\}^{-1}z_{n,j}] \tag{3.19}$$

with the sum as in (3.18). Thus we can consider estimates obtained by maximizing (3.19). We presume theorems similar to those obtained by Dunsmuir and Hannan (1976) can be obtained but have not proved this.

4. Summary and example

We have considered statistical methods based on the likelihood function (3.3), its frequency-domain approximation (3.11) and a likelihood-like function based on only part of the frequency domain (3.19). We have seen how to derive analogues of Fisher's scoring estimator, the maximum likelihood estimator and $C(\alpha)$ tests from these functions.

Asymptotic optimality of these techniques has been established by using some of LeCam's results to show the asymptotic equivalence between the actual problem and an inference problem involving a single observation from a multivariate normal distribution.

4.1. Example

As an example we look at Brillinger's (1973, section 2) analysis of measurement of the 'Chandler wobble' of the Earth's axis of rotation. The data are derived from monthly measurements of the shift of the location of the North Pole over the period 1902–1969. It consists of two series, $\xi_0,\ldots,\xi_{n-1}$ and $\eta_0,\ldots,\eta_{n-1}$ corresponding to shifts along and perpendicular to the Greenwich meridian. The model proposed by Brillinger leads to the lagged covariance matrix:

$$c_m=\begin{pmatrix}a_m & b_m\\ -b_m & a_m\end{pmatrix}, \tag{4.1}$$

where

$$\begin{aligned} a_m &= \sigma^2\, e^{-\beta|m|} \cos(2\pi\gamma m)/(4\beta) + \psi_m\,, \\ b_m &= -\sigma^2\, e^{-\beta|m|} \sin(2\pi\gamma m)/(4\beta)\,, \\ \psi_m &= 2\psi^2 \quad \text{if } m = 0\,, \\ &= -\psi^2 \quad \text{if } m = 1\,, \\ &= 0 \qquad \text{otherwise}\,. \end{aligned}$$

The parameters to be estimated are β, γ, σ and ψ with β, σ, $\psi > 0$. The parameter γ is best regarded as taking values in [0, 1] with the points 0 and 1 being identified with each other; that is, $2\pi\gamma$ is a point on the unit circle. The exact likelihood and its derivatives could, in principle, be calculated using (3.3) and (3.6).

The spectrum defined by (3.7) is

$$f(\lambda) = \begin{pmatrix} f_0(\lambda) & \mathrm{i}f_1(\lambda) \\ -\mathrm{i}f_1(\lambda) & f_0(\lambda) \end{pmatrix} \tag{4.2}$$

where

$$\begin{aligned} f_0(\lambda) &= g(\lambda+\gamma) + g(\lambda-\gamma) + 4\psi^2 \sin(2\pi\lambda)\,, \\ f_1(\lambda) &= g(\lambda+\gamma) - g(\lambda-\gamma)\,, \\ g(\omega) &= \sigma^2(1-e^{-2\beta})/[2\beta\{1-2e^{-\beta}\cos(2\pi\omega) + e^{-2\beta}\}]\,. \end{aligned}$$

Putting

$$z_{n,j} = n^{-1/2} \sum_{k=0}^{n-1} \begin{pmatrix} \xi_k \\ \eta_k \end{pmatrix} e^{2\pi \mathrm{i} jk/n} \tag{4.3}$$

and substituting into (3.11) and (3.12) enables the approximate likelihood and its derivatives to be calculated. The conditions (B0)–(B4) are satisfied provided we extend (B0) to allow $2\pi\gamma$ to take values on the unit circle. This extension does not pose any problems to the theory outlined in Section 2. Thus, provided a preliminary estimator is available, LeCam's conditions are satisfied with $\Delta_n(\theta)$ based on either the exact or approximate likelihood (3.14, 3.16) and with $\Gamma(\theta)$ as in (3.15). Hence the estimator (2.7) has the asymptotic optimality properties we have discussed.

In fact, the computational procedure can be simplified slightly since the matrices (4.2) can be simultaneously diagonalized.

$$\begin{aligned} &2^{-1/2}\begin{pmatrix} 1 & \mathrm{i} \\ 1 & -\mathrm{i} \end{pmatrix} f(\lambda) 2^{-1/2} \begin{pmatrix} 1 & 1 \\ -\mathrm{i} & \mathrm{i} \end{pmatrix} \\ &\qquad = 2\,\mathrm{diag}\{g(\lambda+\gamma) + 4\psi^2\sin^2(2\pi\lambda),\ g(\lambda-\gamma) + 4\psi^2 \sin^2(2\pi\lambda)\}\,. \end{aligned} \tag{4.4}$$

Thus $f(\lambda)$ may be replaced by the right-hand side of (4.4) if ξ_k and η_k are replaced by $2^{-1/2}(\xi_k + i\eta_k)$ and $2^{-1/2}(\xi_k - i\eta_k)$ in (4.3). The resulting expression can then be manoeuvred into the form given by Brillinger. Thus Brillinger's method of regarding the data as a single complex-valued series does lead to optimal estimates. In general, however, this will be true only when the covariances are of the form (4.1) corresponding to a model which is invariant under rotation of the ξ, η axes.

Brillinger used a method of moments estimator as a preliminary estimator and then applied repeated iterations of (2.7) using the version of Δ_n based on the approximate likelihood and $\Gamma(\theta)$ replaced by the approximation, $\Gamma_n(\theta)$, defined by

$$[\Gamma_n(\theta)]_{u,v} = \frac{1}{2n}\sum_{j=0}^{n-1} \mathrm{tr}\left[\{f(j/n, \theta)\}^{-1}\frac{\partial}{\partial\theta_u}f(j/n, \theta)\{f(j/n, \theta)\}^{-1}\frac{\partial}{\partial\theta_v}f(j/n, \theta)\right].$$

Only two iterations were required for the process to converge and presumably the approximate maximum likelihood estimate was obtained. He also used the inverse of $n\Gamma_n(\theta)$ for estimating the variances of the estimates.

The raw data were heavily contaminated by a yearly cycle. Brillinger compensated for this by subtracting the cyclic term derived from the monthly averages. An alternative approach would have been to exclude narrow bands of frequencies around the peaks corresponding to the yearly cycle and its harmonics as in Section 3.5 and hence allow for slow fluctuations in the yearly cycle as well as the cycle itself.

References

Aitchison, J. and Silvey, S. D. (1958). Maximum-likelihood estimation of parameters subject to restraints. *Ann. Math. Statist.* **29**, 813–828.

Bahadur, R. R. (1960). On the asymptotic efficiency of tests and estimators. *Sankya* **22**, 229–252.

Bartlett, M. S. (1955). Approximate confidence intervals III. A bias correction. *Biometrika* **42**, 201–204.

Billingsley, P. (1961). *Statistical Inference for Markov Processes.* University of Chicago Press, Chicago.

Brent, R. P. (1979). Some new algorithms for Toeplitz systems of linear equations. Abstract of paper given at the Applied Mathematics Conference of the Australian Mathematical Society in Leura, NSW.

Brillinger, D. R. (1973). An empirical investigation of the Chandler wobble and two proposed excitation processes. *Bull. Internat. Statist. Inst.* **45**, Book III, 413–436.

Brillinger, D. R. (1980). Analysis of variance and problems under time series models. In: P. R. Krishnaiah, ed., *Handbook of Statistics,* Vol. 1, 237–278. North-Holland, Amsterdam.

Bühler, W. J. and Puri, Prem S. (1966). On optimal asymptotic tests of composite hypotheses with several constraints. *Zeit. Wahrscheinlichkeitstheorie* **5**, 71–88.

Cybenko, G. (1980). The numerical stability of the Levinson–Durbin algorithm of Toeplitz systems of equations. *Siam J. Sci. Stat. Comput.* **1**, 303–319.

Davies, R. B. (1973). Asymptotic inference in stationary Gaussian time-series. *Adv. Appl. Prob.* 5, 469–497.

Deistler, M., Dunsmuir, W. and Hannan, E. J. (1978). Vector linear time series models: corrections and extensions. *Adv. Appl. Prob.* **10**, 360–372.

Dunsmuir, W. and Hannan, E. J. (1976). Vector linear time series models. *Adv. Appl. Prob.* **8**, 339–364.

Dunsmuir, W. and Robinson, P. M. (1981). Parametric estimators for stationary time series with missing observations. *Adv. Appl. Prob.* **13**, 129–146.

Dzhaparidze, K. O. (1977). Estimation of parameters of spectral density with fixed zeroes. *Theory of Prob. Applications.* **22**, 708–729.

Fletcher, R. (1980). *Practical Methods of Optimization.* Wiley, New York.

Gardener, G., Harvey, A. C. and Phillips, G. D. A. (1980). An algorithm for exact maximum likelihood estimation of autoregressive-moving average models by means of Kalman filtering. *Applied Statistics* **29**, 311–322.

Hájek, J. (1972). Local asymptotic minimax and admissibility in estimators. In: LeCam, Neyman and Scott, eds., *Sixth Berkeley Symp. in Math. Stat. Prob.*, Vol. 1, 175–194. University of California Press, Berkeley.

Hannan, E. J. (1970). *Multiple Time Series.* Wiley, New York.

Hannan, E. J., Dunsmuir, W. T. M. and Deistler, M. (1980). Estimation of vector ARMAX models. *J. Multivar. Anal.* **10**, 275–295.

Hosking, J. R. M. (1980). Lagrange-multiplier tests of time-series models. *J. R. Statist. Soc.* **B42**, 170–181.

Hosking, J. R. M. (1981). Lagrange-multiplier tests of multivariate time-series models. *J. R. Statist. Soc.* **B43**, 219–230.

LeCam, L. (1960). Locally asymptotically normal families of distributions. *University of California Publications in Statistics* **3**, 37–98.

LeCam, L. (1969). *Théorie Asymptotique de la Décision Statistique.* University of Montreal Press, Montreal.

LeCam, L. (1974). *Notes on Asymptotic Methods in Statistical Decision Theory.* University of Montreal Press, Montreal.

Moran, P. A. P. (1970). On asymptotically optimal tests of composite hypotheses. *Biometrika* **57**, 47–55.

Neyman, J. (1959). Optimal asymptotic tests of composite statistical hypotheses. In: U. Grenander, ed., *Probability and Statistics*, 213–234. Wiley, New York.

Nicholls, D. F. (1977). A comparison of estimation methods for vector linear time-series models. *Biometrika* **64**, 85–90.

Parzen, E. (1971). Efficient estimation of stationary time series mixed schemes. *Bull. Internat. Statist. Inst.* **44**, Book II, 315–319.

Rao, C. R. (1965). *Linear Statistical Inference and its Applications.* Wiley, New York.

Walker, A. M. (1964). Asymptotic properties of least-squares estimates of parameters of a stationary non-deterministic time-series. *J. Austral. Math. Soc.* **4**, 363–384.

Whittle, P. (1953). The analysis of multiple stationary time-series. *J. R. Statist. Soc.* **B15**, 125–139.

Whittle, P. (1962). Gaussian estimation in stationary time series. *Bull. Inst. Internat. Statist.* **39**, 105–129.

D. R. Brillinger and P. R. Krishnaiah, eds., *Handbook of Statistics, Vol. 3*
© Elsevier Science Publishers B.V. (1983) 93–109

5

Applications of Spectral Analysis in Econometrics

C. W. J. Granger and Robert Engle

1. Beginnings

The forerunners of modern spectral analysis were Fourier series fitting techniques, which assumed a series contained important deterministic cycles of known period, and the periodograms, which assumed the same model but the components had periods that needed to be determined. These models were used by economists, despite the considerable computing costs, the best examples being the works by Moore (1914) and Beveridge (1921, 1922). An account of these and other early applications can be found in Cargill (1974). The main objective of this work was to search for cycles in data with the hope that cycles of similar periods in pairs of series would indicate relationships between these series, an example being sunspots and rainfall and hence wheat prices. In a sense the search for cycles was too successful, for instance Beveridge found evidence of over twenty in his long English wheat price series. This unlikely multiplicity of cycles brought the basic model into some disrepute and undoubtedly this was partly responsible for G. Udny Yule developing the alternative autoregressive and moving average models in the late 1920s and early 1930s. The resulting tension between the time-domain and frequency-domain approaches lasted until quite recently. The reason for the periodogram giving evidence of too many apparent cycles is explained by the low correlation between estimates at adjacent frequencies and the fact that it is an inconsistent estimator of the theoretical spectrum. Smoothing procedures used now to estimate spectra circumvent these problems.

The link between Fourier series, the periodogram and modern spectral methods was pointed out by Davis (1941) in a book well advanced of its time and which received remarkably little attention. By 1959 spectral methods still had not been applied to economic data. However, in that year Oskar Morgenstern, following up a strong suggestion by John von Neumann, initiated a project at Princeton to investigate the usefulness of spectral methods in economics. The project was supervised by John Tukey, who had recently developed the interpretation of cross-spectral techniques, and was staffed by Herman Karreman, Michio Hatanaka and Clive Granger, with Thomas Wonnacott and Michael Godfrey also being involved later. The first report of this project was published in 1961 (Granger, 1961) and the complete report resulted in the book by Granger and Hatanaka (1964). At the same time Marc Nerlove was using these techniques at Stanford to study seasonal adjustment problems (Nerlove, 1964, discussed below); and Hannan (1960) had previously worked on the same problem. Other early writers in this field were Fand (1966), Morgenstern (1961) and Cunnyngham (1983).

Empirical studies in econometrics appear to go through phases where different techniques become particularly popular. Initially single-equation regressions were dominant but were then replaced by the more appropriate, but difficult, simultaneous equation models. In the late 1960s and very early 1970s spectra methods became popular and probably more papers were published using these techniques than using the more classical simultaneous models. By the mid 1970s time-domain time-series techniques came into vogue, due to the appearance of the influential book by Box and Jenkins (1970). Nold (1972) produced a bibliography of applications of spectral methods in economics covering much of the most active period, listing 101 papers by 68 different authors, although some of the references given are only marginally relevant. Recently, spectral techniques have largely been out of favor by applied econometricians although they are still used as one of the bundle of empirical techniques available for analysis of time-series data. The theoretical aspects of the frequency-domain representations remain important when the properties of these various techniques are considered.

2. Applications of the power spectrum

The obvious features of a univariate, power spectrum that can be easily noted are any peaks, such as at the seasonal frequencies, $2\pi k/12$, $k = 1, 2, \ldots, 6$, for monthly data, and any shape that is complicated compared to the simple shapes that arise from a white noise or first-order autoregressive and moving average models. The seasonality question will be considered in Section 4. Economies have been seen to follow swings with alternating periods of prosperity and depression, known as the business cycle. An early and obvious application of spectral techniques was to investigate these swings. It should be emphasized that the business cycle has never been at all regular, or deterministic, and so corresponds to one, or several, frequency bands rather than to particular frequency points. The obvious problem with this topic is that the business cycle corresponds to rather low frequencies and so estimation of this component is difficult, even with monthly or quarterly data, unless very long series are available. The situation is little improved by considering a number of different series from the same economy, as this provides little extra information; most parts of the economy are inclined to move together at low frequencies. Although some evidence was found for certain low-frequency components being especially important (see, for instance, Howrey, 1968 and

Harkness, 1968), in general all low frequencies were usually observed to be important for the levels of major economic variables, and so the business cycle component did not prove to be special or outstanding. The relative importance of low-frequency components compared to all higher-frequency components was found so frequently that a spectrum that steadily declined from low to higher frequencies, except possibly at seasonal frequencies, was called the 'typical spectral shape' in Granger (1966). Unfortunately there are a number of different time-domain models that produce such a spectral shape, including AR(1) with a parameter near one and integrated models of order d, where d can be a fraction,[1] and which includes the random walk model.

Fig. 1 shows a typical spectral shape, that of a composite stock price index (taken from Granger and Morgenstern (1970, p. 142)). The estimated spectrum, after linear trend was removed from the series, is shown with 95% confidence

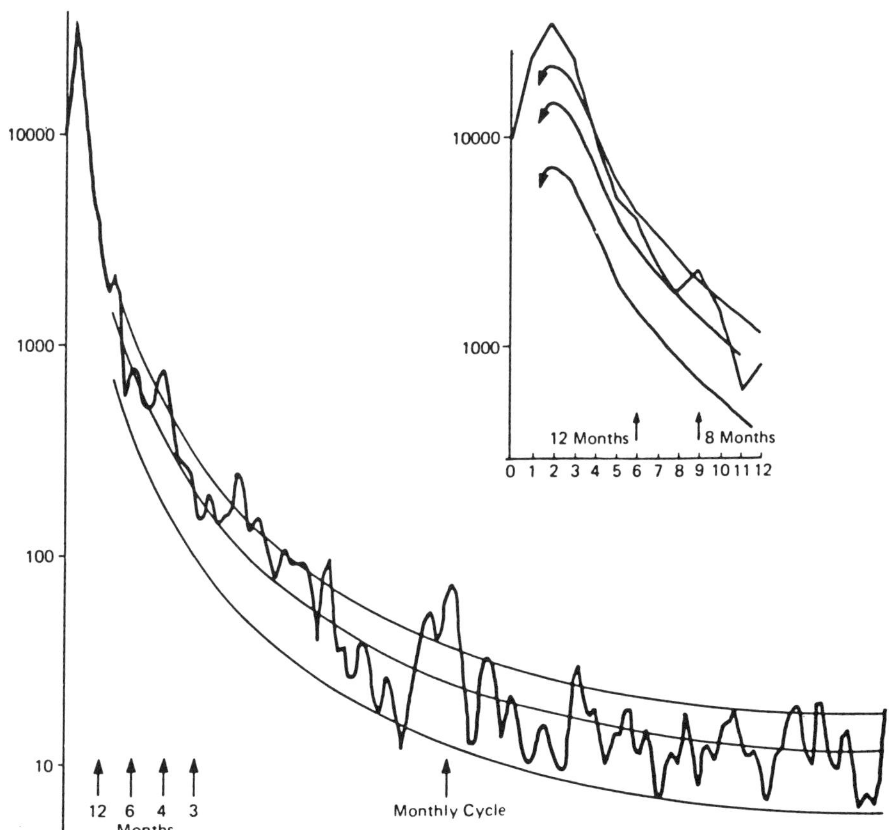

Fig. 1. Power spectrum of the composite weekly SEC stock price index.

[1]Such models can arise from aggregation of simple dynamic models, as shown for instance in Granger (1980).

intervals, with extra detail for low frequencies shown in the insert. A monthly cycle is evident but few other features of clear significance.

Because it is difficult to estimate the spectrum at very low frequencies it is also difficult to distinguish between these models using the estimated power spectrum of the original series. Sometimes it is easier to distinguish between some of these models by looking at the spectrum of the first differenced series. The typical spectral shape was found so frequently that it was used as a method of evaluating a large-scale econometric model by Howrey (1971, 1972). The Klein–Goldberger and Wharton econometric models were used to produce simulated data and the spectra of these data compared to the typical shape. In general, the models passed this not particularly stringent test. The 'typical spectral shape' is of course an oversimplification and actual spectra may have other discernable properties, as Nerlove (1971) found in a study of U.S. price series. He also found difficulty in interpreting these extra properties.

The other obvious use of the power spectrum is to investigate the relevance of a particular model suggested by a theory. For example, a number of economic theories suggests that the change in particular series should be white noises, so that the spectra of these changes will be flat over all frequencies if the theory is correct. This procedure was used by Sargent (1972) to test rational expectations for forward interest rates, by Granger and Morgenstern (1963, 1970) to test the random walk theory for stock market prices and by Labys and Granger (1970) to test the same theory for commodity prices. The method was found useful and occasionally some slight deviations from the predicted spectral shape were found. It would be possible to use a similar method to test other specific time-domain models, but this has not been done, as economic theory does not usually provide sufficiently specific models.

3. Application of the cross spectrum

Potentially the most important technique available in the early period was the cross spectrum and the functions derived from it, the coherence, and the phase and gain diagrams. If x_t and y_t are a pair of jointly stationary series, with spectra $f_x(\omega)$, $f_y(\omega)$ respectively, and cross spectrum $f_{xy}(\omega)$, the derived functions are

$$\text{Coherence:} \quad R_{xy}(\omega)=\frac{|f_{xy}(\omega)|^2}{f_x(\omega)f_y(\omega)},$$

$$\text{Phase:} \quad \phi(\omega)=\arctan\left[\frac{\operatorname{Im} f_{xy}(\omega)}{\operatorname{Re} f_{xy}(\omega)}\right],$$

$$\text{Gain:} \quad G_{xy}(\omega)=\frac{|f_{xy}(\omega)|}{f_x}(\omega).$$

The coherence measures the strength of relationships (squared correlation coefficient) between corresponding frequency components in the two series. As components with different frequencies are necessarily uncorrelated for jointly stationary series, the coherence thus totally measures the (second-order) strength of relationships between the series and has the added advantage that, in theory, its value is not altered by application of the same filters to the individual series. The gain essentially measures the regression coefficient of the ω-frequency component of x_t on the corresponding component of y_t. In the case where one series is leading the other, the phase diagram can be used to measure this lead. Thus, for instance, if $y_t = ax_{t-k} + e_t$, where e_t is independent of x_s for all t, s, the phase diagram for (y_t, x_t) will be a straight line of slope k, even when k is not an integer. These functions have likely useful interpretation for economic variables, particularly if these variables are decomposed into low-frequency parts ('business cycle', 'permanent income'), seasonal components and high frequencies ('transitory income'). Because economic theory predicts that different relationships may hold for different frequencies, spectral techniques may be uniquely suited to uncover these relationships. An extension of this idea discussed below is Band Spectrum Regression.

As an illustration of cross-spectral diagrams, Fig. 2 shows coherence and phase diagrams for wheat spot and medium-future commodity price changes. Coherence is seen to be higher at low frequencies and to decline at middle and higher frequencies. The phase diagram indicates little or no lag between the series and is seen to be highly variable when estimated coherence is very low.

The cross spectrum may also be used to identify or select time-domain models. Because the cross spectrum between two jointly stationary series is a fully general representation of the relation between two series it includes all time-domain transfer function or distributed lag models as special cases. By

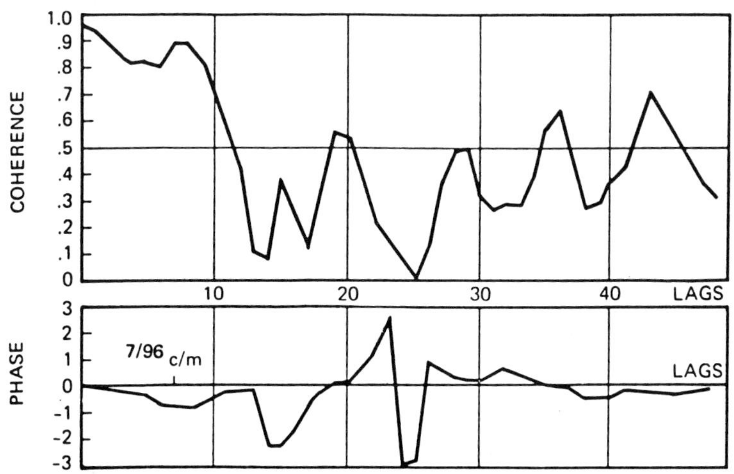

Fig. 2. Cross-spectrum between differences in wheat spot and medium-future price series.

estimating the cross spectrum first, it may be possible to choose a time-domain representation which is supported by the data. Engle (1976) has used this approach to specify the relationship between housing investment and interest rates. He found that the distributed lag weights change sign as would be predicted by an accelerator type of model. Thus spectral methods may be useful in a first exploratory look at economic data to pick acceptable models for further study.

Many studies have applied spectral techniques to economic data as the primary method of analysis and no attempt will be made to summarize them all. Instead, a brief account will be given of results in two fields, the term structure of interest rates and evaluation of leading indicators.

The rates of interest charged on loans depend partly on the length of time the loan will be outstanding and various theories attempt to explain this 'term structure'. Sargent (1968) found that coherences were generally high, particularly between rates of similar term, and that in general the longer (Government) rate leads the shorter rates with the lead longest as the differences in term increase. Granger and Rees (1968) using British data found similar coherence results but with the lags reversed; however, it seems very likely that the data they used were unsatisfactory in quality. Cargill and Meyer (1972) used a different approach; they estimated distributed lag relationships from the observed cross spectrum and found that long rates could not explain short rates but there was "a close relationship between short- and long-term rates and a fast response of the long rate to changes in the short rate". Thus the studies do not agree, as so often seems to happen with empirical work in economics using different data sets and statistical methods. The first two studies have the difficulty that lags are inferred from the phase diagram; this interpretation is only correct if no feedback occurs between the series.

The timing of the long swings in the macroeconomy is very irregular and prediction of turning points, the upturns and downturns, is of considerable interest to governments and companies. One method of prediction is to find series that consistently lead at the turns and the National Bureau of Economic Research has suggested many such leading indicators and also an index of these indicators. A possible way of evaluating the claims made for these indicators, in terms of their consistency and the extent of leads, is by looking at the coherence and phase diagrams at low frequencies from the cross spectrum between the indicator series and a measure of the state of the economy such as the index of industrial production. This has been done by Hatanaka (Chapter 12 of Granger and Hatanaka, 1964) and by Hymans (1973). They found that the indicators did lead, in that the phase diagrams indicate such a lead, but the coherences are often lower than might be hoped for and the leads are less than those suggested by the National Bureau. The best individual indicator seemed to be the layoff rate. Hymans found that the National Bureau's index of leading indicators could be improved by a better choice of weights, with some of the present components given zero weight. Rather similar results have been found recently by Neftci (1979) using time-domain methods. The main criticism of these pieces of work is that the National Bureau choses series that lead at turning points and these series do not necessarily lead at other parts of the cycle, whereas the studies just mentioned assume a constant lead throughout the cycle.

The potentially important partial cross-spectral techniques, in which the relationships between a pair of series is considered in the frequency domain after removal of the effects of one or more other series, have been little used in economics. One application is by Hatanaka (in Granger and Hatanaka, 1964) who considered inventory cycles and the acceleration principle. Using department store data he tried to distinguish between the acceleration principle and two alternative hypotheses but was unable to reach a decisive conclusion. However, he states "the reason for this failure is not in the partial cross-spectral analysis but in the lack of adequate data". A further application is in the Brillinger and Hatanaka (1970) study reported in Section 6.

4. Evaluation of seasonal adjustment procedures

Many economic series contain important seasonal components, as shown by the clear peaks observable in estimated power spectra at the seasonal frequencies. As these peaks usually appear to have finite width, the seasonal component appears not to consist only of deterministic terms. The presence of a strong seasonal in a series is thought by many econometricians to be rather troublesome as it obscures the more economically important business cycle and low frequencies. Thus attempts are made to remove, or reduce, the seasonal component and many techniques for seasonal adjustment have been suggested. Some of these techniques use one- or two-sided linear filters, and so their effects on the power spectrum are easily determined, but other techniques, including the X-11 method used for all U.S. series and many international statistics, are nonlinear and so their effects cannot be completely determined from available theory. The natural method for evaluating seasonal adjustment techniques is spectral analysis of the unadjusted and adjusted series as seasonal questions are easily phrased in the frequency domain. The first use of spectral techniques to investigate the seasonal appears to have been by Hannan (1960), but a more influential paper is that by Nerlove (1964) who considered the effects of the Bureau of Labor Statistics method of seasonal adjustment on seventy-five U.S. employment, unemployment and labor force series. The spectra of the unadjusted series showed, of course, strong peaks at the seasonal frequencies, but, more surprisingly, the adjusted series often had spectra with dips at seasonal frequencies, suggesting in a sense that the adjustment procedure had removed 'too much'. It is certainly true that if a series containing no seasonal is run through many adjustment procedures, something will be taken out. Nerlove also ran the cross spectrum between the adjusted and unadjusted series and found the gain at nonseasonal frequencies was usually substantially lower than the value one, particularly at the higher frequencies. This suggests that the higher-frequency components could have been badly disrupted. The phase diagrams indicated no further problems.

A difficulty with the Nerlove approach is that the gain estimates can be badly affected by 'leakage' from the strong seasonal peaks for the unadjusted data. To circumvent this problem Godfrey and Karreman (1967) constructed non-seasonal data from autoregressive models, added a variety of seasonal components, adjusted the resulting series using a number of different techniques and finally compared the spectra of the adjusted and the original, nonseasonal input series. They also looked at the cross spectra between these series. Ideally, one might suppose that a perfect adjustment technique would remove the added seasonal component, leaving the adjusted series virtually identical to the original nonseasonal component. They also found that the adjusted series had spectra with dips at the seasonal frequency and the coherence between the adjusted and the nonseasonal series was high at frequencies lower than the first seasonal frequency ($2\pi/12$) but was low at all higher frequencies, especially at the seasonal frequencies. Thus, the important low frequencies appeared to be unaffected by the adjustment processes, but all higher frequencies were badly disturbed. The results are potentially very serious when modeling relationships between series and the use of nonlinear adjustment methods has to be justified with some care. However, the dips in the spectrum and the coherence in adjusted series at seasonal frequencies are to be expected whenever the seasonal component is estimated by a regression procedure using a least-squares criterion. A similar feature occurs if a mean or a linear trend is estimated by least squares and subtracted from a series, the resulting series will have a dip at zero frequency. It is seen that an appropriate criterion in spectral terms is less obvious than was originally supposed. This question of what criterion to use and further examples of the use of spectral techniques to evaluate seasonal adjustment techniques can be found in the book edited by Zellner (1979), see particularly the papers by Baron, Wecker, Kuiper, Granger and by Cleveland, Dunn, and Terpenning and in Grether and Nerlove (1970) and Nerlove, Grether and Carvalho (1979).

5. Spectral regression

Because regression methods are the main statistical tool in economics it was natural that spectral methods would be adapted to the specification and estimation of linear regressions. The first analyses examined the estimation of regressions where the disturbances were assumed only to follow some stationary stochastic process. Watson and Hannan (1956) developed bounds for the efficiency of ordinary least squares and Hannan (1963) proposed an estimator for this setup later, called the Hannan Efficient estimator. This is simply generalized least squares and is easily illustrated.

Suppose y and x are $T \times l$ vectors of random variables with the following conditional distributions:

$$E(y \mid x) = x\beta ,$$

$$\mathrm{Var}(y \mid x) = \sigma^2 \Omega .$$

Letting $\epsilon = y - x\beta$, then $\mathrm{Var}(\epsilon) = \sigma^2\Omega$.

The generalized least-squares estimator is

$$\hat{\beta} = \frac{x'\Omega^{-1}y}{x'\Omega^{-1}x} .$$

If Ω is the covariance matrix of a stationary process, then it can always be approximately diagonalized by the matrix of Fourier coefficients as shown by Grenander and Szegö (1958). In this case the GLS estimator can be written approximately as

$$\hat{\beta} \sim \frac{\sum_j f_{xy}(\omega_j) f_\epsilon^{-1}(\omega_j)}{\sum_j f_x(\omega_j) f_\epsilon^{-1}(\omega_j)} .$$

This is Hannan's Efficient estimator which is made feasible by replacing spectra by their estimates.

If Ω can be exactly diagonalized, which could be the case for a process which is a circulant or when the sample size is large, then an exact expression for $\hat{\beta}$ can be obtained which shows that the particular estimates of f_{xy} and f_x required are the periodograms. In fact, a simple way to calculate this estimate is to regress the Fourier transform of y on the Fourier transform of x weighting each observation by $f_\epsilon(\omega_j)^{-1/2}$. This is simply a weighted least-squares problem with complex data.

When Ω and f_ϵ are unknown, these quantities must be estimated from the data. Standard methods of estimation of power spectra can be applied to the residuals from a consistent estimation method to obtain asymptotically efficient estimators which do not require specifying the process of the disturbances. Such procedures can also be iterated to obtain maximum likelihood estimates of β and f_ϵ. The ability to find fully efficient estimators assuming simply that the disturbances are stationary is very attractive as misspecification of the nature of this process may lead to substantial inefficiencies, indeed even worse estimates than ordinary least squares for some cases as shown by Engle (1974). Against this must be set the problems of window design for estimates of f_ϵ and the probable deviation of finite sample performance from asymptotic optimality. Engle and Gardner (1976), using Monte Carlo evidence, established that the finite sample results are quite acceptable except where f_ϵ has a very strong low-frequency peak which is therefore hard to estimate.

The extension of this argument to dynamic regressions where there may be lagged dependent variables was initially considered by Hannan (1965), Amemiya and Fuller (1967), and more recently Espassa (1977) and Engle (1980). In this case the problem becomes nonlinear so that iterative methods must be used although there are several ways to formulate the iterations. Under Gaussian assumptions, an estimator is found by Engle (1980) to be exact maximum likelihood. Spectral simultaneous equations estimation was initially discussed in the static case by Hannan and Terrell (1973) and for dynamic models by Espassa and Sargan (1977).

A variety of econometric studies have used these and closely related methods. For example, a computationally simpler estimation method, the Hannan Inefficient estimator, was used by Sims (1972a,b) and Cargill and Meyer (1972). An approximation to Hannan's Efficient estimator was employed by Sims (1972a), Cargill and Meyer (1972) and more recently by Geweke (1977) and Engle (1979). It would seem that there are substantial opportunities for application of such simple mechanical approaches to dealing with serial correlation in regression.

A direct extension of this approach is to consider estimating models on only a subset of the frequencies. A variety of economic and statistical reasons might be offered for such a choice. For example, there may be measurement errors which are concentrated in some frequencies (Engle and Foley, 1975), or there may be different models which explain short- and long-run behavior, or perhaps seasonality should be excluded for either of these reasons. The first formal statement and application of this approach was called Band Spectrum Regression by Engle (1974). He looked at the consumption function to determine whether the marginal propensity to consume an additional dollar of income appeared to be different for high frequencies (transitory) and low frequencies (permanent). The permanent income hypothesis would suggest a substantial difference but none was observed.

A further justification for running regressions on separate spectral bands is as a specification test. If a model is well specified, the estimates should not be significantly different. Engle (1978) performed such a test on a set of price equations and found some rather significant differences.

Hylleberg (1977) performed a Monte Carlo experiment of the method for eliminating seasonality from a regression. The Band Spectrum Regression performed quite well in this application for most of the situations. This is a particularly important application in the light of the results of Wallis (1974) and Sims (1974) who show that seasonal adjustment can seriously distort parameter estimates in linear regression. This may occur either because there is a different model at the seasonal frequencies or because the independent and dependent variables are separately adjusted.

The extension of BSR to models with lagged dependent variables and simultaneous equations is discussed by Engle (1980) and Espassa and Sargan (1977) although there is a difficulty with the empirical results in the latter as pointed out in the former. The instrumental variables method formulated in Engle (1980) is examined in a simulation experiment by Bunzel and Hylleberg (1980) with again satisfactory results although in several of their situations other approaches perform equally well.

6. Frequency-domain factor analysis

Spectral methods are generally applied to one or two series at a time but there is no reason not to apply them in a multivariate context as well. When there are many series, the calculation and interpretation of the gain, phase and coherence between each pair at each frequency become formidable. Efforts to restrict the generality of the interactions allowed have focused upon principal component or factor analytic models. Both of these methods in their traditional form deal only with contemporaneous correlations among the variables, and similarly the frequency-domain versions depend upon only the cross spectra at each frequency.

The principal component setup was initially proposed by Brillinger (1975) and the factor analytic model appears first in Geweke (1975, 1977) and in collaborative work with Sargent and Sims (1977). A variety of economic applications of the frequency-domain factor model have been published including the Sargent–Sims (1977) model of the macroeconomy, Geweke's (1977) model of production and Singleton's (1980) model of the term structure of interest rates. In each case, both the economic questions asked and the estimation methods are novel. We will develop a simplified version of the Singleton model below.

Consider an $M \times 1$ vector of economic variables y_t which will be the yield to maturity of different length bonds. In the Singleton case these included Government bonds from 3 months to 10 years. These are assumed to be driven by k white noise independent unit variance factors given by the vector Z_t and possibly serially correlated but independent disturbances specific to each maturity given by ϵ_t. The model is simply

$$y_t = \sum_{s=0}^{\infty} A_s Z_{t-s} + \epsilon_t ,$$

where Z_t and ϵ_t have spectral density matrices I and $F_\epsilon(\omega)$ respectively. The spectral density of y is immediately derived to be

$$F_t(\omega) = \tilde{A}(\omega)\tilde{A}(\omega)^\dagger + F_\epsilon(\omega) ,$$

where $\tilde{A}(\omega)$ is the Fourier transform of the series A_s given by

$$\tilde{A}(\omega) = \sum_{s=0}^{\infty} A_s \, e^{i\omega s} .$$

The estimation problem is to find F_ϵ and $\tilde{A}$ at each frequency based upon data F_y at that frequency.

The model described above is however simply the familiar factor analytic model used in cross-sectional studies but with complex covariances.[2] Standard

[2]In fact, all methods of real multivariate analysis have their complex counterpart which may be of potential use in time-series analysis.

estimation techniques adapted for complex arithmetic can therefore be applied for each frequency separately because the frequency bands are independent. If however there are some constraints on the A_s or on F_ϵ implied by the original time-domain formulation, then the analysis may not decompose so easily. In the models discussed here, there are no such constraints; however, in the works of Engle and Watson (1981), these constraints essentially argue for reversion to time-domain methods.

The economic question asked by both Sargent and Sims and by Singleton is, what is the dimension of Z? How many independent noise sources are there in the economy being studied? This question is of interest because the finding of several sources of noise could explain how different economic agents could have different expectations and information sets. The test is based upon sequentially testing increasing values of k as null hypotheses against the unrestricted model and stopping when an acceptable level is achieved.

Singleton finds $k = 2$ for his term structure model. Thus there are two independent information series involved in the formulation of expectations about the yields on different maturity securities. This is consistent with earlier studies which use the short-term rate and inflation as the key determinants of the structure but is also consistent with many other interpretations. As usual, factor analytic methods face difficulties precisely labeling the factors. Nevertheless, the notion that $k = 2$ suggests that a parsimonious model of the multivariate time-series relationship among interest rates is consistent with the data. Such a relationship might be particularly useful for forecasting in the general multivariate problem.

7. Advanced techniques

The traditional spectral techniques deal just with stationary series, linear relationships and second moments, but various extensions removing these assumptions have been proposed and occasionally applied to economic data. For example, Brillinger and Rosenblatt (1967a,b) have provided theory for Fourier transforms of sets of higher lagged moments, expressed most conveniently in terms of lagged cumulants, although interpretation of the resulting functions is not always clear or simple. Godfrey (1965) has estimated the bispectrum being the Fourier transform of their lagged moments, such as $E[X_tX_{t-p}X_{t-q}]$ when X_t has zero mean, for two economic series—a stock price series for a single company (International Telephone and Telegraph) and the 'Federal Float' which is a quantity of cash held within the Federal Reserve System and is used to measure the level of activity of the member banks in the System. He found that a transformation of the stock prices P_t to $\log(P_t + a)$, where a is near zero produces a series that more nearly obeyed a linear model. The Federal Float series, which is strongly seasonal, produced a bispectrum which rejected "the hypothesis that the entire process including the seasonal frequencies is well represented as a linear process". Thus the bispectral results were found useful for detecting nonlinearity but the actual generating mechanism is not always readily discerned from the results.

There are various methods of investigating possible nonstationarity in series. The most obvious procedure is to calculate spectra or other functions for a variety of overlapping time-periods to see if there is an obvious tendency for change. Brillinger and Hatanaka (1970) investigated the permanent income hypothesis by estimating moving spectra, cross spectra and partial cross spectra. They found that a permanent income hypothesis with time-changing horizon was consistent with the data. A more sophisticated approach to nonstationarity is to estimate the harmonizable spectrum. A harmonizable process x_t has Cramer's representation

$$x_t = \int_{-\pi}^{\pi} e^{it\omega}\, dZ(\omega)$$

but now the terms $dZ(\omega)$ are not orthogonal, having

$$E[dZ(\omega)\,\overline{dZ(\lambda)}] = ddF(\omega, \lambda)\,,$$

where $F(\omega, \lambda)$ is a bivariate distribution function, so that covariances are given by

$$E[x_t\bar{x}_{t-k}] = \int\int e^{it(\omega-\lambda)}\, e^{ik\lambda}\, ddF(\omega, \lambda)$$

and are thus dependent on t and so the series will be nonstationary. In fact, harmonizable processes are a very general class and include most nonstationary processes as special cases, including random walks and models with time-varying parameters. Thus, for example, a seasonal frequency component can be correlated with the business cycle component in some process, so that the amplitude of the seasonal could be larger during times of prosperity than during depressions. This cannot occur with a strictly stationary series. Joyeux (1979) has discussed estimation and interpretation of the harmonizable spectrum, and applied the technique to two individual economic series. With new housing starts, it was found that the high- and low-frequency components were intercorrelated, and thus the series is nonstationary. When the method was applied to a personal income series, the high- and low-frequency components were found to be uncorrelated, which agrees with the economic theory that permanent and transitory components of income are independent. Harmonizable processes are a very general class of nonstationary processes with easy interpretations, but they do appear to require rather long series for successful analysis.

8. Conclusions

Spectral methods today find an important place in both theoretical and applied economic research, but in general it is as a companion to time-domain and conventional regression methods rather than as alternatives. Thus time-domain and frequency-domain procedures are not competitors but can help each other. The development of estimators for most time-series problems based on frequency-domain statistics has generally been completed. Applications of these techniques, particularly the more complex, are, however, rather scarce. In part this is due to unfamiliarity, but more important in our opinion is the superiority of time-domain parameterizations for many problems. The frequency domain provides simple ways of estimating models with large numbers of free parameters (which increase with sample size in most cases) while the time domain generally imposes tighter parameterizations. Empirically for the type of data and quantity of data generally available to economists, the time-domain formulations are more satisfactory.

Frequency-domain methods thus are particularly useful in exploratory techniques and in theoretical research into the properties of statistical procedures for stationary data series. There is a broad body of literature on which these research directions can build and it is likely that this basis will continue to develop.

References

Amemiya, T. and Fuller, W. A. (1967). A comparative study of alternative estimators in a distributed lag model. *Econometrica* **35**, 509–529.

Beveridge, Sir W. H. (1921). Weather and harvest cycles. *Economic Journal* **31**, 429–452.

Beveridge, Sir W. H. (1922). Wheat prices and rainfall in western Europe. *J. Royal Stat. Soc.* **85**, 412–478.

Box, G. E. P. and Jenkins, G. M. (1970). *Time Series Analysis, Forecasting and Control.* Holden-Day, San Francisco.

Brillinger, David R. (1975). *Time Series: Data Analysis and Theory.* Holt, Rinehart and Winston, New York.

Brillinger, D. R. and Hatanaka, M. (1970). A permanent income hypothesis relating to the aggregate demand for money (an application of spectral and moving spectral analysis). *Economic Studies Quarterly* **21**, 44–71.

Brillinger, D. R. and Rosenblatt, M. (1967a). Asymptotic theory of estimates of kth-order spectra. In: B. Harris, ed., *Spectral Analysis of Time Series.* Wiley, New York.

Brillinger, D. R. and Rosenblatt, M. (1967b). Computation and interpretation of kth-order spectra. In: B. Harris, ed., *Spectral Analysis of Time Series.* Wiley, New York.

Bunzel, H. and Hylleberg, S. (1979). Seasonality in dynamic regression models: A comparative study of finite sample properties of various regression estimators including band spectrum regression, manuscript Institute of Economics, University of Aarhus.

Cargill, T. F. (1974). Early applications of spectral methods to economic time series. *History of Political Economy* **6**, 1–16.

Cargill, T. F. and Meyer, R. A. (1972). A spectral approach to estimating the distributed lag relationship between long and short term interest rates. *International Economic Review* **13**, 223–238.

Cunnyngham, J. (1963). The spectral analysis of economic time series, Working Paper No. 14, U.S. Bureau of the Census. Government Printing Office, Washington.

Davis, H. T. (1941). *The Analysis of Economic Time Series.* Principle Press, Bloomington, Indiana.

Engle, R. F. (1974a). Band spectrum regression. *International Economic Review* **15**, 1–11.

Engle, R. F. (1974b). Specification of the disturbance of efficient estimation. *Econometrica* **42**, 135–146.

Engle, R. F. (1976). Interpreting spectral analyses in terms of time-domain models. *Annals of Economic and Social Measurement* **5**, 89–109.

Engle, R. F. (1978). Testing price equations for stability across spectral frequency bands. *Econometrica* **46**, 869–881.

Engle, R. F. (1979). Estimation of the price elasticity of demand facing metropolitan producers. *Journal of Urban Economics* **6**, 42–64.

Engle, R. F. (1980). Exact maximum likelihood methods for dynamic regressions and band spectrum regressions. *International Economic Review* **21**, 391–407.

Engle, R. F. and Foley, D. F. (1975). An asset price model of aggregate investment. *International Economic Review* **16**, 625–647.

Engle, R. F. and Gardner, R. (1976). Some finite sample properties of spectral estimators of a linear regression. *Econometrica* **44**, 149–165.

Engle, R. F. and Watson, M. (1981). A one-factor multivariate time series model of metropolitan wage rates. *J. Amer. Statist. Assoc.* **76**, 774–781.

Espassa, A. (1977). *The Spectral Maximum Likelihood Estimation of Econometric Models With Stationary Errors.* Vandenhoeck and Ruprecht, Göttingen.

Espassa, A. and Sargan, J. D. (1977). The spectral estimation of simultaneous equation systems with lagged endogenous variables. *International Economic Review* **18**, 583–605.

Fand, D. (1966). A time series analysis of the "bills-only" theory of interest rates. *Review of Economics and Statistics* **48**, 361–371.

Geweke, J. (1975). Employment turnover and wage dynamics in U.S. manufacturing, unpublished Ph.D. dissertation, University of Minnesota.

Geweke, J. (1977). The dynamic factor analysis of economic time series models. In: D. J. Aigner and A. S. Goldberger, eds., *Latent Variables in Socioeconomic Models,* Chapter 19, 365–383. North-Holland, Amsterdam.

Geweke, J. (1977). Wage and price dynamics in U.S. manufacturing. In: C. Sims, ed., *New Methods in Business Cycle Research.* Federal Reserve Bank, Minneapolis.

Grether, D. and Nerlove, M. (1970). Some properties of 'optimal' seasonal adjustment. *Econometrica* **38**, 682–703.

Godfrey, M. D. (1965). An exploratory study of the bi-spectrum of economic time series. *Applied Statistics* **14**, 48–69.

Godfrey, M. D. and Karreman, H. (1967). A spectrum analysis of seasonal adjustment. In: M. Shubik, ed., *Essays in Mathematical Economics in Honor of Oskar Morgenstern.* Princeton University Press, Princeton.

Granger, C. W. J. (1961). First report of the Princeton economic time series project. *L'Industria* **4**, 194–206.

Granger, C. W. J. (1966). The typical spectral shape of an economic variable. *Econometrica* **34**, 150–161.

Granger, C. W. J. (1980). Long memory relationships and the aggregation of dynamic models. *J. of Econometrics* **14**, 227–238.

Granger, C. W. J. and Hatanaka, M. (1964). *Spectral Analysis of Economic Time Series.* Princeton University Press, Princeton.

Granger, C. W. J. and Morgenstern, O. (1963). Spectral analysis of New York stock market prices. *Kyklos* **16**, 1–27.

Granger, C. W. J. and Morgenstern, O. (1970). *Predictability of Stock Market Prices.* D. C. Heath & Co., Lexington, Mass.

Granger, C. W. J. and Rees, H. J. B. (1968). Spectral analysis of the term structure of interest rates. *Review of Economic Studies* **35**, 67–76.

Grenander, U. and Szegö, G. (1958). *Toeplitz Forms and Their Applications.* University of California Press, Berkeley.

Hannan, E. J. (1960). The estimation of seasonal variation. *The Australian J. of Statistics* **57**, 31–44.

Hannan, E. J. (1963). Regression for time series. In: M. Rosenblatt, ed., *Time Series Analysis.* Wiley, New York.

Hannan, E. J. (1965). The estimation of relationships involving distributed lags. *Econometrica* **33**, 206–224.

Hannan, E. J. and Terrell, R. D. (1973). Multiple equation systems with stationary errors. *Econometrica* **41**, 299–320.

Harkness, J. P. (1968). A spectral analytic test of the long swing hypothesis in Canada. *Review of Economics and Statistics* **50**, 429–436.

Howrey, E. P. (1968). A spectral analysis of the long-swing hypothesis. *International Economic Review* **9**, 228–252.

Howrey, E. P. (1971). Stochastic properties of the Klein–Goldberger model. *Econometrica* **30**, 73–88.

Howrey, E. P. (1972). Dynamic properties of a condensed version of the Wharton model. In: Bittickman, ed., *Economic Models of Cyclical Behavior,* Vol. II, 601–603. National Bureau of Economic Research.

Hylleberg, S. (1977). A comparative study of finite sample properties of band spectrum regression estimators. *Journal of Econometrics* **5**, 167–182.

Hymans, S. (1973). On the use of leading indicators to predict cyclical turning points. *Brookings Papers on Economic Activity,* 339–384.

Joyeux, R. (1979). Harmonizable processes in economics. Ph.D. thesis, Economics Department, University of California, San Diego.

Labys, W. C. and Granger, C. W. J. (1970). *Speculation, Hedging and Commodity Price Forecasts.* D. C. Heath and Co., Lexington, Mass.

Moore, H. L. (1914). *Economic Cycle: Their Law and Causes.* August M. Kelley, New York (1967 reprint).

Morgenstern, O. (1961). A new look at economic time series analysis. In: H. Hegeland, ed., *Money, Growth and Methodology and Other Essays in Economics, in Honor of Johan Akerman.* C. W. K. Gleerue, Lund, Sweden.

Neftci, S. N. (1979). Lead-lag relations, exogeneity and prediction of economic time series. *Econometrica* **47**, 101–114.

Nerlove, M. (1964). Spectral analysis of seasonal adjustment procedures. *Econometrica* **32**, 241–286.

Nerlove, M. (1971). On the structure of serial dependence in some U.S. price series, Conference on the Econometrics of Price Determination, Board of Govenors of the Federal Reserve System.

Nerlove, M., Grether, D. M. and Carvalho, J. L. (1979). *Analysis of Economic Time-Series.* Academic Press, New York.

Nold, F. C. (1972). A bibliography of applications of techniques of spectral analysis to economic time series, Technical Report No. 66, Institute for Mathematical Studies in the Social Sciences, Stanford University.

Sargent, T. J. (1968). Interest rates in the nineteen-fifties. *Review of Economics and Statistics* **50**, 164–172.

Sargent, T. J. (1972). Rational expectations and term structure of interest rates. *Journal of Money, Credit and Banking* **4**, 74–97.

Sargent, T. J. and Sims, C. (1977). Business cycle modeling without much a priori economic theory. In: C. Sims, ed., *New Methods in Business Cycle Research,* 45–110. Federal Reserve Bank, Minneapolis.

Sims, C. (1972a). Are there exogenous variables in short run production relations. *Annals of Economic and Social Measurement* **1**, 17–36.

Sims, C. (1972b). Money, income and causality. *American Economic Review* **62**, 540–552.

Sims, C. (1974). Seasonality in regression. *Journal of the American Statistical Association* **69**, 618–626.

Singleton, K. J. (1980). A latent time series model of the cyclical behavior of interest rates. *International Economic Review* **21**, 559–576.

Wallis, K. (1974). Seasonal adjustment and relations between variables. *Journal of the American Statistical Association* **69**, 18–31.

Watson, G. S. and Hannan, E. J. (1956). Serial correlation in regression analysis II. *Biometrika* **43**, 436–448.

Zellner, A. (1979). Seasonal analysis of economic time series, Economic Research Report, ER-1, U.S. Department of Commerce, Bureau of the Census.

D. R. Brillinger and P. R. Krishnaiah, eds., *Handbook of Statistics, Vol. 3*
 111–123

6

Signal Estimation

E. J. Hannan

1. Introduction

The problem we here consider is that where r sensors (antennae, receivers) record a signal transmitted to all sensors, each of which also receives noise. Examples are tide gauges or pressure recorders which record sea level or seismometers which record the direction of motion of the earth. Sometimes there might be no such easily identified apparatus as in the case of economic measurements, each being affected by a common economic force.

The noise could consist partly of noise internal to the system in which case it would be regarded as independent of the signal, or at least incoherent with it. (See other chapters for definitions of terms used herein.) It could also be regarded as incoherent as between sensors. Noise external to the apparatus need not be incoherent as between sensors, for example because this noise is constituted by a second signal. The noise might even be coherent with the main signal if, for example, this second signal was a refracted form of the main signal.

All techniques below are based on the assumption that the records $y_j(t)$ of the sensors are additively composed of signal and noise and that the records are sampled at a discrete sequence $t = 1, 2, \ldots, T$ of time points, choosing the time unit as the sampling interval, for convenience. It is always assumed that this sampling does not eliminate our ability to measure appropriate characteristics of the signal. Of course, this need not be so. For example, if $y_1(t) = x(t) + \eta_1(t)$ and $y_2(t) = x(t-\tau) + \eta_2(t)$ and we wish to measure τ but only the $y_j(t)$ are observed, then if these are Gaussian all that can be known that is relevant is the cross spectrum $f_{12}(\omega) = \exp \mathrm{i}\tau\omega \cdot f_x(\omega)$, $-\infty < \omega < \infty$. (Throughout we assume spectra to be absolutely continuous with, at least, continuous densities.) However, if only the discrete-time points are used, then frequencies are aliased. (See Section 1.3(iv) of Chapter 16, "Review of Various Approaches to Power Spectrum Estimation".) Thus all that can be known about the spectrum from the data is

$$\sum_{j=-\infty}^{\infty} \mathrm{e}^{\mathrm{i}\tau(\omega+2\pi j)} f_x(\omega + 2\pi j), \quad -\pi < \omega \leq \pi, \tag{1.1}$$

and if τ is not an integer we cannot determine it without further assumptions. In future we implicitly assume that such problems do not arise. In the major applications we have in view this is not a costly assumption since observations may be made as frequently as is desired and will be made sufficiently frequently to ensure that $f_x(\omega)$ is effectively zero outside $(-\pi, \pi)$, in which case (1.1) reduces to $\exp i\tau\omega \cdot f_x(\omega)$. Though $x(t)$ is not observed and hence $f_x(\omega)$ cannot be directly obtained, the cross spectrum $f_{12}(\omega)$ can be estimated, of course, and hence τ can be estimated since $\tau\omega$ is just the argument of the complex number $f_{12}(\omega)$.

In Chapter 1 on "Wiener Filtering", problems of signal estimation are considered based only on the spectra and cross spectra (assumed known or estimated). This apparatus of Wiener filtering has largely been replaced in recent years by Kalman filtering methods, based on, in some ways, more special models, namely

$$y(t) = Hx(t) + Du(t) + \eta(t),\ x(t+1) = Fx(t) + Gu(t) + \zeta(t+1)\,, \tag{1.2}$$

$$E\{\eta(s)\eta(t)'\} = \delta_{st}Q,\ E\{\zeta(s)\zeta(t)'\} = \delta_{st}R,\ E\{\eta(s)\zeta(t)'\} = \delta_{st}S\,. \tag{1.3}$$

Here $y(t)$ is a vector of r components, $x(t)$ is not observed and has n components and $u(t)$ is observed and has s components. The unobserved signal is $x(t)$. There is a large and important theory concerning (1.2), its detailed structure and estimation. We have introduced (1.2) and (1.3) partly as a basis from which to begin to discuss more special models below. We introduce them also because it should be understood that they may provide a good basis from which to construct statistical procedures of the kind dealt with in this chapter. However, as such methods are not usually frequency-domain methods we shall not deal with them here, in any detail. The first part of (1.2) is very general and merely expresses $y(t)$ as composed of a signal component, $Hx(t)$, not directly observed, and noise. Thus for $r = 2$ and the example first given H could be the unit matrix and $x(t)$ could have the two components $x(t)$ and $x(t-\tau)$. It is the second part of (1.2) that is special. Because of that the relations (1.2) correspond to a rational transfer function model. Indeed they can be rewritten as

$$\begin{aligned} y(t) &= Hx(t\,|\,t-1) + Du(t) + \varepsilon(t)\,, \\ x(t+1\,|\,t) &= Fx(t\,|\,t-1) + Gu(t) + K\varepsilon(t)\,, \end{aligned} \tag{1.4}$$

wherein $\varepsilon(t)$ is the linear innovation sequence for $y(t)$ and $x(t\,|\,t-1)$ is the best linear predictor of $x(t)$ from the past of the $y(t)$, $u(t)$ sequences. The rational transfer functions are $D + H\{zI_n - F\}^{-1}G$, $I_n + H\{zI_n - F\}^{-1}K$, which describe the influence, respectively, of the $u(t)$, $\varepsilon(t)$ sequences on $y(t)$. Thus (1.2) and (1.3) lead to finite parameter models which in turn may lead to economical statistical methods. For the reason stated above, we now leave these models and return to the main purposes of the chapter.

In accordance with the purpose of this volume we shall deal with methods based on the Fourier coefficients,

$$w_j(\omega_v) = T^{-1/2} \sum y_j(t)\, e^{it\omega_v}, \quad \omega_v = 2\pi v/T,\ -\tfrac{1}{2}T < v \leq \tfrac{1}{2}T. \tag{1.5}$$

As is well known, some computational advantages can arise from replacing ω_v by $\omega'_v = 2\pi v/T'$, $T' \gg T$, $-\frac{1}{2}T' < v \leq \frac{1}{2}T'$, where T' is highly composite. (See Chapter 19 on "Computer Programming of Spectrum Estimation".) It is also sometimes recommended that $y_j(t)$, on the right in (1.5), be multiplied by a 'taper'. We omit such details here. Of course, modulo rounding errors, computer errors, the transformation (1.5) neither gains nor loses information. Its use is mainly related to stationarity assumptions. Thus if $y(t)$ is a stationary random vector (and some other 'regularity' conditions are satisfied) and ω is a fixed frequency, then the m vectors $w_y(\omega_v)$, composed of the $w_j(\omega_v)$, for m values of ω_v nearest to ω, m fixed, become, as $T \to \infty$, distributed independently and identically, with the probability density function,

$$(\pi^r \det\{2\pi f(\omega)\})^{-1} \exp(-w_y(\omega_v)^* \{2\pi f(\omega)\}^{-1} w_y(\omega_v)). \tag{1.6}$$

More is known than this. For example, when $r = 1$ we consider the quantities $|w_y(\omega_v)|^2/\{2\pi f(\omega_v)\} = z_v$, let us say, and consider the empirical distribution function of these, using $0 < v < \frac{1}{2}T$, i.e. the function $F_T(x)$ which is the proportion of the z_v that are less than or equal to x, $0 \leq x < \infty$. It can be shown, under appropriate conditions, that this function converges uniformly and almost surely to $1 - e^{-x}$, which is what would be expected if (1.6) held for all $0 < v < \frac{1}{2}T$ (which itself will not be true unless $y(t)$ is Gaussian and the matrix function $f(\omega)$ is a constant matrix). In any case, most of the methods of this chapter can be obtained, via a certain amount of 'sleight of hand', by maximum likelihood (ML) acting as if (1.6) did hold for all ω_v. It may then be shown that the methods are asymptotically valid in the following sense. Even if $y(t)$ is not Gaussian, but is reasonably regular, then the asymptotic distribution of the estimates is that which would obtain were the data Gaussian and the true ML estimators were constructed.

2. Regression problems

In (1.2) if $G = 0$ we have

$$y(t) = Du(t) + e(t), \tag{2.1}$$

$$x(t+1 \mid t) = Fx(t \mid t-1) + K\varepsilon(t), \qquad e(t) = Hx(t \mid t-1) + \varepsilon(t). \tag{2.2}$$

Then (2.2) is of the same form as (1.4) with G, $D = 0$ and with $y(t)$ now called $e(t)$. Thus $e(t)$ is generated by a rational transfer function model and methods based on (2.1) and (2.2) can be very successful. However, there are also cases where methods based on the Fourier coefficients can be useful. One reason is that (2.1) may hold only over certain frequencies, as we now explain. Let us transform (2.1) using (1.5). Then

$$w_y(\omega_v) = Dw_u(\omega_v) + w_e(\omega_v), \quad -\tfrac{1}{2}T < v < \tfrac{1}{2}T. \tag{2.3}$$

There may be noise effects so that $u(t)$ is not observed but these noise effects may predominate at certain frequencies so that (2.3) holds, to an adequate approximation, at other frequencies. Again the apparatus through which $y(t)$ is measured may respond effectively to the signal only at certain frequencies. For that matter the model (2.1) might itself be valid only if the sequences involved are composed only of certain frequencies, being nonlinear, for example, outside of this set of frequencies. Thus we now assume that (2.3) holds only for $\omega \in \mathcal{B} \subset (-\pi, \pi)$. It will be convenient to put $\mathcal{B} = \mathcal{B}_- \cup \mathcal{B}_+$ where $\mathcal{B}_+ \subset (0, \pi)$ and $\mathcal{B}_-$ is the reflection of $\mathcal{B}_+$ in the origin.

Returning to (2.3) it is suggested that we use these equations, for $\omega_v \in \mathcal{B}$, by means of a regression procedure. This is made simpler by the assumption discussed below (1.6) whereby the $w_e(\omega_v)$ may be treated as independent with $E\{w_e(\omega_v)\omega_e(\omega_v)^*\} = 2\pi f_e(\omega_v)$. If this function were known, then the ML estimator using $\omega_v \in \mathcal{B}$ would be, *for the case where* $r = 1$,

$$\hat{D} = \left[\sum_{\mathcal{B}} w_y(\omega_v)w_u(\omega_v)^* f_e(\omega_v)^{-1}\right]\left[\sum_{\mathcal{B}} w_u(\omega_v)w_u(\omega_v)^* f_e(\omega_v)^{-1}\right]^{-1}. \tag{2.4}$$

(For the case $r > 1$, the reader may consult the references in the Bibiliographic Notes.) The row vector $\hat{D}$ may be treated, asymptotically, as normal with mean vector D and covariance matrix estimated by $(2\pi)^{-1}$ by the second factor in (2.4). The quantity $w_u(\omega_v)w_u(\omega_v)^* f_e(\omega_v)^{-1}$ may be considered as $(2\pi)^{-1}$ by the signal-to-noise ratio so that (2.4) would be an excellent formula to use. The main problem arises from the need to estimate that function. Unless care is exercised the inaccuracy in $\hat{D}$ will be dominated by the inaccuracy in the estimate of f_e, which should not be so. One procedure would be first to estimate by replacing $f_e(\omega_v)$ by unity in (2.4). Let $\tilde{D}$ be the resulting estimate. Then we may form

$$2\pi\hat{f}_e(\omega) = \frac{1}{m}\sum_{\omega} |w_y(\omega_v) - \tilde{D}w_u(\omega_v)|^2 \tag{2.5}$$

where here and below Σ_ω is a sum over a band of m frequences, ω_v, centered at ω. Of course, (2.5) will be computed for $\omega = 2\pi j/T$ for such j as make $2\pi j/T$ lie in $\mathcal{B}$. Then $\hat{f}_e(\omega)$ is used in (2.4) in place of $f_e(\omega)$. The process should now be iterated, using $\hat{D}$ in place of $\tilde{D}$, until (say) the trace of the matrix in the second factor stabilises. However, m must not be taken too small or else errors in $\hat{f}_e$ *will* be important. Of course, if m is taken too large, then $f_e(\omega)$ will constitute a smoothed estimate of $f_e(\omega)$ and efficiency in estimation will be lost.

An alternative procedure would be to construct an estimate of $f_e(\omega)$ based on a parametric model, such as (2.2). However, it may be difficult to do this if only a band $\mathscr{B}$ that is less than $[-\pi, \pi]$ is to be used. Used with care and in situations where T is not very small, these methods based on (2.4) seem useful.

A particular case is that where $u(t)$ is composed of lagged values of a variable $z(t)$ so that (2.1) may be rewritten as

$$y(t) = \sum_{-a}^{b} \beta(j) z(t-j) + e(t). \tag{2.6}$$

Now $s = b + a + 1$. We may choose to replace $w_u(\omega_v) w_u(\omega_v)^*$ by

$$|w_z(\omega_v)|^2 [\exp \mathrm{i}(k-l)\omega_v]_{-a \leq k, l \leq b}$$

and $w_y(\omega_v) w_u(\omega_v)^*$ by

$$w_y(\omega_v) \overline{w_z(\omega_v)} \{\exp(-\mathrm{i} l \omega_v)\}_{-a < l \leq b}.$$

One context in which (2.6) has been used (but not via (2.4)) is that where

$$y(t) = x(t-\tau) + \eta_1(t), \qquad z(t) = \ + \eta_2(t). \tag{2.7}$$

Now by expanding $\exp \mathrm{i}\tau\omega$ in a Fourier series it is easy to check that, for stationary $x(t)$,

$$x(t-\tau) = \sum_{-\infty}^{\infty} \frac{\sin \pi(j-\tau)}{\pi(j-\tau)} x(t-j), \tag{2.8}$$

the series converging in mean square. If we estimate (2.6), say, by least-squares regression of $y(t)$ on the $z(t-j)$, say for $a = b < \infty$, where we choose a reasonably large, then the vector of estimated coefficients $\hat{\beta}(j)$ will be such that

$$\left\{\sum_{-a}^{a} \hat{\beta}(j) \,\mathrm{e}^{\mathrm{i}j\omega}\right\} f_z(\omega) \doteq f_x(\omega) \,\mathrm{e}^{\mathrm{i}\tau\omega} = f_x(\omega) \sum_{-\infty}^{\infty} \frac{\sin \pi(j-\tau)}{\pi(j-\tau)} \,\mathrm{e}^{\mathrm{i}j\omega},$$

since both sides represent the cross spectrum between $y(t)$ and $z(t)$. Thus if $f_x(\omega)/f_z(\omega)$ is near to constant over (say) a narrow band where $f_z(w)$ is concentrated, then we can hope that $\hat{\beta}(j)$ will be approximately proportional to $\sin \pi(j-\tau)/\{\pi(j-\tau)\}$, which is maximised at $j = \tau$. Thus a first estimate of τ may be given by the j for which $\hat{\beta}(j)$ is greatest in amplitude. Since

$$\sum_{-a}^{a} \hat{\beta}(j) \,\mathrm{e}^{\mathrm{i}(j-\tau)\omega} = \sum_{-\infty}^{\infty} \mathrm{e}^{\mathrm{i}k\omega} \left\{\sum_{-a}^{a} \hat{\beta}(j) \frac{\sin \pi(k-j+\tau)}{\pi(k-j+\tau)}\right\},$$

which should be near to a constant over the band, a better approximation might be got by choosing τ so as to maximise, with respect to τ,

$$\left| \sum_{-\infty}^{\infty} \hat{\beta}(j) \frac{\sin \pi(j-\tau)}{\pi(j-\tau)} \right| .$$

This argument is all rather loose and no precise justification has been given so far but the method seems to work well in simulations. Since the method is suitable only for narrow band signals, least squares may perform as well as (2.4) for such cases as are relevant since $\mathcal{B}$ may now be narrow and $f_e(\omega)$ near to constant over this band. This means that we are weighting according to the power in $z(t)$ over this band which is equivalent to using least squares if the power in $z(t)$ is negligeable outside of the hand. Losses (biases) due to the approximations introduced may be outweighted by the gain due to estimating only a few $\beta(j)$ (for small a) compared to more elaborate procedures. The use of a finite parameter model in this case leads to the consideration of (1.2), once again, as a basis for delay estimation. We shall discuss this topic in the next section.

Of course, the whole apparatus associated with what is often called the linear model may be applied to the $w_y(\omega_v)$, $w_u(\omega_v)$, the problems becoming those in complex multivariate analysis. (See Chapter 20 on "Likelihood Ratio Tests on Covariance Matrices and Mean Vectors of Complex Multivariate Normal Populations and Their Applications in Time Series" and Chapter 15 on "Frequency-Domain Analysis of Multidimensional Time-Series Data".)

3. Delay estimation

We now consider a spatially arranged array of sensors. For brevity we discuss only the case of two-dimensional space and the case where only one measurement is taken at each sensor, though the other cases are important. The array may be passive (i.e. merely recording the signals it receives) or active (i.e. that signal may be a reflected form of a signal transmitted from the array). We consider here only the former case. If the source is not far away, in terms of the diameter of the array, the wave fronts will be circular but we consider only the case where the fronts are linear. We ignore frequency shifts, due to receiver tuning errors or differential Doppler shifts. We consider the case only of a stationary source. Again we consider only 'off-line' estimation situations and do not discuss 'on-line', real-time, calculations. All of the other cases can be treated (though the treatment becomes very complex in general) and we mention them here so as to indicate the richness of the range of phenomena under the heading of this paragraph. We have, at the beginning of the chapter, already referred to assumptions relating to coherence between the noises or of the noises with the signals. One way of handling the former is to model this coherence as due to a second signal (or to more than one other signal). The problem of many signal sources does not seem to have received any proper treatment. It seems to be more difficult than others mentioned here.

Let the signal be propagated in the direction given by the vector ϕ of unit length. Let, now, $p(k)$ be the vector giving the position, relative to a fixed coordinate system, of the kth sensor. If the speed of propagation is c, then the lag in the arrival of the signal at the kth sensor, as compared to the origin of coordinates, is seen to be

$$\langle p(k), \phi\rangle/c \tag{3.1}$$

where we now write $\langle p(k), \phi\rangle$ for the inner product between the two vectors. In general, c will depend on ω. The quantity (3.1), as a function of ω, is known as the phase delay. If $f_x(\omega)$ is the spectrum of the signal at the origin of coordinates, then the matrix of spectra and cross spectra of the outputs of the sensors is

$$f_x(\omega)[\exp\{i\omega\langle p(j)-p(k), \phi\rangle/c(\omega)\}]_{j,k=1,\ldots,r} + f_n(\omega) \tag{3.2}$$

where $f_n(\omega)$ is diagonal with the jth noise spectrum in that place in the diagonal. The quantity $\lambda(\omega) = c(\omega)/\{\omega/2\pi\}$ is the wavelength at frequency ω and $\omega\phi c(\omega)^{-1}$ could be called the wave-number vector or $\kappa(\omega) = \omega/c(\omega)$ the wave number. Thus $\omega\lambda(\omega)/2\pi = c(\omega)$ and $\omega = \kappa(\omega)c(\omega)$. Of course, wave number has the same interpretation in terms of oscillation along the direction of propagation at a fixed time as does angular frequency ω at a fixed point in space, as time varies. The situation may be understood by relating it to a more general model, namely (using p for a point in the plane)

$$x(p, t) = \int_0^{2\pi}\int_0^\infty\int_0^\infty [\cos\{r\kappa\cos(\theta-\psi) - t\omega\}\,d\xi(\kappa, \omega, \psi) \\ + \sin\{r\kappa\cos(\theta-\psi) - t\omega\}\,d\zeta(\kappa, \omega, \psi)]\,.$$

Here p has been represented in polar form, (r, θ). This formula composes $x(p, t)$ linearly from plane waves, in each direction ψ, at each wave number κ and with each frequency ω. The functions ξ and ζ determine the amplitude and phase of the oscillation. To see this, keep κ, ω and ψ fixed. Then when t is fixed also, the integrand is constant along lines orthogonal to the ray in the direction ψ and in that direction is a sinusoidal oscillation with (angular) frequency κ. For (r, θ) fixed and t varying, the integrand represents a temporal oscillation with frequency ω. The cases we treat in the remainder of this chapter are the simple ones where the functions ξ and ζ are concentrated at a particular value of ψ and where also their mass is concentrated along a curve $\kappa = \omega/c(\omega)$ in the κ, ω plane. If $c(\omega) \equiv c$, then $x(p, t)$ is a wave form with linear wave fronts orthogonal to the direction ψ and the wave form propagates without changing its shape. If $c(\omega)$ is not constant, then each constituent frequency component is of this kind but they are propagating with different velocities so that the wave form does change shape as different frequency components move through one another. The general case is one without a predominant direction of motion and consequently without any linear wave fronts.

We now return to the special case where there is only one direction of propagation and $\kappa = \omega/c(\omega)$. In fact, there may also be attenuation of the signal so that the spectrum of what is observed is

$$f_x(\omega)a(\omega)a(\omega)^* + f_n(\omega)\,, \tag{3.3}$$

where

$$a(\omega)^* = (a_j(\omega)\exp\{-\mathrm{i}\omega\langle p(j), \phi\rangle/c(\omega)\})_{j=1,\dots,r}, \quad a_j(\omega) > 0\,.$$

The $a_j(\omega)$ describes the relative attenuation at the jth sensor. What we seek to measure is $c(\omega)^{-1}\phi$, though sometimes attenuation may also be of interest.

The phase $\theta_{jk}(\omega)$ between the output of the jth and kth sensors, i.e. the argument of their cross spectrum, is (from (3.2) or (3.3))

$$\theta_{jk}(\omega) = \omega\langle p(j) - p(k), \phi\rangle/c(\omega)\,, \tag{3.4}$$

which is ω by the relative phase delay. This quantity is estimated by means of

$$\frac{w_j(\omega_v)\overline{w_k(\omega_v)}}{|w_j(\omega_v)\overline{w_k(\omega_v)}|} = \exp \mathrm{i}\hat{\theta}_{jk}(\omega_v), \quad j, k = 1, \dots, r\,. \tag{3.5}$$

It is these quantities that we shall use to estimate $c(\omega)^{-1}\phi$. We call $\sigma_{jk}(\omega)$ the coherence between the output of the jth and kth sensors. We shall parameterise $c(\omega)^{-1}\phi$ by means of a vector τ. Examples are as follows:

(a) $r = 2$, $c(\omega) \equiv c$. Then $\tau = \langle p(1) - p(2), \phi\rangle/c$. Of course, $c^{-1}\phi$ cannot itself be determined for $r = 2$ but only the delay, τ. If ϕ is known this determines c and if c is known the cosine of the angle between ϕ and the line joining $p(1)$ and $p(2)$ is known.

(b) $r = 2$, $c(\omega) = (a + b\omega)^{-1}$. Then τ has two components $a\langle p(1) - p(2), \phi\rangle$, $b\langle p(1) - p(2), \phi\rangle$.

(c) $r \geq 3$, $c(\omega) \equiv c$, $\tau = c^{-1}\phi$, provided the $p(j)$ are not on the same line.

(d) $r \geq 3$, $c(\omega)^{-1}$ is represented by a cubic spline with prescribed knots. Then τ describes the vector ϕ together with the parameters of the spline, again assuming the $p(j)$ not to be on the same line.

Of course, if the $p(j)$ lie on the same line, then $c^{-1}\phi$ cannot be determined since, in the Gaussian case, all that we may know is contained in the knowledge of the $\langle p(j) - p(k), \phi\rangle/c$; while if the $p(j)$ lie along a line the $\binom{r}{2}$ row by 2 column matrix with typical row $p(j) - p(k)$ is singular and conversely. If we knew the $\sigma(\omega)$, we would choose $\hat{\tau}$, the estimate of τ, so as to maximise

$$Q(\tau) = -\sum_{\omega_v \in \mathscr{B}_+}\sum_{j<k} \sigma^{jk}(\omega_v)\sigma_{jk}(\omega_v)\cos\{\hat{\theta}_{jk}(\omega_v) - \theta_{jk}(\omega_v)\}\,. \tag{3.6}$$

This is again ML, after a certain amount of reorganisation of the formula. Of course, (3.6) is a function of τ through the $\theta_{jk}(\omega_v)$ (see (3.6)). Here $\sigma^{jk}(\omega)$ is the typical entry in the matrix that is the reciprocal of the matrix with entries

$\sigma_{jk}(\omega)$. If $r = 2$, then (3.6) becomes

$$\sum_{\omega_v \in \mathscr{B}_+} \left\{ \frac{\sigma^2(\omega_v)}{1 - \sigma^2(\omega_v)} \right\} \cos\{\hat{\theta}(\omega_v) - \theta(\omega_v)\}, \tag{3.7}$$

where the subscripts 1 and 2 have now been eliminated since there are only two records. The minus sign is inserted in (3.6) so that no minus sign appears in (3.7). The physical meaning of these formulae is rather direct. Imagine that, in the two-sensor case, we can steer the array by rotating it about its centre. We do this so that the received signal is strongest, in sum, by rotating the array so that it is (nearly) at right angles to the direction of propagation. This estimates that direction for us. If we cannot steer the array, almost the same effect is achieved by rephasing the outputs of the different sensors. Thus if $\theta(\omega_v)$ is the relative rephasing at frequency ω_v, then the summed output is

$$w_1(\omega_v) + w_2(\omega_v)\, e^{i\theta(\omega_v)}, \tag{3.8}$$

whose squared modulus is

$$|w_1(\omega_v)|^2 + |w_2(\omega_v)|^2 + 2\mathscr{R}\{w_1(\omega_v) w_2(\omega_v)\, e^{-i\theta(\omega_v)}\}.$$

Since $|w_1(\omega_v)\overline{w_2(\omega_v)}|$ can be viewed as estimating $\sigma(\omega_v)\{f_1(\omega_v) f_2(\omega_v)\}^{1/2}$ where f_1 and f_2 are the spectra of the ouptuts (see Chapter 15 on "Frequency-Domain Analysis of Multidimensional Time-Series Data"), we are led to replace this by $\sigma(\omega_v)\{f_1(\omega_v) f_2(\omega_v)\}^{1/2} \cos\{\hat{\theta}(\omega_v) - \theta(\omega_v)\}$, omitting the terms that do not depend on $\theta(\omega_v)$ and a factor 2. When an appropriate weighting by frequencies is introduced, this leads to (3.7). Clearly this is maximised by taking $\theta(\omega_v)$, on average, as near to the $\hat{\theta}(\omega_v)$ as possible remembering that it, $\theta(\omega_v)$, can be varied only by varying τ. Since $\{1 - \sigma^2(\omega_v)\}/\sigma^2(\omega_v)$ is (essentially) proportional to the variance of $\hat{\theta}(\omega_v)$, the weighting is clearly the appropriate one. The numerical approximation to the steering of an array described above is called 'delay and sum beamforming' and the additional detail in (3.6) and (3.7) is due to the appropriate weighting of frequencies and a certain amount of re-arrangement. The word 'beamforming' occurs because of the duality that exists between the problems of arranging an array to optimally receive and to optimally transmit a signal in a given direction.

Under rather general conditions the $\hat{\tau}$ may, asymptotically, be treated as normal with mean τ and variance covariance matrix $T^{-1}V^{-1}$, where V may be estimated as

$$\left[\frac{1}{T} \sum_{\omega_v \in \mathscr{B}_+} \sum_{j<k} \sigma^{jk}(\omega_v) \sigma_{jk}(\omega_v) \frac{\partial \hat{\theta}_{jk}(\omega_v)}{\partial \tau_a} \frac{\partial \hat{\theta}_{jk}(\omega_v)}{\partial \tau_b} \right]_{a,b=1,2,\ldots}. \tag{3.9}$$

Here τ_a is the ath component of τ. Thus in case $\tau = c^{-1}\phi$, then (3.8) becomes

$$\left[\frac{1}{T}\sum_{\mathscr{B}_+}\omega_v^2\sum_{j<k}\sigma^{jk}(\omega_v)\sigma_{jk}(\omega_v)\{p_a(j)-p_a(k)\}\{p_b(j)-p_b(k)\}\right], \tag{3.10}$$

and in case $r=2$ and $\tau=\langle p(1)-p(2), \phi\rangle/c$, this becomes

$$\frac{1}{T}\sum_{\mathscr{B}_+}\omega_v^2\left\{\frac{\sigma^2(\omega_v)}{1-\sigma^2(\omega_v)}\right\}. \tag{3.11}$$

There is no doubt that these procedures would be very effective and that under most circumstances met in practice the asymptotic theory would apply well but $\sigma_{jk}^2(\omega)$ will not be known and care must be exercised in its estimation. The most obvious procedure is the use of

$$\hat{\sigma}_{jk}(\omega)=\frac{|\Sigma_\omega w_j(\omega_v)\overline{w_k(\omega_v)}|}{\{\Sigma_\omega|w_j(\omega_v)|^2\,\Sigma_\omega|w_k(\omega_v)|^2\}^{1/2}}, \tag{3.12}$$

where the same notation is used as in (2.5), and (3.12) is computed for $\omega=2\pi j/T$ and such j as make this value lie in $\mathscr{B}_+$. A main problem is again the possibility that errors in the estimation of $\sigma_{jk}(\omega)$ will dominate. It should be observed that $(m-1)\hat{\sigma}_{jk}^2/\{1-\hat{\sigma}_{jk}^2\}$ is approximately distributed as F with 2 and $2m-2$ degrees of freedom. Hence for values of m likely to be used in practice $\hat{\sigma}_{jk}(\omega)$ does not differ from zero at any reasonable significance level unless it is above 0.30 (i.e. $\hat{\sigma}_{jk}^2>0.09$). It is likely that low values of $\hat{\sigma}_{jk}(\omega)$ will occur at higher frequencies. It is then dangerous to include these frequencies because, as (3.11) shows, they have high weight in the formulae and if a spuriously high value of $\sigma_{jk}^2(\omega_v)$ leads to overweighting, then an inaccurate estimate of τ may be obtained. Of course, in practice, the range of frequencies that should be used may be known because the range over which the signal is present and the range over which the apparatus responds to the signal may be known. It is also known that (3.12) will be biased down if $\theta_{jk}(\omega)$ is not zero and the delay $\langle p(j)-p(k), \phi\rangle/c(\omega)$, relatively large. Usually the value of this delay is approximately known, in which case the $y_j(t)$ can be rephased so that this is small. This should always be done. If no such information is available, then a first estimate of τ could be used to obtain estimates of the delays and the rephasing could be done on this basis.

The calculation of $\hat{\tau}$ may be aided by the use of a fast Fourier transform algorithm. Consider (3.7), for example, for $\tau=\langle p(1)-p(2), \phi\rangle/c$. Let M be the period of $Q(\tau)$, i.e. $M=T/b$ where b is the largest v for which $\omega_v\in\mathscr{B}_+$. Then for $\tau=t/2$, $t=0, \pm1, \ldots, \pm M$, $Q(\tau)$ becomes the real part of

$$\frac{1}{2T}\sum_{1}^{2T}C_v\exp(\mathrm{i}2\pi vt/2T), \tag{3.13}$$

where C_v is $2T\sigma^2(\omega_v)/\{1-\sigma^2(\omega_v)\}\exp\{-\mathrm{i}\hat{\theta}(\omega_v)\}$ for $\omega_v\in\mathscr{B}$ and is otherwise zero. We may evaluate $Q(\tau)$ for $\tau = t/2$ by fast Fourier transformation of the C_v, choose the value of $t/2$ maximising $Q(t/2)$ and then locate the actual maximum by a standard function optimisation routine. In more elaborate cases a similar procedure may be used. For example, when $r = 2$ and $c(\omega) = c\omega^{-1}$, $\omega > 0$, then $\tau = c^{-1}\langle p(1) - p(2), \phi\rangle$ as before and we must break the range of ω into subranges over which ω^2 is nearly linear and proceed as above to evaluate $Q(\tau)$ at a discrete set of points in each subrange.

Finite parameter models such as (1.2) could be used for the present situation and one use is given in Section 2. Such methods are more easily put into a recursive form suitable for real-time calculation but we do not discuss that here.

Often the signal will not be stationary, though it is usually relevant to treat the noise as stationary. The essential point concerning transient signals is that the $w_x(\omega)$ will now vary smoothly with ω and not in the chaotic fashion that the discussion below (1.6) indicates.

Thus if we call $w_x(\omega)$ the Fourier coefficient of $x(t)$, the signal recorded at the origin and $w(\omega)$, $w_n(\omega)$ the vectors of $y_j(t)$ and noise Fourier coefficients, then, approximately

$$w(\omega_v) = w_x(\omega_v)\zeta(\omega_v) + w_n(\omega_v), \tag{3.14}$$

where

$$\zeta_j(\omega_v) = a_j(\omega_v) \exp i\{\omega_v\langle p(j), \phi\rangle/c(\omega_v)\}.$$

We shall, however, here discuss only the case where there is no attenuation. Now put

$$\langle w\rangle(\omega) = \frac{1}{m}\sum_{\omega} w(\omega_v), \tag{3.15}$$

using the same notation as in (2.5) and (3.12). *Then, provided the transient signal is phased so that it is concentrated near $t = 0$,* approximately

$$\langle w\rangle(\omega_v) = w_s(\omega_v)\zeta(\omega_v) + \langle w_n\rangle(\omega_v) \tag{3.16}$$

since w_s and ζ are smooth features of ω. The requirement that $x(t)$ not be rephased is forced by the fact that a substantial rephasing, by $T/2$ for example, would introduce a factor $\exp(i\omega T/2)$ into $w_x(\omega)$ which oscillates at such a high frequency that averaging over a band of m frequencies would reduce $\langle w_x\rangle(\omega_v)$ well below $w_x(\omega_v)$ in magnitude. Of course, for $m = 1$ this would not be so. We are averaging in (3.15) so as to enhance the signal-to-noise ratio since each component of $\langle w_n\rangle(\omega)$ will have a mean square near m^{-1} by the mean square of that component of $w_n(\omega)$. This will be vitiated if $\langle w_x\rangle(\omega)$ is much less than $w_x(\omega)$ in magnitude.

We may choose τ by minimising

$$\sum_{\mathscr{B}_+} [\{\langle w\rangle(\omega_v) - w_x(\omega_v)\zeta(\omega_v)\}^* f_n(\omega_v)^{-1}\{\langle w\rangle(\omega_v) - w_x(\omega_v)\zeta(\omega_v)\}]$$

with respect to τ and the $w_x(\omega_v)$. This is the same as maximising with respect to τ the function

$$Q(\tau) = \sum_{\mathscr{B}_+} |\zeta^*(\omega_v)f_n(\omega_v)^{-1}\langle w\rangle(\omega_v)|^2/\{\zeta^*(\omega_v)f_n(\omega_v)^{-1}\zeta(\omega_v)\}\,. \tag{3.17}$$

The quantities

$$\frac{1}{m_1-1}\sum_{\omega} \{w(\omega_v) - \langle w\rangle(\omega_v)\}\{w(\omega_v) - \langle w\rangle(\omega_v)\}^* = \hat{f}_n(\omega) \tag{3.18}$$

may be used to estimate $f_n(\omega_v)$, where now Σ_ω is a sum over a band of m_1 frequencies centred at ω. Here m_1 is chosen having in mind the smoothness of $f_n(\omega)$ while m in (3.15) reflects the smoothness in $w_x(\omega)$ and $\zeta(\omega)$. *Note that in this treatment no assumption of incoherence between the noise series is required.* An asymptotic theory can be constructed for such methods but we do not go into that here. Of course, the weighting by $f_n(\omega_v)^{-1}$ in (3.17), or for that matter the weighting in (2.4) and (3.6), could be replaced by a priori chosen weight function, for example $f_n(\omega) \equiv I_r$. If $m = 1$ this would have to be done in (3.17), the formula then becoming

$$Q(\tau) = \frac{1}{r}\sum_{\mathscr{B}_+} |\zeta^*(\omega_v)w(\omega_v)|^2 \tag{3.19}$$

since $\zeta^*(\omega_v)\zeta(\omega_v) = r$. For that matter if $f_n(\omega)$ is diagonal, $\zeta^*(\omega_v)f_n(\omega_v)^{-1}\zeta(\omega_v)$ is again independent of τ. In (3.19) or when $f_n(\omega_v)$ is diagonal the calculation of $Q(\tau)$ may again by simplified by using a fast Fourier transform algorithm as in (3.13). Of course, (3.19) may be used in the stationary case also.

We conclude by mentioning that the virtue of methods such as those based on (3.17) or (3.19) is that they lend themselves to the multiple signal case. Thus since (3.19) is valid whether or not the signal is stationary, it could be used to obtain initial estimates for the multiple signal case.

Bibliographic Notes

Section 1

A general reference on time-series methods is Hannan (1970). The chapters on "Wiener Filtering" (Chapter 1), "Likelihood Ratio Tests on Covariance Matrices and Mean Vectors of Complex Multivariate Normal Populations in Time Series" (Chapter 20), "Frequency-Domain Analysis of Multidimensional Time-

Series Data" (Chapter 15), "Review of Various Approaches to Power Spectrum Estimation" (Chapter 16) and "Computer Programming of Spectrum Estimation" (Chapter 19), in this volume, also contain basic information for this chapter. The model (1.2) has an enormous literature surrounding it commencing from Kalman (1960). A special issue of *IEEE, Automatic Control* **AC-19**, No. 6, December 1974 dealt with this model. For results of the type of (1.6) see, for example, Hannan (1970). The result concerning $F_T(x)$, and related results, is given in Chen and Hannan (1980).

Section 2

Techniques of the kind in (2.4) were introduced in Hamon and Hannan (1963). There is quite a large subsequent literature concerning them. See, for example, Engle (1974) and Doran (1976). The method based on (2.7) and (2.8) is introduced in Chan, Riley and Plant (1980).

Section 3

A special issue of *IEEE, Acoustics, Speech and Signal Processing* **ASSP-29**, No. 3, June 1981 is devoted to delay estimation and this volume contains a great deal of information about the subject of Section 3. Formula (3.6) was introduced in Hamon and Hannan (1974). See also Hannan and Thomson (1973). The techniques based on (3.16) were introduced in Cameron and Hannan (1979). The techniques based on (3.6) have also been used by other people. See Carter (1981) for references. We emphasise again the wide range of problems in this area and the narrow range covered in this survey and again refer the reader to the special issue of *IEEE*, **ASSP-29** mentioned above.

References

Cameron, M. A. and Hannan, E. J. (1979). Transient signals. *Biometrika* **66**, 243–258.

Carter, G. Clifford (1981). Time delay for passive sonar signal processing. *IEEE Trans. Acoustics, Speech, Signal Processing* **ASSP-29**.

Chan, Y. T., Riley, J. M. and Plant, J. B. (1980). A parameter estimation approach to time delay estimation and signal detection. *IEEE Trans. Acoustics, Speech, Signal Processing* **ASSP-28**, 8–16.

Chen, Zhao Guo and Hannan, E. J. (1980). The distribution of periodogram ordinates. *J. Time Series Anal.* **1**, 73–82.

Doran, H. E. (1976). A spectral principal components estimator of the distributed lag model. *International Economic Review* **17**, 8–25.

Engle, R. F. (1974). Band spectrum regression. *International Economic Review* **15**, 1–11.

Hamon, B. V. and Hannan, E. J. (1963). Estimating relations between time series, *J. Geophysical Research* **68**, 6033–6041.

Hamon, B. V. and Hannan, E. J. (1974). Spectral estimation of time delay for dispersive and non-dispersive systems. *J. Roy. Statist. Soc. Series C (Applied Statistics)* **23**, 134–142.

Hannan E. J. (1970). *Multiple Time Series*. Wiley, New York.

Hannan, E. J. and Thomson, P. (1973). Estimating group delay. *Biometrika* **60**, 241–253.

Kalman, R. E. (1960). A new approach to filtering and prediction problems. *J. Basic Engineering* **82**, 35–46.

D. R. Brillinger and P. R. Krishnaiah, eds., *Handbook of Statistics, Vol. 3*

7

Complex Demodulation: Some Theory and Applications*

T. Hasan

1. Introduction

Complex demodulation may be viewed in part as a narrow-band filtering technique which lets one look at the *components* of a time series, within a small frequency band of interest, as a function of time. Operationally, it is much like heterodyning which is used, for example, in AM radio to process information carried through the amplitude and phase modulations. The theory of modulation/demodulation is therefore a well-established and often-used technique in communications (e.g. see Brown and Palermo, 1969).

The original motivation for the use of this technique in time-series analysis was provided by Tukey (1961) who pointed out its usefulness for viewing the *components* generating either a peak in the spectrum of a series, or, a frequency of interest, as a narrow-band signal. Since the method of complex demodulation shifts each frequency of interest to zero and then applies a low-pass filter, the author observed that it made sense to look at the resulting low-frequency images of the more or less gross-frequency components of the time series as they would be more evident to the eyes.

This technique has the further advantage of producing statistics which can be used in many data analytic and formal statistical procedures. For example, it may be used to detect the presence of narrow frequency band signals, to examine a stretch of series for stationarity or to estimate the arrival time of a transient signal (Childers and Pao, 1972). Alternatively, one can construct tests based on the complex demodulates in order to formalize the above procedures. The technique may be used to estimate time-dependent spectra (Priestley, 1965) and has proven invaluable in situations requiring estimation of higher-order spectra (e.g. Godfrey, 1965a; Huber et al., 1971). Complex demodulation has also proved useful in pitch detection by use of a modified procedure called 'saphe cracking' (Bogert, Healy and Tukey, 1963). It has been used in the

*This manuscript is part of the author's doctoral dissertation, written at the University of California at Berkeley. The research was supported in part by US Public Health Service Grant USPHS ES01299-14 and National Science Foundation Grant MCS 7801422.

search for a series $X(t)$, driving an observed series $Y(t)$ (Brillinger, 1973). In cases of frequency modulation where the frequency of the dominant spectral peak increases linearly (and slowly) with time, e.g. $\omega_t = \alpha + \beta t$, demodulation with a time-varying frequency has been successful in estimating the slope β (Munk et al., 1963). By 'remodulating' the demodulates we essentially obtain a narrow band-pass filtered version of our original series, denoted $X(t, \omega)$, which can then be used in principal components and canonical analyses of time series (Brillinger, 1975). Finally, the method of complex demodulation has proven to be very useful in estimating the parameters in certain models in earthquake analysis (Bolt and Brillinger, 1979).

In Section 2, we introduce the basics of complex demodulation and present some known results. Section 3 is concerned with formalizing the statistical properties of the demodulates and some statistics based on them. Of special interest are the subsections on spectrum estimation and the setting of approximate confidence intervals. The applications are presented in Section 4, along with a large sample result for estimating the parameters in a class of models of the form

$$X(t) = \sum_{k=1}^{K} R_k^T(t; \theta_k) \cos(\omega_k t + \delta_k) + \epsilon(t), \tag{1.1}$$

where $\epsilon(t)$ is a stationary, mixing time series, $\{\theta_k, \omega_k, \delta_k\}$ are the parameters to be estimated, and where the superscript T denotes the dependence of the amplitude function upon the length of the series.

2. Basics of complex demodulation

In this section we introduce complex demodulation: the methodology, computational procedures, some general results and a discussion on filters.

2.1. Methodology

Let $X(t)$, $t = 1, \ldots, T$, be a realization of the time series of interest. Operationally, complex demodulation requires that we first form a frequency-shifted series

$$Y(t) = X(t) \exp\{-i\omega_0 t\},$$

where ω_0 is the center frequency of the band in which we want to view the time series. We note that complex multiplication is necessary to discriminate between the frequencies $\omega_0 + \delta$ and $\omega_0 - \delta$, where δ is typically small, after the frequency shift. Next, we smooth the series $Y(t)$ by low-pass filtering, that is, by forming

$$W_x^L(t, \omega_0) = \sum_{u=-L}^{L} a(u)Y(t+u), \tag{2.1}$$

where $\{a(u)\}$, $u = -L, \ldots, 0, \ldots, L$, are the nonzero low-pass filter coefficients. We shall assume that $a(u)$ is of the form $h(u/L)/H^L(0)$, where $h(v)$, $-\infty < v < \infty$, is bounded, is of bounded variation and vanishes for $|v| > 1$, and

$$H^L(\lambda) = \sum_{u=-L}^{L} h(u/L) \exp\{i\lambda u\}. \tag{2.2}$$

If the following condition is satisfied,

$$\sum_{u=-\infty}^{\infty} [1 + |u|^\lambda]|a(u)| < \infty$$

for some $\lambda > 0$, then we can define the transfer function of the coefficients $\{a(u)\}$, $A(\lambda)$, as

$$A(\lambda) = \sum_{u=-\infty}^{\infty} a(u) \exp\{i\lambda u\}. \tag{2.3}$$

We note that $A(\lambda)$ takes on the value 1 at zero frequency which is why we have defined $a(u)$ as above. The functions $h(u/L)$ are usually called *data windows* or *tapers*.

$W_x^L(t, \omega_0)$ appearing in (2.1) is called the *complex demodulate at time t and frequency* ω_0. We shall usually suppress the use of ω_0 in the argument in $W_x^L(t, \omega_0)$ which we shall then denote simply as $W_x^L(t)$.

Let us denote the real and imaginary parts of $W_x^L(t)$ by $W_1(t)$ and $W_2(t)$, respectively. Furthermore, we note that since $W_x^L(t)$ is complex-valued, we can also write it as

$$W_x^L(t) = |W_x^L(t)| \exp\{-i\phi_x^L(t)\}, \tag{2.4}$$

where $|W_x^L(t)|$ is the *instantaneous amplitude* and $\phi_x^L(t)$ is the *instantaneous phase.*

As pointed out by Tukey (see Ref. [28]), the term 'instantaneous' statistically implies a stretch of time long enough to provide many degrees of freedom on the frequency band that leads to an (averaged) estimate at time t. It should be clear then that $|W_x^L(t)|$ and $\phi_x^L(t)$ represent, respectively, estimates of the 'average' amplitude and 'average' phase in the frequency band $(\omega_0 \pm \delta)$, evaluated in the neighborhood of each time instant t. Similarly, $C_0|W_x^L(t)|^2$, where C_0 is some constant proportional to the bandwidth of the filter used in demodulating, can be thought of as the estimate of the 'average' power evaluated in the neighborhood of the time-instant time t (see Priestley, 1965).

It is well known that if the series $X(t)$ is wide-sense stationary, that is $EX(t) = c_x$, c_x constant, and $\operatorname{cov}\{X(t+u), X(t)\} = c_{xx}(u)$, then there exists a random measure $\mathrm{d}Z_x(\omega)$ such that

$$X(t) = \int_{-\pi}^{\pi} \exp\{\mathrm{i}\omega t\}\, \mathrm{d}Z_x(\omega)\,. \tag{2.5}$$

By use of representation (2.5), the complex demodulate at frequency ω_0 can be written as

$$\begin{aligned} W_x^L(t) &= \sum_u a(u) \int_{-\pi}^{\pi} \exp\{\mathrm{i}(\omega - \omega_0)(t+u)\}\, \mathrm{d}Z_x(\omega) \\ &= \int_{-\pi}^{\pi} A(\omega - \omega_0) \exp\{\mathrm{i}(\omega - \omega_0)t\}\, \mathrm{d}Z_x(\omega)\,. \end{aligned} \tag{2.6}$$

In order to make certain approximations to the integral in (2.6), we shall assume that $A(\omega)$ corresponds to the transfer function of an ideal low-pass filter centered at ω_0 with bandwidth 2Δ, that is,

$$\begin{aligned} A(\omega) &= 1 \quad \text{for } |\omega \pm \omega_0| \leqslant \Delta\,, \\ &= 0 \quad \text{otherwise}\,, \end{aligned} \tag{2.7}$$

for $-\pi < \omega < \pi$ and Δ small. In this case, $A(\omega)$ does not satisfy (2.3). However, it is still possible to define the output of such a filter as a limit in mean square, so that

$$W_x^L(t) = \int_{\omega_0-\Delta}^{\omega_0+\Delta} \exp\{\mathrm{i}(\omega - \omega_0)t\}\, \mathrm{d}Z_x(\omega)\,. \tag{2.8}$$

We shall make further use of this representation in illustrating some statistical properties of the demodulates.

2.2. Examples

Let us now consider complex demodulation of the following series:

$$X(t) = R(t)\cos(\omega t + \delta)\,, \quad t = 1, \ldots, T\,, \tag{2.9}$$

where $R(t)$ is a known amplitude function. Frequency shifting by ω_0, we obtain

$$\begin{aligned} Y(t) &= X(t)\exp\{-\mathrm{i}\omega_0 t\} \\ &= \tfrac{1}{2}R(t)\{\exp\{\mathrm{i}[(\omega - \omega_0)t + \delta]\} + \exp\{-\mathrm{i}[(\omega + \omega_0)t + \delta]\}\}\,. \end{aligned}$$

If $R(t)$ is slowly varying[1] and if ω is close to ω_0, then the result of low-pass filtering the series $Y(t)$ using coefficients with the transfer function given by (2.7) is

$$W_x^L(t) \cong \tfrac{1}{2}A(0)R(t)\exp\{i[(\omega - \omega_0)t + \delta]\}. \tag{2.10}$$

Since, by definition, $A(0) = 1$, we obtain the following approximate expressions for the instantaneous amplitude and the instantaneous phase:

$$|W_x^L(t)| \cong \tfrac{1}{2}R(t) \tag{2.11}$$

and

$$\arg W_x^L(t) \cong (\omega - \omega_0)t + \delta. \tag{2.12}$$

Usually we will want to plot either $|W_x^L(t)|$ or $\log|W_x^L(t)|$ and $\arg W_x^L(t)$ against time.

We now present a few examples of complex demodulating the series (2.9) for different forms of $R(t)$.

EXAMPLE 1 (Constant). Suppose $R(t) = R$, $t = 1, \ldots, T$, then (2.11) becomes

$$|W_x^L(t)| \cong \tfrac{1}{2}R. \tag{2.13}$$

Expression (2.13) indicates that if we plot the instantaneous amplitude against time, we can expect a near constant plot near R. The instantaneous phase (modulo 2π) will give segments of straight lines with slope $(\omega - \omega_0)$ as indicated by (2.12). Such an appearance of the phase plot, called spiralling, suggests the presence of a periodic component with period near $2\pi/\omega_0$.

EXAMPLE 2 (Beating Waves). Suppose $R(t) = R\cos\eta t$ with η very small and $\eta \ll \omega_0$, then expression (2.11) becomes

$$|W_x^L(t)| \cong R|\cos\eta t|. \tag{2.14}$$

Now the plot of the instantaneous amplitude will have the appearance of fluctuating slowly as $|\cos\eta t|$.

EXAMPLE 3 (Exponential Decay). Suppose $R(t) = \alpha\exp\{-\beta t\}$, then (2.11) becomes

$$|W_x^L(t)| \cong \alpha\exp\{-\beta t\}. \tag{2.15}$$

[1]Priestley (1965) provides a more formal discussion of 'slowly-varying' (or 'evolutionary' as he calls it) processes. Heuristically speaking, for (2.10) to hold even in an approximate sense, it is necessary that the length of the filter be much smaller than the maximum interval over which the underlying process may be treated as approximately stationary. In this case, we can use standard linear filter theory despite the nonstationary character of the input.

Suppose we take the log of the instantaneous amplitude, (2.15) then becomes

$$\log|W_x^L(t)| \cong -\beta t + \log\alpha\,. \tag{2.16}$$

When plotted against time, (2.16) will give us an approximately straight line.

2.3. *Ramifications*

(1) As we saw above in Example 3, the log instantaneous amplitude function is approximately linear in t when $X(t)$ is an exponentially decaying cosinusoid. Therefore, it seems reasonable to fit a least-squares regression line and obtain estimates for α, β. For other forms of $R(t)$ we could also consider some type of curve fitting.

(2) We could also consider fitting a line to a linear segment of the phase plot and thus obtain an estimate of the slope $(\omega-\omega_0)$, call it Δ, appearing in (2.12). We could then complex demodulate the original data again at frequency $\omega_0+\Delta$ and the procedure could be iterated until the phase plot was approximately constant over time (at least in stretches of interest). Once the phase was near horizontal, we would be essentially getting at an estimate of the phase angle δ.

(3) For the instantaneous phase, $\phi_x^L(t)$, it also makes sense to look at the derivative, $\mathrm{d}\phi_x^L(t)/\mathrm{d}t$, to see how the phase angle is changing with time. This might be useful for estimating arrival times for transient signals.

2.4. *Some computational considerations*

There are a variety of computational considerations suggested in Bingham, Godfrey and Tukey (1967). For example, the paper points out that we need approximately $6+4L$ computations per data point. To reduce the number of computations, the authors suggest that

(1) We use *decimation*. That is, since we have low-pass filtered the shifted series, we do not lose much information by computing the complex demodulates at every Dth point, where $D=L/\alpha$; i.e. some fraction of L, instead of at every value of t, $t=1,\ldots,T$. In this case we will need approximately $6+4\alpha$ computations per original data point and we of course expect

$$6+4\alpha \ll 6+4L\,.$$

(2) We do the computations via a fast Fourier transform (FFT) algorithm. That is, first compute the FFT of the entire time series

$$d_x^T\left(\frac{2\pi s}{T}\right)=\sum_{t=1}^{T} X(t)\exp\left\{-\frac{\mathrm{i}2\pi st}{T}\right\} \quad \text{for } s=0,1,\ldots,T-1\,. \tag{2.17}$$

Next, we multiply by a suitable transfer function, $A^T(\lambda)$, which is centered at $2\pi s_0/T\cong\lambda_0$, the frequency of interest, and is zero except over a relatively short band of frequencies. Finally, let us shift the result by $2\pi s_0/T$ and take the inverse Fourier transform. This will yield

$$\frac{1}{N}\sum_{s^*-s_0=0}^{N-1} d_x^T\left(\frac{2\pi(s^*-s_0)}{T}\right)A^T\left(\frac{2\pi(s^*-s_0)}{T}\right)\exp\left\{\frac{\mathrm{i}2\pi(s^*-s_0)t}{N}\right\}, \tag{2.18}$$

where $N = \#\{s^* \in S\}$. We recognize this as the complex demodulate at time t (e.g. compare with (2.6)).

The authors point out that a possible disadvantage of this method is that we have replaced a transverse filter of limited length by a circular filter extending over the entire time series. So we now have to worry about leakage across time rather than frequency.

2.5. *Use of complex demodulation to obtain a band-pass filter and the corresponding Hilbert transform*

We now give a result, well known in the communications theory literature, which indicates how complex demodulation may be used to obtain a band-pass filtered version of a series, $X(t)$, and its corresponding Hilbert transform, $X^H(t)$. Let

$$\begin{aligned} V_1(t) &= W_1(t)\cos\omega_0 t + W_2(t)\sin\omega_0 t\,,\\ V_2(t) &= W_1(t)\sin\omega_0 t - W_2(t)\cos\omega_0 t\,, \end{aligned} \tag{2.19}$$

where $W_1(t)$ and $W_2(t)$ are the real and complex parts of the complex demodulate, $W_x^L(t)$, and ω_0 is the frequency of demodulation. We now have

LEMMA 2.1. *Let $\{a(u)\}$ be a filter with transfer function $A(\omega)$, $-\infty < \omega < \infty$, then the operation carrying the series $X(t)$, $-\infty < t < \infty$, into the series $V_1(t)$ of* (2.19) *is linear and time invariant with transfer function*

$$B(\omega) = \frac{A(\omega - \omega_0) + A(\omega + \omega_0)}{2}. \tag{2.20}$$

The operation carrying the series $X(t)$ into $V_2(t)$ of (2.19) *is also linear and time invariant with transfer function*

$$C(\omega) = \frac{A(\omega - \omega_0) - A(\omega + \omega_0)}{2\mathrm{i}}. \tag{2.21}$$

□

We note that if

$$A(\omega) = \begin{cases} 1 & \text{for } |\omega| \le \Delta,\ \Delta \text{ small}, -\pi < \omega < \pi\,, \\ 0 & \text{otherwise}\,, \end{cases} \tag{2.22}$$

then expression (2.20) becomes

$$B(\omega) = \begin{cases} \frac{1}{2} & \text{for } |\omega \pm \omega_0| \le \Delta,\ -\pi < \omega,\ \omega_0 < \pi\,, \\ 0 & \text{otherwise}\,, \end{cases} \tag{2.23}$$

which is the form of the transfer function for a band-pass filter, centered at ω_0, and (2.21) becomes

$$C(\omega)=\begin{cases}-\tfrac{1}{2}\mathrm{i} & \text{for } |\omega-\omega_0|\leqslant \Delta\,,\\ \tfrac{1}{2}\mathrm{i} & \text{for } |\omega+\omega_0|\leqslant \Delta,\ -\pi<\omega,\ \omega_0<\pi\,,\\ 0 & \text{otherwise}\,,\end{cases} \tag{2.24}$$

which is the form of the transfer function for the Hilbert transform.

2.6. Filters

There is a vast amount of time-series literature written on the subject of filters. The usual considerations that are taken into account to make a choice in this matter will of course also apply when using complex demodulation. We shall, therefore, only briefly discuss some of the filters that have been used in conjunction with complex demodulation.

Bingham, Godfrey and Tukey (1967) discuss filters made up of k successive equal-weight moving averages of length $2L/k+1$. For $k=1$ we have what is known as the simple moving average. Granger and Hatanaka (1964) point out that a second application of the simple moving average of length $2M+1$, where $M\leqslant 1$, leaves the first zero of the transfer function unchanged; and if $M=L$ the effect on the transfer function is that the ratio of the side peak to the main peak becomes approximately 4.5% (compared to 21% for a single application of the moving average). The following rule of thumb is proposed by the authors in using the above procedure: if we wish to demodulate the frequency band $(\omega_0\pm\delta)$, the length of the first moving average should be such that $[2\pi/(2L+1)-\delta]$ is small. Since shifting the frequency of interest ω_0 down to zero has caused what was at the zero-frequency point to be shifted to ω_0, the authors also suggest that the second moving average (of length $2M+1$) be chosen such that $[\omega_0-2\pi/(2M+1)]$ is as small as possible.

Filters for complex demodulation, other than those based on combinations of moving averages, are discussed, for example, in Bloomfield (1976) and Childers and Pao (1972). In analyzing earthquake data where the phenomenon of split peaks often occurs, the use of a raised cosine taper had been suggested by Bolt and Brillinger (1979).

Filtering can also be carried out in the frequency domain, specifically, via the fast Fourier transform (FFT) algorithm discussed earlier. Besides computational efficiency, such a procedure would also allow us to compute the derivative of the instantaneous phase with (see Anderssen and Bloomfield, 1974).

3. Statistical properties

In this section we shall be concerned with certain statistical properties of the complex demodulates and the spectral estimates obtained through them. Unless otherwise specified we assume throughout that $X(t)$, $t=0,\pm1,\ldots$, the time series of interest, is wide-sense stationary. This will enable us to use representation (2.5) to illustrate the properties. We also assume that the well-separated values of the time series are only weakly dependent. This last property, known as *mixing*, may be written as

$$\sum_{u_1} \cdots \sum_{u_k} |\text{cumulant}\{X(t+u_1), \ldots, X(t+u_k), X(t)\}| < \infty . \tag{3.1}$$

Brillinger (1975) discusses in detail the ideas underlying condition (3.1).

3.1. First- and second-order moments of demodulates

Using (2.1) it can be shown that

$$EW_x^L(t) = A(\omega_0)\exp\{-\mathrm{i}\omega_0 t\}c_x \tag{3.2}$$

where $EX(t) = c_x$.

So, for example, if $A(\omega_0)$ corresponds to the transfer function of a simple moving average, with filter coefficients $a(u) = 1/(2L+1)$, $u = -L, \ldots, 0, \ldots, L$, then

$$EW_x^L(t) = \frac{\sin \frac{2L+1}{2}\omega_0}{(2L+1)\sin\frac{\omega_0}{2}} \exp\{-\mathrm{i}\omega_0 t\}c_x . \tag{3.3}$$

Expression (3.3) tells us that for $\omega_0 \neq 0 \bmod 2\pi$, $EW_x^L(t)$ will be approximately zero for large L (as $T \to \infty$), even when $EX(t) = c_x \neq 0$. This is not surprising in view of the fact that the complex demodulate, when suitably normalized, behaves like a finite Fourier transform (which also shares this property), computed in the neighborhood of t. We shall explore this relationship further later in this section.

The result discussed above will of course hold for any filter whose transfer function satisfies (2.3).

To find the covariance between the demodulates we note that

$$W_x^L(t+u) = \int_{-\pi}^{\pi} A(\omega-\omega_0)\exp\{\mathrm{i}(\omega-\omega_0)t\}\exp\{\mathrm{i}(\omega-\omega_0)u\}\,\mathrm{d}Z_x(\omega) . \tag{3.4}$$

Thus

$$\begin{aligned} c_{ww}(u) &= \operatorname{cov}\{W_x^L(t+u), W_x^L(t)\} \\ &= \int_{-\pi}^{\pi} |A(\lambda-\omega_0)|^2 \exp\{\mathrm{i}(\lambda-\omega_0)u\} f_{xx}(\lambda)\,\mathrm{d}\lambda , \end{aligned} \tag{3.5}$$

where $f_{xx}(\lambda)$ is the power spectrum of the series $X(t)$ at frequency λ.

Since $|A(\lambda-\omega_0)|^2$ is bounded by definition and $f_{xx}(\lambda)$ is bounded under the mixing condition (3.1), we can see immediately by the Riemann–Lebesgue lemma that $c_{ww}(u)$ will tend to zero as $u\to\infty$. For example, if $A(\lambda-\omega_0)$ corresponds to an ideal low-pass filter, then

$$c_{ww}(u)=\int_{\omega_0-\Delta}^{\omega_0+\Delta}\exp\{\mathrm{i}(\lambda-\omega_0)u\}f_{xx}(\lambda)\,\mathrm{d}\lambda\,. \tag{3.6}$$

For small Δ, we can approximate this by

$$f_{xx}(\omega_0)\int_{\omega_0-\Delta}^{\omega_0+\Delta}\exp\{\mathrm{i}(\lambda-\omega_0)u\}\,\mathrm{d}\lambda=f_{xx}(\omega_0)\frac{2\sin\Delta u}{u}\,. \tag{3.7}$$

We see that as $u\to\infty$, $c_{ww}(u)$ tends to zero, although the rate of convergence is quite slow. It should be clear that if $X(t)$ corresponds to white noise, then demodulates farther apart than the length of the filter will be independent.

Furthermore, if we take the Fourier transform on both sides of (3.6), we obtain

$$f_{ww}(\lambda)=2f_x(\omega_0)A(\lambda) \tag{3.8}$$

where

$$A(\lambda)=\begin{cases}1 & \text{if } |\lambda|\leq\Delta\,,\\ 0 & \text{otherwise}\,.\end{cases}$$

This shows that the power spectrum of the complex demodulate is box-car shaped at zero frequency, as we would expect, with magnitude proportional to the power spectrum of the original series, $X(t)$, at frequency ω_0.

The covariance between demodulates at *different frequencies* is also of interest. Once again, using (2.6) we see that this covariance is given as

$$\begin{aligned}&\operatorname{cov}\{W_x^L(t,\omega_1),W_x^L(t,\omega_2)\}\\&=\int_{-\pi}^{\pi}\int_{-\pi}^{\pi}A(\lambda_1-\omega_1)\overline{A(\lambda_2-\omega_2)}\exp\{\mathrm{i}(\lambda_1-\omega_1)t\}\exp\{-\mathrm{i}(\lambda_2-\omega_2)t\}\\&\quad\times\delta(\lambda_1-\lambda_2)f_{xx}(\lambda_1)\,\mathrm{d}\lambda_1\,\mathrm{d}\lambda_2,\end{aligned} \tag{3.9}$$

where $\delta(\lambda)$ is the Dirac delta function (e.g. see Papoulis, 1962). Again we note that for an ideal low-pass filter, the transfer function is given by

$$A(\lambda_1-\omega_i)=\begin{cases}1 & \text{if } \omega_i-\Delta\leq\lambda_i\leq\omega_i+\Delta,\ -\pi\leq\lambda_i\leq\pi\,,\\ 0 & \text{otherwise}\end{cases}$$

for $i=1,2$. So for $|\omega_1-\omega_2|>\Delta$,

$$\operatorname{cov}\{W_x^L(t,\omega_1),W_x^L(t,\omega_2)\}=0.$$

We can also show that if $X(t)$ is wide-sense stationary and Gaussian, then

$$\vec{E}\log|W_x^L(t)| = \log|EW_x^L(t)| + O(L_T^{-1}) \tag{3.10}$$

and

$$\overrightarrow{\text{cov}}\{\log_{10}|W_x^L(t+u)|, \log_{10}|W_x^L(t)|\}$$
$$= \frac{1}{2c_x^2}\left[\int_{-\pi}^{\pi}\frac{|A(\omega-\omega_0)|^2}{|A(\omega_0)|^2}\cos(\omega u)f_{xx}(\omega)\,d\omega\right.$$
$$\left. + \text{Re}\int_{-\pi}^{\pi}\frac{A(\omega-\omega_0)A(-\omega-\omega_0)}{A(\omega_0)^2}\exp\{i\omega u\}f_{xx}(\omega)\,d\omega\right]$$
$$= O(L_T^{-1}), \tag{3.11}$$

where $\vec{E}$ and $\overrightarrow{\text{cov}}$ denote the expected value and the covariance derived in a term by term manner from a Taylor expansion. We have subscripted L by T in (3.11) to emphasize that we mean $L\to\infty$ as $T\to\infty$.

3.2. First- and second-order moments of instantaneous power ordinates

As we mentioned in Section 3.1, we can think of $|W_x^L(t)|^2$ as being proportional to the instantaneous power at time t. We shall develop this concept more fully in the next subsection. In the meantime we note that

$$\text{cov}\{|W_x^L(t+u)|^2, |W_x^L(t)|^2\}$$
$$= \int_{-\pi}^{\pi}\int_{-\pi}^{\pi}\int_{-\pi}^{\pi}\int_{-\pi}^{\pi} A(\lambda-\lambda_0)\overline{A(\lambda'-\lambda_0)}A(\mu-\lambda_0)\overline{A(\mu'-\lambda_0)}$$
$$\times\exp\{i[(\lambda-\lambda')+(\mu-\mu')]t\}\exp\{i(\lambda-\lambda')u\}$$
$$\times\text{cov}\{dZ_x(\lambda)\,dZ_x(\lambda'), dZ_x(\mu)\,\overline{dZ_x(\mu')}\}. \tag{3.12}$$

If we assume further that $X(t)$ is Gaussian, we can show that

$$\text{cov}\{|W_x^L(t+u)|^2, |W_x^L(t)|^2\}$$
$$= \int_{-\pi}^{\pi}\int_{-\pi}^{\pi} A(\lambda-\lambda_0)A(-\lambda-\lambda_0)\overline{A(\lambda'-\lambda_0)A(-\lambda'-\lambda_0)}$$
$$\times\exp\{i(\lambda-\lambda')u\}f_{xx}(\lambda)f_{xx}(\lambda')\,d\lambda\,d\lambda' + \int_{-\pi}^{\pi}\int_{-\pi}^{\pi}|A(\lambda-\lambda_0)|^2$$
$$\times|A(\lambda'-\lambda_0)|^2\exp\{i(\lambda-\lambda')u\}f_{xx}(\lambda)f_{xx}(\lambda')\,d\lambda\,d\lambda'. \tag{3.13}$$

So assuming that the spectrum is reasonably smooth over the bandwidth of the spectral window $A(\lambda)$, we obtain approximately,

$$\text{cov}\{|W_x^L(t+u)|^2, |W_x(t)|^2\} \cong 4f_{xx}(\lambda_0)^2\frac{\sin^2\Delta u}{u^2}. \tag{3.14}$$

Expression (3.13) may also be used to show that

$$\operatorname{var}|W_x^L(t)|^2 \cong 4\Delta^2 f_{xx}(\lambda_0)^2 \tag{3.15}$$

and, thus,

$$\operatorname{corr}\{|W_x^L(t+u)|^2, |W_x^L(t)|^2\} \cong \frac{\sin^2 \Delta u}{\Delta^2 u^2} . \tag{3.16}$$

3.3. Spectrum estimation

Perhaps one of the most important applications of complex demodulation is spectrum estimation. This approach is most fruitful when dealing with 'slowly-evolving' or 'evolutionary' (nonstationary) time series because it gives us a spectrum estimate which represents the average power of the process in the neighborhood of each time instant t for each frequency (band) of interest. It also preserves the phase information of the series (at each frequency) which ordinary spectrum estimation suppresses.

The theoretical foundation for carrying out spectral analysis of such processes was developed by Priestley (1965) in his work on 'evolutionary spectra'. Priestley assumes that the spectrum is changing *slowly* over time (a measure of which he defines) and then by using estimates involving only local functions of the time series, he defines some form of 'average' spectrum which essentially has the same type of physical interpretation as the spectrum of a stationary process. His technique for estimating the evolutionary spectrum for discrete-parameter nonstationary processes is essentially identical to the spectrum estimate proposed below using complex demodulates.

We define the *running periodogram* as

$$I_{xx}^L(t, \omega) = B_L^{-1}|W_x^L(t, \omega)|^2 , \tag{3.17}$$

where

$$B_L = 2\pi \sum_{u=-L}^{L} a(u)^2 \tag{3.18}$$

is the *bandwidth parameter* and $\{a(u)\}$ are the low-pass filter coefficients.

It can be easily shown that if $X(t)$ is wide-sense stationary,[2] then

(i) $$EI_{xx}^L(t, \omega_0) = B_L^{-1} \int_{-\pi}^{\pi} |A(\omega - \omega_0)|^2 f_{xx}(\omega)\, d\omega . \tag{3.19}$$

[2]Priestley (1965) gives similar results for the evolutionary spectrum for the case where $X(t)$ is an evolutionary process.

If we also assume that $X(t)$ is Gaussian, then

(ii)
$$\operatorname{cov}\{I_{xx}^L(t,\omega_0), I_{xx}^L(t+u,\omega_0)\}$$
$$= B_L^{-2}\int_{-\pi}^{\pi}\int_{-\pi}^{\pi} A(\omega-\omega_0)A(-\omega-\omega_0)\overline{A(\omega'-\omega_0)A(-\omega'-\omega_0)}$$
$$\times \exp\{\mathrm{i}(\omega-\omega')u\} f_{xx}(\omega) f_{xx}(\omega')\,\mathrm{d}\omega\,\mathrm{d}\omega + B_L^{-2}\int_{-\pi}^{\pi}\int_{-\pi}^{\pi} |A(\omega-\omega_0)|^2$$
$$\times |A(\omega'-\omega_0)|^2 \exp\{\mathrm{i}(\omega-\omega')u\} f_{xx}(\omega) f_{xx}(\omega')\,\mathrm{d}\omega\,\mathrm{d}\omega'\,. \tag{3.20}$$

Furthermore, if $X(t)$ is strictly stationary and mixing, then
(iii) $I_{xx}^L(t,\omega_0)$ is approximately distributed as

$$(\chi_2^2/2) f_{xx}(\omega_0) \quad \text{for } \omega_0 \neq 0 \bmod 2\pi$$

and as

$$\chi_1^2 f_{xx}(\omega) \quad \text{for } \omega_0 = \pm\pi, \pm 3\pi, \ldots.$$

Expression (3.19) is seen to be a weighted average of the power spectrum of the original series, $X(t)$, with weight concentrated in the neighborhood of ω_0, the frequency of demodulation, and relative weight determined by the filter coefficients. If we let the $\{a(u)\}$ correspond to a simple moving average, then

$$A(\omega-\omega_0) = \frac{\sin\left[\frac{2L+1}{2}(\omega-\omega_0)\right]}{(2L+1)\sin\left[\frac{\omega-\omega_0}{2}\right]} \tag{3.21}$$

and in this case

$$EI_{xx}^L(t,\omega_0) = \frac{2L+1}{2\pi}\int_{-\pi}^{\pi}\left[\frac{\sin\left[\frac{2L+1}{2}(\omega-\omega_0)\right]}{(2L+1)\sin\left[\frac{\omega-\omega_0}{2}\right]}\right]^2 f_{xx}(\omega)\,\mathrm{d}\omega\,. \tag{3.22}$$

So, $I_{xx}^L(t,\omega_0)$ will be an *approximately unbiased* estimate of $f_{xx}(\omega_0)$ if $\omega_0 \neq 0 \pmod{2\pi}$ for large L (as $T\to\infty$).

Turning to the covariance term (3.20), we see that it may be reexpressed as

$$\operatorname{cov}\{I_{xx}^L(t,\omega_0), I_{xx}^L(t+u,\omega_0)\}$$
$$= \left| B_L^{-1}\int_{-\pi}^{\pi} A(\omega-\omega_0)A(-\omega-\omega_0)\exp\{\mathrm{i}\omega u\} f_{xx}(\omega)\,\mathrm{d}\omega\right|^2$$
$$+ \left| B_L^{-1}\int_{-\pi}^{\pi} |A(\omega-\omega_0)|^2 \exp\{\mathrm{i}\omega u\} f_{xx}(\omega)\,\mathrm{d}\omega\right|^2. \tag{3.23}$$

The first term appears to be a contribution from near the origin, while the second term is a convolution form over all frequencies except those near the origin.

We are, of course, interested in finding out how far apart the running periodogram ordinates must be in order for them to be approximately independent. This, in general, will depend on both the behavior of the true spectrum, $f_{xx}(\omega)$, near ω_0 and the particular data window employed. Of course for white noise, $I^L_{xx}(t, \omega_0)$ and $I^L_{xx}(t+u, \omega_0)$ will be independent for $|u| \geq 2L+1$.

To see why (iii) holds, assume for now that the low-pass filter used in demodulating the series corresponds to a simple moving average. Then

$$W^L_x(t) = \frac{1}{2L+1} \sum_{u=-L}^{L} X(t+u) \exp\{-\mathrm{i}\omega(t+u)\} .$$

Now let us make the following change of variables

$$T' = 2L + 1 \quad \text{and} \quad u = v - L .$$

Then

$$W^L_x(t) = \frac{1}{T'} \sum_{v=0}^{T'-1} X(t+v-L) \exp\{-\mathrm{i}\omega(t+v-L)\} \tag{3.24}$$

and

$$(2\pi)^{-1} T' |W^L_x(t)|^2 = (2\pi T')^{-1} \left| \sum_{v=0}^{T'-1} X(t+v-L) \exp\{-\mathrm{i}\omega v\} \right|^2 . \tag{3.25}$$

Recall that the *periodogram* of the values $X(0), \ldots, X(T-1)$ is defined as

$$I^T_{xx}(\omega) = (2\pi T)^{-1} \left| \sum_{t=0}^{T-1} X(t) \exp\{-\mathrm{i}\omega t\} \right|^2 . \tag{3.26}$$

We can thus interpret (3.25) to mean that the running periodogram at time t behaves like a periodogram ordinate computed in the neighborhood of t. It is well known (e.g. see Brillinger, 1975) that the distribution of a periodogram ordinate can be approximated by χ^2_2.

The extension of the above argument to the case where the filter coefficients $\{a(u)\}$ correspond to an arbitrary low-pass filter is quite straightforward. Brillinger (1975) defines a *weighted periodogram* as

$$I_{xx}(\omega) = \left[2\pi \sum_t h(t/T)^2 \right]^{-1} \left| \sum_t h(t/T) X(t) \exp\{-\mathrm{i}\omega t\} \right|^2 , \tag{3.27}$$

where $h(v)$ is the data window whose transform is given by (2.2). If we assume that $a(u) = h(u/L)/H^L(0)$, then

$$W^L_x(t) = (H^L(0))^{-1} \sum_{u=-L}^{L} h(u/L) X(t+u) \exp\{-\mathrm{i}\omega_0(t+u)\} . \tag{3.28}$$

Following the same steps as before we see that

$$I_{xx}^{L}(t, \omega_0) = [H_L(0)]^2\Big[2\pi \sum_{u=-L}^{L} h(u/L)^2\Big]^{-1}[H_L(0)]^{-2}\Big|\sum_{v=0}^{2L} h\Big(\frac{v-L}{L}\Big)$$
$$\times X(t+v-L)\exp\{-\mathrm{i}\omega_0 v\}\Big|^2$$
$$= \Big[2\pi \sum_{u=-L}^{L} h(u/L)^2\Big]^{-1}\Big|\sum_{v=0}^{2L} h\Big(\frac{v-L}{L}\Big)$$
$$\times X(t+v-L)\exp\{-\mathrm{i}\omega_0 v\}\Big|^2. \quad (3.29)$$

Estimate of $f_{xx}(\omega)$

We now turn to the problem of obtaining an estimate of the power spectrum at the frequency of demodulation. Let us define

$$\hat{f}_{xx}^{N}(\omega) = \frac{1}{N}\sum_{t=L+1}^{T-l} I_{xx}^{L}(t, \omega) \quad (3.30)$$

where $N = T - 2L$. Then, if $X(t)$ is wide-sense stationary with zero mean,

(i) $$E\hat{f}_{xx}^{N}(\omega_0) = B_L^{-1}\int_{-\pi}^{\pi} |A(\omega-\omega_0)|^2 f_{xx}(\omega)\,\mathrm{d}\omega\,. \quad (3.31)$$

Further, under the Gaussian assumption,

(ii) $$\operatorname{var}\hat{f}_{xx}^{N}(\omega_0) = N^{-2}B_L^{-2}\int_{-\pi}^{\pi}\int_{-\pi}^{\pi} A(\omega-\omega_0)A(-\omega-\omega_0)$$
$$\times \overline{A(\omega'-\omega_0)A(-\omega'-\omega_0)}\,\frac{\sin^2 N[(\omega-\omega')/2]}{\sin^2(\omega-\omega')/2}$$
$$\times f_{xx}(\omega)f_{xx}(\omega')\,\mathrm{d}\omega\,\mathrm{d}\omega + N^{-2}B_L^{-2}\int_{-\pi}^{\pi}\int_{-\pi}^{\pi}|A(\omega-\omega_0)|^2$$
$$\times A(\omega'-\omega_0)|^2\,\frac{\sin^2 N[(\omega-\omega')/2]}{\sin^2(\omega-\omega')/2}\,f_{xx}(\omega)f_{xx}(\omega')\,\mathrm{d}\omega\,\mathrm{d}\omega'\,. \quad (3.32)$$

It can also be shown that

$$\hat{f}_{xx}^{N}(\omega_0) = \int_{-\pi}^{\pi} H^L(\omega_0-\alpha)I_{xx}^{N}(\alpha)\,\mathrm{d}\alpha\,, \quad (3.33)$$

where

$$H^L(\alpha) = \frac{\sum_{r=0}^{2L} \sum_{u=r-L}^{L} a(u)a(u-r)\exp\{-\mathrm{i}\alpha r\}}{\sum_{u=-L}^{L} a(u)^2}. \tag{3.34}$$

The estimate of $\hat{f}_{xx}^N(\omega_0)$ is seen to be a weighted average of periodogram values over frequencies that range over the bandwidth of the spectral window $H^L(\alpha)$. If we assume that the spectrum is flat over the bandwidth of $H^L(\alpha)$, then

$$E\hat{f}_{xx}^N(\omega_0) \cong f_{xx}(\omega_0) \int_{-\pi}^{\pi} H^L(\alpha)\,\mathrm{d}\alpha \tag{3.35}$$

and

$$\operatorname{var} \hat{f}_{xx}^N(\omega_0) \cong f_{xx}^2(\omega_0) \int_{-\pi}^{\pi} [H^L(\alpha)]^2\,\mathrm{d}\alpha\,. \tag{3.36}$$

Turning to the distribution of $\hat{f}_{xx}^N(\omega)$, a common statistical procedure is to approximate the distribution of such a variate by a multiple of a chi-squared variate, $\lambda\chi_\nu^2$, where

$$\nu = \frac{2[E\hat{f}_{xx}^N(\omega)]^2}{\operatorname{var} \hat{f}_{xx}^N(\omega)} \tag{3.37}$$

and

$$\lambda = 1/\nu \tag{3.38}$$

(e.g. see Brillinger, 1975).

From the approximations (3.35) and (3.36), we see that ν, called the *equivalent degrees of freedom* is approximately

$$\frac{2[\int_{-\pi}^{\pi} H^L(\alpha)\,\mathrm{d}\alpha]^2}{\int_{-\pi}^{\pi} H^L(\alpha)^2\,\mathrm{d}\alpha}. \tag{3.39}$$

3.4. Cross-spectral estimates

Suppose we are given a bivariate time series $\{X(t), Y(t), t = 0, \pm 1, \ldots\}$; then, as we shall see in this section, it is quite straightforward to extend the results of the previous subsection to cross-spectral estimates. First we define the following instantaneous cross-spectral estimates.[3]

Instantaneous cross periodogram

$$I_{yx}^L(t, \omega_0) = B_L^{-1} W_y^L(t, \omega_0) \overline{W_x^L(t, \omega_0)} \tag{3.40}$$

[3]See Priestley and Tong (1973) for the theory of 'evolutionary' cross spectra.

Instantaneous cross spectrum

$$f_{yx}^M(t, \omega_0) = \frac{1}{2M+1} \sum_{k=-M}^{M} I_{yx}^L(t+k, \omega_0). \tag{3.41}$$

Note that we have smoothed over $2M+1$ adjacent complex demodulates to obtain the point estimate at time t. Furthermore, we can write (3.41) as

$$f_{yx}^M(t, \omega_0) = \frac{1}{2M+1} \sum_{k=-M}^{M} |W_y^L(t+k)|\,|W_x^L(t+k)| \exp\{i(\phi_y^L(t) - \phi_x^L(t))\} \tag{3.42}$$

where $\phi_x^L(t)$ and $\phi_y^L(t)$ are the instantaneous phases. The quantity

$$\theta_{yx}^L(t, \omega_0) = \phi_y^L(t, \omega_0) - \phi_x^L(t, \omega_0) \tag{3.43}$$

which represents the *instantaneous phase difference* of the demodulates is also of interest (see Burley, 1969, for example).

Instantaneous cross phase

$$\begin{aligned}\phi_{yx}^M(t, \omega_0) = {} & \tan^{-1}\{-\mathrm{Im}[f_{yx}^M(t, \omega_0)]/\mathrm{Re}[f_{yx}^M(t, \omega_0)]\} \\ & + \pi \cdot \mathrm{sgn}\{-\mathrm{Im}[f_{yx}^M(t, \omega_0)]\}\end{aligned} \tag{3.44}$$

for $\mathrm{Re}[f_{yx}^M(t, \omega_0)] \neq 0$. Burley (1969) points out that a crude preliminary estimate of $\phi_{yx}(t, \omega_0)$ may be obtained as

$$\frac{1}{2M+1} \sum_{k=-M}^{M} \theta_{yx}^L(t+k, \omega_0), \tag{3.45}$$

which would be reasonable provided $\theta_{yx}(t, \omega_0)$ is fairly constant.

Instantaneous gain

$$G_{yx}^M(t, \omega_0) = |f_{yx}^M(t, \omega_0)|/f_{xx}^M(t, \omega_0) \tag{3.46}$$

for $f_{xx}^M(t, \omega_0) \neq 0$, where

$$f_{xx}^M(t, \omega_0) = \frac{1}{2M+1} \sum_{k=-M}^{M} I_{xx}^L(t+k, \omega_0).$$

Instantaneous coherence squared

$$|R_{yx}^M(t, \omega_0)|^2 = |f_{yx}^M(t, \omega_0)|^2/|f_{xx}^M(t, \omega_0) f_{yy}^M(t, \omega_0)|. \tag{3.47}$$

If we write (3.47) as

$$\Big\{\sum_k \sum_{k'} |W_y^L(t+k)|\,|W_y^L(t+k')|\,|W_x^L(t+k)|\,|W_x^L(t+k')| \times \cos[\theta_{yx}(t+k,\omega_0) - \theta_{yx}(t+k',\omega_0)]\Big\} \div \Big\{\sum_k |W_x^L(t+k)|^2 \sum_{k'} |W_y^L(t+k')|^2\Big\}, \tag{3.48}$$

then we can see that $|R_{yx}^M(t,\omega_0)|^2$ has a minimum $\leqslant 1$ if $\theta_{yx}(t+k,\omega_0)$ equals $\theta_{yx}(t+k',\omega_0)$ for all k, k', and the equality holds when $|W_y^L(t+k)|/|W_x^L(t+k)|$ is constant over time (see Burley, 1969). Finally, we define the

Instantaneous residual spectrum

$$f_{\epsilon\epsilon}^M(t,\omega_0) = [1 - |R_{yx}^M(t,\omega_0)|^2] f_{yy}^M(t,\omega_0)\,. \tag{3.49}$$

Statistical properties of $f_{yx}^M(t,\omega_0)$

If $\{X(t), Y(t), t = 0, \pm 1, \ldots\}$ is a bivariate time series with $EX(t) = EY(t) = 0$, $\operatorname{cov}\{X(t+u), X(t)\} = c_{xx}(u)$, $\operatorname{cov}\{Y(t+u), Y(t)\} = c_{yy}(u)$, and $\operatorname{cov}\{Y(t+u), X(t)\} = c_{yx}(u)$ and if $\Sigma_u |c_{xx}(u)| < \infty$, $\Sigma_u |c_{yy}(u)| < \infty$ and $\Sigma_u |c_{yx}(u)| < \infty$, then it can be shown that

(i)
$$Ef_{yx}^M(t,\omega_0) = B_L^{-1} \int_{-\pi}^{\pi} |A(\omega - \omega_0)|^2 f_{yx}(\omega)\, \mathrm{d}\omega\,. \tag{3.50}$$

Further, under the Gaussian assumption,

(ii)
$$\begin{aligned} \operatorname{var} f_{yx}^M(t,\omega_0) = {} & B_L^{-2} N^{-2} \int_{-\pi}^{\pi}\int_{-\pi}^{\pi} |A(\omega-\omega_0)|^2 |A(\omega'-\omega_0)|^2 \\ & \times \frac{\sin^2 N[(\omega-\omega')/2]}{\sin^2(\omega-\omega')/2} f_{yx}(\omega) f_{xy}(\omega')\, \mathrm{d}\omega\, \mathrm{d}\omega' \\ & + B_L^{-2} N^{-2} \int_{-\pi}^{\pi}\int_{-\pi}^{\pi} A(\omega-\omega_0) \\ & \times \overline{A(\omega'-\omega_0) A(-\omega'-\omega_0) A(-\omega-\omega_0)} \\ & \times \frac{\sin^2 N[(\omega-\omega')/2]}{\sin^2(\omega-\omega')/2} f_{yx}(\omega) f_{xy}(\omega')\, \mathrm{d}\omega\, \mathrm{d}\omega'\,, \end{aligned} \tag{3.51}$$

where $N = 2M + 1$.

Expression (3.50) shows that the expected value of the proposed estimate of the instantaneous cross spectrum is a weighted average of $f_{yx}(w)$, with weight concentrated around ω_0.

3.5. Bispectrum estimation

For linear systems, it is well known that an input with frequency ω_0 emerges at the output with the same frequency ω_0. However, for nonlinear systems, the output contains the harmonic frequencies, $2\omega_0, 3\omega_0, \ldots$ as well. The bispectrum, which is defined as

$$f_{xxx}(\omega_1, \omega_2) = (2\pi)^{-2} \sum_u \sum_v c_{xxx}(\omega_1, \omega_2) \exp\{-i(\omega_1 u + \omega_2 v)\}\,, \tag{3.52}$$

where $c_{xxx}(\omega_1, \omega_2)$ is the third-order cumulant, is often used to study the nonlinear properties of such systems.

Several papers discuss the use of complex demodulation to estimate the bispectrum. An early reference is Godfrey (1965a) who uses this procedure to study nonlinear properties of an economic time series. Since then, Huber, Kleiner and Gasser (1971), for example, have used it to investigate phase relations in an EEG signal.

Let $W_x^L(t, \omega_1)$, $W_x^L(t, \omega_2)$ and $W_x^L(t, \omega_3)$ be complex demodulates of the series $X(t)$ at frequencies ω_1, ω_2 and ω_3, respectively, where $\omega_1 + \omega_2 + \omega_3 = 0$, then

$$\hat{f}^N_{xxx}(\omega_1, \omega_2) = N^{-1} \sum_{t=L+1}^{T-L} W_x^L(t, \omega_1) W_x^L(t, \omega_2) W_x^L(t, \omega_3) \tag{3.53}$$

is an estimate of the bispectrum, where $N = T - 2L$ and the principal domain is given by

$$0 \leqslant \omega_1 \leqslant \pi\,,$$

$$0 \leqslant \omega_2 \leqslant \pi/2\,,$$

$$0 \leqslant \omega_1 + 2\omega_2 \leqslant \pi\,.$$

Note that the bispectrum of a real-valued series is complex valued.

We shall now investigate the relationship between estimation of the bispectrum via complex demodulation and via the third-order cumulant (assuming $EX(t) = 0$).

The use of (3.53) yields

$$\begin{aligned} \hat{f}^N_{xxx}(\omega_1, \omega_2) &= \frac{1}{N} \sum_r \sum_u \sum_v a(u)a(v)a(u+v-r)\Big[\sum_t X(t+u)X(t+v) \\ &\quad \times X(t+u+v-r)\Big] \exp\{-i[\omega_1(r-v) + \omega_2(r-u)]\} \\ &= \sum_r \sum_u \sum_v a(u)a(v)a(u+v-r)c^N_{xxx}(r-v, r-u) \\ &\quad \times \exp\{-i[\omega_1(r-v) + \omega_2(r-u)]\}\,, \end{aligned} \tag{3.54}$$

where

$$c^N_{xxx}(r-v, r-u) = \frac{1}{N}\sum_t X(t+u)X(t+v)X(t+u+v-r)$$

is the sample third-order cumulant. It can then be shown that

$$\hat{f}^N_{xxx}(\omega_1, \omega_2) = N\int_{-\pi}^{\pi}\int_{-\pi}^{\pi} H^L(\omega_1-\alpha, \omega_2-\beta)I^N_{xxx}(\alpha, \beta)\,\mathrm{d}\alpha\,\mathrm{d}\beta\,, \tag{3.55}$$

where $I^N_{xxx}(\alpha, \beta)$ is the third-order periodogram and

$$H^L(\omega_1-\alpha, \omega_2-\beta) = \sum_r\sum_u\sum_v a(u)a(v)a(u+v-r) \times \exp\{-\mathrm{i}[(\omega_1-\alpha)(r-v)+(\omega_2-\beta)(r-u)]\}\,.$$

We can see from (3.55) that the bispectrum estimate, $\hat{f}^N_{xxx}(\omega_1, \omega_2)$, is essentially a weighted average of the third-order periodogram.

3.6. Approximate confidence intervals for the running periodogram

For a stationary series, the estimated amplitude, or equivalently, power, should be constant within sampling fluctuations. So if we have a series which is (slowly) evolving in time, then complex demodulation not only lets us handle the spectrum of such a series but also lets us inquire into the nature of the nonstationarity. Essentially, this technique is designed to bring out departures from stationarity in a graphical way, instead of parametrizing particular types of departures in advance and then developing formal significance tests intended to have high power against these particular alternatives.

We now suggest a way of setting approximate $100(1-\alpha)\%$ confidence intervals around the estimate of the running periodogram.

We have shown that

$$\frac{I^L_{xx}(t, \omega)}{f_{xx}(\omega)} \mathrel{\dot\sim} \chi^2_2/2\,. \tag{3.56}$$

Thus, we have, approximately,

$$\Pr\left[-\ln\left(1-\frac{\alpha}{2}\right) < I^L_{xx}(t, \omega)/f_{xx}(\omega) < -\ln\left(\frac{\alpha}{2}\right)\right] = 1-\alpha\,. \tag{3.57}$$

Now suppose that $\hat{f}^N_{xx}(\omega)$, defined by expression (3.30), is based on enough degrees of freedom so that $f_{xx}(\omega)$ is well approximated by $\hat{f}^N_{xx}(\omega)$. In this case we have, approximately,

$$\Pr\left[-\ln\left(1-\frac{\alpha}{2}\right) < I^L_{xx}(t,\omega)/\hat{f}^N_{xx}(\omega) < -\ln\left(\frac{\alpha}{2}\right)\right] = 1-\alpha\,. \tag{3.58}$$

It is important to note that the confidence intervals (3.57) and (3.58) hold only for a given *single value of t* and for a given *single value of ω*.

More often than not we will be concerned with $\log_{10} I^L_{xx}(t,\omega)$. In this case the approximate confidence interval becomes

$$\Pr[\log_{10}\gamma_1 + \log_{10}\hat{f}^N_{xx}(\omega) < \log_{10} I^L_{xx}(t,\omega) < \log_{10}\gamma_2 + \log_{10} f_{xx}(\omega)] = 1-\alpha\,,$$

where

$$\gamma_1 = -\ln(1-\alpha_2) \quad \text{and} \quad \gamma_2 = -\ln\left(\frac{\alpha}{2}\right). \tag{3.59}$$

We interpret this confidence interval as follows: under stationarity, we expect $EI^L_{xx}(t,\omega)$ to be in the confidence region for $(1-\alpha)100\%$ of the values between $t = L+1$ and $t = T-L$.

Fig. 1 is an illustration of the use of this procedure. The error series corresponds to Gaussian white noise. We have demodulated here at the known underlying frequency. We also note that the background noise level has been reached at about $t = 600$.

3.7. Statistical properties of the instantaneous phase

So far we have not made any explicit use of the instantaneous phase $\phi^L_x(t)$, which is an important byproduct of complex demodulation. We first of all present some statistical properties of the instantaneous phase.

Under assumptions similar to the ones presented for the instantaneous

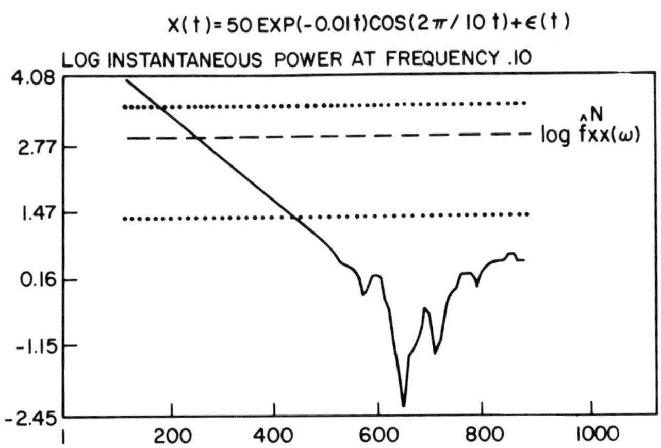

Fig. 1.

amplitude, it can be shown that

$$\vec{E}[\phi_x^L(t)] = \arg EW_x^L(t) + O(L_T^{-1}) \tag{3.60}$$

and

$$\overrightarrow{\text{cov}}\{\phi_x^L(t+u), \phi_x^L(t)\} = \frac{1}{2c_x^2}\left[\int_{-\pi}^{\pi} \frac{|A(\omega-\omega_0)|^2}{|A(\omega_0)|^2}\cos(\omega u) f_{xx}(\omega)\,d\omega - \text{Re}\int_{-\pi}^{\pi} \frac{A(\omega-\omega_0)A(-\omega-\omega_0)}{A(\omega_0)^2} \times \exp\{i\omega u\} f_{xx}(\omega)\,d\omega\right] + O(L_T^{-1})\,. \tag{3.61}$$

It is also instructive to know how the log amplitude and phase vary together *instantaneously* and for this purpose it can be shown that

$$\overrightarrow{\text{cov}}\{\log|W_x^L(t)|, \phi_x^L(t)\} = O(L_T^{-1})\,. \tag{3.62}$$

When $X(t)$ is Gaussian white noise, the plot of the instantaneous phase will be quite irregular in appearance if L is sufficiently small (compared to the total number of data points). As a matter of fact, if $X(t)$ is stationary and mixing, then $\phi_x^L(t)$ will be distributed approximately as a uniform variate on the interval $(0, 2\pi)$ [for $\lambda \neq 0 \pmod{\pi}$]. In order to see this, recall from (3.28) that

$$\begin{aligned} W_x^L(t) &= [H^L(0)]^{-1} \sum_{u=-L}^{L} h\left(\frac{u}{L}\right) X(t+u)\exp\{-i\omega_0(t+u)\} \\ &= [H^L(0)]^{-1} \sum_{v=0}^{2L} h\left(\frac{v-L}{L}\right) X(t+v-L)\exp\{-i\omega_0 v\} \end{aligned} \tag{3.63}$$

for $v = u - L$, which as we saw earlier is essentially a *weighted finite Fourier* transform (except for a normalizing factor). From Brillinger (1975) we know that this transform will be distributed approximately as

$$N^c\left(0, 2\pi N^{-1}\left[\int_0^1 h(v)^2\,dv \Big/ \left(\int_0^1 h(v)\,dv\right)^2\right] f_{xx}(\lambda)\right)$$

(for $\omega_0 \neq 0 \bmod \pi$) where N^c denotes the complex normal distribution and where $N = 2L + 1$. In this case, $W_1(t)$ and $W_2(t)$, the real and complex parts of $W_x^L(t)$, will be approximately independent

$$\text{Normal}\left(0, \pi N^{-1}\left[\int_0^1 h(v)^2\,dv \Big/ \left(\int_0^1 h(v)\,dv\right)^2\right] f_{xx}(\lambda)\right)$$

variates and thus

$$\phi_x^L(t) = \tan^{-1}\left[\frac{W_2(t)}{W_1(t)}\right]$$

will be approximately uniform.

This result is of some interest because it tells us that for a stationary *noise* series (that is $EX(t) = 0$), the phase plot obtained through complex demodulation will look random. On the other hand, when there is a periodic component present in the data, say,

$$X(t) = \rho \cos(\omega t + \phi) + \epsilon(t), \tag{3.64}$$

where $\epsilon(t)$ is a stationary noise series, $X(t)$ will be nonstationary in the mean and in this case we would not expect the phase plot to appear random. In fact, if we complex demodulate the series $X(t)$ given by (3.64) using (for the sake of simplicity) a simple moving average, then

$$W_x^L(t, \omega_0) = \frac{1}{2L+1}\frac{\rho}{2}\left[\exp\{i\phi\}\frac{\sin\frac{2L+1}{2}(\omega_0 - \omega)}{\sin\frac{\omega_0 - \omega}{2}} + \exp\{-i\phi\}\frac{\sin\frac{2L+1}{2}(\omega_0 - \omega)}{\sin\frac{\omega_0 + \omega}{2}}\right]. \tag{3.65}$$

So for ω_0 close to ω, we have

$$W_x^L(t, \omega_0) \cong \frac{\rho}{2}\exp\{i\phi\} \tag{3.66}$$

and we can see immediately that

$$\arg W_x^L(t, \omega_0) \cong \phi, \tag{3.67}$$

as expected. Priestley (1981) gives further results on the distribution of the phase estimate for the nonnull case.

In Fig. 2 we have used

$$X(t) = \cos(2\pi/10t)I_{\{451 \leq t \leq 550\}} + U(t)$$

for $t = 1, \ldots, 1000$, where $I_{\{\cdot\}}$ is an indicator function and where $U(t)$ is uniform white noise. In this example we used a filter corresponding to a simple moving average with L, the number of lags, equal to 50. The phase plot is seen to be nearly horizontal for $t = 400$ to 600 as expected.

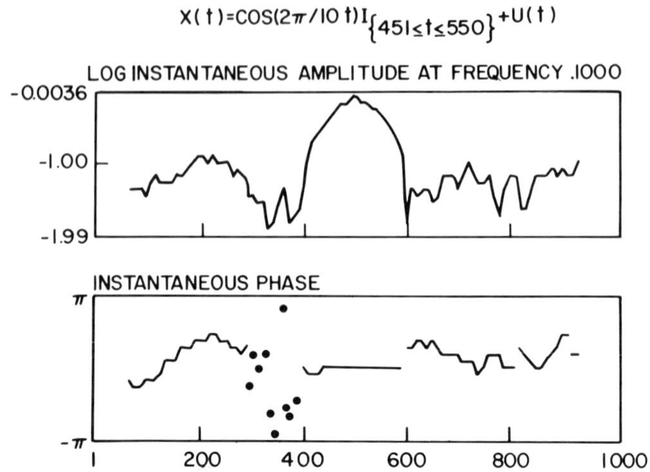

Fig. 2.

The above discussion would seem to suggest that the phase plot might be useful for estimating arrival times of transient signals. We shall see later, however, that the amplitude plot in fact is more sensitive for this particular application. The usefulness of the phase, in the context of the above discussion, is primarily in confirming the presence of a harmonic component in the data.

4. A class of amplitude modulated cosinusoids

In this section we shall be concerned with a general class of time-series models for which complex demodulation seems especially suited. We shall consider models of the form

$$X(t) = \sum_{k=1}^{K} R_k^T(t; \boldsymbol{\theta}_k) \cos(\omega_k t + \delta_k) + \epsilon(t)\,, \tag{4.1}$$

where $\epsilon(t)$ is a stationary, mixing time series, $R_k^T(t, \cdot) \neq 0$ is a fixed function for $k = 1, \ldots, K$, $\{\boldsymbol{\theta}_k^{p\times 1}, \omega_k, \delta_k; k = 1, \ldots, K\}$ are the parameters to be estimated, and where T is the length of the series.

Some examples of $R^T(t; \theta)$ are

(i) the exponential decay model

$$R^T(t; \boldsymbol{\theta}) = \alpha \exp\left(-\frac{\phi t}{T}\right) \quad \text{for } t > 0\,, \tag{4.2}$$

where $\boldsymbol{\theta} = \{\alpha, \phi\}$,

(ii) the Childers–Pao transient model

$$R^T(t; \boldsymbol{\theta}) = \alpha\left(\frac{t}{T}\right)^{\lambda} \exp\left(-\frac{\phi t}{T}\right) \quad \text{for } t > \gamma, \tag{4.3}$$

where $\boldsymbol{\theta} = \{\alpha, \phi, \gamma\}$ $\lambda = 0, 1, \ldots,$

(iii) the linear decay model

$$R^T(t; \boldsymbol{\theta}) = \theta_1 - \theta_2 \frac{t}{T} \quad \text{for } 0 < t < T, \tag{4.4}$$

where $\boldsymbol{\theta} = (\theta_1, \theta_2)$,

(iv) a slowly-varying cosine wave

$$R^T(t; \boldsymbol{\theta}) = \cos(\delta)t \quad \text{for } t > 0 \tag{4.5}$$

with δ small, and finally

(v) constant amplitude

$$R^T(t; \boldsymbol{\theta}) = \theta \quad \text{for } t \geq 0 . \tag{4.6}$$

In practice, once an initial estimate of ω_k, call it $\hat{\omega}_k^{(0)}$, has been obtained, we would complex demodulate at this frequency to:

(a) Check for visual confirmation for the presence of a harmonic component and, if present, get an estimate for the duration.

(b) Obtain initial estimates for the unknown parameters through some kind of curve fitting since the log amplitude in complex demodulation is essentially $\log \hat{R}_k^T(t; \boldsymbol{\theta}_k)$.

The final estimates could then be obtained through nonlinear least squares in the frequency domain (to obtain a more tractable solution since our errors are not uncorrelated) with minimization taking place in some interval, I_k, around $\hat{\omega}_k^{(0)}$ (suggested, for example, by the periodogram). For this reason we can choose to estimate each set of parameters $\{\boldsymbol{\theta}_k, \omega_k, \delta_k\}$, for $k = 1, \ldots, K$, separately; that is, with minimization taking place in disjoint I_k's for each k.

4.1. *Asymptotic normality of estimates*

Let us assume that there exists a continuous function $R(x; \boldsymbol{\theta})$ such that for all $\boldsymbol{\theta} \in \Theta$, Θ compact and where $\boldsymbol{\theta}$ is an interior point of Θ,

$$R_k^T(t; \boldsymbol{\theta}_k) = R_k\left(\frac{t}{T} \boldsymbol{\theta}_k\right), \quad k = 1, \ldots, K. \tag{4.7}$$

Such a choice seems reasonable in discussing asymptotic theory, at least for the exponential decay model, since otherwise our signal would disappear for large t. Note that if we set $t/T = s$, then

$$R_k^T(sT;\boldsymbol{\theta}_k) = R_k(s;\boldsymbol{\theta}_k), \quad 0 \leq s \leq 1 . \tag{4.8}$$

Assuming the dependence upon the length of the series, T, as characterized by expression (4.7), and under further regularity conditions, Hasan (1979, 1982) showed that the vector

$$\begin{bmatrix} T^{1/2}(\hat{\boldsymbol{\theta}}_k^T - \boldsymbol{\theta}_{k0}) \\ T^{1/2}(\hat{\delta}_k^T - \delta_{k0}) \\ T^{3/2}(\hat{\omega}_k^T - \omega_{k0}) \end{bmatrix}_{(p+2)\times 1}, \tag{4.9}$$

where $(\hat{\boldsymbol{\theta}}_k^T, \delta_k^T, \hat{\omega}_k^T)$ are nonlinear least-squares estimates, is asymptotically normal with mean zero and covariance matrix given by

$$2\pi f_{\epsilon\epsilon}(\omega_{k0})\Lambda(\boldsymbol{\theta}_{k0})^{-1}, \tag{4.10}$$

where $f_{\epsilon\epsilon}(\omega)$ is the power spectrum of the error series,

$$\Lambda(\boldsymbol{\theta}_{k0}) = \begin{bmatrix} [\tilde{\gamma}_{lm}(\boldsymbol{\theta}_{k0})]_{p\times p} & [0]_{p\times 2} \\ [0]_{2\times p} & \begin{bmatrix} \gamma(\boldsymbol{\theta}_{k0}) & \gamma_v(\boldsymbol{\theta}_{k0}) \\ \gamma_v(\boldsymbol{\theta}_{k0}) & \gamma_{vv}(\boldsymbol{\theta}_{k0}) \end{bmatrix}_{2\times 2} \end{bmatrix}_{(p+2)\times(p+2)} \tag{4.11}$$

and where

$$\gamma(\boldsymbol{\theta}'_k, \boldsymbol{\theta}''_k) = \int_0^1 R_k(v;\boldsymbol{\theta}'_k)R_k(v;\boldsymbol{\theta}'')\,\mathrm{d}v , \tag{4.12}$$

$$\tilde{\gamma}_{lm}(\boldsymbol{\theta}'_k, \boldsymbol{\theta}''_k) = \frac{\partial^2\gamma(\boldsymbol{\theta}'_k, \boldsymbol{\theta}''_k)}{\partial\theta'_{kl}\partial\theta''_{km}}, \tag{4.13}$$

$$\gamma_{vv}(\boldsymbol{\theta}'_k, \boldsymbol{\theta}''_k) = \int_0^1 v^2 R_k(v;\boldsymbol{\theta}'_k)R_k(v;\boldsymbol{\theta}''_k)\,\mathrm{d}v , \tag{4.14}$$

$$\gamma_v(\boldsymbol{\theta}'_k, \boldsymbol{\theta}''_k) = \int_0^1 v R_k(v;\boldsymbol{\theta}'_k)R_k(v;\boldsymbol{\theta}''_k)\,\mathrm{d}v . \tag{4.15}$$

For $\boldsymbol{\theta}'_k = \boldsymbol{\theta}''_k = \boldsymbol{\theta}_k$, the notation is simplified, for example, to $\gamma(\boldsymbol{\theta}_k)$.

EXAMPLE 4.1. Let

$$X(t) = \theta\cos(\omega t + \delta) + \epsilon(t), \quad \theta \neq 0 ,$$

where $\{\theta, \delta, \omega\}$ are unknown parameters. Trivially, the covariance matrix is given by

$$2\pi f_{\epsilon\epsilon}(\omega_0)\begin{bmatrix} 1 & 0 & 0 \\ 0 & \theta_0^2 & \theta_0^2/2 \\ 0 & \theta_0^2/2 & \theta_0^2/3 \end{bmatrix}^{-1}$$

Related references are Whittle (1952), Walker (1971) and Hannan (1971).

EXAMPLE 4.2. Let

$$X(t) = (\theta_1 - \theta_2 t/T)\cos(\omega t + \delta) + \epsilon(t)\,, \quad \theta_1, \theta_2 \neq 0\,.$$

where $\{\theta_1, \theta_2, \delta, \omega\}$ are the unknown parameters. In this case the covariance matrix is given by

$$2\pi f_{\varepsilon\varepsilon}(\omega_0)\begin{bmatrix} 1 & -\frac{1}{2} & 0 & 0 \\ -\frac{1}{2} & \frac{1}{3} & 0 & 0 \\ 0 & 0 & \theta_1^2 - \theta_1\theta_2 + \theta_2^2/3 & \theta_1^2/2 - \frac{2}{3}\theta_1\theta_2 + \theta_2^2/4 \\ 0 & 0 & \theta_1^2/2 - \frac{2}{3}\theta_1\theta_2 + \theta_2^2/4 & \theta_1^2/3 - \theta_1\theta_2/2 + \theta_2^2/5 \end{bmatrix}^{-1}$$

Finally, we have

EXAMPLE 4.3. Let

$$X(t) = \alpha \exp\{-\phi t/T\}\cos(\omega t + \delta) + \epsilon(t)\,, \quad \alpha \neq 0\,,$$

where $\{\alpha, \phi, \delta, \omega\}$ are the unknown parameters. The covariance matrix is given by

$$2\pi f_{\epsilon\epsilon}(\omega_0)\begin{bmatrix} I_0(\phi_0) & -\alpha_0 I_1(\phi_0) & 0 & 0 \\ -\alpha_0 I_1(\phi_0) & \alpha_0^2 I_2(\phi_0) & 0 & 0 \\ 0 & 0 & \alpha^2 I_0(\phi_0) & \alpha^2 I_1(\phi_0 \\ 0 & 0 & \alpha_0^2 I_1(\phi_0) & \alpha_0^2 I_2(\phi_0 \end{bmatrix}^{-1}$$

$$= 2\pi f_{\epsilon\epsilon}(\omega_0)\alpha_0^{-2} J(\phi_0)^{-1}\begin{bmatrix} \alpha_0^2 I_2(\phi_0) & \alpha_0 I_1(\phi_0) & 0 & 0 \\ \alpha_0 I_1(\phi_0) & I_1(\phi_0) & 0 & 0 \\ 0 & 0 & I_2(\phi_0) & -I_1(\phi_0) \\ 0 & 0 & -I_1(\phi_0) & I_0(\phi_0) \end{bmatrix}$$

where

$$I_\lambda(\phi) = \int_0^1 v^\lambda \exp\{-2\phi v\}\,\mathrm{d}v \quad \text{for } \lambda = 0, 1, 2\,,$$

$$J(\phi) = I_0(\phi)I_2(\phi) - I_1(\phi)^2\,.$$

We now have

$$\operatorname{var} \hat{\alpha} = 2\pi T^{-1} f_{\epsilon\epsilon}(\omega_0) I_2(\phi_0) J(\phi_0)^{-1},$$
$$\operatorname{var} \hat{\phi} = 2\pi T^{-1} f_{\epsilon\epsilon}(\omega_0) \alpha_0^{-2} I_0(\phi_0) J(\phi_0)^{-1},$$
$$\operatorname{var} \hat{\delta} = 2\pi T^{-1} f_{\epsilon\epsilon}(\omega_0) \alpha_0^{-2} I_2(\phi_0) J(\phi_0)^{-1},$$
$$\operatorname{var} \hat{\omega} = 2\pi T^{-3} f_{\epsilon\epsilon}(\omega_0) \alpha_0^{-2} I_0(\phi_0) J(\phi_0)^{-1},$$

etc. We shall discuss this model further in the next subsection.

4.2. *Exponential decay model*

Bolt and Brillinger (1979) have considered the exponential decay model, discussed in Example 4.3, in modeling the impulse response of the earth to a large magnitude earthquake. In their paper the authors present a general algorithm for simultaneously estimating eigenfrequencies, amplitudes, phases and damping coefficients for terrestrial eigenspectra measurements. An important use of complex demodulation in their paper is for obtaining initial estimates of the parameters in the model, especially the decay parameter ϕ, at each frequency of interest.

We shall discuss here how estimates of α and ϕ may be obtained directly from the log amplitude function. It can be shown that

$$Z_x^L(t) = \log|W_x^L(t)| = -\frac{\phi t}{T} + \log \alpha + \left[\alpha \exp\left\{-\frac{\phi t}{T}\right\}\right]^{-1} \operatorname{Re}[\exp\{-\mathrm{i}\phi_x^L(t)\}\eta(t)] + O(L_T^{-1}) \tag{4.16}$$

where

$$\phi_x^L(t) = (\omega - \hat{\omega}^{(0)})t + \delta,$$

$\hat{\omega}^{(0)}$ is the periodogram estimate, and

$$\eta(t) = \sum_u a(t-u)\epsilon(u)\exp\{-\mathrm{i}\hat{\omega}^{(0)}u\}.$$

In a recent paper, Toyooka (1979) discusses the following time-series regression model

$$Y(t) = \sum_{j=1}^{p} \beta_j X_j(t) + c(t)U(t), \tag{4.17}$$

$t = 1, 2, \ldots,$ where the β_j are the unknown parameters, $c(t)$ is a modulation function of the form

$$c(t) = \exp\left[\sum_{i=1}^{q} c_i \gamma_i(t)\right], \tag{4.18}$$

$\gamma_i(t)$ is a bounded function of t for each $i = 1, \ldots, q$, $\{c_i; i = 1, \ldots, q\}$ are additional unknown parameters and, finally, $U(t)$ is a wide-sense stationary process. A two-stage procedure for estimating the parameters $\{\beta_j\}$ and $\{c_i\}$ is proposed and a result showing the large sample behavior of the estimates is also presented.

If we rewrite (4.16) as

$$Z_x^L(t) \cong \sum_{j=1}^{2} \beta_j X_j(t) + \left[\exp \sum_{i=1}^{2} c_i \phi_i(t)\right] U(t), \tag{4.19}$$

where

$$\beta_1 = \log \alpha, \qquad X_1(t) = 1,$$

$$\beta_2 = -\phi, \qquad X_2(t) = t/T,$$

$$c_1 = -\log \alpha, \qquad \phi_1(t) = 1,$$

$$c_2 = \phi, \qquad \phi_2(t) = t/T,$$

then by applying Toyooka's result we can show that

$$\begin{aligned} \operatorname{var} \log \hat{\alpha} &\cong 2\pi T^{-1} f_{uu}(0) \alpha_0^{-2} I_2(\phi_0) J(\phi_0)^{-1}, \\ \operatorname{var} \hat{\phi} &\cong 2\pi T^{-1} f_{uu}(0) \alpha_0^{-2} I_0(\phi_0) J(\phi_0)^{-1}, \end{aligned} \tag{4.20}$$

where

$$I_j(\phi_0) = \int_0^1 v^\lambda \exp\{-2\phi_0 v\}\, dv \quad \text{for } j = 0, 1, 2,$$

$$J(\phi_0) = I_0(\phi_0) I_2(\phi_0) - I_1(\phi_0)^2.$$

We note that the expression of the asymptotic variance for $\hat{\phi}$ obtained here is identical to the expression obtained in Example 4.3. This result may be significant in applications where the ϕ's have to be estimated at a number of frequencies since fitting lines to the log amplitude obtained through complex demodulation should be computationally more efficient than the nonlinear regression method discussed earlier. However, the estimates obtained through the linear fit would probably have larger standard errors since they would presumably be based upon fewer points.

4.3. *The Childers–Pao transient model for visual-evoked responses*

Childers and Pao (1972) consider the transient model

$$X(t) = \sum_{k=1}^{K} \alpha_k t \exp\{-\beta_k t\} \cos(\omega_k t + \delta_k) + \epsilon(t) \tag{4.21}$$

for $t > \tau$, for visual-evoked responses (VER's) monitored by scalp electrodes over the occipital region of the brain. The term transient here refers to the time lag associated with each wavelet before its arrival.

For $K = 1$, the problem is immediately seen to be one of two-phase non-linear regression where the join point has to be estimated. That is, we can rewrite (4.21) as

$$\begin{aligned} X(t) &= \mu + \epsilon(t), \quad t = 1, \ldots, \tau, \\ X(t) &= \alpha t \exp\{-\beta t\} \cos(\omega t + \delta) + \epsilon(t), \quad t = \tau + 1, \ldots, T, \end{aligned} \tag{4.22}$$

where $\{\tau, \mu, \alpha, \beta, \omega, \delta\}$ are parameters to be estimated. We shall find it necessary to reparametrize $\alpha = \alpha/T$ and $\beta = \phi/T$ so that the asymptotics to be discussed later will make sense.

There exists a great amount of literature which deals with the problem of estimating the change point for the mean for linear regression. In most papers it is assumed that the errors are independent and identically distributed normal mean zero variates and that the parameters occur linearly. Further it is assumed in some papers that the join point(s) are smooth, that is, the regression function is continuous at τ.

Clearly the assumptions mentioned above will not be satisfied for the type of data for which complex demodulation is best suited. However, complex demodulation can still be used to obtain a satisfactory estimate of τ. The estimation procedure for the remaining parameters $\{\alpha, \phi, \omega, \delta\}$ is then identical to that of the exponential decay model considered by Bolt and Brillinger. In the exponential decay case, a reasonable estimate of the arrival time is provided by the peak in the graph of the log instantaneous amplitude minus one-half the number of time lags used for filtering (see Hasan, 1979).

If a noise record preceding the arrival of the transient is available,[4] an alternative estimation procedure would be to first set confidence bands in the manner described in Section 3. We of course have some cutoff point in mind for the noise record and again complex demodulation can be helpful in this respect, as long as the onset time is not too close to the beginning of the data. This suggests that a noise record of fair length be collected preceding the signal, if at all possible. We can now take as an estimate of τ the first significant jump out of our confidence bands (that is, one which remains out for some duration of the signal). Using simulated data, Hasan (1979) found that this estimate precedes the (known) arrival time by a random amount, but usually within one-half the filter length. It seems sensible then to adjust this estimate by adding one-half the filter length (or possibly more), since it is probably better to err by overestimating the arrival time than the other way around. [One reason being the

[4]Unfortunately such a record is not always available. For example, for the type of earthquake data considered in the previous section, the seismometer would start clipping (going out of bounds) at the arrival of the signal and by the time it resets itself the decay phenomenon would already be in effect.

anomalous behavior of the estimate of the frequency ω, under the null (signal not present) and alternative (signal present) hypotheses (see Whittle, 1952). This erratic behavior could lead to unreliable results if, having estimated τ, we then proceeded to estimate the parameters of the underlying model over a time period in which the signal was not present.]

Acknowledgments

This research is part of the author's doctoral dissertation, which was written at the University of California, Berkeley, under the supervision of Professor David R. Brillinger. Professor Brillinger's friendship, guidance and encouragement are gratefully acknowledged.

References

[1]* Anderssen, R. S. and Bloomfield, P. (1974). A time series approacn to numerical differentiation. *Technometrics* **16**, 69–75.

[2]* Banks, R. J. (1975). Complex demodulation of geomagnetic data and the estimation of transfer functions. *Geophys. J. R. Astr. Soc.* **43**, 83–101.

[3]* Beamish, D., Hanson, H. W. and Webb, D. C. (1979). Complex demodulation applied to Pi2 geomagnetic pulsations. *Geophys. J. R. Astr. Soc.* **58**, 471–493.

[4]* Bingham, C., Godfrey, M. D. and Tukey, J. W. (1967). Modern techniques of power spectrum estimation. *IEEE Trans. Audio Electroacoust.* **AU-15**, 56–66.

[5]* Bloomfield, P. (1976). *Fourier Analysis of Time Series: An Introduction.* Wiley, New York.

[6]* Bogert, B. P., Healy, M. J. R. and Tukey, J. W. (1963). The frequency analysis of time series for echoes cepstrum, pseudo-autocovariance, cross-cepstrum and saphe-cracking. In: M. Rosenblatt, ed., *Proceedings of the Symposium on Time Series Analysis*, 209–243. Wiley, New York.

[7]* Bolt, B. A. and Brillinger, D. R. (1979). Estimation of uncertainties in eigenspectral estimates from decaying geophysical time series. *Geophys. J. R. Astr. Soc.* **59**, 593–603.

[8]* Brillinger, D. R. (1973). An empirical investigation of the Chandler wobble and two proposed excitation processes. *Bull. Int. Stat. Inst.* **45**, 413–434.

[9]* Brillinger, D. R. (1975). *Time series: Data Analysis and Theory.* Holt, Rinehart and Winston, New York.

[10] Brown, W. M. and Palermo, C. J. (1969). *Random Processes, Communications, and Radar.* McGraw-Hill, New York.

[11]* Burley, S. P. (1969). A spectral analysis of the Australian business cycle. *Australian Econ. Papers* **VIII**, 193–218.

[12]* Childers, D. G. and Durling, A. (1975). *Digital Filtering and Signal Processing.* West, St. Paul.

[13]* Childers, D. G. and Pao, M. (1972). Complex demodulation for transient wavelet detection and extraction. *IEEE Trans. Audio Electroacoust.* **AU-20**, 295–308.

[14]* Cooley, J. W., Lewis, P. A. W. and Welch, P. D. (1969). The applications of the FFT algorithm to the estimation of spectra and cross-spectra. In: *Symposium on Computer Processing in Communications*, 5–20. Polytech. Inst. of Brooklyn.

*Denotes a reference in which complex demodulation is mentioned; however, the reference itself may not be explicitly referred to in this paper.

[15]* Gasser, T. (1977). General characteristics of the EEG as a signal. In: A. Remond, ed., *Data Processing in Electroencephalography and Clinical Neurophysiology*. Elsevier, Amsterdam.

[16]* Godfrey, M. D. (1965a). An exploratory study of the bispectrum of an economic time series. *Appl. Statist.* **14**, 48–69.

[17]* Godfrey, M. D. (1965b). The statistical analysis of stationary processes in economics. *Kyklos* **14**, 373–386.

[18]* Granger, C. W. T. and Hatanaka, M. (1964). *Spectral Analysis of Economic Time Series*. Princeton University Press, Princeton.

[19] Hannan, E. J. (1971). Nonlinear time series regression. *J. Appl. Prob.* **8**, 767–780.

[20]* Hasan, T. (1979). Complex Demodulation, Ph.D. thesis, University of California, Berkeley, 1979.

[21]* Hatanaka, M. and Suzuki, M. (1967). The theory of the pseudo-spectrum and its applications to nonstationary dynamic econometric models. In: M. Shubik, ed., *Essays in Mathematical Economics in Honor of Oskar Morgenstern*. Princeton University Press, Princeton.

[22]* Huber, P. J., Kleiner, B. and Gasser, T. (1971). Statistical methods for investigating phase relations in stationary stochastic processes. *IEEE Trans. Audio Electroacoust.* **AU-19**, 78, 86.

[23]* Koopmans, L. H. (1974). *The Spectral Analysis of Time Series*. Academic Press, New York.

[24]* Meyer, R. A. Jr. (1972). Estimating coefficients that change over time. *Int. Econ. Rev.* **13**, 705–710.

[25] Munk, W. H., Miller, G. R., Snodgrass, F. E. and Barber, F. N. (1963). *Phil. Trans. Roy. Soc.* **A255**, 505–584.

[26] Papoulis, A. (1962). *The Fourier Integral and its Applications*. McGraw-Hill, New York.

[27] Parzen, E. (1961). Mathematical considerations in the estimation of spectra. *Technometrics* **3**, 167–190.

[28]* Priestley, M. B. (1965). Evolutionary spectra and non-stationary processes (with discussion). *J. Roy. Statist. Soc.* **B27**, 204–237.

[29]* Priestley, M. B. (1981). *Spectral Analysis and Time Series*, Vols. 1 and 2. Academic Press, New York.

[30] Priestley, M. B. and Tong, H. (1973). On the analysis of bivariate non-stationary processes. *J. Roy. Statist. Soc.* **B35**, 179–188.

[31] Toyooka, Y. (1979). An asymptotically efficient estimation procedure in time series regression model with a nonstationary error process. Res. Reports on Info. Sciences, Ser. B: Operations Res., Dept. of Info. Sciences, Tokyo Inst. of Tech., No. B-64.

[32]* Tukey, J. W. (1961). Discussion emphasizing the connection between analyses of variance and spectrum analysis. *Technometrics* **3**, 1–29.

[33]* Tukey, J. W. (1967). An introduction to the calculations of numerical spectrum analysis. In: B. Harris, ed., *Adv. Sem. on Spectral Analysis of Time Series*, 25–46. Wiley, New York.

[34] Walker, A. M. (1971). On the estimation of a harmonic component in a time series with stationary independent residuals. *Biometrika* **58**, 21–36.

[35]* Walter, D. O. (1971). The method of complex demodulation. *Advances in EEG Analysis*, No. 27, Suppl., 53–57.

[36] Whittle, P. (1952). The simultaneous estimation of a time series harmonic components and covariance structure. *Trab. Estad.* **3**, 43–57.

[37]* Hasan, T. (1982). Nonlinear time series regression for a class of amplitude modulated cosinusoids. *J. Time. Ser. Anal.* **3**, 109–122.

D. R. Brillinger and P. R. Krishnaiah, eds., *Handbook of Statistics, Vol. 3*

8

Estimating the Gain of A Linear Filter from Noisy Data

*Melvin J. Hinich**

Introduction

Measuring the gain and phase of a linear relationship between two signals is an important task in a variety of scientific investigations. In some applications, one signal called the input is controlled. For example, various test input signals are used to measure the response of a linear amplifier. In other applications, the two signals are stochastic and it is arbitrary which signal is called the input. This is the case for the magnetotelluric application discussed by Bentley (1973) and Clay and Hinich (1981).

Filter response is estimated using simultaneously observed data from both signals. If there is noise in the input and output data, standard estimators of the gain are biased. This bias is a time-series version of errors-in-variables bias in linear statistical models (Kmenta, Chapter 9, 1971). This chapter presents an asymptotically unbiased estimator of filter gain for a certain class of filters.

Let us begin with a brief review of the basics of linear filter theory for continuous time signals. There are many texts on the market that explain linear filters. A clear and rigorous exposition is given in Chapter 2 of Kaplan (1962).

1. Linear filters

A time-invariant linear filter is characterized by a function called the *impulse response*, which we denote $h(t)$. The output $y(t)$ of the filter for an input $x(t)$ is given by the convolution

$$y(t) = \int_{-\infty}^{\infty} h(t')x(t-t')\,\mathrm{d}t' . \tag{1.1}$$

A filter is called *stable* if $|h(t)|$ is integrable. A filter is called *causal* if $h(t) = 0$

*This work was supported by the Office of Naval Research (Statistics and Probability Program) under contract.

for $t<0$, and thus $y(t)$ depends only on $x(t')$ for $t' \leq t$. If $h(t)$ is stable and causal, then it can be shown using Laplace transform mathematics that $H(s) = \int_{-\infty}^{\infty} h(t) \exp(-st)\,dt$ has no singularities in the half-plane Re $s \geq 0$, i.e. $H(s)$ has no poles in the right-hand part of the complex plane.

It is often convenient to express a filter's response in the frequency domain. The *transfer function*

$$H(f) = \int_{-\infty}^{\infty} h(t) \exp(-i2\pi f t)\,dt \tag{1.2}$$

is the Fourier transform of the impulse response. If $h(t)$ is real, then $H(-f) = H^*(f)$ where star denotes complex conjugate.

The *gain* of the filter is $|H(f)|$, its absolute value as a function of frequency. The *phase* response is

$$\phi(f) = \arctan \frac{\mathrm{Im}\, H(f)}{\mathrm{Re}\, H(f)} \tag{1.3}$$

for $-\pi < \phi \leq \pi$. The output due to a complex sinusoidal input $\exp(i2\pi f t)$ is

$$|H(f)| \exp[i(2\pi f t + \phi(f))] = |H(f)| \exp[i2\pi f(t + \phi(f)/2\pi f)]\,. \tag{1.4}$$

For a causal filter, the time shift $-\phi(f)/2\pi f$ is positive (a delay), and $-\phi(f)$ is called the *phase lag*. Since $H(0)$ is real, $\phi(0) = 0$.

A stable causal filter is called *minimum phase lag*[1] if $H(s)$ has no zeros in the half-plane Re $s \geq 0$. The term minimum phase is used for such a filter since its phase lag is less than any other filter with the same gain, provided that $H(s)$ has a finite number of zeros (for Re $s<0$) (Zadeh and Desoer, Section 9.7, 1963). The property of minimum phase filters that is exploited in the estimation method featured in this work is that $\phi(f)$ can be uniquely determined from $\ln|H(f)|$ by means of the Hilbert transform. We will discuss this relationship in some detail after the following discussion about estimating the phase and gain from observations of stochastic input and output signals.

2. Estimating the phase and gain of the transfer function

For many applications, including the magnetotelluric problem that motivated this work, the input signal is stochastic. If $x(t)$ is a stationary stochastic process, then the output $y(t)$ is a stationary stochastic process. Suppose that the

[1]For discrete-time systems, the output of a linear filter whose impulse is $\{h(t_n)\}$ is $y(t_n) = \sum_{m=-\infty}^{\infty} h(t_m)x(t_n - t_m)$. The filter is stable if $\sum_{m=-\infty}^{\infty} |h(t_m)| < \infty$, and is causal if $h(t_m) = 0$ for all $t_m < 0$. Setting the origin so that $t_0 = 0$, a stable causal discrete-time filter is minimum phase lag if its z transform $\sum_{m=0}^{\infty} h(t_m)z^m$ has no zeros on or inside the unit circle $|z| = 1$ in the complex plane.

autocovariance of $x(t)$ is absolutely integrable. Then the spectrum $S_x(f)$ of $x(t)$ exists, and the cross spectrum of $\{x(t), y(t)\}$ is

$$S_{xy}(f) = H(f)S_x(f) \tag{2.1}$$

(Jenkins and Watts, Section 8.4.2, 1968).

The phase of the cross spectrum is defined to be $\phi_{xy}(f) = \arctan[\operatorname{Im} S_{xy}(f)/\operatorname{Re} S_{xy}(f)]$. If $S_x(f) > 0$, it follows from (2.1) that $\phi_{xy}(f) = \phi(f)$, and the gain is $|H(f)| = |S_{xy}(f)|/S_x(f)$. Thus the phase and gain can be consistently estimated from consistent estimates of $S_{xy}(f)$ and $S_x(f)$.

Nowadays spectra are estimated from digital data. Suppose that $x(t)$ and $y(t)$ are sampled after they are filtered using a low-pass filter with cutoff frequency $1/2\tau$, where τ is the length of the sampling interval. The frequency components for $f > 1/2\tau$ must be removed to avoid aliasing of data sampled at frequency τ^{-1}.

Several methods exist for obtaining consistent estimators of $S_x(f)$ and $S_{xy}(f)$ in the principal band $0 < f < 1/2\tau$ from a sample $\{x(n\tau), y(n\tau): n = 1, \dots, N\}$ where $N \to \infty$. Jenkins and Watts (Chapter 7, 1968) advocate the approach that begins with the computation of sample covariances. Estimates based on the discrete Fourier transforms of $\{x(n\tau)\}$ and $\{y(n\tau)\}$ are outlined by Hinich and Clay (1968). An example of a simple estimator for $S_{xy}(f)$ is given in Section 5. Rigorous coverage of spectrum estimation is given by Anderson (Chapter 9, 1971), Koopmans (Chapter 8, 1974), Brillinger (Chapter 5, 1975), and Fuller (Chapter 7, 1976).

There is always some noise in measurements of signals. Suppose we observe

$$\tilde{y}(n\tau) = y(n\tau) + \varepsilon(n\tau)$$

and

$$\tilde{x}(n\tau) = x(n\tau) + u(n\tau)\,, \tag{2.2}$$

where $\{\varepsilon(n\tau)\}$ and $\{u(n\tau)\}$ are stationary noise processes that are *uncorrelated with each other and with the true signals.* This is the errors-in-variables problem that was previously mentioned.

Setting $Ex(n\tau) = Eu(n\tau) = 0$ to simplify notation with no loss of generality, the autocovariance of $\tilde{x}$ is

$$\begin{aligned} c_{\tilde{x}}(m) &= E\tilde{x}(n\tau)\tilde{x}((n+m)\tau) \\ &= Ex(n\tau)x((n+m)\tau) + Eu(n\tau)u((n+m)\tau) \\ &= c_x(m) + c_u(m)\,. \end{aligned} \tag{2.3}$$

Thus the spectrum of the true input signal plus noise is

$$S_{\tilde{x}}(f) = S_x(f) + S_u(f)\,, \tag{2.4}$$

where $S_u(f)$ denotes the spectrum of the noise in $\tilde{x}$. Similarly,

$$S_{\tilde{y}}(f) = S_y(f) + S_\varepsilon(f), \tag{2.5}$$

where $S_\varepsilon(f)$ denotes the spectrum of the noise in $\tilde{y}$. These expressions imply that *consistent* estimators of $S_{\tilde{x}}(f)$ and $S_{\tilde{y}}(f)$ are *inconsistent* estimators of $S_x(f)$ and $S_y(f)$, respectively.

The cross covariance between $\tilde{x}$ and $\tilde{y}$ is

$$\begin{aligned} c_{\tilde{x}\tilde{y}}(m) &= E\tilde{x}(n\tau)\tilde{y}((n+m)\tau) \\ &= Ex(n\tau)y((n+m)\tau) = c_{xy}(m) \end{aligned} \tag{2.6}$$

since the noise processes are assumed to be uncorrelated (with each other) and are uncorrelated with the true signals. It then follows from (2.6) that the cross spectrum of $\{\tilde{x}(n\tau), \tilde{y}(n\tau)\}$ is equal to the cross spectrum of $\{x(n\tau), y(y\tau)\}$, i.e. $S_{\tilde{x}\tilde{y}}(f) \equiv S_{xy}(f)$. This implies that the phase of $S_{\tilde{x}\tilde{y}}(f)$ is $\phi(f)$, and thus $\phi(f)$ can be consistently estimated from a consistent estimate of $S_{\tilde{x}\tilde{y}}(f)$. In other words,

$$\hat{\phi}(f) = \text{arc tan}[\text{Im } \hat{S}_{xy}(f)/\text{Re } \hat{S}_{xy}(f)] \tag{2.7}$$

is a consistent estimator of the phase $\phi(f)$ if $\hat{S}_{xy}(f)$ is a consistent estimator of the cross spectrum of the *observed* signals.

The gain for the observed signals, on the other hand, is *not* equal to $|H(f)|$ since

$$\frac{|S_{\tilde{x}\tilde{y}}(f)|}{S_{\tilde{x}}(f)} = \frac{|S_{xy}(f)|}{S_x(f) + S_u(f)} < \frac{|S_{xy}(f)|}{S_x(f)} = |H(f)|. \tag{2.8}$$

Thus a direct estimator of the gain based upon consistent estimators of $S_{xy}(f)$ and $S_{\tilde{x}}(f)$ will be *inconsistent.* The asymptotic proportional bias of a direct estimator is $-[1+S_x(f)/S_u(f)]^{-1}$, and is thus *frequency dependent* unless the signal and its noise are white. In other words, a direct estimate of the gain will not even have the correct shape when the noise in the input signal is serially correlated. This inconsistency cannot be eliminated by reversing the labels of the observed signals since there is noise in both signals.

If the filter is minimum phase and $\phi(f)$ satisfies some mild restrictions, it is possible to construct a consistent estimator of the natural log of the gain, up to an additive constant. The estimator is derived using Hilbert transform mathematics. Let us now outline the mathematical relationships between $\ln|H(f)|$ and $\phi(f)$ that enable us to derive the estimator and show its consistency. This chapter concludes with a discussion of the problem of estimating the additive constant.

3. Use of the Hilbert transform

If the filter is minimum phase,

$$\ln|H(f)| = \frac{1}{\pi}\int_{-\infty}^{\infty} \frac{\phi(g)}{f-g}\,\mathrm{d}g \tag{3.1}$$

(Papoulis, Chapter 10, 1962). A finite sum approximation of this integral using estimated phases at discrete frequencies yields a noisy estimate of the log gain for modest sample sizes. A statistically more convenient relationship can be derived from the following expression (Solodovnikov, 48–51, 1960): For $f > 0$,

$$\phi(f) = \frac{1}{\pi}\int_{-\infty}^{\infty} \frac{\mathrm{d}}{\mathrm{d}u}[\ln|H(f\,\mathrm{e}^{u})|]\ \ln[\coth(|u|/2)]\,\mathrm{d}u\,. \tag{3.2}$$

Writing $y' = \ln(\tau f) + u$, (3.2) becomes

$$\phi(f) = \frac{1}{\pi}\int_{-\infty}^{\infty} \frac{\mathrm{d}}{\mathrm{d}y'}[\ln|H(\tau^{-1}\,\mathrm{e}^{y'})|]\ \ln[\coth|y' - \ln(\tau f)|/2]\,\mathrm{d}y'\,,$$

and writing further $y = \ln(\tau f)$, we obtain

$$\phi(\mathrm{e}^{y}) = \frac{1}{\pi}\frac{\mathrm{d}}{\mathrm{d}y'}[\ln|H(\tau^{-1}\,\mathrm{e}^{y'})|]\ \ln(\coth|y' - y|/2)\,\mathrm{d}y'\,. \tag{3.3}$$

Now define

$$s(y) = \frac{\mathrm{d}}{\mathrm{d}y}\ln|H(\tau^{-1}\,\mathrm{e}^{y})|\,, \tag{3.4}$$

$$S(a) = \int_{-\infty}^{\infty} s(y)\exp(-\mathrm{i}2\pi a y)\,\mathrm{d}y\,, \tag{3.5}$$

$$\Phi(a) = \int_{-\infty}^{\infty} \phi(f)\exp(-\mathrm{i}2\pi a y)\,\mathrm{d}y\,, \tag{3.6}$$

and

$$Q(a) = \int_{-\infty}^{\infty} \ln[\coth(|y|/2)]\exp(-\mathrm{i}2\pi a y)\,\mathrm{d}y\,. \tag{3.7}$$

Since (3.5)–(3.7) are Fourier transforms, it follows from the convolution (3.3) that $\Phi(a) = \pi^{-1}S(a)Q(a)$, and thus

$$S(a) = \pi Q^{-1}(a)\Phi(a)\,. \tag{3.8}$$

From 612.1 in Campbell and Foster (1948),

$$Q(a) = (2a)^{-1} \tanh(\pi^2 a), \tag{3.9}$$

so that $Q(0) = \pi^2/2$. Thus from (3.8) and (3.9), we obtain the key relationship

$$\begin{aligned} s(y) &= \int_{-\infty}^{\infty} S(a) \exp(i2\pi a y)\, da \\ &= 2\pi \int_{-\infty}^{\infty} a\Phi(a) \coth(\pi^2 a) \exp(i2\pi a y)\, da . \end{aligned} \tag{3.10}$$

Note that a is dimensionless since y is dimensionless.

4. Numerical approximations

In order to estimate $s(y)$ from a finite sample of discrete-time observations of $\tilde{x}$ and $\tilde{y}$, the integrals in (3.6) and (3.10) must be approximated.

Define L to be the integer closest to cN^α, where $0<c<1$ and $0<\alpha<1$. The parameter α plays a vital role in the proof of the consistency of the estimator. Let us approximate the integral in (3.11) by a finite sum using the logarithmically spaced grid $\{y_j = \ln(j/N)\colon j = L, \ldots, N/2\}$, which is the transform of the equally spaced frequency grid $\{f_j = j/N\tau\}$. The grid width at j is

$$\Delta y_j = y_{j+1} - y_j = \ln(1 + 1/j), \tag{4.1}$$

and thus $\Delta y_L > \Delta y_{L+1} > \Delta y_{L+2} > \cdots$. The *largest* width is then $\Delta y_L = \ln(1 + 1/L)$, which is approximately $1/L$ for large L. Since $L \approx cN^\alpha$ for $0<c$, $\alpha<1$, $1/L \to 0$ as $N \to \infty$ and thus all the grid widths to zero as $N \to \infty$. It then follows from (3.11) that

$$\Phi(a) = \sum_{j=L}^{N/2} \phi(f_j) \exp(-i2\pi a y_j)\, \Delta y_j + O(N^{-\alpha}). \tag{4.2}$$

The sum is not periodic in a since the spacing is logarithmic (Hinich and Weber, 1980). Another approximation, using equally spaced y_j, is given by Clay and Hinich (1981).

In many applications the observed signals are high-pass filtered to remove frequency components below some cutoff f_L. If so, define $\phi(f) = 0$ for $0<f<f_L$ and set $\alpha = 1$ and $c = f_L\tau$. Thus $L \approx f_L N\tau$. The gain is estimated only in the band $f_L < f < 1/2\tau$.

Now let us approximate the integral in (3.10) by a finite sum using the equally spaced grid $\{a_m = m/M\colon m = 0, \pm 1, \ldots, \pm N\}$, where M depends on N. In order for the approximation to converge to the integral, the grid width $a_{m+1} - a_m = 1/M$ must go to zero as $N \to \infty$, and N/M must go to infinity so that the grid will span the line in the limit. Setting $M = N^{\beta}$ for $0<\beta<1$, it follows that $1/M \to 0$ and $N/M \to \infty$ as $N \to \infty$. Then

$$s(y) = \frac{2\pi}{M} \sum_{m=-N}^{N} a_m \Phi(a_m) \coth(\pi^2 a_m) \exp(\mathrm{i}2\pi a_m y) + \mathrm{O}(M^{-1}) \tag{4.3}$$

since the integrand in (3.10) is well behaved. Applying (4.2) to (4.3) for $l = L, \ldots, N/2$, we have

$$s(y_l) = \sum_{j=L}^{N/2} d_{l,j}(N) \phi(f_j) \, \Delta y_j + \mathrm{O}(N^{-\gamma}), \tag{4.4}$$

where $\gamma = \min(\alpha, \beta)$ and

$$\begin{aligned} d_{l,j}(N) &= \frac{2\pi}{M} \sum_{m=-N}^{N} a_m \coth(\pi^2 a_m) \exp[\mathrm{i}2\pi a_m (y_l - y_j)] \\ &= \frac{4\pi}{M} \left[\sum_{m=1}^{N} a_m \coth(\pi^2 a_m) \cos\left(2\pi a_m \ln \frac{l}{j}\right) + \frac{1}{2\pi^2} \right]. \end{aligned} \tag{4.5}$$

The bias in this approximation goes to zero as $N \to \infty$ since $0<\gamma<1$.

Recalling that $s(y) = (\mathrm{d}/\mathrm{d}y) \ln|H(f)|$, where $y = \ln(\tau f)$,

$$\ln|H(f_k)| = \int_{-\infty}^{y_k} s(y) \, \mathrm{d}y + \ln|H(0)| \,. \tag{4.6}$$

Thus from (4.4) and (4.6)

$$\begin{aligned} \ln|H(f_k)| &= \sum_{l=L}^{k} s(y_l) \, \Delta y_l + C + \mathrm{O}(N^{-\gamma}) \\ &= \sum_{l=L}^{k} \sum_{j=L}^{N/2} d_{l,j}(N) \phi(f_j) \, \Delta y_l \, \Delta y_j + C + \mathrm{O}(N^{-\gamma}) \\ &= \sum_{j=L}^{N/2} w_{jk}(N) \phi(f_j) \, \Delta y_j + C + \mathrm{O}(N^{-\gamma}), \end{aligned} \tag{4.7}$$

where

$$w_{jk}(N) = \sum_{l=L}^{k} d_{l,j}(N) \, \Delta y_l \tag{4.8}$$

and $C = \ln|H(0)|$. When $\alpha = 1$ and $c = f_L \tau$, however, $C = \ln|H(f_L)|$ since the lower limit of integration in (4.6) is $\ln \tau f_L$. Once again the approximation bias goes to zero as $N \to \infty$.

5. Estimating the log gain

Suppose that we have P disjoint sets of measurements of the input and output signals, $\{\tilde{x}_p(t_p + n\tau),\ \tilde{y}_p(t_p + n\tau)\colon n = 0, \ldots, N-1\}$, where t_p is the starting time of the pth set ($p = 1, \ldots, P$). The discrete Fourier transforms of the pth set are

$$\tilde{X}_p(f_k) = \sum_{n=0}^{N-1} \tilde{x}_p(t_p + n\tau)\exp(-\mathrm{i}2\pi nk/N)$$

and

$$\tilde{Y}_p(f_k) = \sum_{n=0}^{N-1} \tilde{y}_p(t_p + n\tau)\exp(-\mathrm{i}2\pi nk/N)\,, \tag{5.1}$$

where $f_k = k/N\tau$ for $k = 0, \ldots, N/2$. The following property of a discrete Fourier transform of a stationary time series with well-behaved cumulants (Brillinger, Section 4.4, 1975) is used in the consistency proof below: $\{\tilde{X}_p(f_0),\ \tilde{X}_p(f_1), \ldots, \tilde{X}_p(f_{N/2})\}$ are asymptotically independent as $N \to \infty$, and similarly for $\{\tilde{Y}_p(f_k)\colon k = 0, \ldots, N/2\}$. Thus the phase estimators ($k = 0, \ldots, N/2$)

$$\hat{\phi}(f_k) = \arctan[\operatorname{Im} \hat{S}_{xy}(f_k)/\operatorname{Re} \hat{S}_{xy}(f_k)]\,, \tag{5.2}$$

where

$$\hat{S}_{xy}(f_k) = \frac{1}{NP}\sum_{p=1}^{P} \tilde{X}_p(f_k)\tilde{Y}_p^*(f_k)\,, \tag{5.3}$$

are asymptotically independent as $N \to \infty$. When N is large, the mean and variance of $\hat{\phi}(f_k)$ are approximated as follows (Hinich and Clay, 1968):

$$E\hat{\phi}(f_k) = \phi(f_k) + O(P^{-1}) \tag{5.4}$$

and

$$\operatorname{var}\hat{\phi}(f_k) = \frac{1}{2P}[\gamma^{-2}(f_k) - 1] + O(P^{-2})\,, \tag{5.5}$$

where

$$\gamma(f) = \frac{|S_{\tilde{x}\tilde{y}}(f)|}{[S_{\tilde{x}}(f)S_{\tilde{y}}(f)]^{1/2}} \tag{5.6}$$

is the *coherence of the observed signals.* Applying (2.4) and (2.5) to (5.6),

$$\gamma^2(f) = [(1 + r_x(f))(1 + r_y(f))]^{-1}\,, \tag{5.7}$$

where $r_x(f) = S_u(f)/S_x(f)$ and $r_y(f) = S_\varepsilon(f)/S_y(f)$ are the *noise-to-signal* ratios for the $\tilde{x}$ and $\tilde{y}$ signals, respectively. When $\gamma(f_k)$ is small or is near one, then $\hat{\phi}(f_k)$ is approximately unbiased for small values of P, i.e. for $1 \leqslant P \leqslant 4$.

Approximately unbiased and independent estimators of the phases can also be obtained using a smoothed sample cross spectrum computed from a single sample of size $N_s \gg N$. In other words, we can use a standard smoothing procedure for obtaining an estimate of $S_{xy}(f_k)$ to use in expression (5.2), provided the sample size is sufficiently large so that the estimates are approximately uncorrelated across the grid $\{f_k = k/N\tau\}$. The asymptotic variance of $\hat{\phi}(f_k)$ for any of the standard smoothing methods will also be proportional to

$$\gamma^{-2}(f_k) - 1 = r_x(f_k) + r_y(f_k) + r_x(f_k) r_y(f_k) . \tag{5.8}$$

To estimate $\ln|H(f_k)|$, replace $\phi(f_j)$ in expression (4.7) by $\hat{\phi}(f_j)$ for $j = L, \ldots, N/2$. The estimator is thus

$$\text{est} \ln|H(f_k)| = \sum_{j=L}^{N/2} w_{jk}(N) \hat{\phi}(f_j) \, \Delta y_j + C . \tag{5.9}$$

Since the approximation converges to $\ln|H(f_0)|$ as $N \to \infty$ and $f_{k(N)} \to f_0$ for a properly chosen sequence $\{k(N)\}$, and $\hat{\phi}(f_k)$ is asymptotically unbiased as $P \to \infty$, the estimator (5.9) is asymptotically unbiased as P, $N \to \infty$.

It will now be shown that this estimator is asymptotically unbiased for finite values of P if (1) $3\alpha + 4\beta > 6$ and (2) for some $\varepsilon > 0$, var $\hat{\phi}(f) \leqslant f^{\varepsilon}$ for $f \approx 0$. It follows from (5.5) and (5.8) that for sufficiently large P, condition (2) holds if the input and output *signal-to-noise ratios* (r_x^{-1} and r_y^{-1}) go to infinity at least as fast as $f^{-\varepsilon}$ as $f \to 0$. Condition (2) is obviated if $\phi(f) = 0$ for $f < f_L$.

Given a frequency f_0, suppose that $\lim_{N\to\infty} f_{k(N)} = f_0$. As is proven in the Appendix,

$$\lim_{N\to\infty} N^{3\alpha+4\beta-6} \operatorname{var}[\text{est} \ln|H(f_k)|] < 64\pi^2 c^{-3} f_0^2 \int_0^{1/2\tau} f^{-1} \operatorname{var} \hat{\phi}(f) \, \mathrm{d}f . \tag{5.10}$$

The integral in (5.10) is finite if condition (2) holds. If condition (1) holds, then the variance of the estimator of $\ln|H(f_0)|$ goes to zero as $N \to \infty$. Note that the set $\{\alpha, \beta : 3\alpha + 4\beta > 6, \alpha < 1, \beta < 1\}$ is *nonempty*.

The method has been tested using artificial data. Clay and Hinich (1981) present some results using an equally spaced approximation of $\Phi(a)$. Boehl, Bostick and Smith (1977) used a more primitive version of Hilbert transform smoothing (which they invented) on magnetotelluric data. I have tested the logarithmically spaced approach presented above. For the filters tested, the method gave good fits of the log gain (up to an additive constant) for medium sample sizes, such as $P = 100$ and $N = 40$.

Let us now return to the problem of estimating the constant C in (4.7), which is a multiplicative constant in the gain estimator

$$G(f_k) = \exp[\text{est} \ln|H(f_k)|] . \tag{5.11}$$

Suppose that there are frequencies $\{f_h\}$ for which the *coherence* is high (near one). If condition (2) holds, then $f_1, \ldots, f_L$ will be in this set. Since the coherence is high, $r_x(f_h)$ is small, and thus the direct estimator $\hat{S}_{xy}(f_h)/\hat{S}_{\bar{x}}(f_h)$ of the gain is almost unbiased if $\ddot{S}_{\bar{x}}$ is a consistent estimator. One simple method for estimating C is to regress the direct estimates against the $G(f_h)$. The slope of the fit is the estimate of C.

In some applications, it is known that $S_u(f) = 0$ in some band. The direct estimates are then unbiased for frequencies in this band, and can be used to calibrate $G(f)$.

If there are no bands where the coherence is high or $S_u(f) = 0$, then the phase smoothing method only estimates the shape of the gain.

Appendix

Given an f_0 in $(0, 1/2\tau)$, let $\{k(N)\}$ be a sequence such that $f_k = k/N \to f_0$ as $N \to \infty$. Then

$$\lim_{N\to\infty} N^{3\alpha+4\beta-6} \operatorname{var}[\text{est} \ln|H(f_k)|] < 64\pi^2 c^{-3} f_0^2 \int_0^{1/2\tau} f^{-1} \operatorname{var} \hat{\phi}(f)\, df .$$

PROOF. Let $\kappa = 2\alpha + 4\beta - 6$. Since the phases in (5.9) are asymptotically uncorrelated,

$$\lim_{N\to\infty} N^{\kappa} \operatorname{var}[\text{est} \ln|H(f_k)|] = \lim_{N\to\infty} N^{\kappa} \sum_{j=L}^{N/2} w_{jk}^2 (\Delta y_j)^2 \operatorname{var} \hat{\phi}(f_j) . \tag{A1}$$

Recall that $\Delta y_j \geq L^{-1} + O(N^{-1})$, $L \approx cN^{\alpha}$, and $M = N^{\beta}$. Applying the Schwarz inequality to (4.8), we have

$$w_{jk}^2 \leq \sum_{l=L}^{k} d_{l-j}^2 \sum_{l=L}^{k} (\Delta y_l)^2 . \tag{A2}$$

To bound d_{l-j}, note that $a \coth(\pi^2 a) < 2a_N$ for $0 \leq a \leq a_N$ when $a_N = N/M$ is large. Then from (4.5),

$$|d_{l-j}| < \frac{8\pi}{M} \frac{N^2}{M} . \tag{A3}$$

Applying (A3) to (A2),

$$w_{jk}^2 < 64\pi^2 k^2 (N/M)^4 L^{-2} + O(N^{-1}) . \tag{A4}$$

Thus the right-hand side of (A1) is bounded by

$$64\pi^2 c^{-3} \lim_{N\to\infty} f_k^2 \sum_{j=L}^{N/2} \operatorname{var} \hat{\phi}(f_j)\, \Delta y_j = 64\pi^2 c^{-3} f_0^2 \int_{-\infty}^{-\ln 2} \operatorname{var} \hat{\phi}(f)\, dy$$

$$= 64\pi^2 c^{-3} f_0^2 \int_0^{1/2\tau} f^{-1} \operatorname{var} \hat{\phi}(f)\, df .$$

References

Anderson, T. W. (1971). *The Statistical Analysis of Time Series.* Wiley, New York.

Bentley, C. R. (1973). Error estimation in two-dimensional magnetotelluric analyses. *Physics of the Earth and Planetary Interiors* **7**, 423–430.

Boehl, J. E., Bostick, F. X. Jr., and Smith, H. W. (1977). An application of the Hilbert transform to the magnetotelluric method. Univ. of Texas at Austin, Elect. Engr. Rsch. Lab.

Brillinger, D. (1975). *Time Series, Data Analysis and Theory.* Holt, Rinehart and Winston, New York.

Campbell, G. A. and Foster, R. M. (1948). *Fourier Integrals for Practical Applications.* Van-Nostrand, New York.

Clay, C. S. and Hinich, M. J. (1981). Estimating the Earth's impedance function when there is noise in the electric and magnetic signals. In: D. Findley, ed., *Second Applied Time Series Symposium.* Academic Press, New York.

Fuller, W. A. (1976). *Introduction to Statistical Time Series.* Wiley, New York.

Hinich, M. J. and Clay, C. S. (1968). The application of the discrete Fourier transform in the estimation of power spectra, coherence, and bispectra of geophysical data. *Reviews of Geophysics* **6**, 347–363.

Hinich, M. J. and Weber, W. E. (1980). Determination of the Nyquist frequency for unequally spaced data. *ONR Tech. Report* No. 19, Economics Dept., Virginia Tech, Blacksburg.

Jenkıns, G. M. and Watts, D. G. (1968). *Spectral Analysis and its Applications.* Holden-Day, San Francisco.

Kaplan, W. (1962). *Operational Methods for Linear Systems.* Addison-Wesley, Reading.

Kmenta, J. (1971). *Elements of Econometrics.* Macmillan, New York.

Koopmans, L. H. (1974). *The Spectral Analysis of Time Series.* Academic Press, New York.

Papoulis, A. (1962). *The Fourier Integral and Its Applications.* McGraw-Hill, New York.

Solodovnikov, V. V. (1960). *Introduction to the Statistical Dynamics of Automatic Control Systems.* Dover, New York.

Zadeh, L. A. and Desoer, C. A. (1963). *Linear System Theory.* McGraw-Hill, New York.

D. R. Brillinger and P. R. Krishnaiah, eds., *Handbook of Statistics, Vol. 3*
© Elsevier Science Publishers B.V. (1983) 169–183

9

A Spectral Analysis Primer

L. H. Koopmans

1. Introduction

This introductory chapter is intended for readers who are new to the subject of time-series spectral analysis. In it the more important spectral parameters and their interpretations will be introduced and some of the applications of these parameters will be indicated. Methods of estimation and other applications will be given elsewhere in this book. This discussion will summarize material presented in [11] and the reader is referred to this source for more complete treatments of the various topics. This material is also available in various forms and from a variety of viewpoints in [2, 3, 6, 7, 8, 9, 10].

2. A historical perspective

Historically, the above references were all written during what might be called the second blooming of spectral analysis. As we will see below, there are two major methodologies for viewing, analyzing and interpreting time series: the time-domain and spectral-domain methodologies. Theoretical developments of time-series methods have followed a rather cyclical pattern first emphasizing one domain, then the other. These cycles have, in part, followed what were perceived to be the needs of practitioners in the dominant scientific fields of the time. Thus, in the flowering of the hard sciences of physics and astronomy in the late 1800s, time-series analysis was born as a spectral-domain subject because of the central role spectral methods play in these fields.

Time-series spectral analysis is the mathematical equivalent of the decomposition of light into its color components by a prism. The idea that other physical phenomena could be similarly broken down into spectral components and analyzed using the concepts developed from the study of light was (and still is) immensely appealing.

The edge was taken off of the original enthusiasm for spectral-domain methods by the horrendous computational effort required to calculate numerical spectra. At that time, calculations had to be done by hand and, although a number of simplified methods were developed, it was still a major undertaking to analyze time series of the magnitudes usually encountered in the physical sciences.

It was, perhaps, with some relief that interest shifted in the 1920s to the shorter and more manageable series seen in economics and the business world. The first flowering of time-domain methods began at this time. One of the dominant interests of economists and business people is the extrapolation of time series into the future—the important problem of *prediction*. The statistical methods of linear regression developed at that time, with their concomitant dimension of prediction, appeared to be a natural and logical methodology to apply to time series. One simply fits a regression equation to the time series for a block of consecutive times (e.g. days). Replication is achieved by moving the block along the time series, thus picking up consecutive blocks of observations. Once the fitting has been accomplished, one predicts into the future simply by moving the block to the leading edge of the series making the independent or explanatory variables the present and immediate past and putting the dependent or response variable one step into the future. In this way, the usual regression predictor of the response variable becomes the predictor of the next value of the time series. However, the usual assumptions of classical regression are badly violated in this application and the need for mathematical shoring up sparked a large and fruitful development of time-domain methodology. This as well as some more recent developments is detailed in [1].

The seed of spectral-domain methodology was not dead during this period, but only dormant. In fact, as the time-domain methods were probed more deeply, it became increasingly apparent that many of the mathematical properties of these series, even those important to time-domain methods, were more usefully and elegantly expressed in terms of spectral parameters. At the same time (during the period of World War II), the digital computer was under development. The reawakened theoretical interest combined with the realization that many practical problems in engineering as well as in the physical sciences could now be attacked by analyzing spectra on computers sparked the second flowering of spectral analysis. Much of the key work was accomplished in the 1940s and 1950s and was first published in book form in the late 1950s. The key work on the theoretical development of spectral methods was reported by Grenander and Rosenblatt [8] while the computational and engineering applications of spectra were detailed by Blackman and Tukey [2]. John Tukey is widely credited with almost single-handedly sparking the renaissance in the applications of spectral methods.

The other books on spectral analysis mentioned above report subsequent details of explosively expanding theoretical developments and applications from the 1960s and beyond. Important applications of spectral analysis now occur not only in the physical and engineering sciences, but in the natural and social sciences and in medicine. Granger and Hatanaka [7] and others have shown spectral methods to be of use in economics and the business world, the original province of time-domain methods.

The most recent swing of the time domain–frequency domain pendulum to time-domain methods was sparked in 1970 by the important book of Box and Jenkins [5]. By expanding the time-domain methodology and adapting it to the computer, these authors and others have evolved a useful and flexible methodology for carrying out such important functions as prediction (or forecasting). This methodology is currently being rapidly expanded and developed in both its theoretical aspects and its applications. Although the energy of this development tends to make those who think exclusively in the frequency domain feel a little out-of-date and a bit lonely at time-series meetings, the data analyst who must actually work with and try to understand real time series benefits greatly from it. Such individuals, and there is an ever-increasing number of them, must evaluate and learn to use every possible tool available. At this time in history, the choice of tools for analyzing time series is large (but by no means complete!). This volume is intended to bring up to date the catalog of spectral-analysis tools. Since, at first exposure, spectra seem somewhat strange and unnatural, especially when compared to time-domain parameters, there is a tendency to overlook or even avoid the use of spectral methods. In doing so, an important dimension of time-series analysis is lost. It is the goal of this chapter to try to convince you that, if you are familiar with the regression and correlation methods of elementary statistics, spectral analysis will provide a framework for thinking about the structure of one- and two-dimensional time series in precisely these familiar terms. Moreover, the ideas of partial correlation and the methods of multivariate analysis have spectral analogs which make it possible to think constructively about time series of dimension greater than two.

3. The time-series model

The model we cover here describes phenomena for which the generating mechanism can be considered, at least for relatively long stretches of time, to be unchanging with the passage of time. The model itself has this unchanging or *stationary* property beginning with the infinite past and extending into the infinite future. Moreover, it is assumed that the interesting measurable characteristics of the mechanism behave in a manner that can be well described probabilistically. If $X(t)$ represents a numerical characteristic measured at time t, in the model $X(t)$ is viewed as a random variable whose value in the given observed realization has been selected according to a probability distribution. The collection of these random variables for all times $-\infty < t < \infty$, along with their joint probability distributions, is a *stochastic process*. By imposing the physical conception of unchangingness or stationarity on the probability distributions of the stochastic process, this model becomes a *stationary stochastic process*. However, the spectral theory, at least as we will deal with it, involves only the first two moments, the mean and covariance, of the model. Consequently, it is really only necessary to impose the stationarity conditions on these moments. When this is done, the stochastic process is called *weakly* (or *second-order*) *stationary*.

The stationarity assumption implies that the process means $EX(t)$, (where E denotes the expectation operator), are unchanging in time; $EX(t) \equiv_t m$. Only the behavior of the residuals of the process from this constant mean value are of theoretical interest. We can shift to the residuals without changing notation simply by assuming $m = 0$. This will be taken as the value of the mean hereafter. The only process parameter of interest is then the covariance function $R(t_1, t_2) = EX(t_1)X(t_2)$ which describes the stochastic relationship between measured values of the physical phenomena at pairs of time points t_1 and t_2.

The condition of stationarity implies that the physical phenomenon has no relevant time origin; its behavior during one time epoch is the same as it would be for any other. If this condition is imposed on the model, it would imply that the joint behavior of the process at times t_1 and t_2 is precisely the same as it would be for any time translation of these points, $t + t_1$ and $t + t_2$. That is, for all t, t_1 and t_2,

$$R(t_1, t_2) = R(t + t_1, t + t_2) .$$

This being true, by taking $t = -t_1$, we see that $R(t_1, t_2) = R(0, t_2 - t_1)$. That is, the covariance depends on t_1 and t_2 only through the time difference $t_2 - t_1$. The covariance is then completely characterized by the function

$$C(\tau) = R(0, \tau), \quad -\infty < \tau < \infty ,$$

called the *autocovariance function of the process*.

The implication of accepting this model for the physical process under study is that all of the interesting and relevant information about the process is then contained in the values of $C(\tau)$. One such value is $C(0) = EX^2(t)$, the process variance. (Because of the stationarity property, this quantity actually does not depend on t.) The variance represents the average 'energy' or *power* of the process. It has the physical interpretation of a time average of energy because of the property

$$\lim_{T \to \infty} \frac{1}{2T} \int_{-T}^{T} X^2(t)\, \mathrm{d}t = C(0) .$$

The precise meaning of this expression in the stochastic setting and its proof can be found in several of the references given in the introduction.

Without the factor of $1/2T$ in the last displayed expression, power resembles a sum of squares similar to the usual measure of variability seen in the 'analysis of variance'. The representation of the response vector as a linear combination of subcollections of mutually orthogonal vectors makes possible the decomposition of the total sum of squares into a sum of component sums of squares each of which represents the contribution of a different factor in the model. The term 'analysis of variance' actually refers to this decomposition.

Spectral analysis performs precisely this same operation on time series. In the time-series context, the orthogonal vectors of the decomposition are the cosine functions

$$A(\lambda)\cos(\lambda t+\theta(\lambda)), \quad -\infty<t<\infty,$$

where, for given frequency λ (in radians per unit time), $A(\lambda)$ represents the amplitude and $\theta(\lambda)$ the phase of the cosine function. The functions are viewed as being indexed by λ and functions with different values of this index are orthogonal. The fact that these same functions crop up in so many different mathematical contexts is what makes Fourier analysis such a rich field of study. Their appearance in the context of weakly stationary stochastic processes provides the mathematical foundation for the spectral analysis of time series.

4. Spectral representations

The spectral representations we will deal with involve writing the cosine functions in a different form. We first rely on the law of cosines to write $\cos(\lambda t+\theta)=\cos\theta\cos\lambda t-\sin\theta\sin\lambda t$. We then use the representation $e^{i\phi}=\cos\phi+i\sin\phi$ to write $A\cos(\lambda t+\theta)=c\,e^{i\lambda t}+\bar{c}\,e^{-i\lambda t}$, where c is the complex number such that $\theta=\arg(c)$ and $A=2|c|$. It follows that if we let $c(\lambda)=c$ and $c(-\lambda)=\bar{c}$, then a sum of the form

$$\sum_{\lambda\geq 0} A(\lambda)\cos(\lambda t+\theta(\lambda))$$

can equally well be represented as

$$\sum_{\lambda} c(\lambda)\,e^{i\lambda t},$$

where both positive and negative frequencies are involved in this second form. The functions $e^{i\lambda t}$ inherit the orthogonality of the cosine function for different values of λ.

For weakly stationary processes, the 'sum' is actually an integral and the time series has the *spectral representation* (or *decomposition*)

$$X(t)=\int e^{i\lambda t}\,Z(d\lambda).$$

The complex-valued amplitude function $Z(\lambda)$ is a stochastic process and some care is required to properly define this integral. However, intuition is best served by ignoring both mathematical precision and theoretical details. Simply view this expression as representing $X(t)$ as a linear combination of the orthgonal functions $e^{i\lambda t}$. The complex amplitude $Z(d\lambda)$ contains both the amplitude and phase of the cosine in the alternative 'sum' representation given earlier. Consequently, amplitude and phase are random quantities. Thus $X(t)$ can be viewed as being made up of a 'sum' of an infinite number of cosine terms, each of a different color or frequency and with randomly selected amplitude and phase.

The analog of the analysis of variance is now obtained from a similar *spectral representation (decomposition) of the autocovariance function*:

$$C(\tau) = \int e^{i\lambda\tau} F(d\lambda)\,.$$

The function $F(\lambda)$ is called the spectral distribution function or, more simply, the power spectrum of the process. It represents the total power in frequencies to the left of λ. In intuitive terms, the quantity $F(d\lambda) = F(\lambda + d\lambda) - F(\lambda)$ represents the amount of power in the time series at frequency λ. The analysis of variance would correspond to having the total power $C(0)$ equal to the 'sum' of the power contributions at each frequency. This interpretation follows from the spectral representation of the autocovariance function by setting $\tau = 0$:

$$C(0) = \int F(d\lambda)\,.$$

However, the spectral representation of the autocovariance function has an importance beyond this. It tells us how to obtain $C(\tau)$ for all τ if the function $F(\lambda)$ were known. It can be shown (with some difficulty) that $F(\lambda)$ could be recovered if $C(\tau)$ were known completely. That is, these two functions are equivalent parameterizations of the time series. In a sense, they contain the same information about the process. This statement is quite misleading, however, and lies at the root of the unfortunate dichotomization of time-series analysis into separate time-domain and frequency-domain methodologies. One can argue that, since both parameters contain the same information, it is sufficient to study one of them. The time-domain devotees concentrate on the study of $C(\tau)$, while spectrum analysts confine their attention to $F(\lambda)$. The problem with this dichotomized effort is that each parameter displays the time-series information in different ways. Some features of the series are easily detected by looking at $C(\tau)$ but nearly impossible to detect from $F(\lambda)$. The converse is equally true. This is why the practicing time-series analyst *must* be able to operate effectively in both domains. The time-domain tools have the advantage of retaining the time dimension, thus the intuition associated with time-varying phenomena.

Spectral-domain methods, on the other hand, exchange time for frequency and it is necessary to develop new intuition and thought processes in order to interpret the results of spectral analyses for which the goal is the study of $F(\lambda)$.

5. The different types of spectra

Before going on, it is appropriate to look more closely at the kinds of physically relevant spectra the mathematical model is capable of representing. A physical phenomenon that exhibits all of the relevant forms of spectra is light. For example, the spectra of starlight are known to be composed of both lines at distinct frequencies (colors) and a more amorphous blend of energy in bands of frequencies. These are physical realizations of what are called *discrete* or *line spectra* and *continuous spectra*, both of which are representable in the mathematical model.

The power spectrum can be represented as the sum

$$F(\mathrm{d}\lambda) = p(\lambda) + f(\lambda)\,\mathrm{d}\lambda\,,$$

where $p(\lambda)$, called the spectral function, represents the power in the discrete spectrum at frequency λ and $f(\lambda)$, the spectral density function, represents the intensity of the continuous spectrum at λ. There are at most a countable number of points $\lambda_0 = 0$, $\pm\lambda_1$, $\pm\lambda_2, \ldots$ at which $p(\lambda)$ can be positive. The discrete power in any interval of frequencies I is then $\Sigma_{\lambda_i \in I}\, p(\lambda_i)$. The continuous power in I is $\int_I f(\lambda)\,\mathrm{d}\lambda$. The representation of $F(\mathrm{d}\lambda)$, above, admits the possibility of a mixed spectrum in which both continuous and discrete power are present together, as in the starlight example. Pure spectral types would be represented mathematically by taking the function representing the other type to be identically zero. By far the more commonly occurring case in practice, and the one considered almost exclusively in the statistical estimation of spectra, is that of pure continuous spectra. Mixed spectra can be easily reduced to this case by first identifying, estimating and removing the discrete components. Since we will not be concerned with spectral estimation in this chapter, details of this procedure are omitted. However, in the subsequent discussion, we will be concerned primarily with pure continuous spectra. Where the theory does not depend on spectral type, the $F(\mathrm{d}\lambda)$ notation will be retained. Where it does, we will use the spectral density notation.

6. Spectra and linear filters

The relationship between the random amplitude $Z(\mathrm{d}\lambda)$ and the spectrum $F(\mathrm{d}\lambda)$ is important and can be expressed as follows:

$$EZ(\mathrm{d}\lambda)\overline{Z(\mathrm{d}\mu)} = \begin{cases} F(\mathrm{d}\lambda) & \text{if } \mu = \lambda\,, \\ 0 & \text{if } \mu \neq \lambda\,. \end{cases}$$

This expression tells us that the variance of $Z(\mathrm{d}\lambda)$ at frequency λ is $F(\mathrm{d}\lambda)$. (We ignore the measure theoretic niceties and think of $Z(\mathrm{d}\lambda)$ as a complex-valued random variable with zero mean attached to the frequency λ. The variance is then $E|Z(\mathrm{d}\lambda)|^2$.) Moreover, the covariance $EZ(\mathrm{d}\lambda)\overline{Z(\mathrm{d}\mu)}$ is zero if $\mu \neq \lambda$, indicating that the amplitude functions at different frequencies are uncorrelated. Thus problems of describing and dealing with the possibily complicated interrelations of the variables $X(t)$ through the autocovariance function $C(\tau)$—problems, in theory, involving simultaneously the infinity of time dimensions—reduce to an infinite number of uncorrelated and identical finite-dimensional problems—one associated with each frequency. Moreover, except for the minor complication that complex quantities are involved, these finite-dimensional problems will closely resemble familiar problems of statistics.

Perhaps one of the most important applications of the spectral theory, and of these intuitive ideas, is to *linear filters*. The uses of linear filters are woven throughout the entire fabric of time-series analysis. They are used to model physical mechanisms that convert one time series into another. Thus the earth converts the impulse of an earthquake into the complex pattern of waves seen on seismographs in a manner that can be, to a good first approximation, described by a linear filter. Many other physical 'filters' are also well described by linear filters. In other uses, linear filters are designed to perform purposeful transformations of time series. Time-series models, such as the autoregressive and moving average models, familiar in many applications, are defined in terms of linear filters. The construction of a linear predictor of future time-series values is the construction of a special linear filter. The list of applications goes on and on.

Granting their importance, just what are linear filters? A general description of their properties is as follows. (See [11] Chapter 4 for a more careful discussion.) A linear filter L transforms an input time series $X(t)$ into an output time series $Y(t)$, written $Y(t) = L(X(t))$, in such a way that $L(\alpha_1 X_1(t) + \alpha_2 X_2(t)) = \alpha_1 L(X_1(t) + \alpha_2 L(X_2(t))$. Here α_1 and α_2 are real constants which change the scales of the two time series $X_1(t)$ and $X_2(t)$, and the sum indicates addition of the series at each time t. This property accounts for the term '*linear*' in the name of these filters. The separate properties $L(\alpha X(t)) = \alpha L(X(t))$ and $L(X_1(t) + X_2(t)) = L(X_1(t)) + L(X_2(t))$ are called *scale preservation* and the *superposition principle*, respectively. The last property of a linear filter is *time invariance*, which specifies that if $L(X(t)) = Y(t)$, then $L(X(t+h)) = Y(t+h)$ for any h. Intuitively, this simply means that the filter operates in the same fashion no matter what the time origin is—its behavior does not change with time.

The importance of linear filters in the mathematical theory of weakly stationary stochastic processes is connected with the fact that they preserve weak stationarity. That is, if $X(t)$ is a weakly stationary process, then so is $Y(t) = L(X(t))$. Consequently, the behavior of L on $X(t)$ must be observable from the relationships between the parameters of the input and output processes. The relationship between the autocovariance functions of $X(t)$ and $Y(t)$ can be either complicated or simple depending on the specific form of the filter. On the other hand, the relationship between input and output spectra is always simple, regardless of the form of the filter. This is one of the key advantages of the spectral theory.

Without going into details, if the input process has spectral representation

$$X(t) = \int e^{it\lambda} Z_X(d\lambda),$$

the output process $Y(t) = L(X(t))$ has representation

$$Y(t) = \int e^{i\lambda t} D(\lambda) Z_X(d\lambda),$$

where $D(\lambda)$ is called the *transfer function* of L. This function is complex-valued, in general, and can be obtained by applying the filter to the sinusoids $e^{i\lambda t}$ for each λ:

$$L(e^{i\lambda t}) = D(\lambda)\, e^{i\lambda t}.$$

For example, an important special linear filter is the derivative

$$L(X(t)) = \frac{dX(t)}{dt}.$$

Applying L to $e^{i\lambda t}$, we see that

$$\frac{d\, e^{i\lambda t}}{dt} = i\lambda\, e^{i\lambda t}.$$

Thus the transfer function of the derivative is $D(\lambda) = i\lambda$.

To see how input and output spectra are related, we note from the expressions above that the amplitude functions are related by the equation

$$Z_Y(d\lambda) = D(\lambda) Z_X(d\lambda).$$

Forming variances,

$$E|Z_Y(d\lambda)|^2 = |D(\lambda)|^2 E|Z_X(d\lambda)|^2$$

or

$$F_Y(d\lambda) = |D(\lambda)|^2 F_X(d\lambda).$$

Thus the spectra of input and output differ simply by the factor $|D(\lambda)|^2$. In particular, the spectrum of the derivative of $X(t)$ would by $|\lambda|^2 F_X(d\lambda)$.

The condition that the output series have finite power or variance is

$$\int |D(\lambda)|^2 F_X(d\lambda) < \infty.$$

When this condition is satisfied, the filter and input series are said to be *matched.* For example, in order for an input series with pure continuous spectrum and spectral density $f_X(\lambda)$ to match the derivative, we must have

$$\int |\lambda|^2 f_X(\lambda)\, d\lambda < \infty .$$

This clearly imposes a restriction on how much power $X(t)$ can have at high frequencies. If we agree that matching is a necessary constraint, it follows that not all time series can be differentiated.

An intuitive idea of how linear filters operate can be gained by using the polar representations of the complex quantities $D(\lambda)$ and $Z_X(d\lambda)$. Write

$$Z_X(d\lambda) = |Z_X(d\lambda)| e^{i\theta(\lambda)} .$$

Then $|Z_X(d\lambda)|$ represents the random amplitude of the periodic contribution to $X(t)$ at frequency λ and $\theta(\lambda)$ is the random phase, as described in Section 4. Now, writing $D(\lambda) = |D(\lambda)|\, e^{i\phi(\lambda)}$, we see that

$$Z_Y(d\lambda) = |D(\lambda)|\, |Z_X(d\lambda)|\, e^{i(\theta(\lambda)+\phi(\lambda))} .$$

That is, the effect of the filter is to multiply the amplitude at frequency λ by the factor $|D(\lambda)|$ and to shift the phase by $\phi(\lambda)$. These separate components of the transfer function are called the *gain function* and *phase (shift) function*, respectively. The gain and phase-shift functions of the derivative are

$$D(\lambda) = |\lambda|$$

and

$$\phi(\lambda) = \begin{cases} \dfrac{\pi}{2} & \text{for } \lambda > 0 , \\ \dfrac{-\pi}{2} & \text{for } \lambda < 0 . \end{cases}$$

Note that if a linear filter with transfer function $D(\lambda)$ is viewed as modeling a 'black box', whose properties are to be determined from the input and output time series, it is not sufficient to compute the power spectra of input and output. The reason for this is that only the gain function of the filter can be determined from the spectra:

$$|D(\lambda)| = \sqrt{\frac{F_Y(d\lambda)}{F_X(d\lambda)}} .$$

In order to capture the phase shift of the filter as well, we need additional spectral parameters for defining relationships between the two time series. These parameters are discussed next.

7. Spectral parameters for bivariate time series

Two weakly stationary processes $X(t)$ and $Y(t)$ are said to be *stationarily correlated* if the covariance $R_{XY}(t_1, t_2) = EX(t_1)Y(t_2)$ depends only on $t_1 - t_2$. The cross-covariance function $C_{XY}(\tau)$ is then defined to be

$$C_{XY}(\tau) = EX(t+\tau)Y(t).$$

A pair of stationarily correlated weakly stationary processes constitutes a *bivariate* weakly stationary process. The cross-covariance function is the new time-domain parameter which, along with the autocovariance functions $C_X(\tau)$ and $C_Y(\tau)$, completely describes the relevant properties of the bivariate process. The corresponding spectral parameter $F_{XY}(d\lambda)$, called the cross-spectral distribution or, more simply, the *cross spectrum*, satisfies the relation

$$C_{XY}(\tau) = \int e^{i\lambda\tau} F_{XY}(d\lambda).$$

The cross spectrum has discrete and continuous components $p_{XY}(\lambda)$ and $f_{XY}(\lambda)$, called the cross-spectral function and cross-spectral density, for which

$$F_{XY}(d\lambda) = p_{XY}(\lambda) + f_{XY}(\lambda)\, d\lambda.$$

These functions will be nonzero only where the corresponding spectral functions or spectral densities are nonzero for both component processes.

The input and output of a linear filter will always be stationarily correlated. Consequently, we can compute the cross spectrum of such series. It is convenient to use the fact that $F_{XY}(d\lambda)$ is the (complex) covariance of $Z_X(d\lambda)$ and $Z_Y(d\lambda)$:

$$F_{XY}(d\lambda) = EZ_X(d\lambda)\overline{Z_Y(d\lambda)}.$$

If $Y(t) = L(X(t))$ and L has transfer function $D(\lambda)$, then

$$\begin{aligned} F_{XY}(d\lambda) &= EZ_X(d\lambda)\overline{[D(\lambda)Z_X(d\lambda)]} \\ &= \overline{D(\lambda)}EZ_X(d\lambda)\overline{Z_X(d\lambda)} \\ &= \overline{D(\lambda)}F_X(d\lambda). \end{aligned}$$

Thus the transfer function, complete with both gain and phase information, can be computed as

$$D(\lambda) = \frac{\overline{F_{XY}(d\lambda)}}{F_X(d\lambda)}.$$

This, of course, is only one possible use of the cross spectrum. In general, the cross spectrum contains information about the interrelationship between the components of a bivariate time series in much the same way that a covariance measures the linear relationship between two random variables. In fact, this analogy is much closer than one might imagine. In each frequency dimension λ, the cross spectrum *is* essentially the covariance of the two 'random variables' $Z_X(\mathrm{d}\lambda)$ and $Z_Y(\mathrm{d}\lambda)$. The chief difference is that these variables are complex-valued, which makes the covariance complex-valued as well.

Two different real-valued representations of the cross spectrum are in common use, each depending on a particular expression for complex numbers. In order for our notation to agree with that seen in practice, we will take the spectrum to be of continuous type. The cross spectrum is then determined by the cross-spectral density $f_{XY}(\lambda)$. Representing $f_{XY}(\lambda)$ in Cartesian form (with a negative sign) leads to the equation

$$f_{XY}(\lambda) = c(\lambda) - \mathrm{i}q(\lambda),$$

where $c(\lambda)$ and $q(\lambda)$ are the cospectral density (or cospectrum) and quadrature spectral density (or quadspectrum), respectively. Thus one complete list of real-valued spectral parameters for the bivariate process would be $c(\lambda)$, $q(\lambda)$, $f_X(\lambda)$ and $f_Y(\lambda)$.

A second set of parameters is obtained from applying the polar representation $z = r\,\mathrm{e}^{\mathrm{i}\theta}$ to $f_{XY}(\lambda)$, where $r = |z|$ and $\theta = \arg z$. Here, we let

$$\rho(\lambda) = \frac{|f_{XY}(\lambda)|}{\sqrt{f_X(\lambda) f_Y(\lambda)}} \quad \text{and} \quad \psi(\lambda) = \arg f_{XY}(\lambda).$$

These parameters are called the *coherence* and *phase*, respectively. Along with $f_X(\lambda)$ and $f_Y(\lambda)$ they represent an alternate real-valued parameterization of the bivariate process. In the author's view, this parameterization is the more useful one because of its interpretability.

Writing $Z_X(\mathrm{d}\lambda)$ and $Z_Y(\mathrm{d}\lambda)$ in polar form, we have

$$\begin{aligned} f_{XY}(\lambda) &= EZ_X(\mathrm{d}\lambda)\overline{Z_Y(\mathrm{d}\lambda)} \\ &= E|Z_X(\mathrm{d}\lambda)|\,|Z_Y(\mathrm{d}\lambda)|\,\mathrm{e}^{\mathrm{i}(\theta_X(\lambda) - \theta_Y(\lambda))}. \end{aligned}$$

If the phases $\theta_X(\lambda)$ and $\theta_Y(\lambda)$ were constant, the exponential would factor out of the expectation giving

$$\psi(\lambda) = \theta_X(\lambda) - \theta_Y(\lambda).$$

In this case, $\psi(\lambda)$ would represent the phase lead of the $X(t)$ time series over that of the $Y(t)$ series at frequency λ. Since, in general, the phases will be random, this interpretation will not be precisely correct. However, $\psi(\lambda)$ will still represent a weighted stochastic average of the phase differences and it is useful to think of this parameter as the (or an) average phase lead of $X(t)$ over $Y(t)$.

The coherence behaves almost exactly like the absolute value of a correlation coefficient. For example, $0 \leqslant \rho(\lambda) \leqslant 1$, with values near 0 indicating a weak linear relationship at frequency λ and values near 1 a strong relationship. In the time-series context, 'linear' refers to linear filters. That is, $\rho(\lambda)$ measures the degree to which $Y(t)$ can be represented as the output of a linear filter with input $X(t)$. In fact, $\rho^2(\lambda)$ has precisely the interpretation of the coefficient of determination. It is the proportion of the variation (power) of $Y(t)$ at λ that is attributable to its linear relationship with $X(t)$ in the following sense: If $\hat{L}$ is the linear filter that minimizes the power, $E(Y(t)-L(X(t)))^2$ in the 'residual process' among all filters L, then $\rho^2(\lambda)$ is the ratio of the spectral density $f_{\hat{Y}}(\lambda)$ to $f_Y(\lambda)$, where $\hat{Y}(t)=\hat{L}(X(t))$. The process $\hat{Y}(t)$ represents the best approximation to $Y(t)$ as a 'linear function' of $X(t)$ and $\rho^2(\lambda)$ then represents the proportion of the power in $Y(t)$ at frequency λ attributable to $\hat{Y}(t)$. Thus, for example, if $Y(t)$ is exactly a linear function of $X(t)$, $Y(t)=L(X(t))$, and if $D(\lambda)$ is the transfer function of L, then we see from earlier calculations that

$$\begin{aligned}\rho^2(\lambda) &= \frac{|f_{XY}(\lambda)|^2}{f_X(\lambda)f_Y(\lambda)} \\ &= \frac{|\overline{D(\lambda)}f_X(\lambda)|^2}{f_X(\lambda)(|D(\lambda)|^2 f_X(\lambda))} \\ &= 1\,.\end{aligned}$$

Another important property of the absolute value of a correlation coefficient is its invariance under linear transformation. This property also holds for coherence. Thus, if $X(t)$ and $Y(t)$ have coherence function $\rho(\lambda)$ and if $U(t)=L_1(X(t))$ and $V(t)=L_2(Y(t))$, where L_1 and L_2 are arbitrary linear filters, then $\rho(\lambda)$ will also be the coherence of $U(t)$ and $V(t)$ at all frequencies for which the spectral densities $f_U(\lambda)$ and $f_V(\lambda)$ are both positive. These properties make it possible to translate one's intuition about correlation and simple linear regression directly to coherence for a frequency by frequency assessment of the association between two time series.

8. Spectra of multidimensional processes

The discussion of the last section implies that at each frequency the matrix of spectral densities for a bivariate time series,

$$f(\lambda)=\begin{bmatrix} f_X(\lambda) & f_{XY}(\lambda) \\ f_{YX}(\lambda) & f_Y(\lambda) \end{bmatrix}$$

(where $f_{YX}(\lambda)=\overline{f_{XY}(\lambda)}$), behaves like the covariance matrix of two random variables.

Linear filters in two or more dimensions also have a familiar matrix representation in the frequency domain. For example, if a 'black box' with two inputs and two outputs is a linear filter, each input would contribute linearly to each output. If $B_{ij}(\lambda)$ represents the transfer function of the contribution of the ith input to the jth output, the behavior of the filter is completely described by the matrix of transfer functions

$$\boldsymbol{B}(\lambda)=\begin{bmatrix} B_{11}(\lambda) & B_{12}(\lambda) \\ B_{21}(\lambda) & B_{22}(\lambda) \end{bmatrix}.$$

In particular, if $\boldsymbol{Z}(\mathrm{d}\lambda)=(Z_X(\mathrm{d}\lambda), Z_Y(\mathrm{d}\lambda))^T$ is the vector of complex amplitudes of a weakly stationary bivariate input process, the vector output process can be represented as

$$\boldsymbol{W}(t)=\int \mathrm{e}^{\mathrm{i}\lambda t}\boldsymbol{B}(\lambda)\boldsymbol{Z}(\mathrm{d}\lambda),$$

where matrix multiplication has been used inside the integral and the integration has then been carried out coordinatewise. In the same sense, the spectral density matrix of the output process is

$$\boldsymbol{B}(\lambda)\boldsymbol{f}(\lambda)\boldsymbol{B}^*(\lambda),$$

where $*$ denotes the transpose of the coordinatewise complex conjugate of $\boldsymbol{B}(\lambda)$.

Thus writing the amplitude vector of the output process $\boldsymbol{W}(t)$ in the form

$$\boldsymbol{Z}_{\mathrm{out}}(\mathrm{d}\lambda)=\boldsymbol{B}(\lambda)\boldsymbol{Z}_{\mathrm{in}}(\mathrm{d}\lambda),$$

we see that the linear filter behaves exactly like a two-dimensional linear transformation in the frequency domain. Moreover, the spectral density matrix is transformed by a linear filter in exactly the same way that a covariance matrix is by a linear transformation. Again, the only added complication is that the matrix elements are complex-valued.

These ideas and formulas generalize immediately to p-dimensional time series for any finite value of p. Then, the parameters and methods of analyzing the covariance structure of a vector random variable can be applied to the frequency-domain representation of the vector weakly stationary stochastic process. Partial and multiple coherences can be used for the same purposes as are partial and multiple correlation coefficients. However, in the time-series context, phase information is also available in the polar representations of these parameters. Partial and multiple coherences are important parameters in time-series versions of multiple regression.

Methods of multivariate analysis such as canonical correlation analysis and principal component analysis also have time-series analogs which use these and other spectral parameters defined in analogy with the appropriate multivariate parameters. Examples of these methods are given in [4, 6, 11]. The possibilities have by no means been exhausted. It is expected that many fruitful applications of spectral methods to multidimension processes will be made in the future. The third flowering of spectral analysis may well lie in this direction.

References

[1] Anderson, T. W. (1971). *The Statistical Analysis of Time Series.* Wiley, New York.

[2] Blackman, R. B. and Tukey, J. W. (1959). *The Measurement of Power Spectra from the Viewpoint of Communications Engineering.* Dover, New York. Reprinted from *Bell System Tech. J.* **37** (1958).

[3] Bloomfield, Peter (1976). *Fourier Analysis of Time Series: An Introduction.* Wiley, New York.

[4] Bloomfield, Peter and Zlotnik, H. (1979). Spectrum analysis of multiple time series data, in *Proceedings of IMS Special Topics Meeting on Time Series Analysis,* Ames, Iowa. Published by the Institute of Mathematical Statistics.

[5] Box, G. E. P. and Jenkins, G. M. (1970). *Time Series Analysis Forecasting and Control.* Holden-Day, San Francisco.

[6] Brillinger, David R. (1975). *Time Series Data Analysis and Theory.* Holt, Rinehart and Winston, New York.

[7] Granger, C. W. J. and Hatanaka, M. (1964). *Spectral Analysis of Economic Time Series.* Princeton University Press, Princeton.

[8] Grenander, U. and Rosenblatt, M. (1957). *Statistical Analysis of Stationary Time Series.* Wiley, New York.

[9] Hannan, E. J. (1970). *Multiple Time Series.* Wiley, New York.

[10] Jenkins, G. M. and Watts, D. G. (1968). *Spectral Analysis and Its Applications.* Holden-Day, San Francisco.

[11] Koopmans, L. H. (1974). *The Spectral Analysis of Time Series.* Academic Press, New York.

D. R. Brillinger and P. R. Krishnaiah, eds., *Handbook of Statistics, Vol. 3*

10

Robust–Resistant Spectral Analysis

*R. Douglas Martin**

1. Introduction

Resistance and robustness

This tutorial paper describes some methodologies for obtaining *resistant* estimates of spectral densities. Resistance is a term coined by Mosteller and Tukey (1977); roughly speaking it means insensitivity to changes in the data. More specifically, a *resistant* estimate is one which is not affected very much by (i) changing a small fraction of the observed data by possibly large amounts, or (ii) changing all the data by small amounts. Large changes in a small fraction of the data occur either when gross errors are made in recording the data, or when the data by its very nature makes occasional large excursions. All the data may be changed by small amounts, for example, when grouping, rounding or quantization effects are present.

The term resistant is a purely data-oriented word, which has as both an advantage and a disadvantage the fact that technical probability and mathematical statistics issues are not involved. The quality of resistance is to be judged solely in terms of the functional or algorithmic form of the estimate. While this may require technical mathematical tools, in order to establish continuity for example, the probabilistic aspects of statistics are not really required. The advantage is a transparency of concept which is widely accessible. The disadvantage is that without the formal theory of statistics it is not possible to discuss important basic issues such as *consistency* and *efficiency*, let alone discuss *robustness*, which is the probabilistic counterpart of resistance.

The last decade or so has seen a vigorous development of the *theory*, *concepts* and *algorithms* of robust statistical procedures for the case where the observations are *independent*. References to much of the literature, excluding the very most recent publications and preprints, may be found in the recent book by Huber (1981). Work on robustness in the *time-series* setting has lagged

*This research was supported in part by the Office of Naval Research Contract N00014-82-K-0062.

considerably, and this is perhaps understandable in view of the increased technical problems imposed by a serial dependency in the data.

An overview of robust methods for time series, which gives some indication of the problems involved, may be found in Martin (1981). In fact, the difficulties associated with robustness theory for time series are sufficiently technical and so much in a state of current development (cf., for example, Cox, 1981, Papantoni-Kazakos and Gray, 1979, and Boente, Fraiman and Yohai, 1982) that extensive treatment of robustness aspects of spectrum estimation is both premature and beyond the scope of this presentation. Thus the emphasis will be on resistant spectrum estimation, with some occasional discussion of robustness issues. Since situations sometimes encourage use of the terms robustness and resistance interchangeably, and without regard to their distinction (for example, one finds 'robustify' a more pleasant term than 'resistify'), we shall freely alternate use of the terms.

That robustness–resistance is indeed an important issue in the context of spectrum estimation was exposed by Kleiner, Martin and Thomson (1979), who proposed two robust–resistant methodologies (see also Thomson, 1977a,b). Although the techniques and emphasis in this paper differ somewhat from those in Kleiner, Martin and Thomson (1979), the latter provided considerable inspiration for this tutorial. See also Martin and Thomson (1982), which overlaps somewhat with this paper.

The time-series setup

We shall restrict attention in this paper to the special case of *discrete time series* (or *stochastic processes*) x_t, $t = 0, \pm 1, \pm 2, \ldots$, corresponding to observations at equispaced time intervals $t' = 0, \pm\Delta, \pm 2\Delta, \ldots$. It is of no importance here whether the x_t are observations of an inherently discrete-time proces , or are the samples of a continuous time process. In the latter case we simply assume that the sampling rate $f_s = 1/\Delta$ is sufficiently high that *aliasing errors* are of no real concern (Bloomfield, 1976; Brillinger, 1981). For notational convenience the observation, or sampling, time interval Δ is suppressed throughout by assuming $\Delta = 1$.

Furthermore, we assume that the time series x_t of interest has finite variance and is reasonably well modeled by the *wide-sense stationarity* (WSS) assumptions that for all $t = 0, \pm 1, \pm 2, \ldots$

$$Ex_t \equiv \mu \tag{1.1}$$

and

$$\operatorname{cov}(x_t, x_{t+l}) = C(l), \quad l = 0, \pm 2, \ldots . \tag{1.2}$$

That is, the mean Ex_t is constant and the covariance between x_t and x_{t+l} depends only upon l and not upon t. Given the above *second-order* description of x_t in the time domain, the corresponding *frequency-domain* description is given by the *spectrum* or *spectral density*

$$S(f) = \sum_{l=-\infty}^{\infty} C(l)\,\mathrm{e}^{-\mathrm{i}2\pi fl}\,. \tag{1.3}$$

It is easy to check that $S(f)$ is symmetric and periodic, with period unity. Correspondingly, the $C(l)$ are *Fourier coefficients* which may be obtained from $S(f)$ as

$$C(l) = \int_{-1/2}^{1/2} S(f)\,\mathrm{e}^{\mathrm{i}2\pi fl}\,\mathrm{d}f\,. \tag{1.4}$$

The relationships (1.3) and (1.4) are often referred to as (discrete-parameter) *Fourier transform* pairs.

A more basic and intuitive interpretation of the spectral density is obtained from the so-called *spectral representation* theorem which roughly states that X_t may be expressed as the *random Fourier series*

$$x_t \approx \mu + \sum_{l=-N/2}^{N/2} A_l\,\mathrm{e}^{\mathrm{i}2\pi lft}\,, \tag{1.5}$$

where A_l are random complex coefficients with the properties: (i) $EA_l \equiv 0$, and $\mathrm{cov}(A_l, A_m) = 0$ for $l \neq m$, (ii) $S(l/N) = E|A_l|^2$. Thus the value of the spectrum at frequency $f_l = l/N$ is the variance (in the absolute value sense) of the lth random coefficient. A precise statement of the spectral representation theorem and associated details may be found in many standard references (see, for example, Grenander, 1981).

The basic problem to be treated here is that of obtaining resistant estimates of $S(f)$. The paper is organized as follows: Section 2 discusses time-series outlier types and corresponding nonnormality of the probability models, illustrated by several examples, and introduces a basic additive outlier (AO) model. Section 3 discusses the lack of resistance–robustness of conventional autoregressive (AR) spectral density estimates, and shows how to robustify the conventional Ar estimates. Section 4 is a central one in which the lack of resistance–robustness of smoothed periodogram estimates is discussed and a robust methodology is presented. Section 5 briefly mentions the robustness–resistance properties of the methods described in Section 4. Several examples are presented in Section 6, and Section 7 briefly mentions some open problems.

2. Time-series outliers and nonnormality

Qualitative features

Time series occurring in practice often exhibit anomalies in the form of *outliers* of various types. For the time being let us use the term *outlier* quite loosely in the time-series setting, and attempt some degree of preciseness a little later on. In this spirit we note that outliers come in many forms, such as *bumps* or *patches* differing from the remainder of the record, *spikes* of varying shape and frequency, *level shifts* of varying frequency and magnitude, and *gross errors or malfunctions in recording.*

Even though we are focussing our attention on the rather narrow class of discrete time-series data well approximated by the wide-sense stationarity assumption, diversity of qualitative features in observed data records are considerable. Specific kinds of features and anomalies are often associated with particular subject matter contexts. For example, economic time series often exhibit isolated outliers or outlier 'patches' whose onsets are associated with specific events such as strikes, oil embargos, etc. In speech analysis and synthesis, one encounters rather periodic structure, including spikes, in the record (or 'waveform') for *voiced* sounds (see, for example, Rabiner and Schaffer, Chapter 3, 1978).

In one engineering problem, researchers gathering large volumes of underwater acoustic data on several recorders simultaneously are concerned about *level shifts* due to the physical process itself, *latch-ups* consisting of a string of constant values due to a measuring instrument saturation or malfunction, and *isolated outliers* due to various causes. Another engineering problem is concerned with communication at extra low frequencies (ELF) in the range 70–100 hertz. In this frequency range, electrical activity due to storms, both local and distant, produces huge spikes in the data of varying amplitude and frequency of occurrence (see Figs. 1 and 2 of Evans and Griffiths, 1974).

An example of a data set containing many outliers due to a physical reverberation phenomenon consists of measurements of so-called '*glint*' noise shown in Fig. 1 (see, for example, McGrew, 1972, for a discussion of the

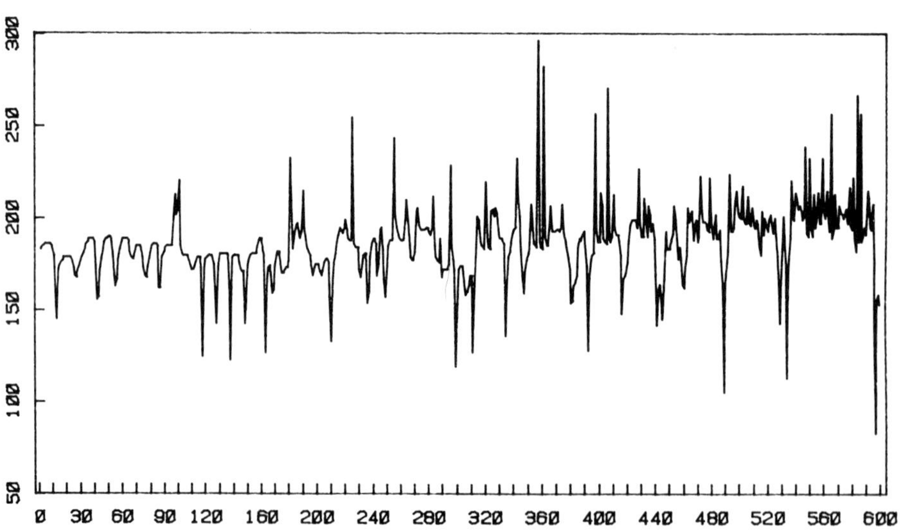

Fig. 1. Glint noise data.

physical basis for a glint noise model). The ordinate in Fig. 1 displays apparent angle of an aircraft target, as observed by a radar receiver, versus angle of rotation of the target. The frequently occurring glint spikes, which indicate an apparent angle quite different from the true angle, are due to constructive interference of the reverberation-like composite return of the reflected signal from different positions of the complex structure of the aircraft.

One can also find special anomalies in time-series data arising in geophysics, oceanography and other engineering and applied science problems. The diversity may at first appear overwhelming, and it might seem that specially tailored techniques would be required for each particular problem. Indeed, it is true that specialized techniques have been devised for certain problems. For example, *intervention analysis*, a structured dummy-variable technique, has been proposed by Box and Tiao (1975) for dealing with situations where a known cause, such as a strike in an economic time series, may cause a special effect, which may in some instances exhibit an outlier-like character. This approach assumes that the intervention effect has a parametric structure which is either known, or easily guessed, except for a few parameter values which are estimated along with other model parameters by Gaussian maximum-likelihood estimation. Brillinger (1973) proposed a variant of this technique involving cross-spectral analysis.

On the other hand, the intervention analysis technique is hardly of universal applicability. First of all, even when the time of onset of a special event indicates the possibility of outliers, it may be difficult to specify a parametric structure for the effect. Second, many time-series outliers have times of occurrence which are not specified by the occurrence of other events, and it may be that they are not even easily detected by the eye in a plot of the data (see Fig. 4D of Kleiner, Martin and Thomson, 1979, or Fig. 6 of Martin, Samarov and Vandaele, 1981).

Thus there is a need for time-series spectrum estimation (as well as parameter estimation) methodologies which are insensitive to outliers, and which do not rely on very special knowledge about either the location of outliers or a probability model which generates the outliers. Fortunately, many of the outlier types described above have an important common feature: they occur a small to moderate fraction of the time, and except for these times, the series exhibits what appears to be a moderately homogeneous and Gaussian stochastic behavior. The resistant spectrum estimation techniques which we shall describe are well suited for series having this general character. For, like other robust techniques in the non time-series setting, they are constructed so as to provide estimates which represent the bulk of the data.

The additive outlier model

Subsequently we shall discuss what constitutes a time-series outlier in rather precise qualitative terms. Partly for reasons having to do with notation, it is convenient to introduce an explicit outlier generating model at this time. The model is a so-called *additive outlier* (AO) model which allows for a particularly harmful type of nonnormal probability model which generates outliers. In this model the observations are labeled y_t, with y_t related to the process x_t of interest by

$$y_t = x_t + v_t , \tag{2.1}$$

where v_t accounts for outliers, either in isolation or in clumps.

Any perturbation of the process x_t, whose spectrum we are interested in estimating, by local disturbances, gross outliers and other aberrations can be expressed in the form (2.1) by simply defining v_t to be the difference between the observations y_t and the *core* process x_t. In general, this produces a complicated dependency between v_t and x_t that, along with the serial dependency needed to adequately describe patchy outliers, greatly complicates any attempt to produce theoretical results. For the latter purpose, one would usually for convenience require that the v_t be independent and identically distributed (i.i.d.). However, the only basic attribute of the v_t which we wish to stress here is that the v_t are assumed to be zero much of the time: $P(v_t = 0) = P(y_t = x_t) = 1 - \gamma$, and typically we will have γ not too large, say $0.01 \leq \gamma \leq 0.25$. Accordingly, the process x_t is observed perfectly about $100(1 - \gamma)\%$ of the time, and is corrupted by outlier effects about $100\gamma\%$ of the time.

From now on we shall assume that the observed data is y_t, with general versions of (2.1) accounting for outliers. The situation where x_t is observed perfectly is then the special case where $v_t \equiv 0$.

Univariate and bivariate outliers

Outliers in time series are associated with distinct nonnormal or non-Gaussian features of the data. Sometimes this is clearly reflected in heavy-tailed behavior of the univariate or marginal distribution function $F(y; t)$, or density $f(y; t)$. Although density estimates are not very useful for revealing a heavy-tailedness, except in very large sample sizes, the presence of outliers is sometimes clearly reflected in normal QQ-plots.[1] For example, the glint noise data segment (which is part of a longer record) displayed in Fig. 1 yields the QQ-plot of Fig. 2. The heavy-tailed nature of the marginal distribution, which is associated with the glint spikes, is quite evident.

Another segment of the same glint noise record is displayed in Fig. 3, and a corresponding normal QQ-plot is shown in Fig. 4. Although the spikes are still quite evident in the time record, the QQ-plot of Fig. 4 gives a much weaker indication of nonnormality than that of Fig. 2. Since the data would nonethe-

[1] A (normal) QQ-plot is a plot of the order statistic values $y(i)$ versus the normal quantiles $y(p_i) = \Phi^{-1}(p_i)$, $p_i = (i - 1/2)/n$, $1 \leq i \leq n$, where Φ is the standard normal distribution function. See, for example, Gnanadesikan (1977).

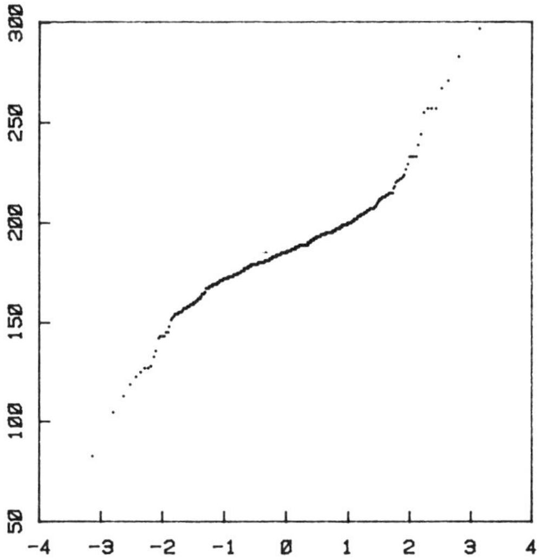

Fig. 2. Normal QQ-plot for glint noise data.

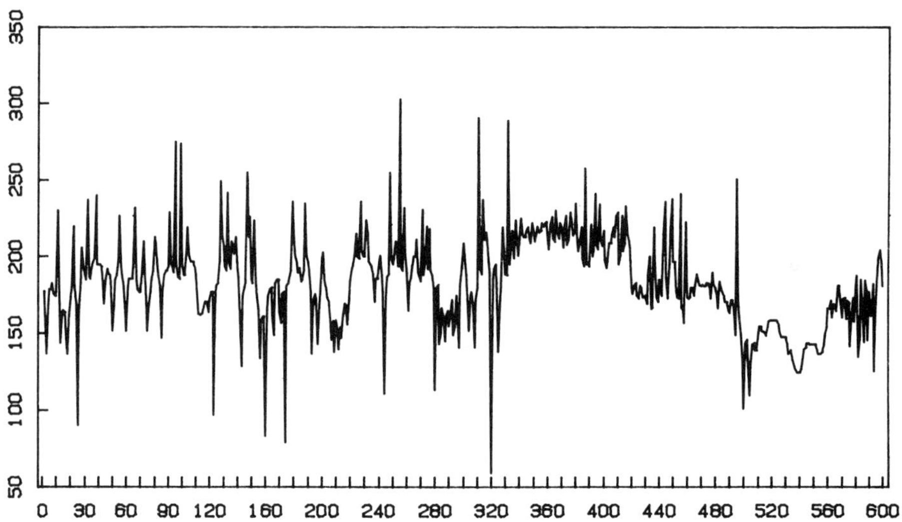

Fig. 3. Glint noise data.

less appear to be strongly non-Gaussian, one might expect this aspect to be more clearly evident in the higher-order distributions for the data. Although QQ-plots do not extend easily and naturally to multivariate situations, one can easily get some idea of the lag-1 bivariate distribution of the data by looking at

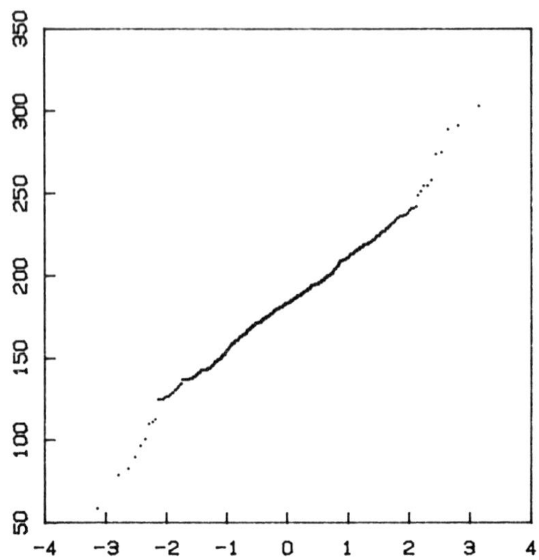

Fig. 4. Normal QQ-plot for glint noise data.

scatter plots, i.e., plots of y_t versus y_{t-1}. Fig. 5 displays such a plot for the glint noise segment of Fig. 3. The nonnormality is more strongly evident here than in the univariate QQ-plot.

Outliers more generally

Although time-series outliers which are hidden in a marginal view of the data will sometimes be exposed in a bivariate view, this need not always be the case.

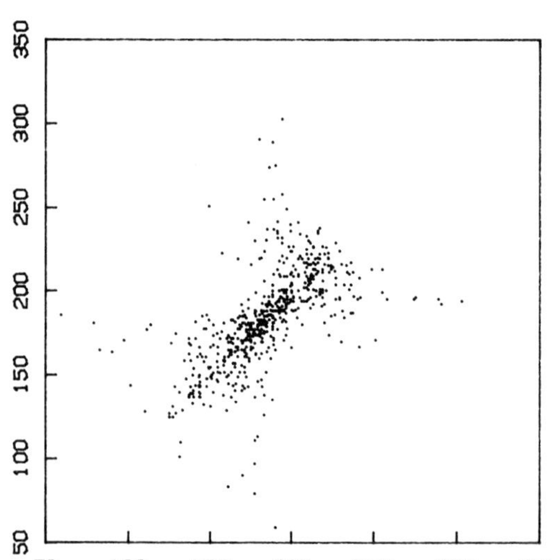

Fig. 5. Lag-1 scatter plot of glint noise data.

For example, the artificial data in Fig. 6 produces the lag-1 scatter plot shown in Fig. 7, which reveals no outliers. Clearly, a univariate QQ-plot will also reveal nothing. On the other hand, the lag-2 scatter plot of y_t versus y_{t-2} in Fig. 8 clearly reveals two outliers (N.B. a single outlier in the original series typically results in a pair of outliers in a scatter plot).

This example suggests that we should routinely plot lag-l scatter plots for $l = 1, 2, \ldots, L$, with L chosen large enough that the lag-L plot shows no

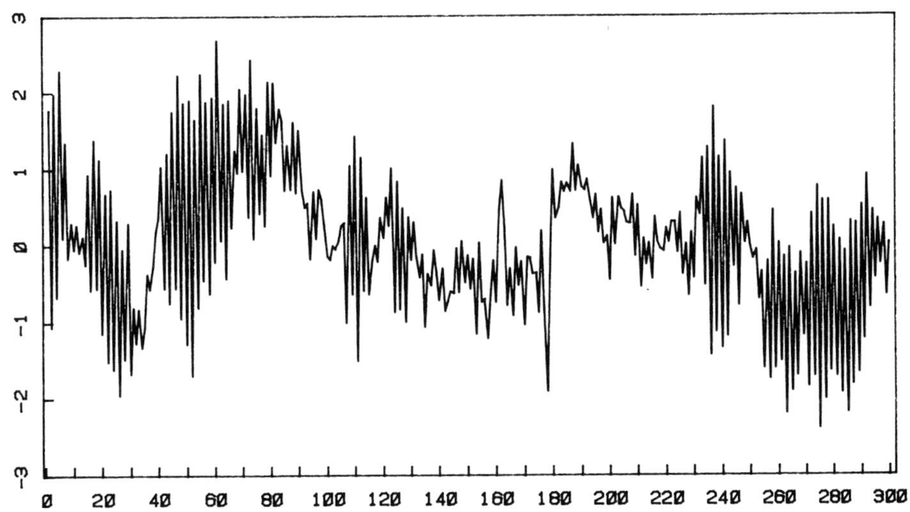

Fig. 6. Artificial data.

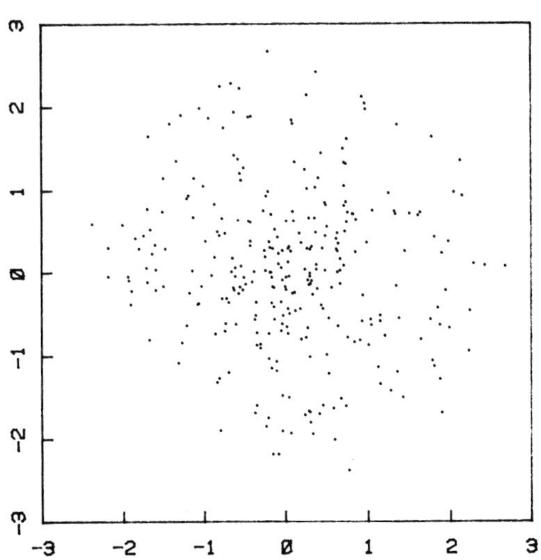

Fig. 7. Lag-1 scatter plot of artificial data.

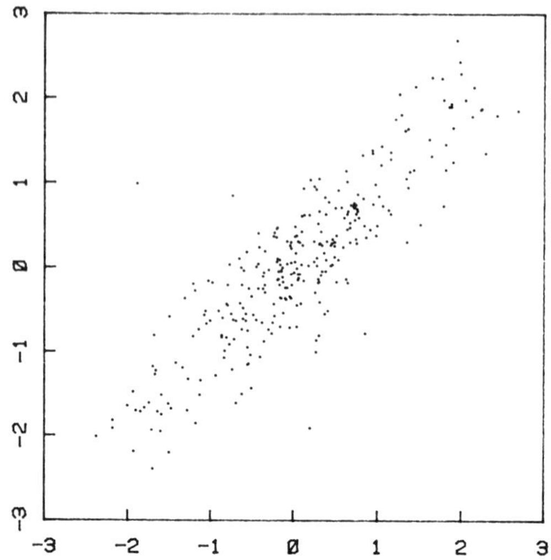

Fig. 8. Lag-2 scatter plot of artificial data.

interesting correlation situation. Although the suggestion is hardly new, it is seldom followed in practice.

The preceding example also begs the obvious question, "*What is a time-series outlier anyway*?" The following comments provide a natural definition in terms of conditional densities for prediction.

Imagine a probability model giving rise to outliers which are transparent in a bivariate view. Denote the bivariate density for the series at times t_1 and t_2

$$f(y_1, y_2) = f(y_2 | y_1) f(y_1)$$

with marginal density $f(y_1)$ and conditional density $f(y_2 | y_1)$. Outliers which do not appear in a marginal view must fall within the area of concentration of $f(y_1)$. But if such outliers are exposed in a bivariate view, they must fall outside the area of concentration of $f(y_1, y_2)$, which is say roughly ellipsoidal. Hence an outlier y_2 at time t_2 must fall outside the region of concentration of $f(\cdot | y_1)$.

Now let us generalize to the higher-order case. An outlier y_{M+1} at time t_{M+1} is a data point which falls outside the region of concentration of $f(y_{M+1} | y_M, \ldots, y_1)$ for some M and $t_1, \ldots, t_M$, where $y_1, \ldots, y_M$ are the data points corresponding to the latter time points. Since selection of nonconsecutive time points would be very difficult in practice, one naturally restricts attention to the case of consecutive time points: $t_{M+1} = t$, $t_M = t - 1, \ldots, t_1 = t - M$. In order to obtain a fairly transparent interpretation of what this general definition means, we shall use a zero-mean Gaussian series as a frame of reference.

For a Gaussian series we would have

$$f(y_t | y_{t-1}, \ldots, y_{t-M}) = N(y_t; \hat{y}_t^{t-1}, s_M), \tag{2.2}$$

where $N(y; \mu, \sigma^2)$ is the Gaussian density with mean μ, variance σ^2, and where

$$\hat{y}_t^{t-1} = E(y_t \mid y_{t-1}, \ldots, y_{t-M}) \tag{2.3}$$

is the conditional-mean predictor of y_t given $y_{t-1}, \ldots, y_{t-M}$, and s_M^2 is the associated conditional mean-squared-error (MSE) of prediction. It is well known that in the Gaussian case, $\hat{y}_t^{t-1}$ is linear:

$$\hat{y}_t^{t-1} = \sum_{l=1}^{M} a_{Ml} y_{t-l}, \tag{2.4}$$

where the coefficients $\boldsymbol{a}_M^{\mathrm{T}} = (a_{M1}, \ldots, a_{MM})$ satisfy the normal equations

$$\boldsymbol{C}\boldsymbol{a}_M = \boldsymbol{g} \tag{2.5}$$

with $C_{ij} = \operatorname{cov}(y_i, y_j)$, $1 \leq i, j \leq M$ and $g_i = \operatorname{cov}(y_0, y_i)$, $1 \leq i \leq M$. For a series with nonzero mean, an intercept coefficient γ would have to be included in (2.4).

Now since $N(y_t; \hat{y}_t^{t-1}, s_M^2) = N(y_t - \hat{y}_t^{t-1}; 0, s_M^2)$, we have y_t *is an outlier if and only if the prediction residual*

$$r_t = y_t - \hat{y}_t^{t-1} \tag{2.6}$$

is large relative to s_M for some M. In practice, it seldom appears to be necessary to choose M very large in order to expose outliers that have any substantial consequence on estimation procedures. Often M in the range 2 to 4 will be adequate, assuming the data is not oversampled.

In the event that the time series is non-Gaussian, it sometimes turns out that we are justified in using the approximation

$$f(y_t \mid y_{t-1}, \ldots, y_{t-M}) \approx \frac{1}{s_t} g\left(\frac{y_t - \hat{y}_t^{t-1}}{s_t}\right) \tag{2.7}$$

for some non-Gaussian density g which is approximately symmetric. In such cases the interpretation of an outlier as a point y_t for which $r_t = y_t - \hat{y}_t^{t-1}$ is large relative to the scale measure s_t still holds. Here one needs to be careful about details concerning the non-Gaussian model. For example, if $y_t \equiv x_t$ is a perfectly observed autoregression of order M, then (2.7) holds with equality, $s_t \equiv s_M$ is the scale of the *innovations*, g is the innovations density and the conditional mean predictor $\hat{y}_t^{t-1}$ is linear (assuming the conditional expectation exists). On the other hand, if the *additive outlier* model (2.1) holds, then the conditional-mean predictor will be a nonlinear function of the data. In particular, $\hat{y}_t^{t-1}$ will be a *robust–resistant* predictor having the general character described in the next paragraph. Further details will be provided in Section 4.

Robust prediction residuals diagnostics

Now the scale $\sigma_r = \sqrt{\text{var}\, r_t}$, of r_t is often much smaller than $\sigma_x = \sqrt{\text{var}\, x_t}$. Correspondingly, outliers which may be small on the scale of σ_x may be quite large on the scale of σ_r. Thus we would hope that with well-chosen M, and with $\boldsymbol{a}_M$ replaced by a good estimate $\hat{\boldsymbol{a}}_M$, outliers which do not appear in univariate, or even bivariate, looks at the data would be well exposed through examination of the prediction residuals r_t.

The only problem is that the predictors $\hat{y}_t^{t-1} = \Sigma_{l=1}^{M} \hat{a}_{Ml} y_{t-l}$ may be severely degraded by the presence of outliers in the predictor variables y_{t-1}, $y_{t-2}, \ldots, y_{t-M}$, and also by the effect on $\hat{\boldsymbol{a}}_M$ of the outliers in all the data. This can be remedied by replacing the predictor variables by suitable 'cleaned' data values $\hat{y}_{t-1}, \ldots, \hat{y}_{t-M}$ having the following character: if y_u is not an outlier, then $\hat{y}_u = y_u$; whereas if y_u is a gross outlier, then $\hat{y}_u$ is some form of interpolate based on other $\hat{y}_t$. The resulting robust–resistant predictor is

$$\hat{y}_t^{t-1} = \sum_{l=1}^{M} \hat{a}_{Ml} \hat{y}_{t-l} \tag{2.8}$$

and outliers in the y_t should be in clear evidence in the residuals $r_t = y_t - \hat{y}_t^{t-1}$, even when they are not evident in the marginal distribution of y_t. Details for constructing such cleaned values, and for estimating $\boldsymbol{a}_M$ robustly as well are given in Section 4.

3. Autoregressive spectral density estimates

Standard autoregressive estimates

There now exists a large literature which discusses and promotes so-called *autoregressive* (*AR*) *spectral density* estimates (cf. Parzen, 1974; Jones, 1981). The basic idea is straightforward. Suppose that in fact the observed data was generated by a stationary pth-order autoregression

$$x_t = \varphi_1 x_{t-1} + \cdots + \varphi_p x_{t-p} + \varepsilon_t, \quad t = 0, \pm 1, \pm 2, \ldots, \tag{3.1}$$

where the ε_t have zero mean, variance σ_ε^2, and are white noise, i.e. $E\varepsilon_t\varepsilon_{t+l} = 0$, $l \neq 0$. Stationarity corresponds to the condition that all roots of the characteristic polynomial $g(z) = 1 - \varphi_1 z^{-1} - \cdots - \varphi_p z^{-p}$ lie inside the unit circle.

Now we need two facts. The first is that for any linear-time-invariant system described by a finite-difference equation with input w_t and output x_t, *the transfer function* $H(f)$ is defined as

$$H(f) = \left. \frac{x_t}{w_t} \right|_{w_t = e^{i2\pi ft}} . \tag{3.2}$$

In words, $H(f)$ is the (complex) ratio of the output to the input when the input is the complex exponential $e^{i2\pi ft}$, $t = 0, \pm 1, \pm 2, \ldots$, of frequency f (see, for example, Rabiner and Schaffer, 1978). Second, if the difference equation in question is driven by (i.e. has input) a wide-sense stationary process ζ_t with spectral density $S_\zeta(f)$, then the output process x_t is wide-sense stationary with spectral density

$$S_x(f) = |H(f)|^2 S_\zeta(f) . \tag{3.3}$$

See, for example, Rosenblatt (1962).

It is straightforward to check that for the autoregression (3.1) we have

$$H(f) = \frac{1}{D(f)} = \frac{1}{1 - \sum_{l=1}^{p} \varphi_l \, e^{i2\pi fl}} . \tag{3.4}$$

Since the input ε_t is white noise, we have $S_\varepsilon(f) \equiv \sigma_\varepsilon^2$, and (3.3) gives

$$S_X(f) = \frac{\sigma_\varepsilon^2}{|D(f)|^2} \tag{3.5}$$

for the spectral density of an autoregression.

An autoregressive spectral density estimate then is of the form

$$\hat{S}_X(f) = \frac{\hat{\sigma}_\varepsilon^2}{|\hat{D}(f)|^2} , \tag{3.6}$$

where

$$\hat{D}(f) = 1 - \sum_{l=1}^{p} \hat{\varphi}_l \, e^{i2\pi fl} \tag{3.7}$$

with $\hat{\sigma}_\varepsilon^2, \hat{\varphi}_1, \ldots, \hat{\varphi}_l$ estimates of $\sigma_\varepsilon^2, \varphi_1, \ldots, \varphi_p$ based on the observed data $x_1, \ldots, x_n$. These estimates have traditionally been one of several seemingly minor variants of least-squares estimates.

As it happens, there has been considerable discussion in the literature concerning which variant of least squares should be used. One consideration has been the desire to obtain estimates $\hat{\varphi}_1, \ldots, \hat{\varphi}_p$ which correspond to a *stationary* autoregression. Solutions of the so-called Yule–Walker equations

$$Z^{\mathrm{T}} Z \hat{\boldsymbol{\varphi}} = Z^{\mathrm{T}} \boldsymbol{x} , \qquad \hat{\boldsymbol{\varphi}}^{\mathrm{T}} = (\hat{\varphi}_1, \ldots, \hat{\varphi}_p) , \tag{3.8}$$

$$\boldsymbol{x}^{\mathrm{T}} = (x_2, x_3, \ldots, x_n, 0, \ldots, 0) , \qquad Z_{ij} = \begin{cases} x_{i-j+1} & 0 < i - j + 1 \leq n, \quad 1 \leq i \leq n + p - 1, \quad 1 \leq j \leq p , \\ 0 & \text{otherwise,} \end{cases} \tag{3.9}$$

have this property. However, the zero's in the 'design matrix' Z are quite obviously a source of potential harmful 'end effects'. A striking example of how harmful this choice can be, may be found in Section 5 of Martin (1980). Much attention, beginning with Burg's proposal (1972), has been given to estimation alternatives to (3.8) and (3.9) which do not suffer from end effects, yet still yield estimates in the region of stationarity.[2]

The robust autoregressive parameter estimates described in the next subsection are not guaranteed to correspond to stationary autoregressions. However, this has not proved to be a practical limitation in our applications, and the estimates do handle undesirable end effects in a natural way (cf. Martin, 1980).

Robust autoregressive estimates

Robustification of autoregressive spectral density estimates is easily accomplished by using robust estimates $\hat{\sigma}_\varepsilon^2, \hat{\varphi}_1, \ldots, \hat{\varphi}_p$ in (3.6) and (3.7). One way to obtain robust $\hat{\varphi}$'s is through the use of *generalized M estimates* (GM estimates) or *bounded-influence autoregression* estimates (BIAR estimates) as they are sometimes called. These estimates are obtained by solving a vector equation of the form

$$\sum_{t=p}^{n-1} W(\zeta_t)\zeta_t\psi\left(\frac{y_{t+1} - \zeta_t^{\mathrm{T}}\hat{\varphi}}{V(\zeta_t)\hat{s}_\varepsilon}\right) = 0\,, \tag{3.10}$$

where $\zeta_t^{\mathrm{T}} = (y_t, y_{t-1}, \ldots, y_{t-p+1})$, W and V are certain weight functions, ψ is a bounded and continuous psi function commonly encountered in robust estimation (see Huber, 1981), and $\hat{s}_\varepsilon$ is a robust scale estimate (for the innovations ε_t) obtained by solving an additional equation along with (3.10).

Two particular classes of BIAR estimates correspond to the choices:

(i) $$V(\zeta_t) \equiv 1, \quad W(\zeta_t) = w(\zeta_t^{\mathrm{T}}\hat{C}^{-1}\zeta_t) \tag{3.11}$$

with the nonnegative and continuous weight function w falling off sufficiently rapidly that $W(\zeta_t)\zeta_t$ is bounded and continuous, and

(ii) $$V(\zeta_t) = W(\zeta_t)\,. \tag{3.12}$$

Here $\hat{C}$ is a $p \times p$ robust covariance matrix estimate.

The general notion that bounded-influence estimation is important from the robustness point of view in the ordinary regression context is due to Hampel

[2]It is quite unfortunate that the term *maximum entropy* has come into widespread use in this context, since the term is used to impugn additional 'special' qualities to the autoregressive spectral density estimate which are quite undeserved.

(1975) and also Mallows (1976). The choice of weights (i) yields a *Mallows-type BIAR* estimate; the choice (ii) is referred to as the *Hampel–Krasker–Welsch (HKW)*-type estimate (Krasker and Welsch, 1979). See also Maronna, Bustos and Yohai (1979).

Details concerning computations of the Mallows-type BIAR estimate, as well as some examples, are given in Martin (1980). See also Martin and Zeh (1978).

Recently Bustos and Yohai (1982) have introduced a quite different and very appealing class of robust estimates for general autoregressive-moving average (ARMA) processes. These estimates are called *RA estimates* because they are based on robust residual autocorrelations.

Comments

If one insists on computing a pure autoregressive spectral density estimate, then one will naturally want to compute a robust variant in situations where outliers are at all a possibility. In this event one of the above robustifications of the ordinary estimate might prove useful. However, computation of only a pure autoregressive spectral density estimate, robust or otherwise, is not recommended. Even when the oft-cited limitation of shortness of the data record is used to justify use of AR estimates an alternative such as least-squares or robust nonlinear trignometric regression will often be viable. In general, the robust prewhitening approach discussed in the next section is recommended, with computation of a robust AR-type estimate occasionally being a useful adjunct.

As we shall see near the end of the next section, robust estimates such as BIAR estimates (or RA estimates) are quite useful in providing initial robust estimates for our robust spectrum estimation methodology.

4. Smoothed periodogram estimates

Standard periodogram based estimates

The usual nonparametric approach for obtaining spectral density estimates for data records which are not too long is as follows. Let $f_k = k/n$. First the discrete Fourier transform (DFT)

$$Y(f_k) = \frac{1}{n} \sum_{t=1}^{n} w_t y_t \, e^{i2\pi f_k t}, \quad k = 0, \ldots, n-1, \tag{4.1}$$

is computed from the observed data $y_1, \ldots, y_n$, where $\{w_t\}$ is a *data window* or *taper* which is used to mitigate leakage due to end effects (Bloomfield, 1976; Brillinger, 1981). The computation (4.1) is invariably carried out using a fast Fourier transform (FFT), the particular variety chosen depending upon whether or not n is a power of 2. This reduces the computational complexity from $O(n^2)$ to $O(n \log n)$.

Next the *periodogram*

$$\hat{S}_y(f_k) = |Y(f_k)|^2 \tag{4.2}$$

is computed, and finally the *smoothed periodogram estimate*

$$\bar{S}_y(f_k) = \sum_{l=-L}^{L} b_l \hat{S}(f_{k+l}) \tag{4.3}$$

is computed, where the frequency-domain smoothing weights $b_l = b_{n,l}$ are chosen to achieve a good compromise between bias and variability. Actually the b_l also depend upon k, but only so as to make reasonable adjustments for $k = 0$ and $[n/2]$, (cf. Bloomfield, 1976). For appropriately chosen sequences $\{b_{n,l}\}$, $\bar{S}_y(f_k) = \bar{S}_y(f_k, b_{n,l})$ is a consistent estimate when $y_t \equiv x_t$ with x_t a WSS process having spectrum $S_x(f)$. See, for example, Brillinger (1981) and Grenander (1981).

For unusually large data sets, or where real-time processing is essential, time-domain smoothing of periodograms on contiguous or overlapping segments is often used (Welch, 1967). However, we shall not deal with this situation here (see Martin and Thomson, 1982).

One may consult Bloomfield (1976), Brillinger (1981) and Koopmans (1974) for more specific information on smoothed periodogram estimates.

Lack of robustness of the standard estimate

If in fact the time series contains outliers generated by an AO model

$$y_t = x_t + v_t\,, \tag{4.4}$$

then the smoothed periodogram estimate can be extremely unreliable. This is particularly true in regions of the spectrum which have relatively low amplitude. This is intuitively clear from the fact that outliers in the data results in inflated variance for the data, and they can affect the DFT $Y(f_k)$ at all frequencies.

First of all we note that bias is a problem. For suppose v_t is a wide-sense stationary series with spectrum $S_v(f_k)$ and that v_t is independent of x_t. Then under reasonable assumptions $\bar{S}_y(f_k) = \bar{S}_y(f_k, b_{n,l})$ will provide a consistent estimate of $S_y(f)$, $f \in [0, \frac{1}{2})$, with $S_y(f)$ given by

$$S_y(f) = S_x(f) + S_v(f)\,. \tag{4.5}$$

Furthermore, central limit theorem behavior of $Y(f_k)$ implies that the inflated variance σ_y^2 of y_t due to the v_t is propagated to $Y(f_k)$. The large sample variance expression is

$$\mathrm{VAR}_\infty \bar{S}_y(f) = S_y^2(f) = S_x^2(f) + 2S_v(f)S_x(f) + S_v^2(f)\,. \tag{4.6}$$

Thus variability may also be a problem.

The case of mutually independent v_t, where $S_v(f) \equiv \sigma_v^2$ provides a convenient case to examine. The bias is then just σ_v^2, and the variance is inflated by the additive term σ_v^4, which can be quite large, as well as by the cross-product terms.

Strong narrow-band components of $S_x(f_k)$ are unlikely to be obscured by small to moderate outliers. For example, Fig. 9 displays the spectrum for both an x_t process (regarded as Gaussian) having a strong narrow-band component, and white noise v_t process, which contains outliers, with the two processes having a common variance of unity. Think of the v_t process as having the contaminated normal distribution with degenerate central component

$$\mathrm{CND}(\gamma, \sigma^2) = (1-\gamma)N(0,0) + \gamma N(0, \sigma^2)$$

with $\gamma = 0.1$, and $\sigma^2 = 10$, and where $N(0,0)$ is the degenerate distribution with all of its mass at the origin.[3] Thus v_t is zero 90% of the time and is large relative to the values of x_t 10% of the time. Fig. 10 shows the spectrum of

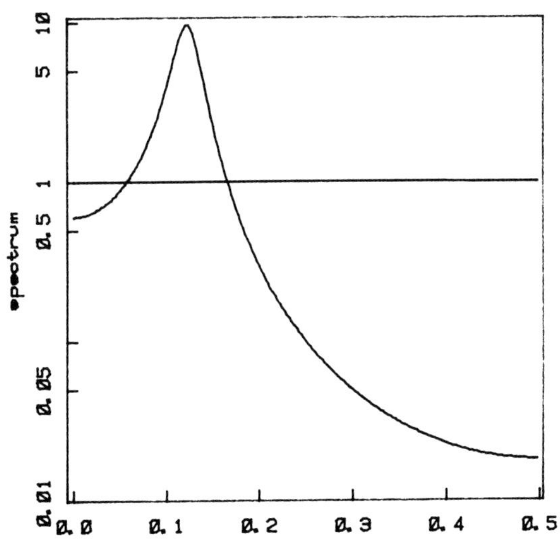

Fig. 9. Spectra of AR2 and outliers processes.

[3] $N(u, \sigma^2)$ denotes the Gaussian distribution with mean u and variance σ^2.

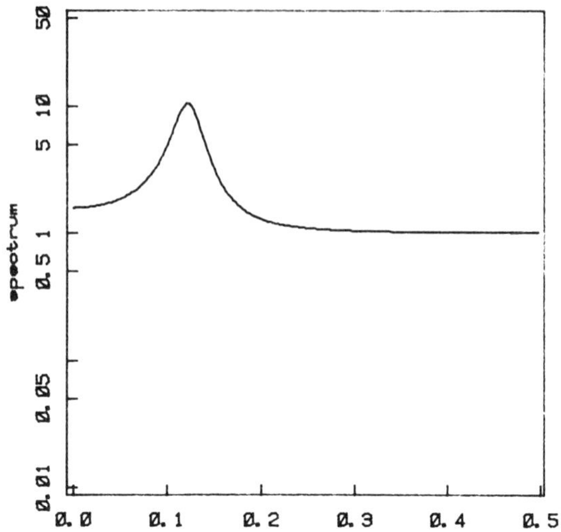

Fig. 10. Spectrum of sum of AR2 and outliers processes.

$y_t = x_t + v_t$, and the dominant peak is still clearly visible (note, however, the range of values on the ordinate).[4]

On the other hand, interesting peaks in the low-amplitude range of the spectrum can be completely obscured by a smallish fraction of relatively small outliers. Fig. 11 displays the spectrum of an x_t process which contains two low-amplitude peaks in addition to the same dominant peak of Figs. 9 and 10,

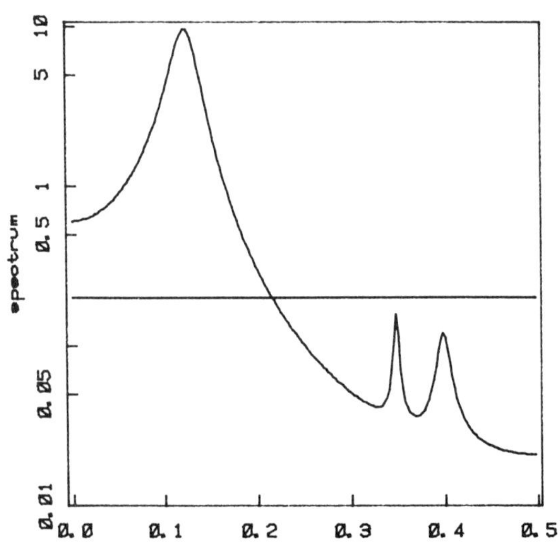

Fig. 11. Spectra of interesting and outliers processes.

[4]It should be noted that throughout we use the convention of folding the spectrum over so that we are really plotting $2S(f)$, and correspondingly the area under the spectrum from 0 to $\frac{1}{2}$ is the variance σ^2.

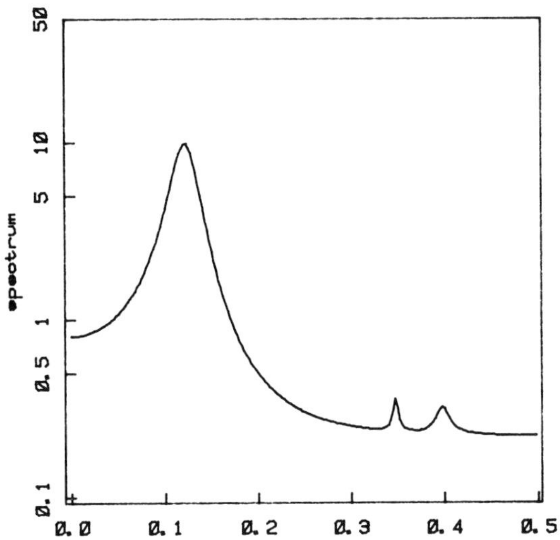

Fig. 12. Spectrum of sum of interesting process and outliers process.

along with the spectrum of a white-noise process v_t. We again have in mind a Gaussian x_t process and a v_t process having the CND(γ, σ^2) distribution with $\gamma\sigma^2 = \text{var } v_t$. Here var $x_t = 1$ while var $v_t = 2$. In Fig. 12 we display the spectrum of $y_t = x_t + v_t$, and see that the potentially interesting low-amplitude peaks are somewhat obscured by the additive v_t process.

It should be noted that while Fig. 12 reflects the expected value of a periodogram-based spectrum estimate for the given additive outlier situation, it does not give an indication of the increased variability caused by additive outliers. The latter can be substantial, and we return to this example in Section 6.

A basic approach to robust spectrum estimation

The essential ingredients of a general robustification procedure for complicated situations consists of *cleaning* the data to remove or down-weight outliers, followed by applications of a standard nonrobust procedure. This is often carried out in an iterative manner. The trick, of course, is to construct a good 'data-cleaning' algorithm. Subsequently I shall describe what I consider to be good *robust filter* and *smoother* algorithms for carrying out the data-cleaning phase. These algorithms, which will be called *filter cleaners* and *smoother cleaners*, are based on fairly low-order autoregressive approximations say, $p = 2$–6, for convenience. When a good smoother-cleaner is imbedded in an overall iteration loop of the general form described later in this section, it produces at the final iteration both cleaned data, which we label $\hat{x}_t^n$ since they are estimates of x_t in the AO model based on all the data $y_1, \ldots, y_n$, and robust estimates $\hat{\varphi}_1, \ldots, \hat{\varphi}_p$ of the low-order autoregressive approximation.

The robust–resistant spectrum estimate is constructed as follows. Prewhiten the cleaned data $\hat{x}_t^n$ using the estimates $\hat{\varphi}_1, \ldots, \hat{\varphi}_p$, thereby forming the *outlier-free* residuals

$$r_t = \hat{x}_t^n - \sum_{l=1}^{p} \hat{\varphi}_j \hat{x}_{t-j}^n \,. \tag{4.7}$$

Compute the standard smoothed-periodogram estimate

$$\hat{S}_r(f_k) = \sum_{j=-L}^{L} b_j \hat{S}_r(f_{k+j}) \tag{4.8}$$

based on the r_t. Finally, compute the estimate of the spectrum of x_t as

$$\overline{S_x}(f_k) = \frac{\overline{S_r}(f_k)}{|\hat{D}(f)|^2} \tag{4.9}$$

with $\hat{D}(f)$ given by (3.7), but with $\hat{\varphi}_1, \ldots, \hat{\varphi}_p$ computed from the iterative procedure using the smoother cleaner to be described shortly. This estimate is of the prewhitened form suggested some time ago by Blackman and Tukey (1958), except that cleaned data is used instead of the original data. As such, it has the virtue of *leakage* control cited by Blackman and Tukey, as well as being robust–resistant.

Note that $\hat{S}_x(f_k)$ differs from a pure autoregressive spectral density estimate by virtue of having $S_r(f_k)$ in the numerator rather than a robust estimate $\hat{\sigma}_\varepsilon^2$ (which might well be computed as $\hat{\sigma}_\varepsilon^2 = \text{ave}(r_t^2)$ in the present situation). In general, the version (4.9) is to be strongly preferred over the pure autoregressive version, even if a good order-selection rule (e.g. Akiake's AIC, 1973 or Parzen's CAT, 1974) is used in conjunction with the latter. The reason is that the nonparametric component $\overline{S_r}(f_k)$ of the estimate is more likely to pick up fine structure in the low-amplitude region of the spectrum than is the pure AR estimate.

Robust filter cleaners and smoother cleaners

The data cleaner to be used is essentially a *robust filter* or *smoother* operating in a special mode. The terminology used here parallels that of the engineering literature: a *filter* uses the data $y_1, \ldots, y_t$ to form an estimate $\hat{x}_t$ of x_t; a *smoother* uses the entire data record $y_1, \ldots, y_n$ to construct an estimate $\hat{x}_t^n$ of x_t. A robust–resistant filter or smoother is one which is not affected much by large changes in a small fraction of the data, or by small changes, e.g. rounding errors, in all the data.

A robust *filter* will be called a *filter cleaner* when it is constructed so as to leave good data points unaltered, while replacing 'sufficiently large' outliers by one-sided predictions. A robust *smoother* will be called a *smoother cleaner* when constructed so as to leave good data unaltered while replacing 'sufficiently large' outliers by interpolators based on all the remaining data. As in Section 2, 'sufficiently large' is gauged in terms of robust prediction residuals. For data sets with a small fraction of outliers, the above types of data cleaners will leave most of the data unaltered. The filter and smoother cleaners described below have the properties just described.

Both the filter cleaner and the smoother cleaner utilize the following state-variable formulation of a pth-order autoregressive *approximation* to x_t:

$$\boldsymbol{X}_t = \boldsymbol{\Phi}\boldsymbol{X}_{t-1} + \boldsymbol{\varepsilon}_t\,, \tag{4.10}$$

where

$$\boldsymbol{X}_t^{\mathrm{T}} = (x_t, x_{t-1}, \ldots, x_{t-p+1})\,, \tag{4.11}$$

$$\boldsymbol{\varepsilon}_t^{\mathrm{T}} = (\varepsilon_t, 0, \ldots, 0)\,, \tag{4.12}$$

$$\boldsymbol{\Phi} = \begin{pmatrix} \varphi_1 & \varphi_2 & \vdots & \varphi_p \\ 1 & 0 & \vdots & 0 \\ 0 & 1 & \vdots & \vdots \\ 0 & \vdots & 1 & 0 \end{pmatrix}. \tag{4.13}$$

Filter cleaners

Let $\hat{\boldsymbol{\Phi}}$ be a robust estimate of $\boldsymbol{\Phi}$ based on robust estimates $\hat{\varphi}_1, \ldots, \hat{\varphi}_p$ and let $\hat{\sigma}_\varepsilon^2$ be a robust estimate of $\sigma_\varepsilon^2 = \operatorname{var} \varepsilon_t$. The filter cleaner computes robust estimates $\hat{\boldsymbol{X}}_t$ of the vector $\boldsymbol{X}_t$ according to the following recursions, with $\boldsymbol{\Phi}$ and σ_ε^2 replaced by $\hat{\boldsymbol{\Phi}}$ and $\hat{\sigma}_\varepsilon^2$:

$$\hat{\boldsymbol{X}}_t = \boldsymbol{\Phi}\hat{\boldsymbol{X}}_{t-1} + \boldsymbol{\rho}_t s_t \psi\left(\frac{y_t - \hat{y}_t^{t-1}}{s_t}\right), \tag{4.14}$$

where

$$\boldsymbol{\rho}_t = \frac{\boldsymbol{m}_t}{s_t^2}$$

and

$$\hat{y}_t^{t-1} = \hat{x}_t^{t-1} = (\boldsymbol{\Phi}\hat{\boldsymbol{X}}_{t-1})_1 \tag{4.15}$$

is the prediction of y_t and x_t based on $y_1, \ldots, y_{t-1}$ through $\hat{\boldsymbol{X}}_{t-1}$; $\boldsymbol{m}_t$ is the first column of the matrix M_t, which is also computed recursively as

$$M_{t+1} = \boldsymbol{\Phi} P_t \boldsymbol{\Phi}^{\mathrm{T}} + Q\,, \tag{4.16}$$

$$P_t = M_t - w(y_t - \hat{y}_t^{t-1})\frac{\boldsymbol{m}_t \boldsymbol{m}_t^{\mathrm{T}}}{s_t^2}\,. \tag{4.17}$$

The functions ψ and w are described below, and Q is a matrix with all zero entries except for the 1–1 element which is σ_ε^2, and

$$s_t^2 = m_{11,t}\,. \tag{4.18}$$

The filter-cleaned estimate $\hat{x}_t$ is just the first component of $\hat{\boldsymbol{X}}_t$:

$$\hat{x}_t = (\boldsymbol{X}_t)_1 \,. \tag{4.19}$$

The psi function ψ and weight function w are the crucial ingredients of the robust filter. One reasonably good choice for ψ is the two-part Hampel redescending function

$$\psi_{HA}(t) = \begin{cases} t, & |t| \leq a, \\ \dfrac{a}{b-a}(b-t)\,, & a < t \leq b, \\ -\dfrac{a}{b-a}(b+t), & -b \leq t \leq -a, \\ 0, & |t| > b, \end{cases} \tag{4.20}$$

with $a = 2.5$ and $b = 5.5$ reasonable choices for the constants.

Psi functions which, like ψ_{HA}, are zero outside of a finite symmetric interval are referred to as *rejection-type* psi functions.

In earlier papers (cf. Martin, 1979, 1981) it was suggested that the appropriate form for w is $w(t) = \psi'(t)$. The reason for this suggestion was that the robust filter is then an *approximate conditional-mean* (ACM) filter when x_t are Gaussian, the v_t are i.i.d. with distribution F_v, s_t^2 is modified somewhat, and ψ is appropriately specified in terms of F_v. Of course, the conditional-mean filter is optimal in the sense of minimizing the mean-squared-error, as is well known (Jazwinski, 1970; Meditch, 1967). Unfortunately the choice $w = \psi'$ is a discontinuous function for piecewise linear psi functions such as $\psi = \psi_{HA}$, and this goes against the minimal requirement that a robust estimate be a continuous function of the data (Hampel, 1971, 1974). A better choice of w might be

$$w(t) = \psi(t)/t \,. \tag{4.21}$$

In the case of ψ_{HA} this gives

$$w_{HA}(t) = \begin{cases} 1, & |t| < a \\ \dfrac{a}{b-a}\left(\dfrac{b}{|t|} - 1\right), & a \leq |t| \leq b \\ 0, & |t| > b \end{cases} \tag{4.22}$$

which embodies the general features desirable for a good w:

(i) $w(0) = 1$ and $w(t) \approx 1$ for a range of values 'near' the origin

(ii) $w(t) = 0$ for all sufficiently large values of $|t|$.

It is still not clear whether we should be quite satisfied with functions w such as (4.22), which are nonnegative. The ACM filter approximation quoted from above suggests that ψ' should in fact take on negative values for certain reasonable shapes of ψ. If this is a good feature, then smooth w functions should be constructed accordingly. This is an issue which needs further study.

The filter-cleaner recursions (4.14)–(4.19) have a familiar form. When $\psi(t) \equiv t$, $w(t) \equiv 1$ and $s_t^2 = m_{11,t} + \sigma_v^2$, with σ_v^2 the variance of the observation noise, the filter cleaner becomes the usual Kalman filter recursion. The Kalman filter provides the exact conditional-mean estimate when the v_t have the Gaussian distribution $F_v = N(0, \sigma_v^2)$. Note that when $\sigma_v^2 = 0$, as in (4.18), the Kalman filter is just the identity map. The corresponding feature of the appropriately tuned filter cleaner is that it tends to leave outlier free data points unaltered. This mode of behavior is in evidence in the examples of Section 6.

The robust filter (4.14)–(4.19) differs from the simple robust filter of Kleiner, Martin and Thomson (1979) in two important aspects. First of all, (4.14) is a *vector* recursion, and the structure of $\boldsymbol{\rho}_t$ is such that good data points which follow an outlier are used to improve the estimate of x_t at the outlier position. This can yield improved estimates of x_t at times subsequent to the occurrence of an outlier.

Second, the current filter incorporates the data-dependent scale s_t whose values satisfy $s_{t+1} > s_t$ if a gross outlier occurs at time t. This is a particularly important feature when using a redescending psi function, such as ψ_{HA}. For if a fixed scale s were used, the filter could lose track of the data never to regain it for the duration of a fixed length data set $y_1, \ldots, y_n$. The fact that outliers cause s_t to increase in value improves the ability of the filter to regain tracking of the data.

Smoother cleaners

Now that the details of the filter cleaner have been given, the *smoother-cleaner algorithm* is easily described. Denoting the smoother-cleaned estimate of the vector $\boldsymbol{X}_t = (x_t, x_{t-1}, \ldots, x_{t-p+1})^{\mathrm{T}}$ by $\hat{\boldsymbol{X}}^n$, we have the backward-time recursion

$$\hat{\boldsymbol{X}}_t^n = \hat{\boldsymbol{X}}_t + P_t \Phi^{\mathrm{T}} M_{t+1}^{-1} (\hat{\boldsymbol{X}}_{t+1}^n - \Phi \hat{\boldsymbol{X}}_t)\,, \tag{4.23}$$

where the $\hat{\boldsymbol{X}}_t$, $1 \leq t \leq n$, are the filter-cleaned values obtained from the forward-time recursions (4.14)–(4.18), with $\hat{\boldsymbol{X}}_n^n = \hat{\boldsymbol{X}}_n$ as the initial condition for the backward recursion. The smoother-cleaned data values are the first components of the $\hat{\boldsymbol{X}}_t^n$:

$$\hat{x}_t^n = (\hat{\boldsymbol{X}}_t^n)_1\,. \tag{4.24}$$

As in the case of the filter cleaner, there is an approximate optimality result. If $w(t) = \psi'(t)$ and ψ are appropriately specified in terms of F_v, and s_t^2 is modified somewhat, then $\hat{\boldsymbol{X}}_t^n$ and $\hat{x}_t^n$ are approximate conditional-mean estimates (Martin, 1979).

Translation equivariant versions

In actuality, a realistic AO model would differ from (4.4) by virtue of incorporating a location parameter μ:

$$y_t = \mu + x_t + v_t\,. \tag{4.25}$$

To account for this, our data cleaners are made *translation-equivariant* by a simple two-step procedure. First, a robust location M-estimate (Huber, 1967) $\hat{\mu}$ is used to center the original time series:

$$\tilde{y}_t = y_t - \hat{\mu}\,. \tag{4.26}$$

Then the smoother cleaner is used on $\tilde{y}_t$ to produce $\tilde{x}_t^n$. Finally, the cleaned data is recentered:

$$\hat{x}_t^n = \tilde{x}_t^n + \tilde{\mu}\,. \tag{4.27}$$

Implementation of filter- and smoother-type cleaners

In order to implement the filter cleaner and smoother cleaner, one needs robust estimates $\hat{\varphi}$ and $\hat{\sigma}_\varepsilon^2 = \hat{s}_\varepsilon^2$ of $\varphi^{\mathrm{T}} = (\varphi_1, \ldots, \varphi_p)$ and σ_ε^2. Initial estimates may be obtained using *bounded-influence autoregression* (BIAR) estimates as described in Section 3, equations (3.10).

A basic iteration strategy

The Mallows-type BIAR estimates (3.10) and (3.11) may be conveniently computed via the iterated-weighted-least squares (IWLS) methodology:

$$\sum_{t=p}^{n-1} \zeta_t W(\zeta_t) w_t^j \cdot (y_{t+1} - \zeta_t^{\mathrm{T}} \hat{\varphi}^{j+1}) = 0, \quad j = 1, \ldots, \mathrm{NIT}, \tag{4.28}$$

with j denoting the iteration number, NIT the total number of iterations, and where the weights w_i^j are given by

$$w_t^j = \frac{\psi\left(\dfrac{y_{t+1} - \zeta_t^{\mathrm{T}} \hat{\varphi}^j}{\hat{s}_\varepsilon^j}\right)}{y_{t+1} - \zeta_t^{\mathrm{T}} \hat{\varphi}^j}\,. \tag{4.29}$$

In the example of Section 6, the IWLS estimate (4.28) and (4.29) is obtained by

(i) letting $\hat{\varphi}^0$ be the least-squares estimate,
(ii) computing NITH iterations using Huber's psi function

$$\psi_H(t) = \begin{cases} t, & |t| \leq K, \\ K \cdot \mathrm{sgn}(t), & |t| > K, \end{cases} \tag{4.30}$$

with $K = 1.5$, and

(iii) computing NITB additional iterations, starting with $\hat{\varphi}^{\mathrm{NITH}}$, $\hat{s}_{\varepsilon}^{\mathrm{NITH}}$, using Tukey's redescending bisquare function

$$\psi_B(t) = \begin{cases} t(1-(t/c)^2)^2, & |t| < c, \\ 0, & |t| \geqslant c, \end{cases} \tag{4.31}$$

with $c = 6.0$.

Typically NITB = 1 or 2 regardless of the value of NITH. For data sets where the fraction of outliers is small NITH = 4 will suffice. However, for data which has a large fraction of outliers (≈30%), larger values of NITH seem to be required.

The reason for the iteration strategy where ψ_H is used first with a fair number of iterations, followed by 1 or 2 iterations with ψ_B, is that (3.10) has an essentially unique root when $\psi = \psi_H$, but not when $\psi = \psi_B$. We wish to make use of the *rejection character* of ψ_B without risk of computing an extraneous root. With NITB = 1 the solution is essentially unique, and we are willing to seek improvement with small risk by going to NITB = 2. We have observed cases with higher-order autoregressions where further iteration with ψ_B leads to an extraneous root.

Another basic iteration

After computing $\hat{\varphi}$ and $\hat{s}_\varepsilon$ as just described, one then computes smoother-cleaned values $\hat{x}_t^n$. At this point one can initiate another iteration procedure whereby least-squares estimates are used to obtain new φ and s_ε estimates based on cleaned data $\hat{x}_t^n$. These new parameter estimates may then be used to compute new cleaned data, and so on. However, the implicit estimating equation here, namely

$$\sum \hat{\zeta}_t \cdot (\hat{X}_t^n - \hat{\zeta}_t^{\mathrm{T}} \hat{\varphi}) = 0\,, \tag{4.32}$$

where $\hat{\zeta}_t^{\mathrm{T}} = (\hat{X}_{t-1}^n, \ldots, \hat{X}_{t-p}^n)$, is highly nonlinear and again multiple roots are a real possibility. Fortunately, we have found that the initial BIAR estimates $\hat{\varphi}$ and $\hat{s}_\varepsilon$ are sufficiently good that one, or at most two, iterations will suffice here.

5. Robustness–resistance properties of the spectrum estimate

A robust–resistant procedure has the following two properties: (i) when the data is 'good'—i.e. Gaussian—the procedure should be almost as good as the conventional (often optimal) procedure presuming normality, and (ii) when outliers are present the procedure should still work well, and in particular work much better than the conventional procedure.

Resistance

The data-oriented *resistance* version of these properties has the following features in the context of the data-cleaning step of computing the spectral-density estimate. When the data is 'good' it contains no outliers, and the data cleaner should have the property that it alters only a tiny fraction of the data. Correspondingly, the final estimate is close to that obtained by the conventional spectral density estimate without data cleaning.

Now suppose that 'good' data is altered by adding a not-too-large fraction of arbitrarily large outliers. The new spectral-density estimate should not differ much from the original estimate. This is done through the interpolation feature of the filter cleaner and smoother cleaner: at gross outlier positions the output of the data cleaner is an interpolate (an extrapolate or one-sided interpolate for the filter cleaner, and a two-sided interpolate for the smoother cleaner).

The two features just described are evident in the examples presented in the next section.

The second property just described is the resistance property. From a general point of view the resistance property is a consequence of the following fact about the filter cleaner and smoother cleaner. Viewed as a mapping from R^n to R^n, the filter cleaner and hence the smoother cleaner are *bounded* and *continuous* functions of the data if both ψ and w are bounded and continuous, and the centering–recentering steps are ignored. The reader may easily check this claim.

Robustness

For parametric problems where consistency of parameter estimates can be assured by an (often unrealistic) symmetry assumption, the two properties are taken to be (i) high efficiency in terms of variance at the Gaussian model, and (ii) high efficiency at strategically selected non-Gaussian distributions. If one allows for the realistic assumption that non-Gaussian departures involve difficult-to-describe asymmetries, then finite-sample and asymptotic bias will occur, and bias control will be an important aspect of the robustness problem.

Correspondingly, for the nonparametric problem of obtaining robust spectrum estimate, we should provide evidence that (i) the bias, although nonzero, is negligibly small when the v_t contain outliers, and (ii) the efficiency, in terms of variance, is quite high when the v_t are identically zero, and reasonably high when the v_t contain outliers. One way of collecting the needed evidence would be through careful and extensive *finite-sample size* Monte Carlo studies, an unattractive task in the spectrum analysis context. Another approach would be to obtain the asymptotic bias and variance of our procedure, thereby providing approximate bias and variance for largish sample sizes. Although useful analytic expressions for the desired quantities are not likely to be obtained, some headway on the problem seems possible through use of Mallow's (1980) recent contribution on the *linear part* of a nonlinear smoother.

The linear part concept

The *linear part* of a nonlinear smoother, *NL*, is defined in the context of a general AO model

$$y_t = \mu + x_t + v_t, \tag{5.1}$$

where μ is a location parameter, x_t is a stationary Gaussian process and v_t is an i.i.d., possibly non-Gaussian process. Let $(NLy)_t$ define a smoother intended to provide a good estimate of x_t. We are interested here primarily in the case $(NLy)_t = \hat{x}_t^n$. The linear part L of NL is the projection of $(NLy)_t$ onto $H\{x_t\}$, the linear space spanned by the x_t. In particular, $(Lx)_t$ is defined by the coefficients a_j which minimize the mean-squared-error

$$E\left\{\left[(NLy)_t - \sum a_j' x_{t-j}\right]^2\right\}. \tag{5.2}$$

In words, the linear part L describes the best approximation to the *nonlinear* smoother *NL* based on *linear* smoothing of the *unobservable* process x_t, using a quadratic loss function.

Under certain assumptions, Mallows (1980) has proved that the following decomposition holds:

$$(NLy)_t = (Lx)_t + R_t, \tag{5.3}$$

where the projection $(Lx)_t = \sum a_j x_{t-j}$ and the residual process R_t are uncorrelated with respect to one another. The frequency-domain, spectral-densities version of this is

$$S_{NL}(f) = |H_L(f)|^2 S_x(f) + S_R(f), \tag{5.4}$$

where $S_{NL}(f)$, $S_x(f)$ and $S_R(f)$ are the spectral densities of the smoothed process, the Gaussian process x_t, and the residual process R_t, respectively; $H_L(f)$ is the *linear part transfer function* defined by

$$H_L(f) = \sum_{j=-\infty}^{\infty} a_j\, e^{i2\pi fj}, \quad -\tfrac{1}{2} \leq f \leq \tfrac{1}{2}. \tag{5.5}$$

Although our filter cleaners and smoother cleaners do not exactly satisfy Mallow's assumptions, the decomposition can be proved for asymptotically stationary versions of them as well.

Now the desirable characteristics of a good data cleaner in our context are (i) L should be nearly the identity operator, and (ii) the residual process R_t should be 'small'. Property (i) manifests itself by virtue of the a_j being close to zero for $j \neq 0$, and $a_0 \approx 1$. Correspondingly, the linear part transfer function

$$H_L(f) = \sum a_j \, e^{i2\pi fj}, \quad -\tfrac{1}{2} \leq f \leq \tfrac{1}{2}, \tag{5.6}$$

will be a good approximation to the identity operator transfer function

$$H_L(f) \equiv 1, \quad -\tfrac{1}{2} \leq f \leq \tfrac{1}{2}. \tag{5.7}$$

Smoothers with 'small' R_t will have var $R_t = \int_{-1/2}^{1/2} S_R(f)\,df$ small, where $S_R(f)$ is the spectrum of R_t.

That these properties are achievable with a good smoother cleaner is illustrated by some computations of the linear part transfer function and residuals spectrum given in Section 6 of Martin and Thomson (1982).

It should be noted that we may gauge the degree of robustness of both the prewhitening coefficient estimates $\hat{\varphi}$ and the final spectrum estimate by consideration of the approximation

$$\hat{x}_t^n = (Lx)_t + R_t \approx x_t + R_t, \tag{5.8}$$

with x_t and R_t uncorrelated. Since the $\hat{\varphi}$ are essentially obtained by applying least squares to the $\hat{x}_t^n$ (at the final iteration described in Section 4), we have an additive noise problem for which the asymptotic bias and variability of $\hat{\varphi}$ can be computed. When L is indeed close to the identity operator and the variance of R_t is small relative to that of x_t, $\hat{\varphi}$ will have a small asymptotic bias, and high efficiency.

The estimate $\hat{S}_x(f)$ of (4.9) is also based on the $\hat{x}_t^n$, and the prewhitening step plays the same role it always has with good data. The second-order considerations leading to (4.5) and (4.6) apply, and so we have $S_{x_t^n} = S_{Lx}(f) + S_R(f)$. When $(Lx)_t \approx x_t$ and var $R_t \ll$ var x_t, the $\hat{S}_x(f)$ of (4.9) will be a good estimate of $S_x(f)$.

The linear part calculations done to date indicate that for good smoother cleaners, $(Lx)_t$ is *exceedingly* close to x_t and var R_t is *exceedingly* small relative to var x_t when $v_t \equiv 0$ in (4.4), i.e. when no outliers are present (cf. Section 6 of Martin and Thomson, 1982). The robustness of both $\hat{\varphi}$ and $\hat{S}_x(f)$ derives from the fact that the above statement holds with the adjective '*exceedingly*' only weakened somewhat when outliers are present.

6. An example

To illustrate the robust spectrum estimation procedure, we use the model which gives the x_t and v_t spectra displayed in Figs. 11 and 12. The 'interesting' process x_t is a Gaussian process comprised of the sum of three second-order autoregressions having different variances, with $\sigma_x^2 = \text{var } x_t \approx 1$. The v_t process is an i.i.d. (or white noise) sequence having the contaminated normal distribution CND (γ, σ^2), with $\gamma = 0.05$ and $\sigma^2 = 4$, so that var $v_t = 0.2$ as in Fig. 11. This yields a fairly small fraction of contamination, and the scale $\sigma = 2$ of the v_t process is not very much larger than the scale $\sigma_x \approx 1$ of the x_t process.

Fig. 13 shows a segment of length 300 which is part of a realization of length 1000 of the 'interesting' x_t process, while Fig. 14 shows the corresponding segment of the sum $y_t = x_t + v_t$ of the interesting process and the outliers process. Only one outlier is in clear evidence at a casual glance. A closer look reveals 6 or 7 more outliers (recall the discussion of Section 2 on the nature of time-series outliers). There are actually 11 nonzero v_t's in the segment used here.

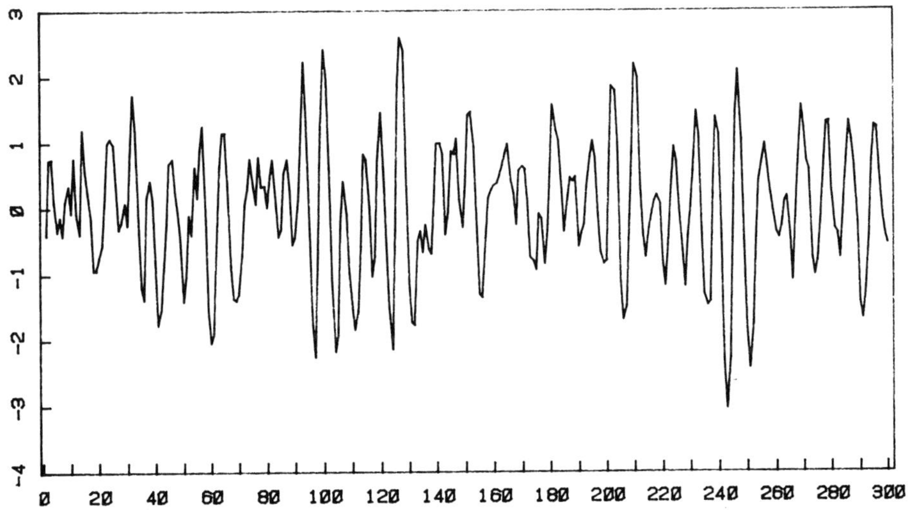

Fig. 13. Portion of interesting process.

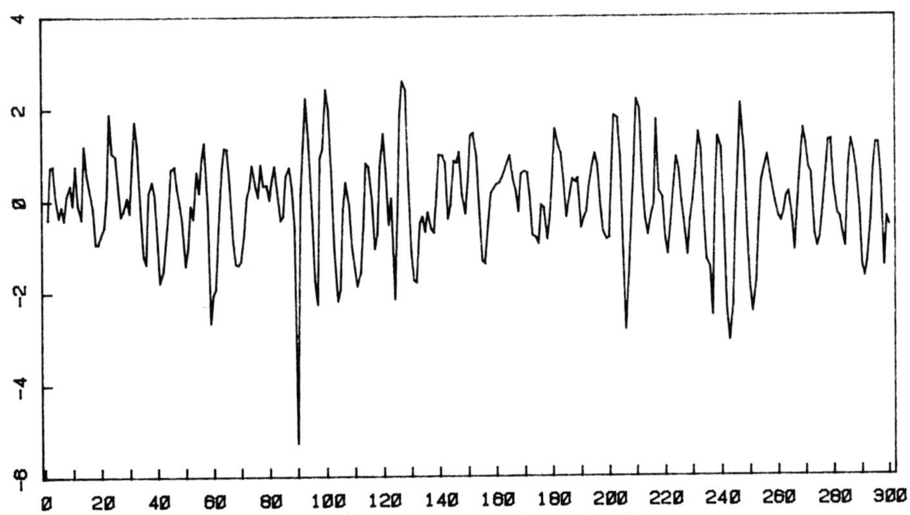

Fig. 14. Portion of the sum of interesting and outliers processes.

In Figure 19 we show the data of Figure 14, along with the smoother-cleaned data. This display clearly reveals 8 of the 11 nonzero v_t's. Figure 20 shows the behavior of the data-dependent scale s_t used in the filter-cleaner (4.14).

Now Figs. 15 and 16 show conventional smoothed-periodogram estimates based on the entire series of length 1000 from which the segments of Figs. 13 and 14 were obtained. The estimate of Fig. 15 clearly displays the two low-amplitude peaks, centered at the proper frequencies. In Fig. 16 we see the effects of the outliers: one of the low-amplitude peaks is nearly obliterated while the other is both broadened and shifted to the right of its original center frequency of $f \approx 0.4$. In addition, the weak bump at about $f = 0.22$ in Fig. 15 has become a more annoying artifact in Fig. 16.

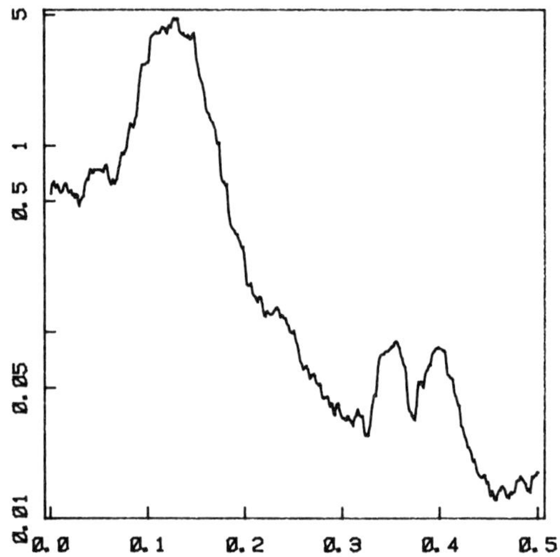

Fig. 15. Smoothed periodogram for outlier-free data.

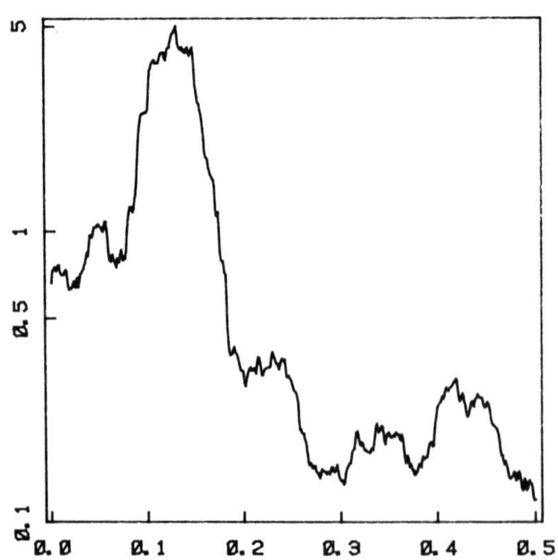

Fig. 16. Smoothed periodogram for data containing additive outliers.

One should also recall that the variability in the vicinity of the peaks in Fig. 15 is about $S^2_{\tilde{x}}(f) \approx (0.05)^2$, whereas the variability in the same region of Fig. 16 is roughly $S^2_{\tilde{v}}(f) = (0.2)^2$, which is a sixteen-fold increase.

Robust spectrum estimates for the good data and the outlier contaminated data are displayed in Figs. 17 and 18. Notice (i) how close the robust and conventional estimates are in the outlier-free case, and (ii) the improved quality of the robust estimate when outliers are present—the original peaks are now clearly evident and centered at the proper frequencies; the main blemish is the annoying artifact at about $f = 0.22$.

The robust spectrum estimates were computed as described in Section 4. The

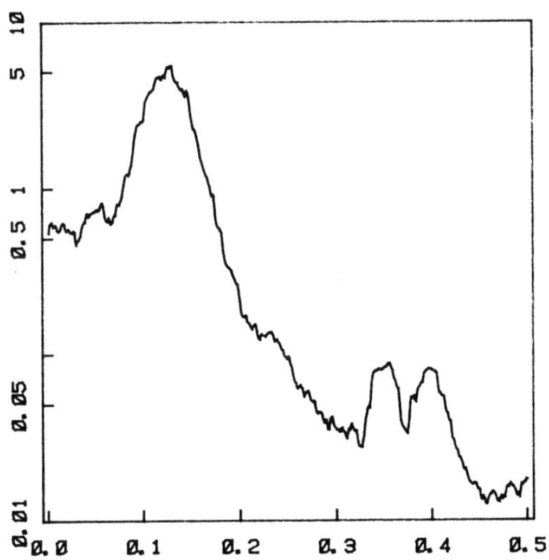

Fig. 17. Robust spectrum estimate for outlier-free data.

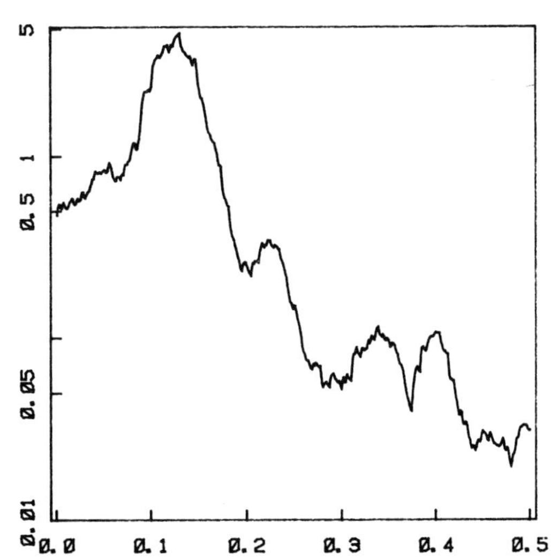

Fig. 18. Robust spectrum estimate for data containing additive outliers.

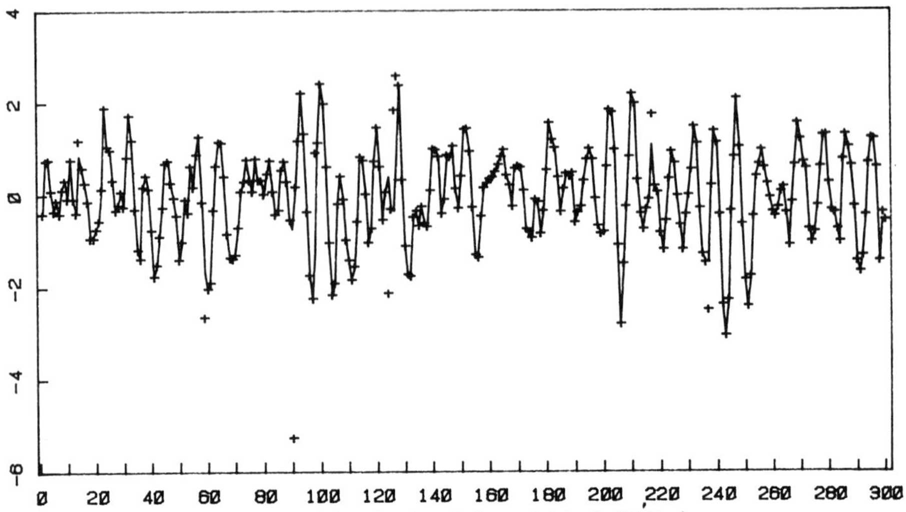

Fig. 19. Data (plus signs) and cleaned data (solid line).

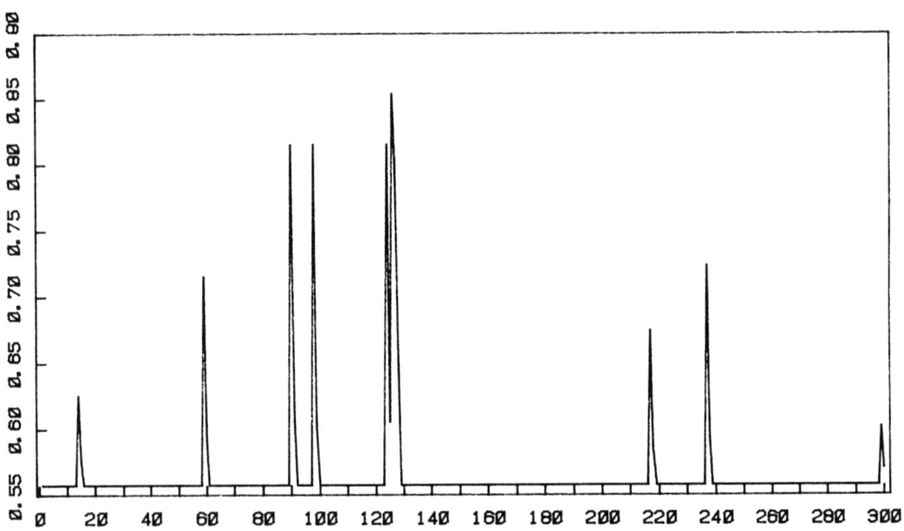

Fig. 20. Data-dependent observation prediction scale $S(t)$.

preliminary BIAR estimate used NITH = 4 and NITB = 1. In effect, zero iterations of the type described in conjunction with (4.32) were used. The data was simply cleaned with the smoother cleaner using the BIAR estimates $\hat{\varphi}$ and $\hat{s}_\varepsilon$, and the final estimate computed from (4.9) using the above $\hat{\varphi}$ to carry out the prewhitening step (4.7).

Further examples using the robust spectrum estimation procedure on both artificial data and real data may be found in Kleiner, Martin and Thomson (1979) and Martin and Thomson (1982). The latter reference includes the glint noise data of Figures 1 and 3.

7. Concluding remarks

The use of robust–resistant spectrum estimation methodologies is still in its infancy. Hopefully the weight of evidence presented here, and in the associated references, will result in widespread use of a robust–resistant spectrum estimate as an important adjunct to computation of a standard estimate.

Of course, a number of detailed issues needs to be studied more carefully in order to provide more secure guidelines for algorithm construction. For example, some combination of theory, Monte Carlo, and linear part calculations are needed to fine tune the choice of psi functions ψ, and tuning constants used in conjunction with the psi functions. Among the other issues that might be raised we would include (i) construction of spectrum estimates which are robust and also consistent when the nominal Gaussian model holds, (ii) extension of the methodology to deal with bivariate time series, (iii) extension of the methodology to deal with unequally spaced data, and (iv) special techniques for large data sets. Some suggestions on these points may be found in Section 8 of Martin and Thomson (1982). However, detailed studies remain to be carried out.

Meanwhile, the robust–resistant spectrum estimation technique presented here appears to be a quite useful exploratory tool. In particular, it is a tool having the power to detect and resolve relatively low-amplitude peaks in the spectrum which may be of scientific interest, but which are easily obscured by the presence of outliers.

References

Akaike, H. (1973). Information theory and an extension of the maximum likelihood principle. In: B. N. Petrov and F. Csáki, eds., *2nd International Symposium on Information Theory*, 267–281. Akademiai Kiado, Budapest.

Blackman, R. B. and Tukey, J. W. (1959). *The Measurement of Power Spectra from the Viewpoint of Communications Engineering*. Dover, New York. Reprinted from *Bell System Tech. J.* **37** (1958).

Bloomfield, P. (1976). *Fourier Analysis of Time Series: An Introduction*. Wiley, New York.

Boente, G., Fraiman, R. and Yohai, V. J. (1982). Qualitative robustness for general stochastic processes. Tech. Report No. 26, Department of Statistics, University of Washington, Seattle.

Box, G. E. P. and Tiao, G. C. (1975). Intervention analysis with applications to economic and environmental problems. *Journal of the American Statistical Association* **70**, 70–79.

Brillinger, D. R. (1973). A power spectral estimate insensitive to transients. *Technometrics* **15**, 559–563.

Brillinger, D. R. (1981). *Time Series: Data Analysis and Theory* (Expanded edition). Holden-Day, San Francisco.

Burg, J. P. (1972). The relationship between maximum entropy spectra and maximum likelihood spectra. *Geophysics* **37**, 375–376.

Bustos, O. and Yohai, V. J. (1982). Robust estimates for ARMA models. Submitted to *J. Amer. Statist. Assoc.*

Cox, Dennis (1981). Metrics on stochastic process and qualitative robustness. Technical Report No. 3. Department of Statistics, University of Washington.

Evans, J. E. and Griffiths, A. S. (1974). Design of a sanguine noise processor based upon world wide extremely low (ELF) recordings. *IEEE Trans. in Comm.* **22**(4), 528–539.

Gnanadesikan, R. (1977). *Methods for Statistical Data Analysis of Multivariate Observations*. Wiley, New York.

Grenander, U. (1981). *Abstract Inference*. Wiley, New York.

Hampel, F. R. (1971). A general qualitative definition of robustness. *Annals Math. Stat.* **42**, 1887–1895.

Hampel, F. R. (1974). The influence curve and its role in robust estimation. *Journal of American Statistical Association* **69**, 383–393.

Hampel, F. R. (1975). Beyond location parameters: Robust concepts and methods. *Proceedings of I.S.I. Meeting, 40th Session*, Warsaw.

Hogg, R. V. (1979). An introduction to robust estimation. In: R. L. Launer and G. N. Wilkinson, eds., *Robustness in Statistics*, 1–18. Academic Press, New York.

Huber, P. J. (1964). Robust estimation of a location parameter. *Ann. Math. Statist.* **35**(1) (March), 73–101.

Huber, P. J. (1981). *Robust Statistics*. Wiley, New York.

Jazwinski, A. H. (1970). *Stochastic Processes and Filtering Theory*. Academic Press, New York.

Jones, R. H. (1981). Fitting a continuous-time autoregression to discrete data. In: D. F. Findley, ed., *Applied Time Series Analysis II*. Academic Press, New York.

Kleiner, B., Martin, R. D. and Thomson, D. J. (1979). Robust estimation of power spectra. *Journal of the Royal Statistical Society, Series B* **41**(3), 313–351.

Koopmans, L. H. (1974). *The Spectral Analysis of Time Series*. Academic Press, New York.

Krasker, W. S. and Welsch, R. E. (1979). Efficient bounded-influence regression estimation using alternative definitions of sensitivity. Tech. Rep. No. TR-3, CCREMS, M.I.T.

Mallows, C. L. (1976). On some topics in robustness, Bell Labs. Tech. Memo, Murray-Hill, New Jersey (talks given at NBER Workshop on Robust Regression, Cambridge, MA, May 1973, and at the ASA-IMS Regional Meeting, Rochester, New York, May 21–23, 1975).

Mallows, C. L. (1980). Some theory of nonlinear smoothers. *The Annals of Statistics* **8**, 695–715.

Maronna, R. A., Bustos, O. and Yohai, V. (1979). Bias and efficiency robustness of general M-estimates with random carriers. In: Th. Gasser and M. Rosenblatt, eds., *Smoothing Techniques for Curve Estimation*. Springer-Verlag, New York.

Martin, R. D. and J. Zeh (1978). Robust generalized M-estimates for autoregressive parameters, including small-sample behavior. Tech. Report No. 214, Dept. of Electrical Eng., University of Washington, Seattle.

Martin, R. D. (1979). Approximate conditional-mean type smoothers and interpolators. In: Th. Gasser and M. Rosenblatt, eds., *Smoothing Techniques for Curve Estimation*, 117–143. Springer-Verlag, New York.

Martin, R. D. (1980). Robust estimation of autoregressive models. In: D. R. Brillinger and G. C. Tiao, eds., *Directions in Time Series*, 228–254. Institute of Mathematical Statistics Publication, Haywood, California.

Martin, R. D. (1981). Robust methods for time series. In: D. F. Findley, ed., *Applied Time Series II*. Academic Press, New York.

Martin, R.D., Samarov, A. and Vandaele, W. (1981). Robust methods for ARIMA models, to appear in proceedings of ASA-CENSUS-NBER Conference on Applied Time Series Analysis of Economic Time Series, Oct. 13–15, 1981. Also Tech. Report No. 21, Department of Statistics, University of Washington, Seattle.

Martin, R. D. and Thomson, D. J. (1982). Robust-resistant spectrum estimation. *Proc. IEEE* **70**(9), 1097–1114.

McGrew, C. A. (1972). Modeling the angular glint of an aircraft target, TM-FASE-022-03-72, Westinghouse Electric Corporation, Systems Develop. Div., Baltimore, MD.

Meditch, J. S. (1967). Orthogonal projections and discrete optimal linear smoothing. *SIAM Jour. Control* **5**(1), 74–89.

Mosteller, W. F. and Tukey, J. W. (1977). *Data Analysis and Regression*. Addison-Wesley, Reading.

Papantoni-Kazakos, P. and Gray, R. M. (1979). Robustness of estimators on stationary processes. *Ann. Probab.* **7**, 989–1002.

Parzen, E. (1974). Some recent advances in time-series modelling. *IEEE Transactions on Automatic Control* **AC-19**, 723–730.

Rabiner, L. R. and Schaffer, R. W. (1978). *Digital Processing of Speech Signals.* Prentice-Hall, Englewood Cliffs, New Jersey.

Rosenblatt, M. (1962). *Random Processes.* Oxford University Press, New York.

Thomson, D. J. (1977a). Spectrum estimation techniques for characterization and development of WT4 waveguide, Part I. *Bell System Technical Journal* **56**, 1769–1815.

Thomson, D. J. (1977b). Spectrum estimation techniques for characterization and development of WT4 waveguide, Part II. *Bell System Technical Journal* **56**, 1983–2005.

Welch, P. D. (1967). The use of the fast Fourier transform for estimation spectra: A method based on time averaging over short, modified periodograms. *IEEE Trans. On Aud. and Elec.* **AU-15**, 70–74.

D. R. Brillinger and P. R. Krishnaiah, eds., *Handbook of Statistics, Vol. 3*
© Elsevier Science Publishers B.V. (1983) 221–247

11

Autoregressive Spectral Estimation*

Emanuel Parzen

Introduction

The problem of spectral analysis of time series is clearly of great interest to the many applied scientists who use spectral analysis in their scientific research. It should be of great interest to statisticians because it embodies prototypes of two of the great problems of modern statistics: functional inference and modeling. A problem of statistical inference usually assumes three ingredients: a sample of observations, a parameter which indexes the family of possible joint probability densities of the sample, and a formula for the probability density of the sample

$$f(\text{sample} \mid \text{parameter}) .$$

Classical statistical inference assumes the parameter $\theta = (\theta_1, \dots, \theta_k)$. Functional inference assumes the parameter is a function, such as $f(\omega)$, $0 \leq \omega \leq 1$.

The *parameter estimation* problem seeks to form optimal estimators (denoted $\hat{\theta}$) of the parameter. A typical *model identification* problem seeks to find the smallest number of significantly nonzero components θ_j of the parameter θ.

Estimation of a function often has similar features to model identification, since a function can be parametrized exactly by a countable infinity of parameters. However, in practice, one can only efficiently estimate a finite number of parameters. Therefore, to estimate a function, one must use the smallest finite number of parameters which provide an adequate approximation of the function.

The goals of functional inference and model identification are in my view best pursued simultaneously. One seeks methods of statistical inference which are finite-parametric and non-(or infinite-) parametric. One achieves this goal by using finite parameter models (which theory indicates might be exact

*This research was supported by the Office of Naval Research (Contract N00014-81-MP-10001, ARO DAAG29-80-C0070).

methods) in ways that enable them to be interpreted also as approximating models.

Autoregressive spectral estimation is one of the new techniques for spectral analysis developed in the last two decades. Its theory and applications are extremely extensive. This article aims to provide an overview (rather than a detailed account) of the main ideas. A comprehensive bibliography guides the reader to articles in which case studies and comparisons of autoregressive spectral estimators are described.

The spectral density is defined in Section 1. Infinite-order AR and MA representations of a stationary time series are introduced in Section 2. Entropy as a motivation for autoregressive schemes is discussed in Section 3. Alternate parametrizations of an autoregressive scheme are outlined in Section 4. Algorithms for computing the coefficients of autoregressive spectral densities are stated in Section 5. Criteria for determining the orders of autoregressive schemes are mentioned in Section 6. Suggestions for empirical spectral analysis are outlined in Section 7. The final section provides a guide to the literature of autoregressive spectral estimation by listing the articles which correspond to some important developments.

1. Correlations and spectral functions of a stationary time series

The theory of time series discusses separately discrete-parameter time series $\{Y(t),\ t=0, \pm 1, \pm 2, \ldots\}$ and continuous-parameter time series $\{Y(t),\ -\infty < t < \infty\}$. Only the former case is discussed in this article. Discrete-parameter series usually arise by observing a continuous-parameter time series at equispaced times. The frequency variable ω of a pure harmonic

$$Y(t) = A \cos 2\pi\omega t + B \sin 2\pi\omega t \tag{1.1}$$

observed at $t = 0, \pm 1, \ldots$ can be assumed to vary in either $-0.5 \leqslant \omega \leqslant 0.5$ or $0 \leqslant \omega \leqslant 1$. The first interval is usually adopted, but the second interval will be adopted in this article because it is more convenient for developing isomorphisms between spectral analysis and nonparametric data modeling using quantile and density-quantile functions (see Parzen, 1979).

The definitions and notation we adopt for the functions used to describe a zero-mean *stationary* Gaussian discrete-parameter time series $Y(t)$, $t = 0, \pm 1, \ldots$ are as follows.

A 'time-domain' specification of the probability law of $Y(\cdot)$ is provided by the *covariance function*

$$R(v) = E[Y(t)Y(t+v)], \quad v = 0, \pm 1, \pm 2, \ldots; \tag{1.2}$$

or by the variance $R(0)$ and the *correlation function*

$$\rho(v) = \frac{R(v)}{R(0)} = \operatorname{corr}[Y(t), Y(t+v)]\,. \tag{1.3}$$

To define spectral (frequency) domain specification of the probability law of $Y(\cdot)$, we first assume summability of $R(\cdot)$ and $\rho(\cdot)$. The Fourier transforms of $R(v)$ and $\rho(v)$ are called the *power spectrum* $S(\omega)$ and *spectral density function* $f(\omega)$ respectively, and are defined by

$$S(\omega) = \sum_{v=-\infty}^{\infty} \mathrm{e}^{-2\pi i v\omega} R(v), \quad 0 \leq \omega \leq 1; \tag{1.4}$$

$$f(\omega) = \sum_{v=-\infty}^{\infty} \mathrm{e}^{-2\pi i v\omega} \rho(v), \quad 0 \leq \omega \leq 1\,. \tag{1.5}$$

The *spectral distribution function* is defined by

$$F(\omega) = \int_0^{\omega} f(\omega')\,\mathrm{d}\omega', \quad 0 \leq \omega \leq 1\,. \tag{1.6}$$

For data analysis, one actually computes a modified spectral distribution function denoted $F_+(w)$ and defined for $0 \leq \omega \leq 0.5$:

$$F_+(\omega) = 2F(\omega), \quad 0 \leq \omega \leq 0.5\,. \tag{1.7}$$

A correlation function $\rho(v)$ has a mathematical property of being a *positive-definite function* which, without assuming summability, guarantees the existence of: (1) a spectral distribution function $F(\omega)$, and (2) the spectral representation of the correlation function $\rho(v)$ given by

$$\begin{aligned}\rho(v) &= \int_0^1 \mathrm{e}^{2\pi i v\omega}\,\mathrm{d}F(\omega)\\ &= \int_0^{0.5} \cos 2\pi v\omega\,\mathrm{d}F_+(\omega)\,.\end{aligned} \tag{1.8}$$

The notion of an *ergodic* time series is not given a rigorous definition in this article but its intuitive meaning is important for us. We call a time series ergodic when the parameters of its probability law possess consistent estimators (and thus can be determined with probability one, given a sample of infinite length). An example of a nonergodic stationary Gaussian zero-mean time series is

$$Y(t) = A \cos 2\pi\omega t + B \sin 2\pi\omega t\,,$$

where A and B are independent $N(0, \sigma^2)$ random variables. One can infer the values of A and B in the sample observed, but one cannot infer the value of

σ^2. This time series has correlation function

$$\rho(v) = \cos 2\pi\omega v, \quad v = 0, \pm 1, \pm 2, \ldots, \tag{1.9}$$

and does not possess a spectral density.

Parzen (1982) proposes that it is useful in practice to distinguish qualitatively between three types of time series, which we call

no memory: white noise,
short memory: stationary and ergodic,
long memory: nonstationary or nonergodic.

A *no-memory* or *white noise* time series is a stationary Gaussian time series satisfying either of the equivalent conditions:

$$\rho(v) = 0 \quad \text{for } v > 0; \qquad f(\omega) = 1,\ 0 \leq \omega \leq 1\,. \tag{1.10}$$

A *short-memory* time series is a stationary time series possessing a summable correlation function $\rho(v)$ and a spectral density $f(\omega)$ which is bounded above and below in the sense that the dynamic range of $f(\omega)$

$$\mathrm{DR}(f) = \max_{0 \leq \omega \leq 1} f(\omega) \div \min_{0 \leq \omega \leq 1} f(\omega) \tag{1.11}$$

satisfies $1 < \mathrm{DR}(f) < \infty$.

A *long-memory* time series is one which is neither no memory nor short memory; alternatively, a long-memory time series is one which is nonstationary or nonergodic. It usually has components representing cycles or trends. An example of a long-memory time series is (1.1) where A and B are independent $N(0, \sigma^2)$ random variables.

For a short-memory time series, one can define the *inverse-correlation* function

$$\rho i(v) = \int_0^1 \mathrm{e}^{2\pi \mathrm{i} v\omega} f^{-1}(\omega)\, \mathrm{d}\omega \div \int_0^1 f^{-1}(\omega)\, \mathrm{d}\omega \tag{1.12}$$

and the *cepstral-correlation function*

$$\gamma(v) = \int_0^1 \mathrm{e}^{2\pi \mathrm{i} v\omega} \log f(\omega)\, \mathrm{d}\omega\,. \tag{1.13}$$

It should be noted that the inverse-correlation function is positive definite. However, the cepstral-correlation function is not. These new types of correlation functions are introduced because they may provide more parsimonious parametrizations in the sense that they decay to 0 faster than does the correlation function. Statistical inference from a sample of the probability law of a time series often achieves greatest statistical efficiency by using the most parsimonious parametrizations. Thus to form estimators $\hat{f}(\omega)$ on the spectral density $f(\omega)$ from a raw estimator $\tilde{f}(\omega)$, greater precision may be attained by first forming estimators $\{f^{-1}(\omega)\}^\wedge$ and $\{\log f(\omega)\}^\wedge$ of the inverse or logarithm of the spectral density. Autoregressive spectral estimation may be regarded as an approach to estimating $f(\omega)$ by first estimating $f^{-1}(\omega)$.

2. Filter models of a short-memory stationary time series

A short-memory zero-mean Gaussian stationary time series $Y(t)$ has a property of fundamental importance: it can be linearly transformed to a white noise time series, denoting $Y^\nu(t)$ or $\varepsilon(t)$, by a filter that is invertible. The definition of $Y^\nu(t)$ is provided by the theory of minimum mean-square error prediction.

The definitions and notation of prediction theory that we adopt are as follows:

$$Y^{\mu,m}(t) = E[Y(t) \mid Y(t-1), \ldots, Y(t-m)] \tag{2.1}$$

denotes the memory m one-step ahead predictor, while

$$Y^{\nu,m}(t) = Y(t) - Y^{\mu,m}(t) \tag{2.2}$$

is the prediction error, and

$$\sigma_m^2 = E[|Y^{\nu,m}(t)|^2] \div E[|Y(t)|^2] \tag{2.3}$$

is the mean-square prediction error measured in units of the variance $R(0)$ of $Y(t)$. Corresponding infinite-memory notation is

$$Y^{\mu}(t) = E[Y(t) \mid Y(t-1), Y(t-2), \ldots], \tag{2.4}$$

$$Y^{\nu}(t) = Y(t) - Y^{\mu}(t), \tag{2.5}$$

$$\sigma_\infty^2 = E[|Y^{\nu}(t)|^2] \div E[|Y(t)|^2]. \tag{2.6}$$

Explicit formulas for the foregoing predictors, prediction errors, and mean-square prediction errors can be obtained from the correlation function $\rho(v)$. The autoregressive coefficients $\alpha_m(1), \ldots, \alpha_m(m)$ of order m are defined by

$$-Y^{\mu,m}(t) = \alpha_m(1)Y(t-1) + \cdots + \alpha_m(m)Y(t-m), \tag{2.7}$$

$$Y^{\nu,m}(t) = Y(t) + \alpha_m(1)Y(t-1) + \cdots + \alpha_m(m)Y(t-m). \tag{2.8}$$

A predictor is characterized by the condition that the prediction error is orthogonal (normal) to the predictor variables:

$$E[Y^{\nu,m}(t)Y(t-k)] = 0, \quad k = 1, \ldots, m. \tag{2.9}$$

By substituting (2.8) into (2.9), one obtains the famous Yule–Walker equations, defining $\alpha_m(0) = 1$.

$$\sum_{j=0}^{m} \alpha_m(j)\rho(j-k) = 0, \quad k = 1, \ldots, m. \tag{2.10}$$

One obtains σ_m^2 by

$$\sigma_m^2 = E[Y^{\nu,m}(t)Y(t)] \div E[|Y(t)|^2] = \sum_{j=0}^{m} \alpha_m(j)\rho(j). \tag{2.11}$$

For a short-memory time series, these equations also hold with $m = \infty$.

The time series of infinite-memory prediction errors $Y^\nu(t)$ is always a white noise series called the *innovations.* It provides a transformation of the time series $Y(t)$ to a white noise time series $Y^\nu(t)$, which we write

$$Y(t) + \alpha_\infty(1)Y(t-1) + \cdots = Y^\nu(t), \tag{2.12}$$

and call an AR(∞) or infinite-order autoregressive scheme for $Y(t)$. An MA(∞) or infinite-order moving average scheme representation is

$$Y(t) = Y^\nu(t) + \beta_\infty(1)Y^\nu(t-1) + \cdots, \tag{2.13}$$

whose coefficients $\beta_\infty(k)$ can be determined recursively from $\alpha_\infty(j)$ by $\beta_\infty(0) = 1$, and for $k > 0$

$$\beta_\infty(k) + \sum_{j=1}^{k} \alpha_\infty(j)\beta_\infty(k-j) = 0. \tag{2.14}$$

The AR(∞) and MA(∞) representations have important implications for spectral analysis since they provide formulas for the spectral density function $f(\omega)$ alternative to the formula that $f(\omega)$ is the Fourier transform of $\rho(v)$. One can show that

$$f(\omega) = \sigma_\infty^2 |h_\infty(e^{2\pi i\omega})|^2, \tag{2.15}$$

$$f^{-1}(\omega) = \sigma_\infty^{-2} |g_\infty(e^{2\pi i\omega})|^2, \tag{2.16}$$

defining

$$h_\infty(z) = \sum_{j=0}^{\infty} \beta_\infty(j)z^j, \qquad g_\infty(z) = \sum_{k=0}^{\infty} \alpha_\infty(k)z^k. \tag{2.17}$$

These infinite series converge in general in mean square on the unit interval. In order to guarantee pointwise convergence at each ω in $0 \leq \omega \leq 1$, one must make an additional smoothness condition such as a Lipshitz condition on $f(\omega)$, which is implied in turn by the condition

$$\sum_{v=-\infty}^{\infty} |v|\,|\rho(v)| < \infty . \tag{2.18}$$

In spectral analysis we usually assume at least the existence of a continuous second derivative, which is implied by the condition

$$\sum_{v=-\infty}^{\infty} |v|^2\,|\rho(v)| < \infty . \tag{2.19}$$

The notion of a time series $Y(\cdot)$ being an *autoregressive scheme of order p*, denoted AR(p), can be defined in terms of prediction theory as follows: $Y(\cdot)$ is an AR(p) if and only if the memory p prediction errors $Y^{\nu,p}(\cdot)$ is a white noise time series and $\alpha_p(p) \neq 0$. The spectral density of $Y^{\nu,p}(\cdot)$ can be expressed in terms of the autoregressive transfer function of order p

$$g_p(z) = \alpha_p(0) + \alpha_p(1)z + \cdots + \alpha_p(p)z^p \tag{2.20}$$

by

$$f_{Y^{\nu,p}}(\omega) = \frac{1}{\sigma_p^2}\,|g_p(e^{2\pi i\omega})|^2 f(\omega) . \tag{2.21}$$

If the time series $Y(\cdot)$ is in the fact AR(p), then its spectral density equals the function

$$f_p(\omega) = \sigma_p^2 |g_p(e^{2\pi i\omega})|^{-2} , \tag{2.22}$$

which we call, in general, the *approximating autoregressive spectral density of order p*. A time series $Y(\cdot)$ can be regarded as approximated by an AR(p) if

$$\bar{f}_p(\omega) = \frac{f(\omega)}{f_p(\omega)} \tag{2.23}$$

can be regarded as not 'significantly' different from the constant 1. In this way a test of the hypothesis that a time series $Y(\cdot)$ is AR(p) can be converted to a test of the hypothesis that the prediction error time series is white noise.

The sequence of approximating autoregressive spectral densities $f_m(\omega)$, $m = 1, 2, \ldots$ may be shown to converge as m tends to ∞ to $f(\omega)$ at each ω in $0 \leq \omega \leq 1$ under suitable conditions (see especially Nevai, 1979). Sufficient conditions are that $f(\omega)$ has finite dynamic range (and therefore is bounded above and below) and has a continuous derivative. When an estimator, denoted $\hat{f}_m(\omega)$, of $f_m(\omega)$ is used as an estimator of $f(\omega)$, one has to take into account two kinds of errors, called respectively bias and variance. Bias is a measure of the deterministic difference between $f_m(\omega)$ and $f(\omega)$, while variance is a measure of the stochastic distance between $\hat{f}_m(\omega)$ and $f_m(\omega)$. As m increases, bias decreases, while variance increases. This is an example of the fundamental problem of empirical spectral analysis which is how to achieve an optimal balance between bias and variance. When one uses autoregressive spectral estimation, this problem reduces to a question of determining the order m of the approximating autoregressive scheme, which is discussed in Section 6.

It should be noted that basic references for the mathematical properties (especially convergence) of autoregressive transfer functions $g_m(z)$ are Geronimus (1960) and Grenander and Szegö (1958).

3. Entropy interpretation of autoregressive spectral densities

The use of autoregressive spectral densities as exact models, and as approximating models, for true spectral densities is often questioned by sceptical statisticians on the ground that their use in general is ad hoc and without theoretical justification. A possible answer to this criticism is provided by entropy concepts. This section reviews these concepts in order to state their application to spectral estimation.

The notion of entropy in statistics is usually first defined for a discrete distribution with probability mass function $p(x)$. The entropy of this distribution, denoted $H(p)$, is defined by

$$H(p) = -\sum_x p(x) \log p(x) . \tag{3.1}$$

For the distribution of a continuous real-valued random variable X, with probability density function $f(x)$, entropy is defined (analogously or formally) by

$$H(f) = -\int_{-\infty}^{\infty} f(x) \log f(x) \,\mathrm{d}x . \tag{3.2}$$

A concept closely related to entropy is information divergence $I(f; g)$ between two probability densities $f(x)$ and $g(x)$, defined by

$$I(f; g) = \int_{-\infty}^{\infty} \left\{-\log \frac{g(x)}{f(x)}\right\} f(x) \,\mathrm{d}x . \tag{3.3}$$

One should note that $I(f; g)$ equals minus the *generalized entropy* $H(f \mid g)$ defined by

$$H(f \mid g) = \int_{-\infty}^{\infty} \left\{-\frac{f(x)}{g(x)} \log \frac{f(x)}{g(x)}\right\} g(x) \,\mathrm{d}x . \tag{3.4}$$

Another fundamental concept is *cross-entropy* defined by

$$H(f; g) = \int_{-\infty}^{\infty} \{-\log g(x)\} f(x)\, dx . \tag{3.5}$$

Note that $H(f) = H(f; f)$.

Information divergence is expressed in terms of cross-entropy and entropy by

$$I(f; g) = H(f; g) - H(f) . \tag{3.6}$$

IMPORTANT INFORMATION INEQUALITY:

$$I(f; g) \geq 0 \tag{3.7}$$

with equality if and only if $f = g$; consequently

$$H(f) \leq H(f; g) . \tag{3.8}$$

PROOF. Apply Jensen's inequality which states for an arbitrary function $h(x)$

$$\int_{-\infty}^{\infty} \{\log h(x)\} f(x)\, dx \leq \log \int_{-\infty}^{\infty} h(x) f(x)\, dx \tag{3.9}$$

with equality if and only if $h(x) = 1$ for almost all x with respect to the probability density $f(x)$.

Some applications of entropy in probability and statistical modeling are as follows.

The method of maximum likelihood parameter estimation can be described abstractly as follows. One introduces a parametric family of probability densities $f_\theta(x)$, indexed by a vector parameter $\theta = (\theta_1, \ldots, \theta_k)$. Suppose there is a true parameter value $\bar{\theta}$ in the sense that the true probability density $f(x) = f_{\bar{\theta}}(x)$. Then $\bar{\theta}$ satisfies

$$H(f) = H(f; f_{\bar{\theta}}) = \min_{\theta} H(f; f_\theta) . \tag{3.10}$$

To estimate $\bar{\theta}$ from data, one forms an estimator $\tilde{H}(f; f_\theta)$ of $H(f; f_\theta)$ and defines an estimator $\hat{\theta}$ of $\bar{\theta}$ by

$$\tilde{H}(f; f_{\hat{\theta}}) = \min_{\theta} \tilde{H}(f; f_\theta) . \tag{3.11}$$

The estimator $\tilde{H}(f; f_\theta)$ could be of the form

$$\tilde{H}(f; f_\theta) = H(\tilde{f}; f_\theta) \tag{3.12}$$

for a suitable raw estimator $\tilde{f}(x)$ of $f(x)$.

The parametric families of probability densities $f_\theta(x)$ are often derived axiomatically using a maximum entropy principle.

THEOREM 1. *Fix k functions $T_j(x)$, $j = 1, 2, \ldots, k$, and k real numbers $\tau_1, \tau_2, \ldots, \tau_k$ such that there exists probability densities $f(x)$ satisfying*

$$\int_{-\infty}^{\infty} T_j(x) f(x)\, \mathrm{d}x = \tau_j, \quad j = 1, \ldots, k. \tag{3.13}$$

The density with maximum entropy $H(f)$ among these densities is of the form

$$\log f_\theta(x) = \sum_{j=1}^{k} \theta_j T_j(x) - \Psi(\theta_1, \ldots, \theta_k), \tag{3.14}$$

where

$$\Psi(\theta_1, \ldots, \theta_k) = \log \int_{-\infty}^{\infty} \mathrm{d}x \exp \sum_{k=1}^{k} \theta_j T_j(x), \tag{3.15}$$

and $\theta_1, \ldots, \theta_k$ are chosen to satisfy

$$\int_{-\infty}^{\infty} T_j(x) f_\theta(x)\, \mathrm{d}x = \tau_j, \quad j = 1, \ldots, k. \tag{3.16}$$

PROOF. The theorem may be proved by calculus of variations arguments. A quick proof is to verify that for any $f(x)$ satisfying the moment constraints (3.13)

$$H(f; f_\theta) = \Psi(\theta_1, \ldots, \theta_k) - \sum_{j=1}^{k} \theta_j \tau_j = H(f_\theta), \tag{3.17}$$

and therefore

$$H(f) \leq H(f; f_\theta) = H(f_\theta). \tag{3.18}$$

Thus the maximum entropy is achieved by $f_\theta(x)$.

Natural exponential models. A parametric family of probability densities $f_\theta(x)$ is said to obey a natural exponential model when it is of the form (3.14). Thus natural exponential models are maximum entropy probability densities.

To extend entropy concepts to short-memory stationary zero-mean Gaussian time series, define the information divergence for a sample $\{Y(t), t = 1, 2, \ldots, T\}$ as a function of the true probability density f of the sample, and a model g for f. We define

$$I(f; g) = \lim_{T \to \infty} I_T(f; g), \tag{3.19}$$

$$I_T(f; g) = \frac{-1}{T} E_f\left[\log \frac{g(Y(1), \ldots, Y(T))}{f(Y(1), \ldots, Y(T))}\right]. \tag{3.20}$$

It should be noted that we are using the notation f and g with a variety of meanings. For a Gaussian zero-mean stationary time series, the probability density of the sample is specified by the spectral densities $f(\omega)$ of the true distribution and $g(\omega)$ of the model. The arguments of the information divergence $I(f; g)$ indicate spectral densities in the following discussion. Pinsker (1963) derives the following very important formula:

$$I(f; g) = \tfrac{1}{2}\int_0^1 \left\{\frac{f(\omega)}{g(\omega)} - \log \frac{f(\omega)}{g(\omega)} - 1\right\} \mathrm{d}\omega . \tag{3.21}$$

Since $u - \log u - 1 \geq 0$ for all u, I has two of the properties of a distance: $I(f; g) \geq 0$, $I(f; f) = 0$. However, I does not satisfy the triangle-inequality.

We define the cross-entropy of spectral density functions $f(\omega)$ and $g(\omega)$ by

$$H(f; g) = \tfrac{1}{2}\int_0^1 \left\{\log g(\omega) + \frac{f(\omega)}{g(\omega)}\right\} \mathrm{d}\omega . \tag{3.22}$$

The entropy of f is

$$H(f) = H(f; f) = \tfrac{1}{2}\int_0^1 \{\log f(\omega) + 1\}\, \mathrm{d}\omega . \tag{3.23}$$

Information divergence can be expressed

$$I(f; g) = H(f; g) - H(f) . \tag{3.24}$$

Hence

$$H(f) \leq H(f; g) . \tag{3.25}$$

An approximating autoregressive spectral density of order m, denoted $\bar{f}_m(\omega)$, to a spectral density $f(\omega)$ is defined by

$$H(f; \bar{f}_m) = \min_{f_m} H(f; f_m) , \tag{3.26}$$

where the minimization is over all f_m of the form

$$f_m(\omega) = \sigma_m^2 |g_m(\mathrm{e}^{2\pi \mathrm{i}\omega})|^{-2} , \tag{3.27}$$

$$g_m(z) = 1 + \alpha_m(1)z + \cdots + \alpha_m(m)z^m . \tag{3.28}$$

One may verify that

$$H(f; f_m) = \tfrac{1}{2}\left\{\log \sigma_m^2 + \frac{1}{\sigma_m^2}\int_0^1 g_m|(\mathrm{e}^{2\pi \mathrm{i}\omega})|^2 f(\omega)\, \mathrm{d}\omega\right\} . \tag{3.29}$$

The coefficients $\bar{\sigma}_m^2, \bar{\alpha}_m(1), \ldots, \bar{\alpha}_m(m)$ of the minimum cross-entropy approximating autoregressive spectral density satisfy

$$\begin{aligned}\bar{\sigma}_m^2 &= \int_0^1 |\bar{g}_m(e^{2\pi i\omega})|^2 f(\omega)\, d\omega \\ &= \int_0^1 \bar{g}_m(e^{2\pi i\omega}) f(\omega)\, d\omega = \sum_{j=0}^m \hat{\alpha}_m(j)\rho(j) \\ &= \min_{g_m} \int_0^1 |g_m(e^{2\pi i\omega})|^2 f(\omega)\, d\omega ,\end{aligned} \tag{3.30}$$

$$\int_0^1 \bar{g}_m(e^{2\pi i\omega})\, e^{-2\pi ik\omega} f(\omega)\, d\omega = \sum_{j=0}^m \bar{\alpha}_m(j)\rho(j-k) = 0, \quad k = 1, 2, \ldots, m. \tag{3.31}$$

Further,

$$H(f; \bar{f}_m) = \tfrac{1}{2}\{\log \bar{\sigma}_m^2 + 1\} = H(\bar{f}_m). \tag{3.32}$$

The autoregressive spectral density $\bar{f}_m(\omega)$ can be derived axiomatically using a maximum entropy principle.

THEOREM 2. *The spectral density with maximum entropy among all spectral densities $f(\omega)$ satisfying the constraints*

$$\int_0^1 e^{2\pi i\omega j} f(\omega)\, d\omega = \rho(j), \quad j = 1, 2, \ldots, m, \tag{3.33}$$

for m specified correlation coefficients $\rho(1), \ldots, \rho(m)$ is $\bar{f}_m(\omega)$.

PROOF. It may be verified that $\bar{f}_m(\omega)$ satisfies the constraints (3.33) and (3.32) holds for any $f(\omega)$ satisfying (3.33). Since

$$H(f) \leq H(f; \bar{f}_m) = H(\bar{f}_m), \tag{3.34}$$

it follows that $\bar{f}_m$ has maximum entropy among all spectral densities satisfying the constraints (3.33).

The maximum entropy principle provides a motivation or justification for the use of autoregressive spectral estimators. However, the maximum entropy principle provides no insight into how to identify an optimal order m, or even what are the effects of different methods of estimating the parameters σ_m^2, $\alpha_m(1), \ldots, \alpha_m(m)$. It provides no guidance for how to learn from the data whether the time series is nonstationary (long memory) or stationary (short memory), or whether the best time series model is AR, MA, or ARMA. It is a principle for deriving probability models, rather than statistically fitting models to data. Further, the maximum entropy principle justifies autoregressive estimators only for short-memory time series. Autoregressive estimators are justified for long-memory time series by the fact that a pure harmonic $Y(t) = A\cos 2\pi\omega t + B\sin 2\pi\psi t$ satisfies $Y(t) - \phi Y(t-1) + Y(t-2) = 0$, where $\phi = 2\cos 2\pi\omega$.

4. Parametrizations of autoregressive spectral estimators

There are many ways for forming autoregressive spectral estimators, because there are four equivalent ways of parametrizing them.

A. *Autoregressive coefficients*

Consider coefficients $\alpha_m(1), \ldots, \alpha_m(m)$ such that $g(z) = 1 + \alpha_m(1)z + \cdots + \alpha_m(m)z^m$ satisfies $g(z) \neq 0$ for complex z such that $|z| \leq 1$. We call $g(z)$ a minimum phase filter transfer function. The autoregressive coefficients determine σ_m^2 by

$$1 = \sigma_m^2 \int_0^1 |g_m(\mathrm{e}^{2\pi\mathrm{i}\omega})|^{-2}\,\mathrm{d}\omega\,.$$

One computes the correlation coefficients $\rho(1), \ldots, \rho(m)$ by

$$\rho(j) = \int_0^1 \exp(2\pi\mathrm{i}\omega j)\sigma_m^2 |g_m(\mathrm{e}^{2\pi\mathrm{i}\omega})|^{-2}\,\mathrm{d}\omega\,.$$

B. *Correlations*

A set of m coefficients $\rho(1), \ldots, \rho(m)$, such that the matrix [with $\rho(0) = 1$, $\rho(-v) = \rho(v)$]

$$\begin{bmatrix} \rho(0) & \rho(-1) & \ldots & \rho(1-m) \\ \rho(1) & \rho(0) & \ldots & \rho(2-m) \\ & & & \\ \rho(m-1) & \rho(m-2) & \ldots & \rho(0) \end{bmatrix}$$

is positive definite, is correlation coefficients of a time series. They determine autoregressive coefficients by solving the Yule–Walker equations [with $\alpha_m(0) = 1$]

$$\sum_{j=0}^{m} \alpha_m(j)\rho(j-k) = 0\,, \quad k = 1, \ldots, m\,.$$

One computes σ_m^2 by $\sigma_m^2 = \sum_{j=0}^{m} \alpha_m(j)\rho(j)$.

C. Partial correlations

Consider coefficients $\pi(1), \ldots, \pi(m)$ satisfying $|\pi(j)| < 1$, $j = 1, 2, \ldots, m$. They represent partial correlation coefficients defined theoretically by

$$\pi(j) = \operatorname{corr}[Y(t), Y(t-j) | Y(t-1), \ldots, Y(t-j+1)] .$$

Partial correlation coefficients determine autoregressive coefficients and residual variances by a recursive algorithm called the Levinson–Durbin recursion (see Levinson, 1947; Durbin, 1960): for $k = 1$

$$\alpha_1(1) = -\pi(1), \qquad \sigma_1^2 = 1 - \pi^2(1) ,$$

while for $k = 2, 3, \ldots, m$

$$\alpha_k(k) = -\pi(k) = \frac{-1}{\sigma_{k-1}^2} \sum_{j=0}^{k-1} \alpha_{k-1}(j)\rho(k-j) ,$$

$$\sigma_k^2 = \sigma_{k-1}^2\{1 - \pi^2(k)\} ,$$

$$\alpha_k(j) = \alpha_{k-1}(j) - \pi(k)\alpha_{k-1}(k-j) .$$

Autoregressive coefficients determine partial correlation coefficients by the recursion (see Barndorff-Nielssen and Schon, 1973)

$$\alpha_{k-1}(j) = \{1 - \pi^2(k)\}^{-1}\{\alpha_k(j) + \pi(k)\alpha_k(k-j)\} .$$

D. Residual variances

Consider coefficients $\sigma_1^2, \sigma_2^2, \ldots, \sigma_m^2$ satisfying

$$1 > \sigma_1^2 > \sigma_2^2 > \cdots > \sigma_m^2 > 0 ,$$

and m coefficients representing sign $\pi(1), \ldots,$ sign $\pi(m)$. The σ's represent residual variances; they determine partial correlation coefficients by a formula noted by Dickenson (1978)

$$\pi(k) = \operatorname{sign} \pi(k)\left[1 - \frac{\sigma_k^2}{\sigma_{k-1}^2}\right]^{1/2} .$$

5. Empirical autoregressive spectral estimation

Given a sample $\{Y(t), t = 1, 2, \ldots, T\}$ of a zero-mean Gaussian stationary time series whose spectral density $f(\omega)$ is to be estimated by an autoregressive spectral density estimator

$$\hat{f}_m(\omega) = \hat{\sigma}_m^2 |\hat{g}_m(e^{2\pi i\omega})|^{-2},$$

$$\hat{g}_m(z) = 1 + \hat{\alpha}_m(1)z + \cdots + \alpha_m(m)z^m,$$

we define the order identification problem to be the choice of order m, and the parameter estimation problem to be the choice of algorithm for computing the coefficients $\hat{\alpha}_m(1), \ldots, \hat{\alpha}_m(m)$ and the residual variance $\hat{\alpha}_m^2$.

For a sample $Y(1), \ldots, Y(t)$ of a zero-mean Gaussian stationary time series, an approximation for the joint probability density function $f_\theta(Y(1), \ldots, Y(T))$ indexed by a parameter θ is obtained as follows. We assume that the time series $Y(t)$ has been divided by $\{R(0)\}^{1/2}$ so that its covariance function equals its correlation function. Then

$$-2 \log f_\theta(Y(1), \ldots, Y(T)) = \log(2\pi)^T \det K_\theta + Y_T^* K^{-1} Y_T,$$

where $*$ denotes complex conjugate transpose, $Y_T^* = (Y(1), \ldots, Y(T))$, $K_\theta = EY_T Y_T^*$ is a covariance matrix with (s, t)-element equal to $\rho_\theta(s-t)$. The subscript θ on $\rho_\theta(v)$ and $f_\theta(\omega)$ indicates that they are functions of the parameters θ, which are to be estimated.

The covariance matrix K_θ is a Toeplitz matrix. Asymptotically, as T tends to ∞, all T by T Toeplitz matrices have the same eigenvectors $\exp(-2\pi itj/T)$, $j = 0, 1, \ldots, T-1$. The corresponding eigenvalues are $f_\theta(j/T)$. An approximation for likelihood function frequently adopted is therefore

$$\begin{aligned} -\frac{1}{T} \log f_\theta(Y(1), \ldots, Y(T)) &= \tfrac{1}{2} \log 2\pi + \tfrac{1}{2} \int_0^1 \left\{ \log f_\theta(\omega) + \frac{\tilde{f}(\omega)}{f_\theta(\omega)} \right\} d\omega \\ &= \tfrac{1}{2} \log 2\pi + H(\tilde{f}; f_0), \end{aligned}$$

where $\tilde{f}(\omega)$ is the sample spectral density defined by

$$\tilde{f}(\omega) = \sum_{t=1}^{T} |Y(t) \exp(-2\pi it\omega)|^2 \div \sum_{t=1}^{T} Y^2(t).$$

Maximum likelihood estimators $\hat{\theta}$ are asymptotically equivalent to the estimators $\hat{\theta}$ minimizing the sample cross-entropy $H(\tilde{f}; f_\theta)$.

If the parametric model $f_\theta(\omega)$ is assumed to be the spectral density of an AR(p), then estimators $\hat{\sigma}_p^2$, $\hat{\alpha}_p(1), \ldots, \hat{\alpha}_p(p)$ of the coefficients satisfy Yule–Walker equations corresponding to the sample correlation function

$$\begin{aligned} \hat{\rho}(v) &= \int_0^1 e^{2\pi i\omega v} \tilde{f}(\omega)\, d\omega \\ &= \sum_{t=1}^{T-v} Y(t)Y(t+v) \div \sum_{t=1}^{T} Y^2(t). \end{aligned}$$

The sample correlation function $\hat{\rho}(v)$ can be computed, using the fast Fourier transform, by

$$\hat{\rho}(v) = \frac{1}{Q}\sum_{k=0}^{Q-1} \exp\left(2\pi i \frac{k}{Q} v\right)\tilde{f}\left(\frac{k}{Q}\right),$$

which holds for $0 \leq v < Q - T$.

It should be noted that we are assuming the time series $Y(t)$ to be zero mean, or more generally to have been detrended by subtraction of $\hat{\mu}(t)$, an estimator of the mean-value function $\mu(t) = E[Y(t)]$. When $\mu(t) = \mu$, a constant, we take $\hat{\mu} = \bar{Y}$. When $\mu(t)$ is a function with period d (as might be the case with $d = 12$ for monthly time series), one might take for $\hat{\mu}(t)$ the mean of $Y(s)$ for all $s = t$ modulo d.

By recursively solving the Yule–Walker equations, one can determine sequences of

(1) estimated residual variances

$$1 > \hat{\sigma}_1^2 > \hat{\sigma}_2^2 > \cdots > \hat{\sigma}_m^2 \ldots,$$

(2) estimated partial correlations

$$\hat{\pi}(1), \hat{\pi}(2), \ldots, \hat{\pi}(m), \ldots,$$

(3) estimated autoregressive coefficients

$$\hat{\alpha}_m(0) = 1, \hat{\alpha}_m(1), \ldots, \hat{\alpha}_m(m),$$

(4) autoregressive spectral density estimators

$$\hat{f}_m(\omega) = \hat{\sigma}_m^2 \left| \sum_{j=0}^{m} \hat{\alpha}_m(j) \exp 2\pi i j\omega \right|^{-2},$$

(5) residual spectral densities

$$\tilde{f}_m(\omega) = \frac{\tilde{f}(\omega)}{\hat{f}_m(\omega)}.$$

By forming a smoothed version $\hat{\hat{f}}_m(\omega)$ of $\tilde{f}_m(\omega)$, one can obtain a final estimator $\hat{f}(\omega)$ of the unknown spectral density:

$$\hat{f}(\omega) = \hat{f}_m(\omega)\hat{\hat{f}}_m(\omega).$$

When $\tilde{f}(\omega)$ is tested for white noise, and found not to be significantly different from white noise, then

$$\hat{f}(\omega) = \hat{f}_m(\omega)\,,$$

and the autoregressive spectral density estimator is the final estimator.

The important question of criteria for choosing the orders of approximating spectral densities is discussed in the next section.

Computing estimators of autoregressive coefficients by solving Yule–Walker equations is called *stationary autoregression* because the autoregressive coefficients obtained are guaranteed to correspond to a stationary time series. When $\hat{\sigma}_m^2$ in the foregoing analysis is tending to approximate 0, we consider the time series to be long memory; experimental evidence indicates that more reliable estimators of the spectral density, and also of the autoregressive coefficients, are provided by *least-squares autoregression*, which solves the normal equations

$$\begin{bmatrix} \hat{K}(0,0) & \dots & \hat{K}(0,m) \\ \hat{K}(1,0) & \dots & \hat{K}(1,m) \\ \dots & \dots & \dots \\ \hat{K}(m,0) & \dots & \hat{K}(m,m) \end{bmatrix} \begin{bmatrix} 1 \\ \hat{\alpha}_m(1) \\ \dots \\ \hat{\alpha}_m(m) \end{bmatrix} = \begin{bmatrix} \hat{\sigma}_m^2 \\ 0 \\ \dots \\ 0 \end{bmatrix}$$

for a suitable estimator $\hat{K}(i, j)$ of

$$K(i, j) = E[Y(t-i)Y(t-j)]\,.$$

Possible estimators (for $i, j = 0, 1, \dots, m$) are least-squares forward algorithm

$$\hat{K}(i, j) = \frac{1}{T-M} \sum_{t=0}^{T-m-1} Y(t+m-i)Y(t+m-j)\,,$$

or least-squares forward and backward algorithm

$$\hat{K}(i, j) = \frac{1}{2(T-M)} \sum_{t=0}^{T-m-1} \{Y(t+m-i)Y(t+m-j) + Y(t+i)Y(t+j)\}\,.$$

When several harmonics are present in the data, whose frequencies are close together, least-squares autoregressive coefficient estimators are more effective than Yule–Walker autoregressive coefficient estimators in providing autoregressive spectral estimators which exhibit the split peaks one would like to see in the estimated spectral density.

An important and popular algorithm for estimation of AR coefficients was introduced by Burg in 1967 (see Burg, 1967, 1968). For references to descriptions of Burg's algorithm, see Kay and Marple (1981).

6. Autoregressive order determination

The problem of determining the orders of approximating autoregressive schemes is an important example of the problem of estimating a function by using the smallest finite number of parameters which provide an adequate approximation of the function. The true spectral density is denoted $f(\omega)$ or $f_\infty(\omega)$. An approximation $\bar{f}(\omega)$ is defined by assuming a family of densities $f_{\theta_1,\ldots,\theta_m}(\omega)$ which are functions of m scalar parameters $\theta_1, \ldots, \theta_m$. The parameter values $\bar{\theta}_1, \ldots, \bar{\theta}_m$ which minimize the cross-entropy $H(f; f_{\theta_1,\ldots,\theta_m})$ define a best approximating spectral density $\bar{f}_m(\omega) = f_{\bar{\theta}_1,\ldots,\bar{\theta}_m}(\omega)$. An estimator of f_m is $\hat{f}_m(\omega) = f_{\hat{\theta}_1,\ldots,\hat{\theta}_m}(\omega)$, where $\hat{\theta}_1, \ldots, \hat{\theta}_m$ minimizes $H(\tilde{f}; f_{\theta_1,\ldots,\theta_m})$.

To evaluate the properties of $\hat{f}_m(\omega)$ as an estimator of $f_\infty(\omega)$, one must distinguish two kinds of error. The *model approximation* or *bias* error is

$$B(m) = I(f_\infty; \bar{f}_m) .$$

The *parameter estimation* error or *variance* is

$$V(m, T) = EI(\bar{f}_m; \hat{f}_m) .$$

As m tends to ∞, $B(m)$ tends to 0 and $V(m, T)$ tends to ∞. The optimal value $\hat{m}$ minimizes $EI(f_\infty; \hat{f}_m)$ as measured by

$$C(m) = B(m) + V(m, T) .$$

In practice, one forms an estimator $\hat{C}(m)$ of the terms in $C(m)$ which depend on m.

One calls $\hat{C}(m)$ a criterion function for order determination. It should be plotted, and interpreted as a function, not just examined for its minimum value. It is useful to define a best value of m (at which $\hat{C}(m)$ is minimized) and a second best value of m (at which $\hat{C}(m)$ achieves its lowest relative minimum).

One also has to define a value $\hat{C}(0)$ of the criterion function at $m = 0$. If

$$\hat{C}(m) > \hat{C}(0), \quad \text{for } m = 1, 2, \ldots,$$

then the optimum order is 0, and the time series is considered to be not significantly different from white noise. Further research is required on the properties of order determining criteria as tests for white noise.

Tests for white noise provide an alternative approach to order determination since an autoregressive estimator $\hat{f}_m(\omega)$ is regarded as an adequate fit (or smoother) if the residual spectral density $\tilde{f}(\omega) \div \hat{f}_m(\omega)$ is not significantly different from the sample spectral density of white noise.

A widely used order determining criterion is that introduced by Akaike (1974). It should be emphasized that Akaike's criterion had a different conceptual basis than the one outlined above; it seeks to determine the order of an exact autoregressive scheme which the time series is assumed to obey. Then one can raise the objection against it that it does not consistently estimate the order, which is done by a criterion due to Hannan and Quinn (1979). Our point of view is that the approximating autoregressive scheme need only have the property that $\tilde{f}(\omega) \div \hat{f}_m(\omega)$ is just barely not significantly different from the sample spectral density of white noise.

Akaike's order determining criterion AIC is defined by

$$\text{AIC}(m) = \log \hat{\sigma}_m^2 + \frac{2m}{T}, \quad m \geq 1 .$$

Possible definitions for AIC(0) are 0 or $-1/T$. The Hannan and Quinn criterion is

$$\text{AICHQ}(m) = \log \hat{\sigma}_m^2 + \frac{m}{T} \log \log T .$$

Parzen (1974, 1977) introduced an approximating autoregressive order criterion called CAT (criterion autoregressive transfer function), defined by

$$\text{CAT}(m) = \frac{1}{T} \sum_{j=1}^{m} \left(1 - \frac{j}{T}\right) \hat{\sigma}_j^{-2} - \left(1 - \frac{m}{T}\right) \hat{\sigma}_m^{-2}, \quad m \geq 1 ,$$

$$\text{CAT}(0) = -\left(1 + \frac{1}{T}\right).$$

In practice, CAT and AIC lead in many examples to exactly the same orders. It appears reassuring that quite different conceptual foundations can lead to similar conclusions in practice.

7. Suggestions for empirical spectral analysis

The basic aim of spectral analysis is to obtain an estimated spectral density which does not introduce spurious spectral peaks, and resolves close spectral peaks. To arrive at the final form of spectral estimator in an applied problem, autoregressive spectral estimators can be used to identify the memory type of a time series (long, short, or no memory) and the type of the whitening filter of a short-memory time series (AR, MA, or ARMA). An empirical time-series spectral analysis should involve the following stages.

A. Preprocessing

To analyze a time-series sample $Y(t)$, $t = 1, \ldots, T$, one will proceed in stages which often involve the subtraction of or elimination of strong effects in order to see more clearly weaker patterns in the time-series structure. The aim of preprocessing is to transform $Y(\cdot)$ to a new time series $\tilde{Y}(\cdot)$ which is short memory. Some basic preprocessing operations are memoryless transformation (such as square root and logarithm), detrending, 'high-pass' filtering, and differencing. One usually subtracts out the sample mean $\bar{Y} = (1/T)\Sigma_{t=1}^{T} Y(t)$; then the time series actually processed is $Y(t) - \bar{Y}$.

B. Sample Fourier transform by data windowing, extending with zeroes, and fast Fourier transform

Let $Y(t)$ denote a preprocessed time series. The first step in the analysis could be to compute successive autoregressive schemes using operations only in the time domain. An alternative first step is the computation of the sample Fourier transform

$$\tilde{\psi}(\omega) = \sum_{t=1}^{T} Y(t)\exp(-2\pi i\omega t), \quad \omega = \frac{k}{Q}, \; k = 0, 1, \ldots, Q-1, \tag{7.1}$$

at an equispaced grid of frequencies in $0 \leqslant \omega \leqslant 1$. We call Q the spectral computation number. One should always choose $Q \geqslant T$, and we recommend $Q \geqslant 2T$. Prior to computing $\tilde{\psi}(\omega)$, one should extend the length of the time series by adding zeroes to it. Then $\tilde{\psi}(\omega)$ can be computed using the fast Fourier transform.

If the time series may be long memory, one should compute in addition a sample 'tapered' or 'data-windowed' Fourier transform

$$\tilde{\psi}_W(\omega) = \sum_{t=1}^{T} Y(t) W\left(\frac{t}{T}\right) \exp(-2\pi i\omega t)\,. \tag{7.2}$$

C. Sample spectral density

The sample spectral density $\tilde{f}(\omega)$ is obtained essentially by squaring and normalizing the sample Fourier transform:

$$\tilde{f}(\omega) = \frac{|\tilde{\psi}(\omega)|^2}{\frac{1}{Q}\sum_{k=0}^{Q-1}\left|\tilde{\psi}\left(\frac{k}{Q}\right)\right|^2}, \quad \omega = \frac{k}{Q}, \quad k = 0, 1, \ldots, Q-1\,. \tag{7.3}$$

D. Sample correlation function

The sample correlation function $\hat{\rho}(v)$ is computed (using the fast Fourier transform).

E. Autoregressive analysis

The Yule–Walker equations are solved to estimate innovation variances $\hat{\sigma}_m^2$, to which are applied order determining criteria (AIC, CAT) to determine optimal orders $\hat{m}$ and also to test for white noise. The value of $\hat{\sigma}_{\hat{m}}^2$ and the dynamic range of the autoregressive spectral estimator $\hat{f}_{\hat{m}}(\omega)$ are used to determine the memory type of the time series. Two orders (called the best $\hat{m}$ and second best $\hat{m}(2)$) are determined as candidates as optimal orders corresponding to the absolute minimum and lowest relative minimum of the criterion function.

F. ARMA analysis

When a time series is classified as short memory, an approximating AR scheme of order $4\hat{m}$ is inverted to form MA(∞) coefficients which are used to estimate covariance matrix of $Y(t-j)$ and $Y^{\nu}(t-k)$. A subset regression procedure is then used to determine a 'best-fitting' ARMA scheme, and the corresponding ARMA spectral density estimator. One will be able to identify moving average schemes and ARMA schemes which are barely invertible, and require a long AR scheme for adequate approximation. The long AR spectral estimator introduces spurious spectral peaks when compared to the MA or ARMA estimator.

G. Nonstationary autoregression

When a time series is classified as long memory, more accurate estimators of autoregressive coefficients are provided by minimizing a least-squares criterion or by Burg estimators. When several harmonics are present in the data, whose frequencies are close together, least-squares autoregressive coefficient estimators are more effective than Yule–Walker autoregressive coefficient estimators in providing autoregressive spectral estimators which exhibit the split peaks one would like to see in the estimated spectral density.

H. Long-memory analysis

In the long-memory case, one may want to represent $Y(t)$ as $S(t)+N(t)$, a long-memory signal plus short-memory noise. An approach to this problem may be provided by treating the sample spectral density values $\tilde{f}(k/Q)$ as a data batch to be studied by nonparametric data modeling methods using quantile functions (see Parzen, 1979). The details of such methods are under development.

I. Nonparametric kernel spectral density estimator

An estimator $\hat{f}(\omega)$ of the spectral density is called *parametric* when it corresponds to a parametric model for the time series (such as an AR or ARMA model), *nonparametric* otherwise. A general form of the nonparametric estimator is the kernel estimator

$$\hat{f}(\omega)=\sum_{v=-\infty}^{\infty} k\left(\frac{v}{M}\right)\hat{\rho}(v)\,\mathrm{e}^{-2\pi i\omega v}, \quad 0\leqslant\omega\leqslant 1\,.$$

The problem of determining optimum truncations points M has no general solution; one approach is to choose $M = 4\hat{m}$ to obtain a preliminary smoothing of the sample spectral density.

J. Inverse correlations and cepstral correlations

Estimators of $\rho i(v)$ and $\gamma(v)$ are computed and used to form nonparametric kernel estimators of $f^{-1}(\omega)$ and $\log f(\omega)$, which may provide additional insights into the peaks and troughs to be given significance in the final estimator of the spectrum.

Extensive comparisons of different methods of spectral estimation are given in Pagano (1980), Beamish and Priestley (1981), and Kay and Marple (1981). It seems clear that autoregressive spectral estimators can give superior results when properly used. One should: determine two best orders; compute autoregressive coefficients by Yule–Walker equations and by least squares since when the time series is long-memory autoregressive spectral estimators are most accurate when based on least-squares estimators of autoregressive coefficients; use approximating autoregressive schemes to determine if an ARMA scheme fits better.

The end of the story of the search for the perfect spectral estimator seems attainable if one does not think of spectral estimation as a nonparametric procedure which can be conducted independently of model identification.

8. A bibliography of autoregressive spectral estimation

The references aim to provide a comprehensive list of the publications in English which are directly concerned with developing the theory and methods of autoregressive spectral estimation.

This section lists some of the publications which contributed to the development of AR spectral estimation.

Yule (1972) introduces autoregressive schemes to model disturbed periodicities as an alternative to Schuster periodogram analysis and its spurious periodicities; Yule–Walker (1931) equations relate autoregressive coefficients and correlations of a stationary time series.

Wold (1938) introduces infinite-order autoregressive and moving average representations of a stationary time series; rigorous conditions are given by Akutowicz (1957).

Mann and Wald (1943) derive asymptotic distribution of estimators of autoregressive coefficients.

Levinson (1947) and Durbin (1960) derive recursive methods of solving Yule–Walker equations which subsequently lead to fast algorithms for calculation of high-order AR schemes.

Whittle (1954) seems to be the first to use autoregressive schemes to estimate a spectral density. He used a low-order model in a case where high-order models are indicated by order determining criterion (Akaike, 1974, p. 720).

Grenander and Rosenblatt (1956) criticize attempts to apply low-order autoregressive schemes, and develop theory of nonparametric spectral density estimation, as do Bartlett, Parzen, and Tukey and Blackman.

Parzen (1964), Schaerf (1964) and Parzen (1968) discuss autoregressive spectral estimation as a method for empirical time-series analysis; no theory is given.

Burg (1967, 1968) publishes his pioneering work on MEM (maximum entropy method of spectral estimation) and his method of calculating their coefficients.

Akaike (1969a,b, 1970) derives asymptotic variance formulas for autoregressive spectral estimators, and states FPE (final predictor error) criterion for order determination; precursor of FPE in Davisson (1965).

Parzen (1969) derives heuristically a formula for the asymptotic variance of AR spectral estimators, confirmed by Kromer (1969) and Berk (1974); an order determining criterion is proposed.

Kromer (1969) in an unpublished Ph.D. thesis presents first rigorous analysis of asymptotic distribution of autoregressive spectral estimators, especially their bias; consistency is proved only in an iterated limit mode of convergence.

Berk (1974) provides first proof of consistency of autoregressive spectral estimators.

Carmichael (1976) in an unpublished Ph.D. thesis provides alternative proof of consistency of autoregressive estimators, and extends technique to general problems of density estimation.

Akaike (1973, 1974, 1977) introduces AIC for model order criterion and relates it to entropy maximization principles.

Parzen (1974, 1977) introduces CAT for AR order determination based on concept of finite parameter AR schemes as approximations to infinite parameter AR schemes.

Hannan and Quinn (1979) derive a modification of AIC which provides consistent estimators of the AR order, when exact model is assumed to be a finite order AR.

Huzii (1977), Shibata (1977) and Bhansali (1980) discuss rigorously the convergence of AR spectral estimators and inverse correlations.

Childers (1978) and Haykin (1979) contain very useful collections of papers.

Pagano (1980), Beamish and Priestley (1981) and Kay and Marple (1981) provide illuminating reviews of AR spectral estimators and comparisons with alternative methods.

References

Akaike, H. (1969a). Fitting autoregressive models for prediction. *Ann. Inst. Statist. Math.* **21**, 243–247.

Akaike, H. (1969b). Power spectrum estimation through autoregression model fitting. *Ann. Inst. Statist. Math.* **21**, 407–419.

Akaike, H. (1970a). A fundamental relation between predictor identification and power spectrum estimation. *Ann. Inst. Statist. Math.* **22**, 219–223.

Akaike, H. (1970b). Statistical predictor identification. *Ann. Inst. Statist. Math.* **22**, 203–217.

Akaike, H. (1973). Information theory and an extension of the maximum likelihood principle. In: B. N. Petrov and F. Csáki, eds., *2nd Int. Symp. on Information Theory,* 267–281. Akademiai Kiado, Budapest.

Akaike, H. (1974). A new look at the statistical model identification. *IEEE Trans. Autom. Contr.* **AC-19**, 716–723.

Akaike, H. (1977). On entropy maximization principle In: P. R. Krishnaiah, ed., *Applications of Statistics*, 27–41. North-Holland, Amsterdam.

Akaike, H. (1979). A Bayesian estension of the minimum AIC procedure of autoregressive model fitting. *Biometrika* **66**, 237–242.

Akutowicz, E. J. (1957). On an explicit formula in linear least squares prediction. *Math. Scand.* **5**, 261–266.

Baggeroer, A. B. (1976). Confidence intervals for regression (MEM) spectral estimates. *IEEE Trans. Inform. Theory* **IT-22**, 534–545.

Barndorf-Nielsen, O. and Schon, G. (1973). On the parametrization of autoregressive models by partial autocorrelations. *J. Multivariate Analysis* **3**, 408–419.

Beamish, M. and Priestley, M. B. (1981). A study of autoregressive and window spectral estimation. *Appl. Statist.* **30**, 41–58.

Berk, K. N. (1974). Consistent autoregressive spectral estimates. *Ann. Statist.* **2**, 489–502.

Bhansali, R. J. (1980). Autoregressive and window estimates of the inverse correlation function. *Biometrika* **67**, 551–566.

Bhansali, R. J. and Dowham, D. Y. (1977). Some properties of the order of an autoregressive model selected by a generalization of Akaike's EPF criterion. *Biometrika* **64**, 547–551.

Brillinger, D. R. (1981). *Time Series: Data Analysis and Theory* (expanded edition). Holden-Day, San Francisco [see p. 512].

Burg, J. P. (1967). Maximum entropy spectral analysis. In: *Proc. 37th Meeting Society of Exploration Geophysicists* (Oklahoma City, OK), Oct. 31, 1967.

Burg, J. P. (1968). A new analysis technique for time series data. *NATO Advanced Study Institute on Signal Processing.* Enschede, The Netherlands.

Carmichael, J. P. (1976). The autoregressive method. Unpublished Ph.D. thesis, Statistical Science Division, State University of New York at Buffalo.

Carmichael, J. P. (1978). Consistency of the autoregressive method of density estimation, Technical Report, Statistical Science Division, State University of New York at Buffalo.

Chatfield, C. (1979). Inverse autocorrelations. *J. R. Statist. Soc.* **A142**, 363–377.

Childers, D. G. (1978). *Modern Spectrum Analysis.* IEEE Press, New York.

Cleveland, W. S. (1972). The inverse autocorrelations of a time series and their applications. *Technometrics* **14**, 277–293.

Davisson, I. D. (1965). The prediction error of a stationary Gaussian time series of unknown covariance. *IEEE Trans. on Information Theory* **IT-11**, 527–532.

Dickenson, Bradley W. (1978). Autoregressive estimation using residual energy ratios. *IEEE Transactions on Information Theory* **IT-24**, 503–505.

Durbin, J. (1960). The fitting of time series models. *Rev. Inst. de Stat.* **28**, 233–244.

Durrani, T. S. and Arslanian, A. S. (1982). Windows associated with high resolution spectral estimators. Submitted for publication.

Fuller, W. A., Hasen, D. P. and Goebel, J. J. (1981). Estimation of the parameters of stochastic difference equations. *Annals of Statistics* **9**, 531–543.

Geronimus, Y. L. (1960). *Polynomials Orthogonal on a Circle and Interval.* Pergamon Press, New York.

Gersch, W. (1970). Spectral analysis of EEG's by autoregressive decomposition of time series. *Math. Biosci.* **7**, 205–222.

Gersch, W. and Sharpe, D. R. (1973). Estimation of power spectra with finite-order autoregressive models. *IEEE Trans. Automat. Contr.* **AC-18**, 367–369.

Grandell, J., Hamrud, M. and Toll, P. (1980). A remark on the correspondence between the maximum entropy method and the autoregressive models. *IEEE Trans. Inform. Theory* **IT-26**, 750–751.

Grenander, U. and Rosenblatt, M. (1957). *Statistical Analysis of Stationary Time Series*. Wiley, New York.

Grenander, U. and Szegö, G. (1958). *Toeplitz Forms and their Applications*. University of California Press, Berkeley.

Griffiths, L. J. and Prieto-Diaz, R. (1977). Spectral analysis of natural seismic events using autoregressive techniques. *IEEE Trans. Geosci. Electron.* **GE-15**, 13–25.

Hannan, E. J. (1970). *Multiple Time Series*. Wiley, New York [see p. 334].

Hannan, E. J. and Quinn, B. G. (1979). The determination of the order of an autoregression. *Journal of the Royal Statistical Society* **41**, 190–195.

Haykin, S. S., ed. (1979). *Nonlinear Methods of Spectral Analysis*. Springer-Verlag, New York.

Hsu, F. M. and Giordano, A. A. (1977). Line tracking using autoregressive spectral estimates. *IEEE Trans. Acoustics, Speech, Signal Process.* **ASSP-25**, 510–519.

Huzii, M. (1977). On a spectral estimate obtained by an autoregressive model fitting. *Ann. Inst. Statist. Math.* **29**, 415–431.

Jones, R. H. (1974). Identification and autoregression spectrum estimation. *IEEE Trans. Automat. Contr.* **AC-19**, 894–898.

Jones, R. H. (1975). Fitting autoregression. *J. Amer. Statist. Assoc.* **70**, 590–592.

Jones, R. H. (1976). Autoregression order selection. *Geophys.* **41**, 771–773.

Kaveh, M. and Cooper, G. R. (1976). An empirical investigation of the properties of the autoregressive spectral estimator. *IEEE Trans. Inform. Theory* **IT-22**, 313–323.

Kay, S. M. (1978). Improvement of autoregressive spectral estimates in the presence of noise. *Rec. 1978 Int. Conf. Acoustics, Speech, and Signal Processing*, 357–360.

Kay, S. M. and Marple, S. L., Jr. (1979). Sources of and remedies for spectral line splitting in autoregressive spectrum analysis. *Rec. 1979 Int. Conf. Acoustics, Speech, and Signal Processing*, 151–154.

Kay, S. M. (1979). The effects of noise on the autoregressive spectral estimator. *IEEE Trans. Acoust., Speech, Signal Process.* **ASSP-27**, 478–485.

Kay, S. M. (1980). Noise compensation for autoregressive spectral estimates. *IEEE Trans. Acoust., Speech, Signal Process.* **ASSP-28**, 292–303.

Kay, S. M. and Marple, Stanley L., Jr. (1981). Spectrum analysis—A modern perspective. *Proceedings of the IEEE* **69**, 1380–1419.

Kromer, R. E. (1969). Asymptotic properties of the autoregressive spectral estimator. Ph.D. thesis, Statistics Department, Stanford University.

Lacoss, R. T. (1971). Data adaptive spectral analysis methods. *Geophysics* **36**, 661–675.

Landers, T. E. and Lacoss, R. T. (1977). Some geophysical applications of autoregressive spectral estimates. *IEEE Trans. Geosci. Electron.* **GE-15**, 26–32.

Levinson, N. (1947). The Wiener (root mean square) error criterion in filter design and prediction. *J. Math. Phys.* **25**, 261–278.

Mann, H. B. and Wald, A. (1943). On the statistical treatment of stochastic difference equations. *Econometrica* **11**, 173–220.

Marple, S. L., Jr. (1980). A new autoregressive spectrum analysis algorithm. *IEEE Trans. Acoust., Speech, Signal Process.* **ASSP-28**, 441–454.

McClave, J. (1975). Subset autoregression. *Technometrics* **17**, 213–220.

Morf, M., Vieira, A., Lee, D. T. L. and Kailath, T. (1978). Recursive multichannel maximum entropy spectral estimation. *IEEE Trans. on Geoscience Electronics* **GE-16**, 85–94.

Nevai, Paul G. (1979). An asymptotic formula for the derivatives of orthogonal polynomials. *SIAM J. Math. Anal.* **10**, 472–477.

Pagano, M. (1980). Some recent advances in autoregressive processes. In: D. R. Brillinger and G. C. Tiao, eds., *Directions in Time Series*, 280–302. Institute of Mathematical Statistics. (Comments by H. T. Davis.)

Parzen, E. (1964). An approach to empirical time series. *J. Res. Nat. Bur. Standards* **68B**, 937–951.

Parzen, E. (1967a). *Time Series Analysis Papers*. Holden-Day, San Francisco. Includes Parzen (1964).

Parzen, E. (1967b). The role of spectral analysis in time series analysis. *Review of the International Statistical Institute* **35**, 125–141.

Parzen, E. (1968). Statistical spectral analysis (single channel case) in 1968. *Proceedings of NATO Advance Study Institute on Signal Processing*. Enschede, The Netherlands.

Parzen, E. (1969). Multiple time series modelling. In: P. R. Krishnaiah, ed., *Multivariate Analysis II*, 389–409. Academic Press, New York.

Parzen, E. (1974). Some recent advances in time series modeling. *IEEE Trans. Automat. Contr.* **AC-19**, 723–730.

Parzen, E. (1977). Multiple time series: determining the order of approximating autoregressive schemes. In: P. Krishnaiah, ed., *Multivariate Analysis IV*, 283–295. North-Holland, Amsterdam.

Parzen, E. (1979a). Forecasting and whitening filter estimation. *TIMS Studies in the Management Sciences* **12**, 149–165.

Parzen, E. (1979b). Nonparametric statistical data modeling. *Journal of the American Statistical Assoc.* **74**, 105–131.

Parzen, E. (1980). Time series modeling, spectral analysis, and forecasting. In: D. R. Brillinger and G. C. Tiao, eds., *Directions in Time Series Analysis*. Institute of Mathematical Statistics.

Parzen, E. (1982). Time series model identification and prediction variance horizon. In: D. Findley, ed., *Applied Time Series Analysis II*, 415–447. Academic Press, New York.

Pinsker, M. (1963). *Information and Information Stability of Random Variables*. Holden-Day, San Francisco.

Priestley, M. B. (1981). *Spectral Analysis and Time Series*. Academic Press, London.

Rice, J. (1979). On the estimation of the parameters of a power spectrum. *J. Multivariate Analysis* **9**, 378–392.

Sakai, H. (1979). Statistical properties of AR spectral analysis. *IEEE Trans. Acoust., Speech, Signal Processing* **ASSP-27**, 402–409.

Schaerf, Mirella C. (1964) Estimation of the covariance and autoregressive structure of a stationary time series. Ph.D. thesis, Statistics Department, Stanford University.

Shibata, R. (1976). Selection of the order of an autoregressive model by Akaike's Information Criterion. *Biometrika* **63**, 117–126.

Shibata, R. (1980). Asymptotically efficient selection of the order of the model for estimating parameters of a linear process. *Ann. Statist.* **8**. 147–164.

Shibata, R. (1981). An optimal autoregression spectral estimate. *Annals of Statistics* **9**, 300–306.

Shore, J. E. (1981). Minimum cross-entropy spectral analysis. *IEEE Transactions on Acoustics, Speech, and Signal Processing* **ASSP-29**, 230–237.

Shore, J. E. and Johnson, R. W. (1980). Axiomatic derivation of the principle of maximum entropy and the principle of minimum cross-entropy. *IEEE Transactions of Information Theory* **IT-26**, 26–37.

Smylie, D. E., Clarke, G. K. C. and Ulrych, T. J. (1973). Analysis of irregularities in the earth's rotation. In: B. A. Bolt, ed., *Methods in Computational Physics*, Vol. 13, 391–430. Academic Press, New York.

Thomson, D. J. (1977). Spectrum estimation techniques for characterization and development of WT4 waveguide, Part I. *Bell Syst. Tech. J.* **56**, 1769–1815.

Tong, H. (1977). More on autoregressive model fitting with noisy data by Akaike's information criterion. *IEEE Trans. Inform. Theory* **1T-23**, 404–410.

Ulrych, T. J. and Bishop, T. N. (1975). Maximum entropy spectral analysis and autoregressive decomposition. *Rev. Geophysics Space Phys.* **13**, 183–200.

Ulrych, T. J. and Clayton, R. W. (1976). Time series modelling and maximum entropy. *Phys. Earth Planetary Interiors* **12**, 188–200.

VanDenBos, A. (1971). Alternative interpretation of maximum entropy spectral analysis. *IEEE Trans. Inform. Theory* **IT-17**, 493–494.

Walker, G. (1931). On periodicity in series of related terms. *Proc. Roy. Soc. London, Series A* **131**, 518–532.

Whittle, P. (1954). The statistical analysis of a seiche record. *J. Marine Res.* **13**, 76–100.

Whittle, P. (1963). On the fitting of multivariate autoregressions, and the approximate canonical factorization of a spectral density matrix. *Biometrika* **50**, 129–134.

Wiener, N. (1949). *The Extrapolation, Interpolation, and Smoothing of Stationary Time Series* Viley, New York. Includes Levinson (1946).

Wiggins, R. A. and Robinson, E. A. (1965). Recursive solution to the multichannel filtering problem. *Journal of Geophysical Research* **70**, 1885–1891.

Wold, H. (1938). *A Study in the Analysis of Stationary Time Series.* Almquist and Wiksell, Stockholm.

Yule, G. U. (1927). On a method of investigating periodicities in disturbed series, with special reference to Wolfer's sunspot numbers. *Philosophical Trans. Royal Soc. London, Series A* **226**, 267–298.

D. R. Brillinger and P. R. Krishnaiah, eds., *Handbook of Statistics, Vol. 3*
© Elsevier Science Publishers B.V. (1983) 249–273

12

Threshold Autoregression and Some Frequency-Domain Characteristics

J. Pemberton and H. Tong

1. Introduction

It may be said that the basic idea underlying the frequency-domain analysis of a linear system is the principle of superposition. Specifically, when *probed* by a linear combination of cosinusoids, a linear system responds with a linear combination of cosinusoids of the same frequencies. This property is both the strength and weakness of the assumption of linearity. The strength lies in the simplicity of its frequency-domain analysis, which may be accomplished either by the 'window' method (see, e.g., Jenkins and Watts, 1968) or through fitting a parametric *linear* time-series model (see, e.g., Akaike, 1974; Parzen, 1974). On the other hand, its weakness lies in its lack of structure, by which we mean that many frequency-domain phenomena frequently observed in science and engineering cannot be properly explained if linearity is assumed. Notable phenomena are *limit cycles* (i.e. sustained oscillation of the same frequency), *synchronization, subharmonics, higher harmonics, jump resonance, time irreversibility* and *amplitude-frequency dependency*. Many of these have a long history and have been associated with many eminent scientists and engineers (see, e.g., Minorsky, 1962).

Again, we may perform a frequency-domain analysis of a nonlinear system either by the 'window' method, relying principally on the theory of higher-order spectra (see, e.g., Brillinger, 1965) or through fitting an appropriate parametric *nonlinear* time-series model. We describe a method based on the latter approach and we supplement it by a diagnostic check based on the former approach, similar in spirit to Jones (1974).

Now, by an appropriate parametric nonlinear time-series model in the present context, we mean those models the structure of which is rich enough to capture the frequency-domain phenomena listed in the opening paragraph. It has been demonstrated by Tong (1978, 1980a,b), Tong and Lim (1980) and Pemberton and Tong (1980) that the new class of nonlinear time-series models first proposed by Tong (1977b) constitutes one such class of appropriate models.

2. Some motivation

The class of threshold autoregressive time-series models, TAR, that we are about to describe is based on two fundamental notions, namely the time *delay* and the *threshold* value in the state space. In this section we give some motivation for these.

In ecology, solar physics, control engineering, etc., difference-delay equations and differential-delay equations play an increasingly important role. For example, the now classic logistic-delay equation due originally to Hutchinson (1948) in the field of ecology

$$\frac{\mathrm{d}x(t)}{\mathrm{d}t} = x(t)(a - bx(t-T)) \tag{2.1}$$

is based explicitly on the notion of a time *delay* T which, in ecological terms, reflects the development time of the species. For some recent references in this field, see May (1980). Lim and Tong (1981) have discussed a statistical approach. Another example comes from solar physics. Recently, Yoshimura (1978) has developed a magnetohydrodynamic model for the sunspot activity.

$$\begin{aligned} \frac{\partial \psi}{\partial t} &= \left[\frac{1-\mu^2}{r^2}\frac{\partial^2}{\partial \mu^2} + \frac{\partial^2}{\partial r^2}\right]\psi + f_1(\Phi(t-T))R\Phi\,, \\ \frac{\partial \Phi}{\partial t} &= \left[\frac{1-\mu^2}{r^2}\frac{\partial^2}{\partial \mu^2} + \frac{\partial^2}{\partial r^2}\right]\Phi + f_2(\Phi(t-T))G\psi\,, \end{aligned} \tag{2.2}$$

where f_1 and f_2 are two 'low-pass' functions,

ψ = poloidal field ,
Φ = toroidal field ,
R = regeneration operator ,
G = generation operator ,
T = delay time ,

and μ and r are two of the three components of a spherical co-ordinate system. This equation may be viewed as a mathematical formulation of Babcock's model (see Tong, 1980a). Again, it is the delay parameter that holds the key to an understanding of the apparent cyclical phenomenon of the annual sunspot numbers.

In many natural phenomena, a qualitative change may take place as a result of an accumulation of many quantitative changes. The qualitative change often takes the form of a 'phase transition' when a certain critical value, i.e. a *threshold* value, is crossed. For example, it is known in animal population studies that some animal may change its reproductive behaviour when its population density exceeds a certain critical level. In other words, there is some self-regulation which perhaps ensures a near optimal exploitation of the natural resources (see, e.g., Solomon, 1969; Lim and Tong, 1981). Another example relates to Babcock's model mentioned earlier, according to which an eruption of the internal magnetic field takes place when a critical toroidal field strength is exceeded. In Yoshimura's mathematical formulation, this critical level is taken care of by the 'truncation' points of the low-pass functions f_1 and f_2. Whittle (1954) has discussed the following mathematical model in relation to the interesting nonlinear phenomenon of higher harmonics in his pioneering statistical study of the seich record:

$$L(x_t) = \begin{cases} 0 & \text{if } x_t \geq h\,, \\ K & \text{if } x_t < h\,, \end{cases} \tag{2.3}$$

where L is a linear differential operator, x_t is the water level, K and h are some constants. Here, h represents the critical water level. The physical idea is, of course, very closely related to an electrical relay system:

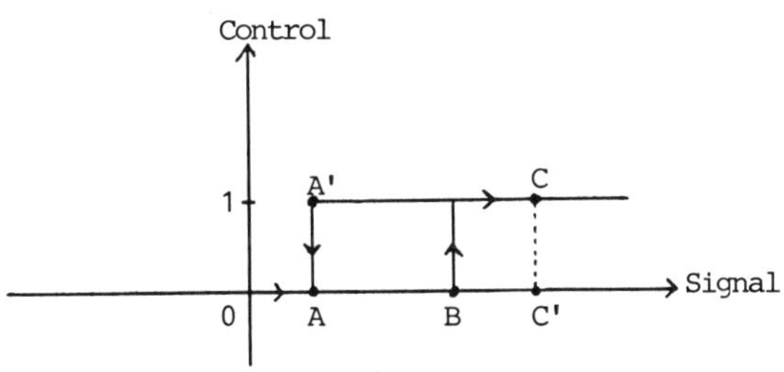

Fig. 2.1.

The system has two control actions which are 'on' and 'off'. When the signal is at the origin, the control action is in the 'off' mode and will remain there until the increasing signal reaches B, at which point the control action is switched to the 'on' mode. On the other hand, if the signal is decreased from the point C, the control action remains in the 'on' mode until the signal reaches the point A′, at which the control action switches to the 'off' mode. The points A and B are the threshold values. There are many other interesting examples of thresholds from diverse fields. They include models of a brain (Lindgren, 1981), neuron firing (Brillinger and Segundo, 1979), antigen-antibody dynamics (Waltman and Butz, 1977), and hydrology (Sugawara, 1962), etc. These references are directly relevant to the development of TAR models.

Leaving aside motivation from the natural science, we would just mention that a TAR model may also arise quite naturally from a Bayesian data-analytic point of view. Specifically, Tong (1981) has considered a simplest nonlinear time-series model for the time series $\{X_n: n = 0, \pm 1, \pm 2, \ldots\}$ in the form

$$E[X_n \mid X_{n-1} = x] = \mu(x)x\,, \tag{2.4}$$

where μ is a 'smooth' function. Suppose that we consider approximating the nonlinear model (2.4) by a linear model in the form

$$E[X_n \mid X_{n-1} = x] = \theta x\,, \tag{2.5}$$

where

$$\theta \sim N(C, V)\,. \tag{2.6}$$

Let L denote the loss function given by

$$L(\theta) = h[1 - \exp\{-\tfrac{1}{2}k^{-1}(\theta - \mu)^2\}]\,, \tag{2.7}$$

where μ is of course the most desirable value of θ (we have suppressed the argument x for brevity), k represents the relative tolerance to difference between μ and θ, and h quantifies the maximum loss. We may now decide whether (2.5) is an acceptable approximation of (2.4) or not by evaluating the expected loss of making the decision. Let D denote the decision space and let δ denote an element of D. It turns out that if the uncertainty, V, of belief of the value of δ is an increasing function of δ, the optimal decision, i.e. one which minimises the expected loss, is to adopt the *linear* approximation (2.5) for as long as $\mu - C$ is no greater than $\{(1+\gamma^2)^{1/2} - 1\}\gamma^{-1}$, where γ is equal to $\{k + V(0)\}^{-1/2}$. Repeating the same linear approximation process, we may conclude that the nonlinear model (2.4) is adequately approximated by a (usually) *small* number of locally linear models, the exact number being determined by the relative tolerance k, and the degree of uncertainty at zero action $V(0)$. It is noteworthy that this kind of discontinuous decision process is intimately related to catastrophe theory (see Smith, Harrison and Zeeman, 1981), thus vindicating the belief of a link between TAR modelling and catastrophe theory which was expressed in Tong and Lim (1980).

3. A simplest TAR model

TAR models in discrete time were first mentioned by Tong (1977b) and reported briefly in Tong (1978, 1980a). A comprehensive account is now available in Tong and Lim (1980). We start our discussion with a time series $\{X_t\colon t = 0, \pm 1, \pm 2, \ldots\}$ which is generated by the simplest form of TAR, the so-called *self-exciting threshold autoregressive model*, which consists of two piecewise linear autoregressive models each of first order, i.e. AR(1). We denote this by SETAR (2;1,1), the first numeral denoting the number of submodels and the numerals preceded by the semicolon denote the respective orders of the linear submodels. Specifically,

$$X_t = \begin{cases} a_1^{(1)} X_{t-1} + \eta_t & \text{if } X_{t-1} \leq r\,, \\ a_1^{(2)} X_{t-1} + \eta_t & \text{if } X_{t-1} > r\,, \end{cases} \tag{3.1}$$

where r is a real constant, called the *threshold* parameter, $\{\eta_t\}$ is a sequence of independent identically distributed random variables, and $a_1^{(1)}$ and $a_1^{(2)}$ are real constants. Trivially, model (3.1) includes the usual AR(1) as a special case by setting r equal to $-\infty$. We also note that $\{X_t\}$ as defined by (3.1) is clearly Markov. Appealing to some standard results on the ergodicity of Markov process, we may establish the following:

THEOREM 3.1. *Suppose that the distribution of η_t is absolutely continuous, with finite mean. A sufficient condition for the existence of an invariant measure for $\{X_t\}$ given by* (3.1) *is that* $|a_1^{(i)}| < 1$, $i = 1, 2$.

REMARK. Intuitively, the sufficient condition ensures a drift back to the 'centre'.

PROOF. Simply note that

$$E_x[|a_1^{(i)}X_{t-1} + \eta_t| - |X_{t-1}|] \leq (|a_1^{(i)}| - 1)|x| + E(|\eta_t|),$$

where E_x denotes that conditional expectation given that $X_{t-1} = x$. Here,

$$i = \begin{cases} 1 & \text{if } x \leq r, \\ 2 & \text{if } x > r. \end{cases}$$

Since $|a_1^{(i)}| < 1$, $i = 1, 2$, it holds that

$$E_x[|a_1^{(i)}X_{t-1} + \eta_t| - |X_{t-1}|] \leq |x| - 1,$$

for $x \notin K$, where

$$K = \{x: |x| \leq [1 + E(|\eta_t|)]/[2 - \max(|a_1^{(1)}|, |a_1^{(2)}|)]\}.$$

Also, the conditional expectation is finite $\forall x \in K$. Hence, using Theorem 3.1 of Tweedie (1975, p. 390), we have proved our theorem.

If we consider $\{X_t: t = 0, 1, 2, \ldots\}$, with X_0 having the distribution determined by the invariant measure, then $\{X_t: t = 0, 1, 2, \ldots\}$ is strictly stationary. We shall henceforth adopt this convention. We also note that Theorem 3.1 is easily generalised to the case of more than two linear submodels.

We may now rewrite (3.1) in the more conventional form

$$X_t = \begin{cases} a_0^{(1)} + a_1^{(1)}X_{t-1} + \varepsilon_t^{(1)} & \text{if } X_{t-1} \leq r, \\ a_0^{(2)} + a_1^{(2)}X_{t-1} + \varepsilon_t^{(2)} & \text{if } X_{t-1} > r, \end{cases} \tag{3.1'}$$

where $\{\varepsilon_t^{(i)}\}$ $i = 1, 2$ is a sequence of independent identically distributed random variables with zero mean. We have now allowed $\varepsilon_t^{(1)}$ and $\varepsilon_t^{(2)}$ to have different variances.

Jones (1978) has discussed methods of calculating stationary distributions and joint moments. For example, he has considered

$$X_t = \begin{cases} 0.5 + X_{t-1} + \varepsilon_t & \text{if } -2.0 < X_{t-1} \leq 1.5 , \\ -1.5 + \varepsilon_t & \text{otherwise} , \ \varepsilon_t \sim N(0, 1) , \end{cases} \tag{3.2}$$

and has calculated, among other things, the autocovariances of various lags which are shown in Table 3.1.

Table 3.1
Autocovariances of model (3.2)

Lag	Autocovariance	Lag	Autocovariance
0	2.21893	7	0.25×10^{-2}
1	0.58111	8	0.13×10^{-2}
2	0.04384	9	0.29×10^{-3}
3	−0.0720	10	-0.6×10^{-4}
4	−0.0468	11	-0.1×10^{-3}
5	−0.0135	12	-0.2×10^{-4}
6	0.48×10^{-3}		

The Fourier transform of these autocovariances up to lag 14, i.e. the spectral density function, is shown in Fig. 3.1. It has a maximum at a *nonzero* frequency, in contrast with the case of a first-order *linear* autoregressive model.

For this simple model, it is possible to use the standard 'cobweb' type argument to investigate stability of the systematic part (i.e. var ε_t set equal to zero). As we can see from Figs. 3.2, 3.3 and 3.4, which are reproduced from Lim (1981), there may be an array of possibilities, namely *divergence*, *stable limit points* and *stable periodic recursions*. The same 'cobweb' type argument may be applied to the case of more than two piecewise linear submodels, all of the first order.

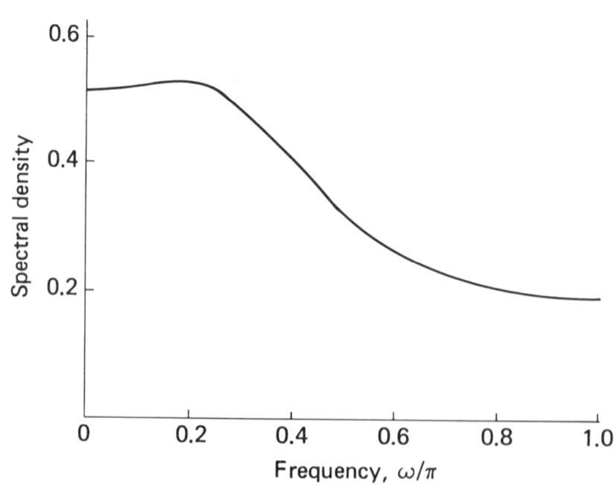

Fig. 3.1.

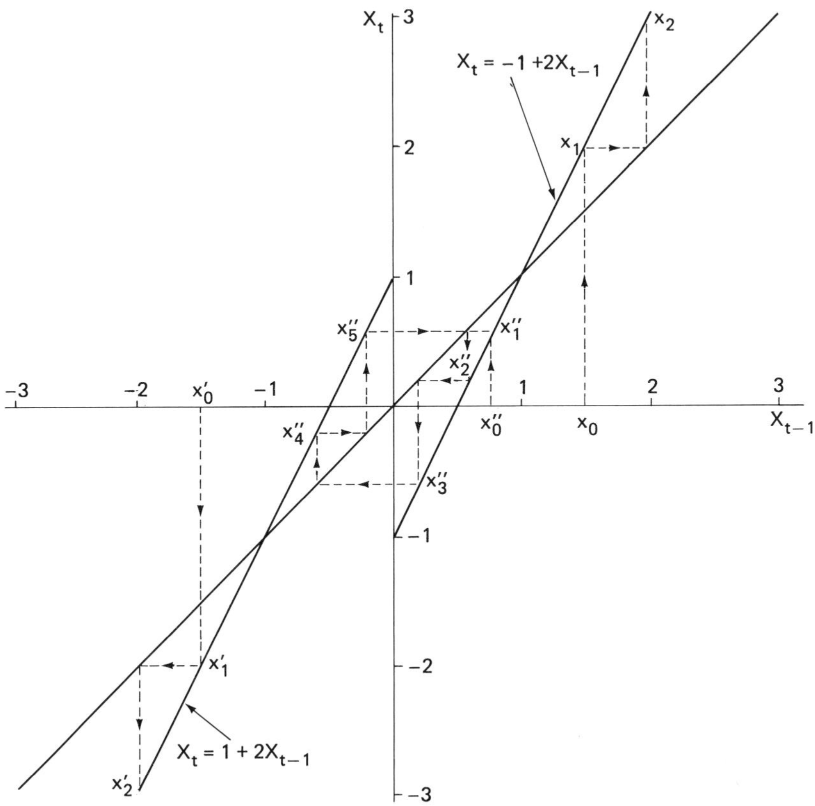

Fig. 3.2. Initial values less than 1 in modulus lead to the periodic recursion (0.6, 0.2, −0.6, −0.2). ±1 are stationary points. Initial values greater than 1 in modulus lead to divergence.

However, the case involving higher-order lags than the first is much more complex. The question of ergodicity, joint distributions and stability for non-linear time-series models involving higher-order lags seems to await further investigation in general.

4. Some generalisation and some frequency-domain characteristics

A SETAR (2;1,1) may be easily generalised to a SETAR $(l; k_1, \ldots, k_l)$ as follows:

Let

$$(-\infty, r) \cup [r_1, r_2) \cup \cdots \cup [r_{l-1}, \infty) \tag{4.1}$$

define a partition of the real line. Let

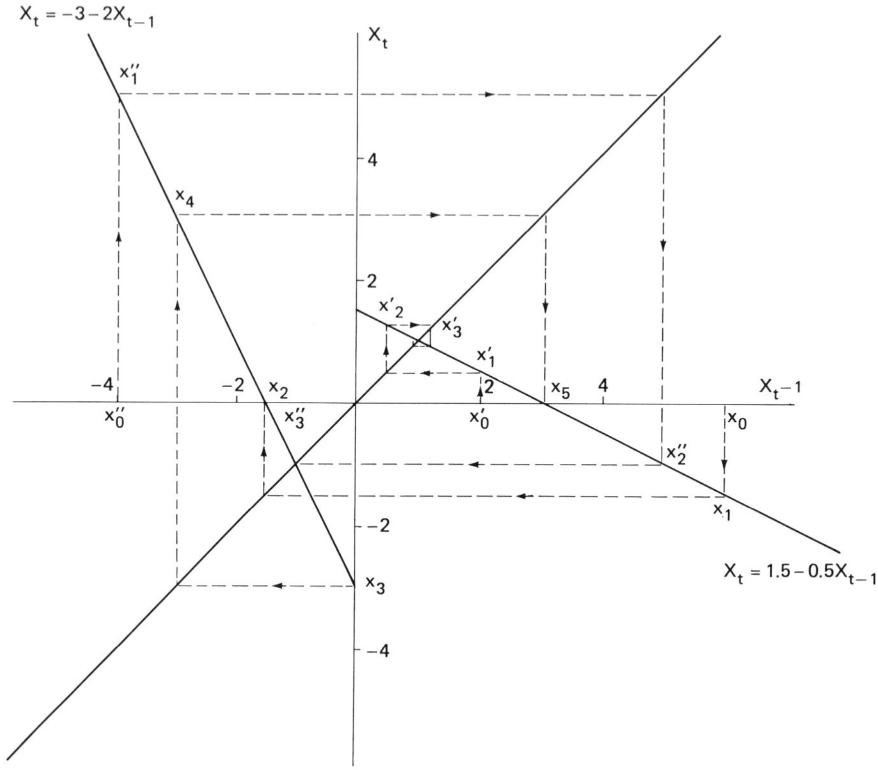

Fig. 3.3. Initial values of the form $3n$, ($n = 0, 1, \ldots$), lead to the periodic recursion $(-3, 3, 0)$. Other initial values lead to the limit points ± 1. Note that a cycle of period 3 does not necessarily imply a chaotic state for SETAR (cf. Li and Yorke, 1975).

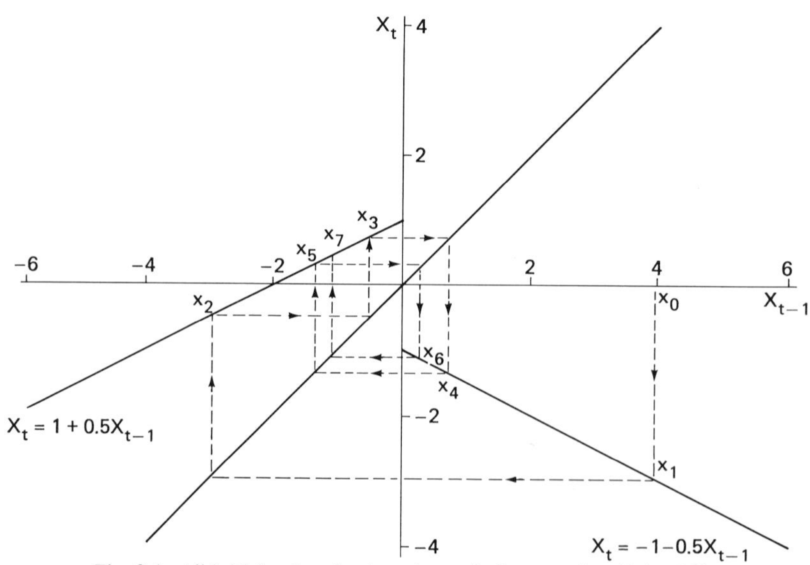

Fig. 3.4. All initial values lead to the periodic recursion $(0.4, -1.2)$.

$$X_t = a_0^{(j)} + \sum_{i=1}^{k_j} a_i^{(j)} X_{t-i} + \varepsilon_t^{(j)}, \tag{4.2}$$

conditional on $X_{t-d} \in [r_{j-1}, r_j)$, $j = 1, 2, \ldots, l$, where $r_0 = -\infty$ and $r_l = \infty$. The $\{\varepsilon_t^{(j)}\}$, $j = 1, \ldots, l$, are allowed to have different variances. $\{X_t\}$ is then said to follow a SETAR $(l; k_1, \ldots, k_l)$. Here, the conditioning variable is X_{t-d} which is a member of $\{X_t\}$ itself; this explains the term 'self-exciting'. There are $l-1$ threshold parameters, namely $r_1, \ldots, r_{l-1}$, and d is the usual delay parameter. On the other hand, if the conditioning variable is defined by an exogeneous variable, say Y_{t-d}, then we may have what is now termed an *open-loop threshold autoregressive system*, or TARSO in short. Specifically,

$$X_t = a_0^{(j)} + \sum_{i=1}^{m_j} a_i^{(j)} X_{t-i} + \sum_{i=0}^{m_j'} b_i^{(j)} Y_{t-i} + \varepsilon_n^{(j)} \tag{4.3}$$

conditional on $Y_{t-d} \in [r_{j-1}, r_j)$.

It is now clear that the main abstract idea behind the above models is the explicit stipulation of a conditioning variable, whose concrete expression may assume a variety of forms (see Tong and Lim, 1980, for further details). In the rest of this section, we allow ourselves the freedom of choice of the concrete expression. The choice is almost invariably suggested by analogy with nonlinear vibrations.

We now demonstrate that SETAR and TARSO models have very rich structure. Specifically, we exhibit their frequency-domain response to mainly deterministic excitations. The characteristics to be demonstrated are well known in nonlinear vibrations.

(i) *Jump resonance*

The output amplitude of a nonlinear system, unlike the linear case, may have a 'resonance jump' at different frequencies depending on whether the input frequency (of constant amplitude) is monotonically increasing or monotonically decreasing (see Fig. 4.1a–c).

The time plots of Figs. 4.2a and 4.2b clearly show that our SETAR can capture this engineering notion. The engineering terminology of a 'hard spring' and a 'soft spring' is an indication of the mode of the 'restoring force' of the system. Figs. 4.2a and 4.2b correspond respectively to the SETAR (2;9,3),

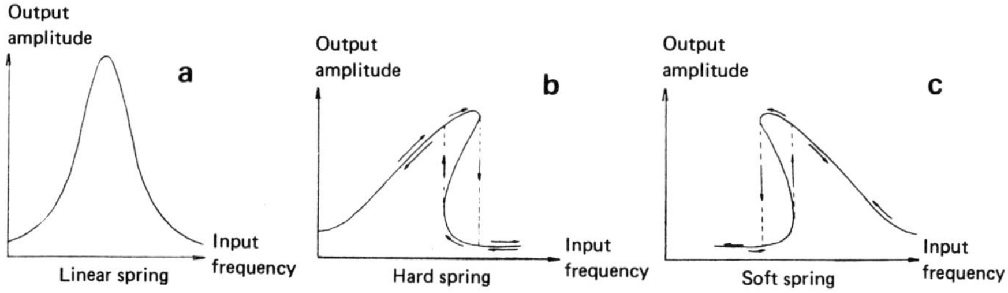

Fig. 4.1.

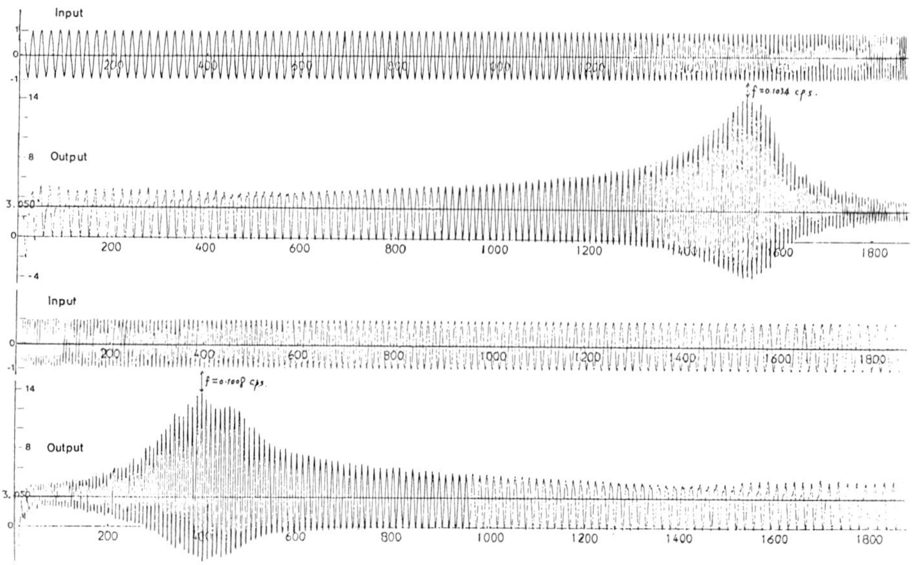

Fig. 42a. Jump pnenomenon of SETAR(2;9,3), $d = 5$ (hard spring type).

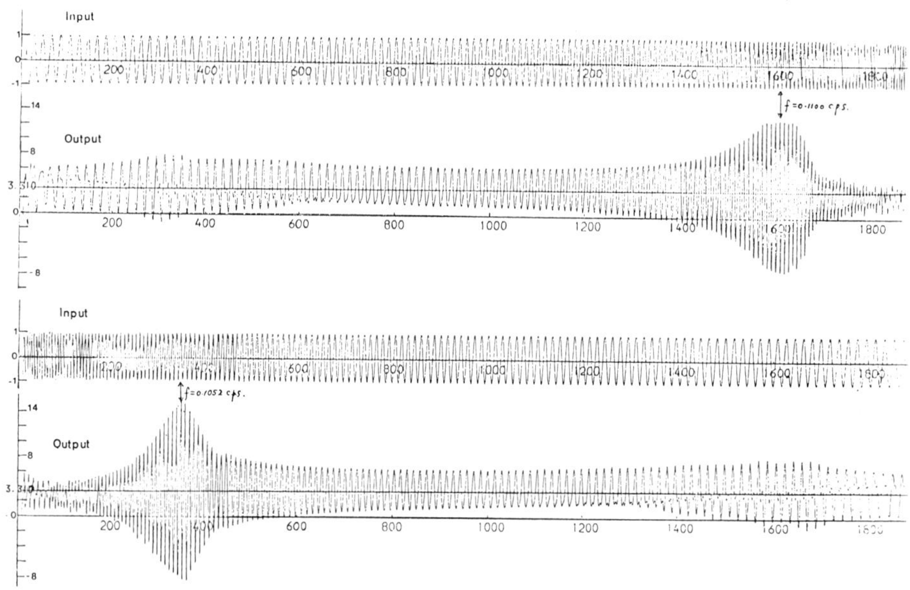

Fig. 42b. Jump phenomenon of SETAR(2;3,8), $d = 6$ (soft spring type).

$d = 5$, and SETAR (2;3,8), $d = 6$ given below. (White noise inputs are replaced by sinusoids in this exercise.)

$$X_t = \begin{cases} 0.4655 + 1.1448X_{t-1} - 0.4801X_{t-2} + 0.1273X_{t-3} \\ \qquad - 0.3580X_{t-4} + 0.2565X_{t-5} - 0.0781X_{t-6} - 0.0493X_{t-7} \\ \qquad + 0.2186X_{t-8} + 0.0526X_{t-9} + \text{input} \quad \text{if } X_{t-5} \leqslant 3.05\,, \\ 1.1940 + 1.1181X_{t-1} - 0.5017X_{t-2} - 0.0594X_{t-3} + \text{input} \\ \qquad\qquad\qquad\qquad\qquad\qquad \text{if } X_{n-5} > 3.05\,, \end{cases}$$

$$X_t = \begin{cases} 1.3003 + 1.3243X_{t-1} - 0.7023X_{t-2} - 0.0750X_{t-3} + \text{input} \\ \qquad\qquad\qquad\qquad\qquad\qquad \text{if } X_{t-6} \leq 3.31\,, \\ 0.2004 + 1.2112X_{t-1} - 0.6971X_{t-2} + 0.6191X_{t-3} \\ \quad - 1.0178X_{t-4} + 0.9967X_{t-5} - 0.7688X_{t-6} + 0.6119X_{t-7} \\ \quad - 0.0551X_{t-8} + \text{input} \qquad \text{if } X_{t-6} > 3.31\,. \end{cases}$$

The output amplitude of a nonlinear system may also have a resonance jump at different amplitudes depending on whether the input amplitude (of constant frequency) is monotonically increasing or monotonically decreasing. Fig. 4.3 corresponds to the time plots of the following threshold model:

$$X_t = \begin{cases} X_{t-1} + 2(Y_t - Y_{t-1}) & \text{if } |Y_{t-1} - Y_{y-2}| \geq 10\,, \\ X_{t-1} + 0.1(Y_t - Y_{t-1}) & \text{if } |Y_{t-1} - Y_{t-2}| < 10\,. \end{cases}$$

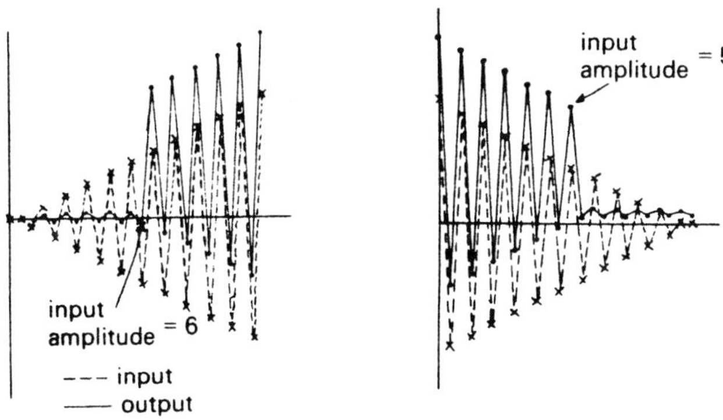

Fig. 4.3.

(ii) *Amplitude-frequency dependency*

It is well known that, unlike a linear system, the output signal may show different frequencies of oscillations for different amplitudes. The time plots of Figs. 4.4a and 4.4b correspond respectively to the two SETAR (2;3,3), $d = 1$, given below.

$$X_t = \begin{cases} 1.6734 - 0.8295X_{t-1} + 0.1309X_{t-2} - 0.0276X_{t-3} + \varepsilon_t \\ \qquad\qquad\qquad\qquad\qquad\qquad \text{if } X_{t-1} > 0.5\,, \\ 1.2270 + 1.0516X_{t-1} - 0.5901X_{t-2} - 0.2149X_{t-3} + \varepsilon_t \\ \qquad\qquad\qquad \text{if } X_{t-1} \leq 0.5, \text{ var } \varepsilon_t^{(i)} = 0.003^2,\ i = 1, 2\,, \end{cases}$$

$$X_t = \begin{cases} 0.15 + 0.85X_{t-1} + 0.22X_{t-2} - 0.70X_{t-3} + \varepsilon_t \qquad \text{if } X_{t-1} \leq 3.05\,, \\ 0.30 - 0.80X_{t-1} + 0.20X_{t-2} + 0.70X_{t-3} + \varepsilon_t \\ \qquad\qquad\qquad \text{if } X_{t-1} > 3.05, \text{ var } \varepsilon_t^{(i)} = 0.003^2,\ i = 1, 2\,. \end{cases}$$

Note that Fig. 4.4a shows the tendency of high frequency of oscillations when the amplitudes are high. Fig. 4.4b shows the reverse tendency. The size of the noise variance does not seem very critical.

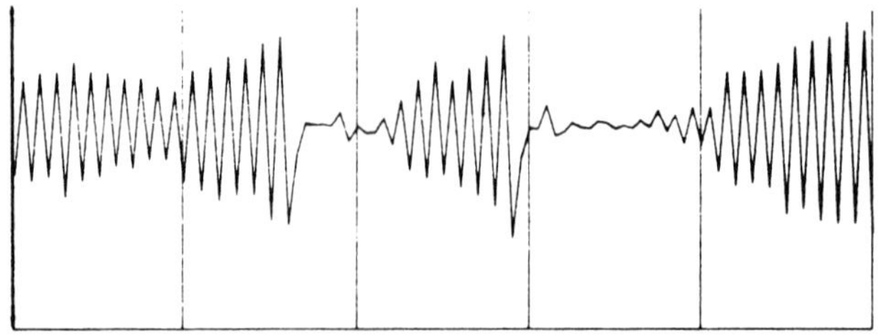

Fig. 4.4a. Amplitude–frequency dependency.

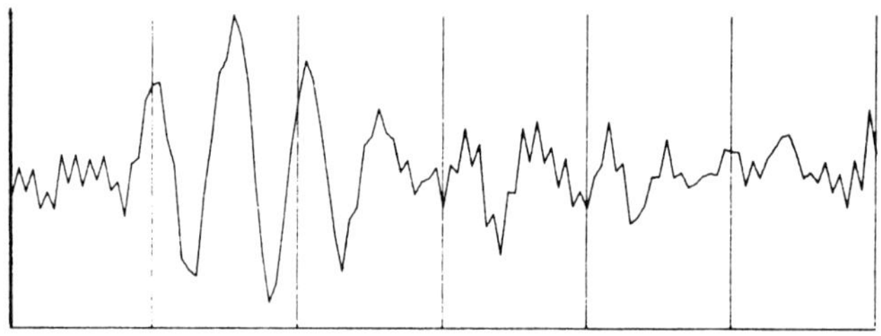

Fig. 4.4b. Amplitude-frequency dependency.

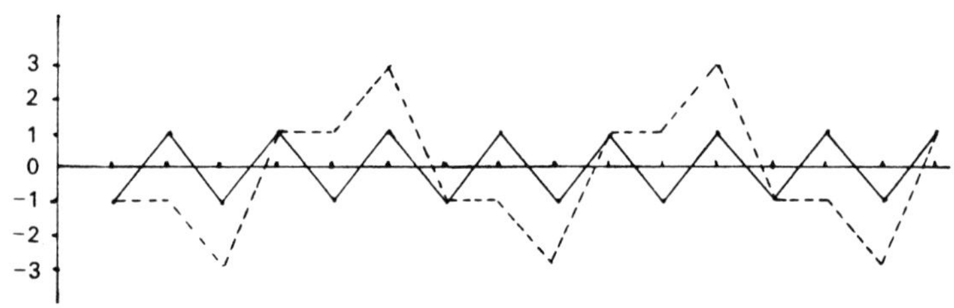

Fig. 4.5. (Input ——, Output ---).

(iii) *Subharmonics*

By a subharmonic it is usually meant an output oscillation at a fraction of the input oscillation frequency. The time plots of Fig. 4.5 correspond to the following simple SETAR (3;0,1,0) with a periodic input $\{Y_t\}$:

$$X_t = \begin{cases} 2X_{t-1} + Y_t & \text{if } |X_{t-1}| \leqslant 2\,, \\ Y_t & \text{if } |X_{t-1}| > 2\,, \end{cases}$$

where

$$Y_t = \begin{cases} -1 & \text{if } t \text{ is odd}\,, \\ 1 & \text{if } t \text{ is even}\,. \end{cases}$$

(iv) *Higher harmonics*

By a higher harmonic it is usually meant an output oscillation at a multiple of the input oscillation frequency. The time plots of Fig. 4.6 correspond to the following simple TAR model with a periodic input $\{Y_t\}$:

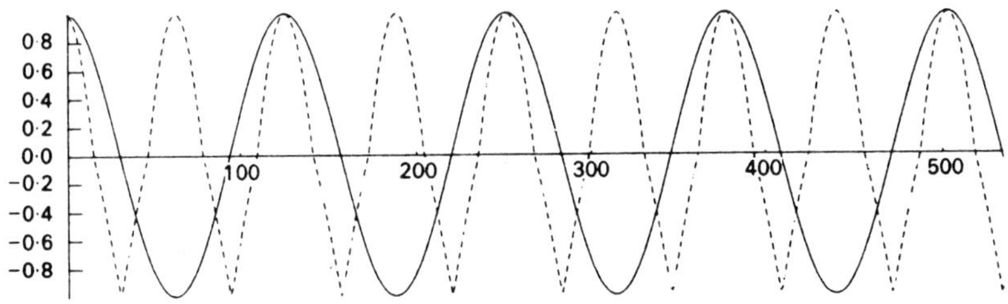

Fig. 4.6. (Input ——, Output ---).

$$X_t = \begin{cases} -(2+\sqrt{2})Y_t - (1+\sqrt{2}) & \text{if } -1 < Y_t \leq -1/\sqrt{2}\,, \\ -\sqrt{2}Y_{t-1} & \text{if } -1/\sqrt{2} < Y_t \leq 0\,, \\ \sqrt{2}Y_{t-1} & \text{if } 0 < Y_t \leq 1/\sqrt{2}\,, \\ (2+\sqrt{2})Y_t - (1+\sqrt{2}) & \text{if } 1/\sqrt{2} < Y_t \leq 1\,. \end{cases}$$

(v) *Limit cycle and strange attractor*

At present, this is the most important property of TAR models and we give a slightly more detailed discussion here.

By a limit cycle in nonlinear vibrations/nonlinear differential equations, it is usually meant sustained unforced oscillations under damping, the classic example being the thermionic value investigated mathematically by Van der Pol and Poincaré (see, e.g., Tong, 1980a). One of the earliest recognitions of its importance in frequency-domain analysis of time series seems to be the example due to Whittle described in Section 2, which anticipated the development of TAR models in a truly remarkable way.

In discrete time, we may formalise by considering a point transformation F from an interval J into itself. Let $F^0(x)$ denote x and $F^{n+1}(x)$ denote $F(F^n(x))$ for $n = 0, 1, \ldots$. Following Li and Yorke (1975), we say that x is a periodic point with period p if $x \in J$ and $x = F^p(x)$ and $x \neq F^k(x)$ for $1 \leq k < p$. In this case $\{F^n(x)\}$ is a periodic sequence. We say that $z \in J$ is *asymptotically periodic* if there is a periodic point x for which

$$F^n(z) - F^n(x) \to 0 \quad \text{as } n \to \infty\,.$$

If the same periodic sequence $\{F^n(x)\}$ is approached independently of the choice of z in some neighborhood, $\{F^n(x)\}$ may be called a discrete-time (stable) *limit cycle*, or, in more modern terminology, a *periodic attractor*. A periodic attractor of infinite period is called a *chaotic state* or a *strange attractor*. We now reproduce some of the examples in Lim (1981).

A limit cycle is demonstrated by the following point transformation

$$X_t = \begin{cases} 0.8023 + 1.0676X_{t-1} - 0.2099X_{t-2} + 0.1712X_{t-3} & \\ \qquad - 0.4528X_{t-4} + 0.2237X_{t-5} - 0.0331X_{t-6} & \text{if } X_{t-2} \leq 3.05\,, \\ 2.2964 + 1.4246X_{t-1} - 1.0795X_{t-2} - 0.0907X_{t-3} & \text{if } X_{t-2} > 3.05\,. \end{cases}$$

The phase planes shown in Figs. 4.7a and 4.7b correspond to two different initial

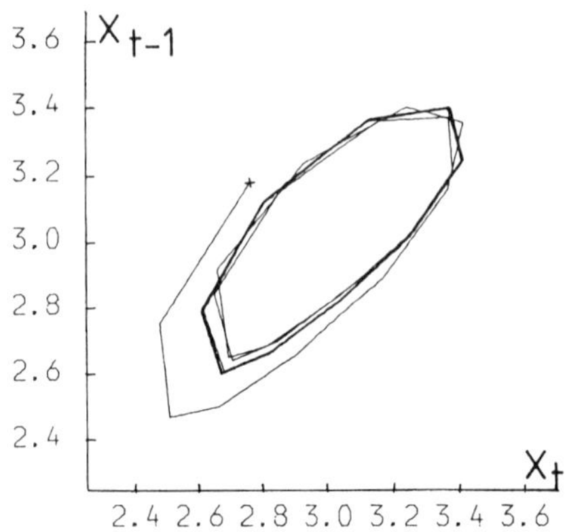

Fig. 4.7a.

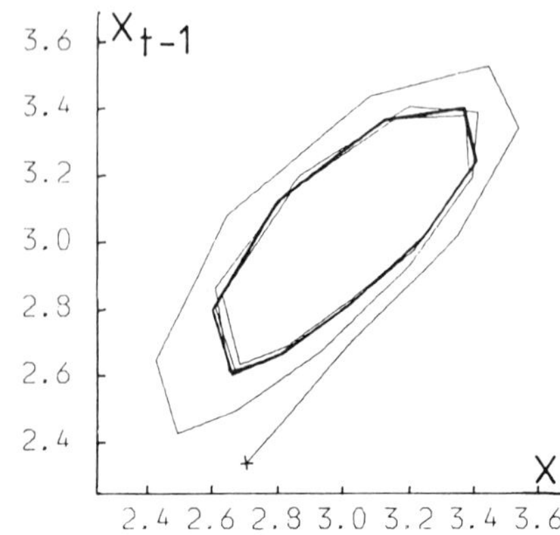

Fig. 4.7b.

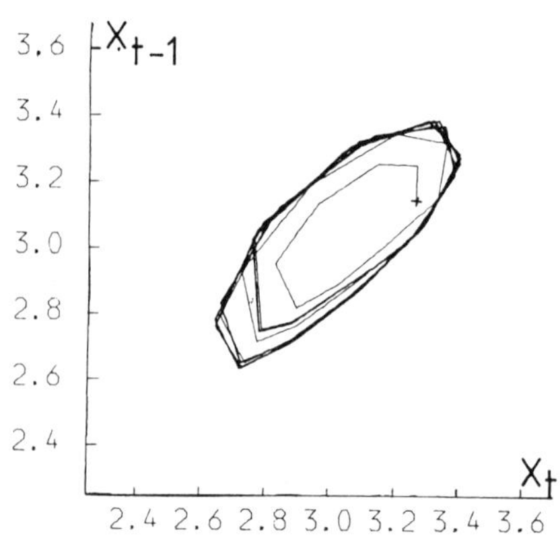

Fig. 4.7c.

values and the same limit cycle is obtained, which has period 9. Beside the aforementioned limit cycle, the experiment shows that the model admits another stable limit cycle (Fig. 4.7c), which has period 35, consisting of 4 subcycles.

Fig. 4.8 demonstrates an unstable limit cycle for the point transformation:

$$X_t = \begin{cases} -0.1331 + 1.2689X_{t-1} - 0.0102X_{t-2} - 0.3789X_{t-3} - 0.1534X_{t-4} \\ \qquad - 0.1313X_{t-5} + 0.1837X_{t-6} - 0.4308X_{t-7} + 0.6265X_{t-8} \\ \qquad - 0.0520X_{t-9} \qquad \text{if } X_{t-5} \leq 2.5563\,, \\ \\ 0.9019 + 1.2829X_{t-1} - 0.9523X_{t-2} + 0.6925X_{t-3} - 0.8224X_{t-4} \\ \qquad + 0.5990X_{t-5} - 0.3584X_{t-6} + 0.3072X_{t-7} - 0.4053X_{t-8} \\ \qquad + 0.5123X_{t-9} - 0.1880X_{t-1} \qquad \text{if } X_{t-5} > 2.5563\,. \end{cases}$$

The solid line shows that $|X_t|$ increases unboundedly with t and the dotted line shows that $X_t \to 2.81$ as t increases.

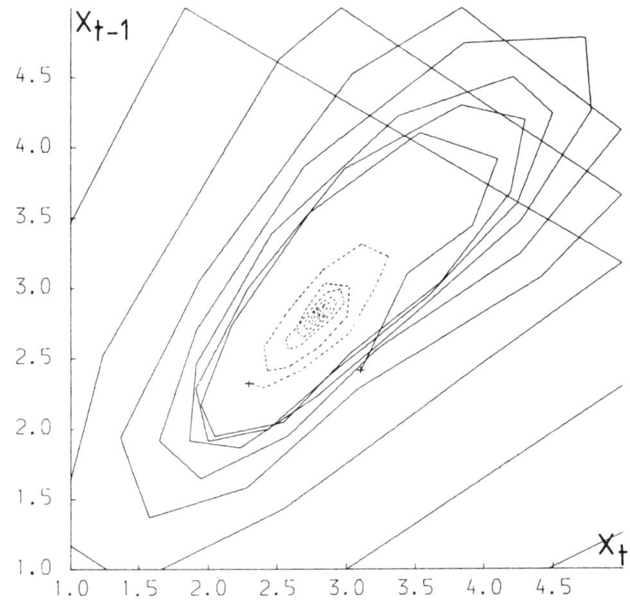

Fig. 4.8. Unstable limit cycle.

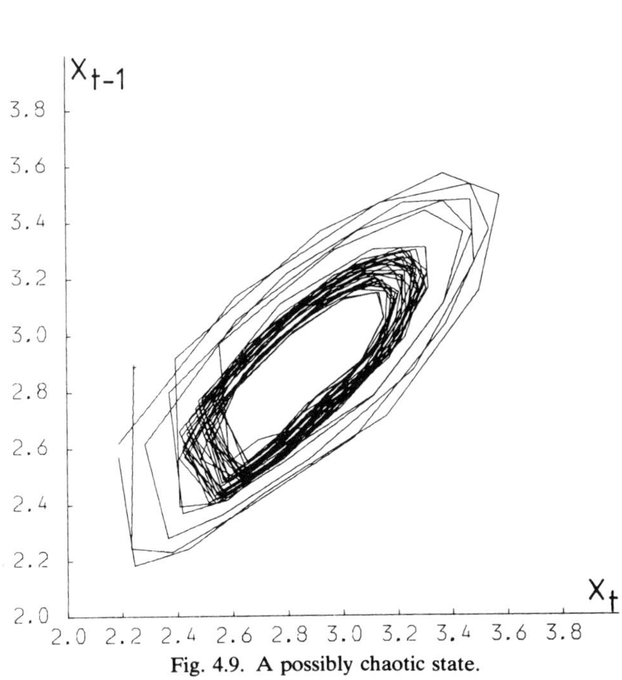

Fig. 4.9. A possibly chaotic state.

Fig. 4.9 illustrates a chaotic state obtained by the following point transformation:

$$X_t = \begin{cases} 0.5890 + 1.1114X_{t-1} - 0.1232X_{t-2} - 0.1430X_{t-3} \\ \qquad\qquad\qquad\qquad\qquad\qquad\qquad \text{if } X_{t-1} \leq 2.5563\,, \\ 0.9333 + 1.1918X_{t-1} - 0.7569X_{t-2} + 0.2723X_{t-3} - 0.3867X_{t-4} \\ \quad + 0.1679X_{t-5} - 0.0812X_{t-6} + 0.0728X_{t-7} - 0.0399X_{t-8} \\ \quad + 0.2149X_{t-9} + 0.0162X_{t-10} \qquad \text{if } X_{t-1} > 2.5563\,. \end{cases}$$

(vi) *Synchronization*

The phenomenon of *synchronization*, also known as *frequency entrainment*, was the first to be studied among many other nonlinear phenomena and was apparently observed for the first time by C. Huygens (1629–1695) during his experiments with clocks. (He was apparently the inventor of the pendulum clock.) He observed that two clocks which were slightly out of step when hung on a wall became in step when placed on a piece of soft wood. It has since been observed in electrical, mechanical, acoustical, electroacoustical, electronics and control systems. Names like Lord Rayleigh, J. H. Vincent, H. G. Moller, E. V. Appleton, van der Pol, A. Andronov and J. J. Stoker have been closely connected with it. In control systems, this phenomenon is usually associated with relays, i.e. *piecewise linear responses.* Currently, there also seems to be a considerable interest in this phenomenon in physiological systems (see, e.g., Hyndeman et al., 1971).

Consider a nonlinear system, say an electron tube, oscillating with a self-excited (i.e. a limit cycle) frequency ω_0, called the *autofrequency*. Suppose that it is then excited by an extraneous periodic oscillation of frequency ω, called the *heterofrequency*. 'Beats' of the two frequencies may be observed. The frequency of the beats decreases as ω approaches ω_0, but this happens only up to a certain value of the difference $|\omega - \omega_0|$ after which the beats disappear *suddenly* and the output oscillates with frequency ω. There is thus a nontrivial zone, $\{\omega\colon \omega_0 - \Delta' < \omega < \omega_0 + \Delta\}$, in which the autofrequency is 'entrained' by the heterofrequency (Fig. 4.10).

Intuitively, we may think of a nonlinear system as possessing a number of autofrequencies (or natural frequencies) whose values may be located by probing the system with some external excitation of various frequencies

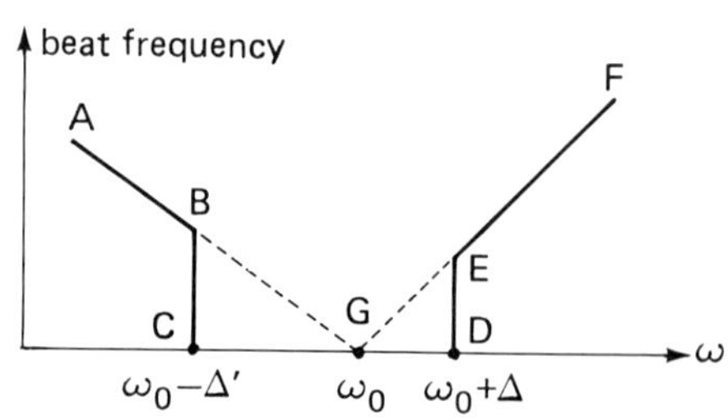

Fig. 4.10. Zone of entrainment (ABCDEF for non-linear case; ABGEF for linear case).

(heterofrequencies) until no beat is observed. Now, the classic set of annual Canadian lynx data (1821–1934) has been much analysed using mainly linear methodology (see, e.g., Tong, 1977a and the discussions therein). It is generally agreed that the data exhibit an approximate $9\frac{1}{2}$-year cycle. Comments have been made about this apparently peculiar timekeeping of the species (op. cit.). Now, the following is the systematic part of a SETAR (2;8,3) model fitted to the data by Tong and Lim (1980). (For a more thorough discussion, see Lim, 1981 and Lim and Tong, 1981.)

$$X_t = \begin{cases} 0.5239 + 1.0359X_{t-1} - 0.1756X_{t-2} + 0.1753X_{t-3} \\ \quad - 0.4339X_{t-4} + 0.3457X_{t-5} - 0.3032X_{t-6} \\ \quad + 0.2165X_{t-7} + 0.0043X_{t-8} \qquad \text{if } X_{t-2} \leqslant 3.1163\,, \\ 2.6559 + 1.4246X_{t-1} - 1.1618X_{t-2} - 0.1094X_{t-3} \\ \qquad\qquad\qquad\qquad \text{if } X_{t-2} > 3.1163\,. \end{cases} \tag{4.4}$$

Driving this system with periodic signals of period 7, 8, 9, 10 and 11 in succession reveals that beats occur except when the periods are 9 and 10. By

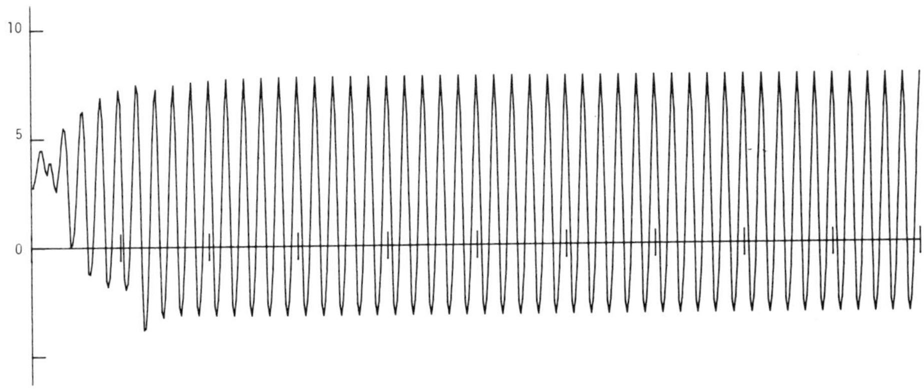

Fig. 4.11. Input signal is periodic with period 10 units of time. Output signal is shown and no beat is observable. Similar output signal is obtained when input signal has period 9 units of time.

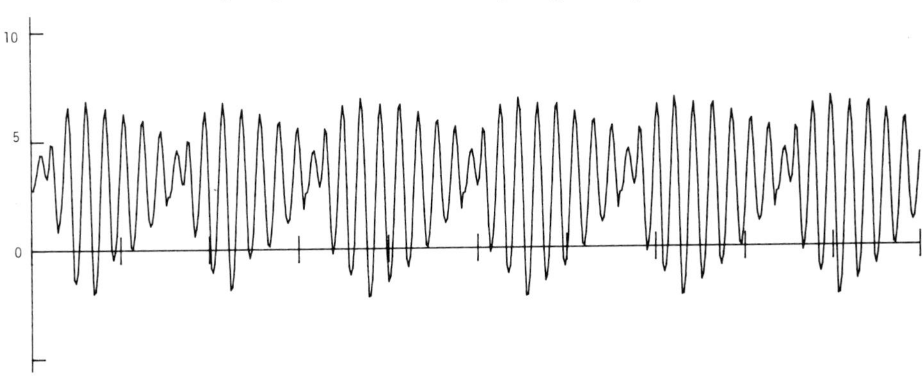

Fig. 4.12. Input signal is periodic with same amplitude as for Fig. 4.11 but with period 11 units of time. Output signal is shown and beats are clearly observable. Beats are also observed when input signal has period 7 or 8 units of time.

adopting the aforementioned interpretation, it seems plausible that the inbuilt regulating mechanism of the lynx population is such that it does not give rise to a unique periodicity but rather it may well 'switch' between two adjacent periodicities, namely 9 and 10 (see, e.g., Figs. 4.11 and 4.12).

5. Diagnostic checks of SETAR models from frequency-domain viewpoint

Earlier we reported some threshold models, the identification of which was fully described in Tong and Lim (1980). Our method is basically one of extending the commonly used least-squares approach to different subsamples defined by the thresholds. Sampling properties are discussed in Lim (1981). We omit the details here.

We now propose to study the appropriateness of the fitted TAR models through their frequency-domain behaviour. Specifically, we compare the second- and third-order properties of the fitted models with those of the data.

We illustrate our approach with the Canadian lynx data (1821–1934), logarithmically transformed.

First, Fig. 5.1 shows the estimated spectral density functions (s.d.f.) reported by Tong (1977a). They were obtained by the usual window method and through a linear AR(11) model. Now, a theoretical study of the s.d.f. of a SETAR (2;1,1) has been completed by Jones (1978), from which it is clear that the theoretical expressions are usually too unwieldly for practical use. We therefore resort to the simulation method. We generally generate artificial data with

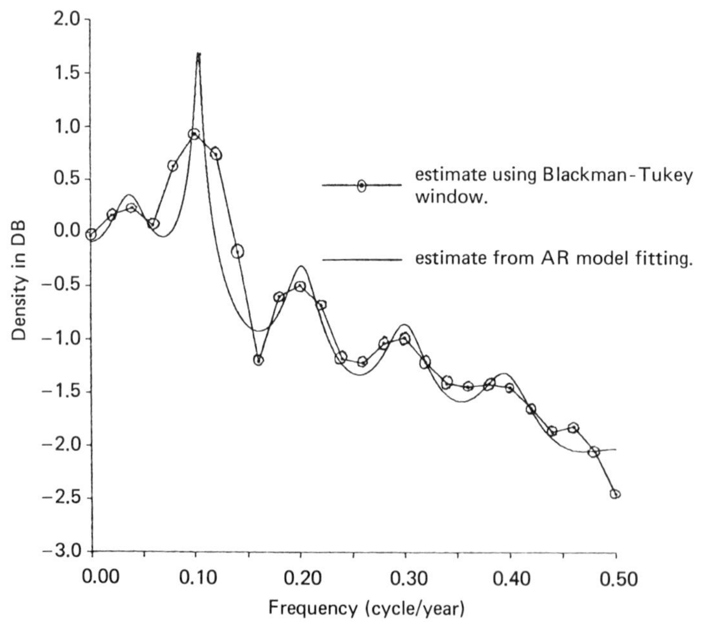

Fig. 5.1. Spectral density functions.

the fitted SETAR model using Gaussian random numbers as the input. The first one thousand data are generally discarded. Fig. 5.2 shows the s.d.f. of the SETAR (2;8,3) model (eq. (4.4) with var $\varepsilon_t^{(1)} = 0.0255$, var $\varepsilon_t^{(2)} = 0.0516$) obtained in this way. A Parzen window with a lag parameter of 200 is applied to a sample of 10,000 data. The dominant peak at approximately 1 cycle per 9.5 years is clearly visible. Its harmonic at approximately 2 cycles per 9.5 years is also visible. The general agreement with the estimates shown in Fig. 5.1 seems good. Fig. 5.3 shows that the autocovariance functions of the fitted SETAR model agree well with the observed data up to lag 20 and then damp out at a faster rate thereafter.

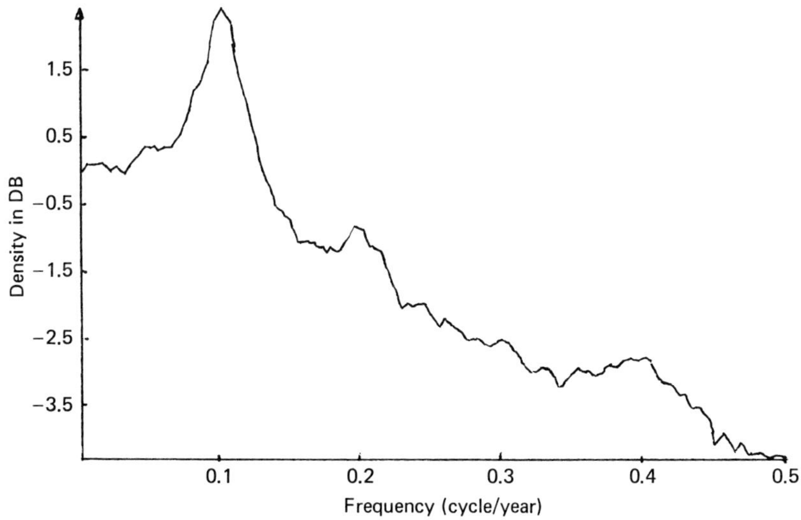

Fig. 5.2.

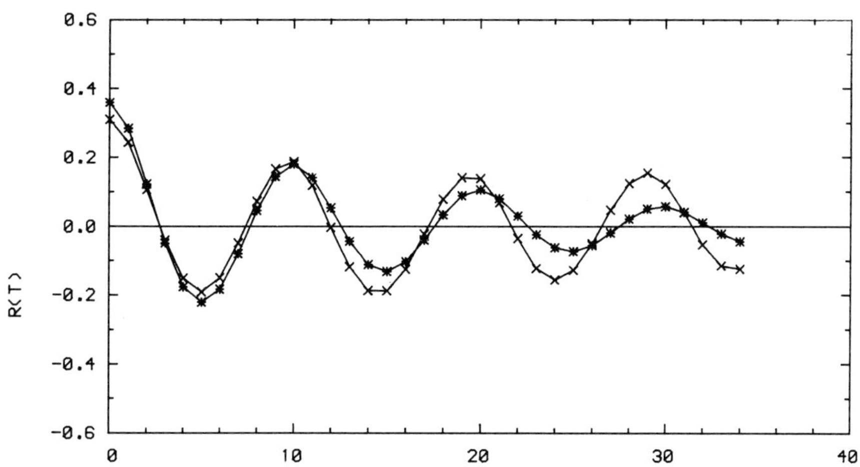

Fig. 5.3. Autocovariance functions: × = data; * = SETAR.

Next, we turn to the bispectrum. Our main reference for its computation and interpretation is Brillinger and Rosenblatt (1967). Immediate details are available in Subba Rao (Chapter 14), and the necessary computer programs are kindly made available by Mr. M. M. Gabr of UMIST. Figs. 5.4 and 5.5 show the modulus of the bispectrum estimate for the (log) lynx observations and the bispectrum estimate of the fitted SETAR model, respectively. In each case, a product of the Parzen window with lag parameter 20 is used. The agreement seems very reasonable although a larger lag parameter for the latter is probably more appropriate, but will involve a much longer computation. (The data have been normalised to have unit third central moment.) It is known that bispectral analysis is useful for the study of nonlinearity and non-Gaussianity, although the result of Pemberton and Tong (1981) shows that some care is needed when using it for the former. Now, one important symptom of nonlinearity and non-Gaussianity is time irreversibility. As has been discussed by Brillinger and Rosenblatt (1967), the argument of the bispectrum is useful in this respect. Specifically, a strictly stationary time series is time reversible (i.e. the probability structure of X_t is the same as that of X_{-t}) if and only if the imaginary parts of all the higher-order spectra (i.e. bispectra, trispectra, etc.) are identically zero. Now, Table 5.1 seems to support the general belief that the (log) lynx data are time irreversible and Table 5.2 shows that the fitted SETAR model has captured this reasonably well. Of particular note is the obvious discontinuity between positive and negative values. It seems quite instructive to

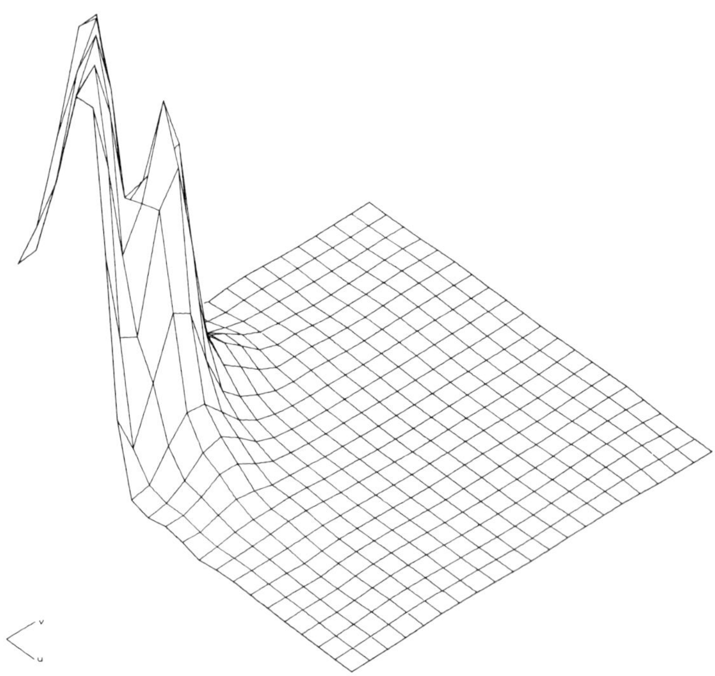

Fig. 5.4. Bispectral density function estimates (log lynx data)—modulus.

Table 5.1
Estimated bispectral argument (log lynx data)

↑ ω_2														
1.00	−0.00													
0.95	−0.00	−0.62	−1.28											
0.90	0.00	0.33	−1.08	−1.63	−1.96									
0.85	−0.00	1.22	0.37	−1.43	−1.95	−2.37	−2.88							
0.80	−0.00	1.27	0.98	−0.91	−1.78	−2.35	−2.89	2.80	2.26					
0.75	−0.00	1.15	1.29	−1.05	−1.65	−2.19	−2.86	2.78	2.22	1.71	0.97			
0.70	−0.00	1.02	1.59	2.83	−2.12	−2.20	−2.78	2.74	2.15	1.58	0.85	0.52	0.58	
0.65	−0.00	0.84	1.66	2.49	−3.09	−2.73	−2.92	2.67	2.10	1.61	0.92	0.50	0.58	0.58
0.60	−0.00	0.72	1.49	2.26	2.79	3.07	3.07	2.60	2.05	1.74	1.52	1.30	0.73	
0.55	−3.14	1.22	1.50	1.77	2.30	2.77	2.88	2.60	2.02	1.71	1.70	1.99		
0.50	−3.14	2.50	1.83	1.67	1.67	2.05	2.56	2.54	2.07	1.62	1.55			
0.45	−3.14	2.87	2.09	1.80	1.69	1.68	1.97	2.31	2.19	1.81				
0.40	−3.14	2.71	2.06	1.82	1.75	1.75	1.84	2.11	2.24					
0.35	−3.14	1.78	1.65	1.62	1.62	1.68	1.82	1.99						
0.30	−3.14	2.38	1.60	1.54	1.50	1.50	1.63							
0.25	−3.14	2.91	1.96	1.60	1.53	1.48								
0.20	−3.14	2.96	2.57	1.83	1.60									
0.15	3.14	2.91	2.77	2.42										
0.10	3.14	2.89	2.79											
0.05	3.14	2.99												
0.00	0.00													
ω_1 (in units of π)	0.00	0.05	0.10	0.15	0.20	0.25	0.30	0.35	0.40	0.45	0.50	0.60	0.55	0.65

Table 5.2
Estimated bispectral argument (threshold model)

ω_2 ↓														
1.00	0.00													
0.95	−0.00	−0.39	−0.60											
0.90	0.00	−0.43	−0.73	−1.08	−1.61									
0.85	−0.00	−0.53	−0.88	−1.22	−1.69	−2.32	−3.05							
0.80	−0.00	−0.72	−1.18	−1.51	−1.90	−2.42	−3.07	2.63	2.20					
0.75	−0.00	−0.98	−1.56	−1.85	−2.13	−2.54	−3.09	2.62	2.18	1.80	1.31			
0.70	−0.00	−1.20	−1.84	−2.18	−2.35	−2.64	−3.10	2.62	2.14	1.75	1.31	0.78	0.49	
0.65	−0.00	−1.42	−2.30	−2.52	−2.58	−2.76	−3.11	2.64	2.11	1.70	1.33	0.91	0.58	0.42
0.60	−3.14	2.28	2.75	−2.03	−2.86	−2.92	3.12	2.69	2.14	1.68	1.35	1.10	0.81	
0.55	−3.14	2.05	1.82	2.15	2.86	3.09	2.99	2.69	2.23	1.71	1.36	1.22		
0.50	−3.14	2.10	1.69	1.64	1.83	2.48	2.68	2.58	2.31	1.88	1.49			
0.45	−3.14	2.16	1.67	1.51	1.56	1.84	2.22	2.36	2.32	2.15				
-0.40	−3.14	2.01	1.58	1.40	1.38	1.51	1.81	2.11	2.26					
0.35	−3.14	1.80	1.43	1.27	1.20	1.25	1.47	1.81						
0.30	−3.14	2.04	1.46	1.25	1.15	1.10	1.19							
0.25	−3.14	2.50	1.72	1.36	1.21	1.12								
0.20	−3.14	2.77	2.17	1.60	1.33									
0.15	3.14	2.88	2.55	2.07										
0.10	3.14	2.94	2.77											
0.05	3.14	3.02												
0.00	0.00													
ω_1	0.00	0.05	0.10	0.15	0.20	0.25	0.30	0.35	0.40	0.45	0.50	0.55	0.60	0.65

(in units of π)

Table 5.3
Estimated bispectral argument of a deterministic signal. Only signs are recorded

ω_2	0.05													0.70
0.95	1	−1												
0.90	1	1	−1	−1										
0.85	1	1	1	−1	−1	−1								
0.80	1	1	1	1	−1	−1	−1	1						
0.75	1	1	1	1	1	−1	−1	−1	1	1				
0.70	1	1	1	1	1	1	−1	−1	1	1	1	1	1	1
0.65	1	1	1	1	1	1	1	−1	1	1	1	1	1	
0.60	1	1	1	1	1	1	1	1	1	1	1	1		
0.55	1	1	1	1	1	1	1	1	1	1	1			
0.50	1	1	1	1	1	1	1	1	1	1				
0.45	1	1	1	1	1	1	1	1	1					
0.40	1	1	1	1	1	1	1	1						
0.35	1	1	1	1	1	1	1							
0.30	1	1	1	1	1	1								
0.25	1	1	1	1	1									
0.20	1	1	1	1										
0.15	1	1	1											
0.10	1	1												
0.05	1													

ω_1 (in units of π)

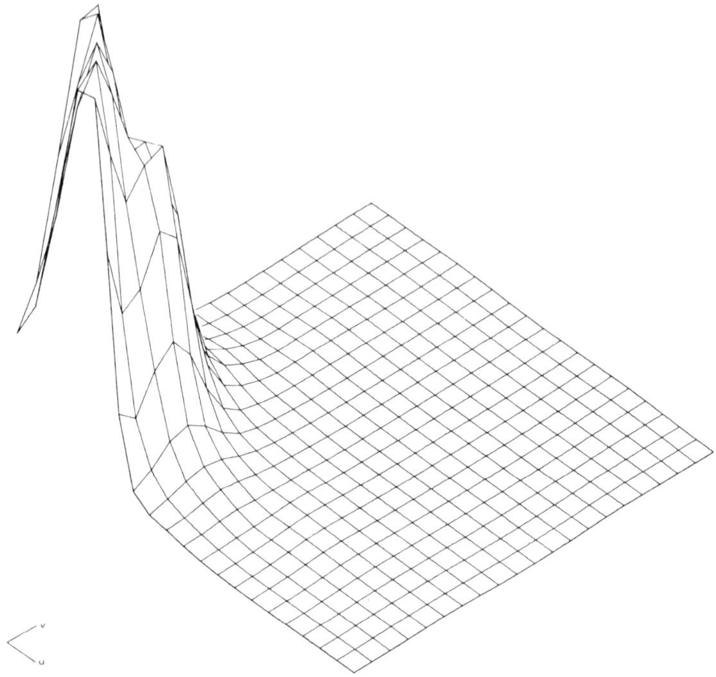

Fig. 5.5. Bispectral density function estimate through SETAR (2;8,3)—modulus.

compare a similar pattern observed (Table 5.3) for the *amplitude modulated* deterministic sequence which consists of repetitions of the basic sequence $\{1, -5, 5, 5, -6\}$. A more systematic study of this type of pattern recognition may be quite useful.

Acknowledgement

We are most grateful to Mr. M. M. Gabr for his generosity in allowing us access to his computer package for the estimation of bispectrum.

References

Akaike, H. (1974). A new look at the statistical model identification. *I.E.E.E.* **AC-19**, 716–723.

Brillinger, D. R. (1965). An introduction to polyspectrum. *Ann. Math. Statist.* **36**, 1351–1374.

Brillinger, D. R. and Rosenblatt, M. (1967). Computation and interpretation of kth order spectra. In: B. Harris, ed., *Spectral Analysis of Time Series*, 153–188. Wiley, New York.

Brillinger, D. R. and Segundo, J. P. (1979). Empirical examination of the threshold model of neuron firing. *Biol. Cybern.* **35**, 213–220.

Doob, J. L. (1953). *Stochastic Processes.* Wiley, New York.

Hutchinson, G. E. (1948). Circular causal system in ecology. *Annals of New York Academy of Sciences* **50**, 221–246.

Hyndeman, B. W., Kitney, R. I. and Sayers, B. McA. (1971). Spontaneous rhythms in physiological control systems. *Nature* **233**, 339–341.

Jenkins, G. M. and Watt, D. G. (1968). *Spectral Analysis and Its Applications*. Holden-Day, San Francisco.

Jones, D. A. (1978). Non-linear autoregressive processes. *Proc. Roy. Soc. London* **A360**, 71–95.

Jones, R. H. (1974). Identification and autoregressive spectrum estimation. *I.E.E.E.* **AC-19**, 894–898.

Li, T-Y. and Yorke, J. A. (1975). Period three implies chaos. *Amer. Math. Monthly* **82**, 988–992.

Lim, K. S. (1981). On threshold time series modelling. Unpublished Ph.D. thesis, University of Manchester, U.K.

Lim, K. S. and Tong, H. (1981). A statistical approach to difference-delay equation modelling in ecology—two case studies. Technical Report No. 146, Dept. of Mathematics, UMIST.

Lindgren, G. (1981). Contribution to the discussion of a paper by D. R. Cox. *Scan. Journ. of Statist.* **8**, 21.

May, R. M. (1980). Non-linear phenomena in ecology and epidemiology. *Annals of New York Academy of Sc.* **357**, 267–281.

Minorsky, N. (1962). *Non-linear Oscillations*. Van-Nostrand, New York.

Parzen, E. (1974). Some recent advances in time series modelling. *I.E.E.E.* **AC-19**, 723–730.

Pemberton, J. and Tong, H. (1980). Threshold autoregression and synchronization. A simulation study. Technical Report No. 133, Dept. of Mathematics, UMIST.

Pemberton, J. and Tong, H. (1981). A note on the distributions of non-linear autoregressive stochastic models. *Journal of Time Series Analysis* **2**, 49–52.

Smith, J. Q., Harrison, P. J. and Zeeman, E. C. (1981). The analysis of some discontinuous decision processes. *Europ. J. of Oper. Res.* **I**, 30–43.

Solomon, M. E. (1969). *Population Dynamics*. E. Arnold, London.

Sugawara, M. (1962). On the analysis of run-off structure about serveral Japanese rivers. *Jap. J. Geophy.* **2**, 1–76.

Tong, H. (1977a). Some comments on the Canadian lynx data. *J. R. Statist. Soc.* **A140**, 432–435, 448–468.

Tong, H. (1977b). Discussion of a paper by A. J. Lawrance and N. T. Kottegoda. *J. R. Statist. Soc.* **A140**, 34–35.

Tong, H. (1978). On a threshold model. In: C. H. Chen, ed., *Pattern Recognition and Signal Processing*. Sijthoff and Noordhof, The Netherlands.

Tong, H. (1980a). A view on non-linear time series model building. In: O. D. Anderson, ed., *Time Series*. North-Holland, Amsterdam.

Tong, H. (1980b). On the structure of threshold time series models. Technical Report No. 133, Dept. of Mathematics, UMIST.

Tong, H. (1981). Discontinuous decision processes and threshold autoregressive time series modelling. *Biometrika* **69**, 274–276.

Tong, H. and Lim, K. S. (1980). Threshold autoregression, limit cycles and cyclical data—with discussion. *J. R. Statist. Soc.* **B42**, 245–292.

Tweedie, R. L. (1975). Sufficient conditions for ergodicity and recurrence of Markov chains on a general state space. *Stoch. Proc. and Their Appl.* **3**, 383–403.

Waltman, P. and Butz, E. (1977). A threshold model of antigen-antibody dynamics. *J. Theor. Biol.* **65**, 499–512.

Whittle, P. (1954). The statistical analysis of a seiche record. *Sears Foundation Jour. of Marine Research* **13**, 76–100.

Yoshimura, H. (1978). Non-linear astrophysical dynamos: multiple-period dynamo wave oscillations and long-term modulations of the 22 year solar cycle. *Astrophy. J.* **226**, 706–719.

D. R. Brillinger and P. R. Krishnaiah, eds., *Handbook of Statistics, Vol. 3*
© Elsevier Science Publishers B.V. (1983) 275–291

13

The Frequency-Domain Approach to the Analysis of Closed-Loop Systems

M. B. Priestley

1. Introduction

The frequency-domain analysis of linear systems is the natural engineering approach to the study of such systems. In its simplest form it is based on 'probing' the system with sine wave inputs of different frequencies and measuring the amplitudes and phases of the corresponding outputs. This technique provides estimates of the system's transfer function at a number of 'spot' frequencies, and later developments, based on stochastic inputs and cross-spectral analysis, may be regarded simply as more sophisticated statistical versions of the sine wave input method. However, the frequency-domain analysis was developed primarily for 'open-loop' systems, i.e. where there is no feedback loop reconnecting the output to the input. The case of 'closed-loop' systems (where a feedback loop is present) can be treated also by frequency-domain methods, but here the estimation of the system's transfer function raises severe statistical problems. One of the first papers to draw attention to this feature is that of Akaike (1967), who pointed out that, in the case of a closed-loop system with a linear feedback controller, there are *two* transfer functions involved in the relationship between input and output (namely, that of the system and that of the feedback loop), and consequently there is an inherent problem of '*nonidentifiability*'. Akaike later suggested that the problem of closed-loop systems is probably best dealt with via a time-domain analysis in which the system and controller are modelled by fitting a (joint) ARMA model to the input and output (see, for example, Akaike, 1968, 1976). There seems little doubt that this type of time-domain analysis provides a more statisfactory approach to the study of closed-loop systems—particularly in view of recent refinements in multivariate ARMA model fitting techniques. Nevertheless, the frequency-domain analysis is not without interest, and it illuminates certain features of closed-loop systems which lie 'beneath the surface' in the time-domain analysis. Moreover, being essentially nonparametric in character, it can be used as a form of supplementary analysis to that of time-domain model fitting, and provides a useful check on the adequacy of the fitted ARMA model.

In this paper we consider the case of discrete time single input/single output closed-loop systems, and investigate the behaviour of an iterative scheme for estimating both the system's transfer function and the structure of the noise disturbance present in the output. It may be noted that there is now an extensive literature on the subject of closed-loop systems. In addition to the papers of Akaike cited above, we would refer to the contributions by Box and MacGregor (1974), Caines and Chan (1975), Chatfield (1975, Chapter 9), Diprose (1968, 1978), Gustavson, Ljung and Soderstrom (1976), Harris (1976), Priestley (1969, 1981, Chapter 10) and Wellstead and Edmunds (1975).

Before describing the analysis of closed-loop systems in detail, we first review briefly some background material.

2. Identification of linear systems

A discrete time system with noise-infected output, shown schematically in Fig. 1, may be described by the well-known model

$$Y_t = \sum_{u=0}^{\infty} a_u U_{t-u} + N_t, \quad t = 0, \pm 1, \pm 2, \ldots . \tag{2.1}$$

Here, U_t denotes the 'input' (at time t), Y_t the 'output' and N_t the 'noise' component of the output. (To comply with the condition of physical realisability, the summation over u in (2.1) must not involve negative values of u.) A familiar estimation problem associated with this model is as follows: Given a set of observations of the input $\{U_1, U_2, \ldots, U_N\}$, together with the corresponding values of the output $\{Y_1, Y_2, \ldots, Y_N\}$, estimate the unknown weight function $\{a_u\}$, or, equivalently, the unknown transfer function $A(\omega) = \Sigma_{u=0}^{\infty} a_u \, e^{-iu\omega}$. This type of problem arises, in one form or another, in many different fields; in particular it has important applications in econometric and control engineering problems. In the latter situation, however, one's aim is usually to devise a suitable feedback loop connecting the output to the input in order to 'control' the output Y_t. In this case, the form of an optimum 'controller' will depend not only on the form of the unknown transfer function $A(\omega)$, but also on the unknown structure of the noise N_t. (See, for example, Aoki, 1967 and Box and Jenkins, 1962.) Accordingly, if we assume that N_t is a

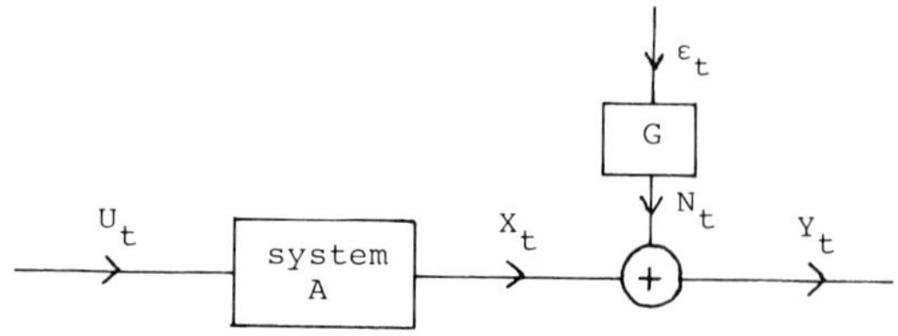

Fig. 1.

linear stationary process, and thus write (2.1) in the form

$$Y_t = \sum_{u=0}^{\infty} a_u U_{t-u} + \sum_{v=0}^{\infty} g_v \varepsilon_{t-v}, \tag{2.2}$$

where ε_t is an uncorrelated (i.e. white noise) stationary process, the estimation problem now becomes: given $(U_1, U_2, \ldots, U_N)$ and $(Y_1, Y_2, \ldots, Y_N)$, estimate $\{a_u\}$ and $\{g_v\}$—or equivalently, estimate $A(\omega)$ and $G(\omega) = \Sigma_{v=0}^{\infty} g_v \, \mathrm{e}^{-iv\omega}$. Box and Jenkins (1963, 1970) considered this problem and suggested the following approach. Assume first that both $A(\omega)$ and $G(\omega)$ are *rational functions* with polynomials of *known degrees* in both the numerators and denominators. The unknown coefficients in these polynomials are then estimated by 'searching' the parameter space until $\Sigma_t \hat{\varepsilon}_t^2$ is minimised, $\{\hat{\varepsilon}_t\}$ being the residuals from the fitted model. This technique may be quite useful in cases when the polynomials are known, a priori, to have fairly low degrees, but it would be extremely difficult to apply this method in more complex situations involving a large number of parameters. However, these authors have suggested an alternative iterative approach (Box and Jenkins, 1966) which proceeds as follows:

(1) First estimate $A(\omega)$ by minimising $\Sigma_t (Y_t - \Sigma_u a_u U_{t-u})^2$, i.e. first assume that $G(\omega) \equiv 1$ and apply standard least-squares theory.

(2) Having estimated $A(\omega)$, then estimate $G(\omega)$ by fitting a model to N_t using, for example, autocorrelation analysis of $\hat{N}_t = Y_t - \Sigma_u \hat{a}_u U_{t-u}$.

(3) Using the estimated *form* of $G(\omega)$, adjust, if necessary, the initial estimate of $A(\omega)$ and the values of the parameters in $G(\omega)$ using a 'search' technique similar to that described above.

This alternative procedure is certainly appealing, since stage 1 may be performed by standard cross-spectral analysis techniques—or, if $A(\omega)$ may be assumed to be rational, by multiple regression techniques. Similarly, stage 2 involves nothing more than the standard model fitting techniques of time-series analysis. However, if this procedure is to produce reliable results at the end of stage 2 it is clearly desirable that the initial estimate of $A(\omega)$ must be 'fairly reliable', so that the estimated *form* of $G(\omega)$ will in turn be 'fairly reliable'. (Note that stage 3 is, of course, essentially the same as the previously mentioned search procedure, using the algebraic forms of $A(\omega)$ and $G(\omega)$ given by stages 1 and 2.)

The basic strategy in the above procedure is first to estimate $A(\omega)$ 'ignoring' $G(\omega)$, and then estimate $G(\omega)$ 'allowing for' $A(\omega)$. Clearly then, if this method is to work successfully, one would like to be able to appeal to some form of '*orthogonality*' property between the functions $A(\omega)$ and $G(\omega)$. (In fact, these considerations are relevant to the general problem of analysing 'residuals' along the lines suggested by Cox and Snell (1968)—see Priestley, 1968.) In this discussion we suggest a possible definition of 'orthogonality', and examine the case when the observations $\{U_t\}$ and $\{Y_t\}$ are taken with a feedback loop *already in existence.* As Box and Jenkins (1966) observe, data collected from an industrial plant (with a view to designing optimal control) will generally have been recorded whilst the plant was operating under some crude form of manual control, or during the operation of some pilot control scheme. In such cases it is important to allow for the existence of feedback between Y_t and U_t.

3. Closed-loop systems: Basic assumptions and notation

We retain our previous model, given by (2.2), but superimpose on this a 'linear feedback plus noise' loop. Schematically, our model may now be described as below.

As before, U_t denotes the (observed) input, X_t the uncorrupted (noise-free) output, N_t the noise in the output and Y_t the observed output. We assume that Y_t is fed back through a linear controller, with unknown frequency response function $a(\omega)$, the output of which, Z_t, is corrupted by a noise component n_t. The noise process n_t is assumed also to be a stationary linear process of the form $n_t = \sum_{k=0}^{\infty} \gamma_k \eta_{t-k}$, where η_t is an uncorrelated (white noise) stationary process and $\Gamma(\omega) = \sum_{k=0}^{\infty} \gamma_k \, e^{-i\omega k}$ is an unknown function. The processes $\{\varepsilon_t\}$ and $\{\eta_t\}$ are assumed to be uncorrelated, i.e. $\operatorname{cov}\{\varepsilon_s, \eta_t\} = 0$, all s, t. For convenience, we assume (w.l.o.g.) that $g_0 = \gamma_0 = 1$,

$$E[\varepsilon_t] = E[\eta_t] = 0\,,$$

$$E[\varepsilon_t^2] = \sigma_\varepsilon^2, \qquad E[\eta_t^2] = \sigma_\eta^2\,.$$

The two noise processes, N_t and n_t, may be regarded as the outputs of linear 'boxes' with transfer functions $G(\omega)$ and $\Gamma(\omega)$ and white noise inputs ε_t and η_t respectively, and we have adopted this convention in Fig. 2.

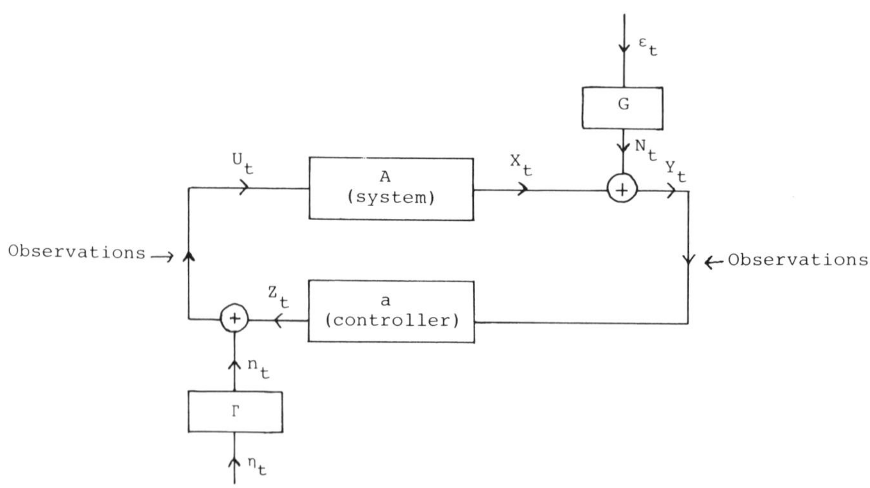

Fig. 2.

The scheme of Fig. 2 may be described by the symbolic equations

$$Y = AU + G\varepsilon\,, \tag{3.1}$$

$$U = aY + \Gamma\eta\,. \tag{3.2}$$

Here, A represents the operator $A(B)$, where $A(z) = \Sigma_{u=0}^{\infty} a_u z^u$ and B denotes the shift operator $BU_t = U_{t+1}$. The operators G, a and Γ are similarly defined, and we have suppressed the suffix t in U_t, Y_t, ε_t and η_t. In addition, we use the same symbol to denote both the transfer function and the corresponding operator, since, for example, $A(\omega) \equiv A(z)$ on making the substitution $z = e^{-i\omega}$.

This notation enables one to derive expressions for spectra and cross spectra almost immediately by *formally* equating (say) $f_{UU}(\omega)$ (the spectral density function of U_t) with $E(UU^*)$, and $f_{YU}(\omega)$ (the cross-spectral density function of Y_t and U_t) with $E(YU^*)$, and replacing each operator by its corresponding transfer function. (Here, $*$ denotes the complex conjugate, but we will also use the notation A^*, $|A|^2$ to denote the functions $A(1/z)$ and $\{A(z)A(1/z)\}$ respectively, even when z does not lie on the unit circle.) These formal manipulations are easily verified using the spectral representations of the relevant processes.

Since the stochastic nature of the system is determined entirely by the 'external' processes ε and η, it is convenient to express both Y and U in terms of these two processes. From (3.1) and (3.2) we obtain (assuming $aA \neq 1$),

$$(1 - aA)U = aG\varepsilon + \Gamma\eta\,, \tag{3.3}$$

$$(1 - Aa)Y = G\varepsilon + A\Gamma\eta\,. \tag{3.4}$$

Using now the assumption that ε and η are uncorrelated, together with the above device for evaluating spectra and cross spectra, one immediately obtains the following expressions for $f_{YU}(\omega)$, $f_{UU}(\omega)$ and $f_{YY}(\omega)$—when these functions exist.

$$2\pi f_{YU}(\omega) = \frac{A|\Gamma|^2\sigma_\eta^2 + a^*|G|^2\sigma_\varepsilon^2}{|1 - Aa|^2}\,, \tag{3.5}$$

$$2\pi f_{UU}(\omega) = \frac{|\Gamma|^2\sigma_\eta^2 + |a|^2|G|^2\sigma_\varepsilon^2}{|1 - Aa|^2} \tag{3.6}$$

and

$$2\pi f_{YY}(\omega) = \frac{|A|^2|\Gamma|^2\sigma_\eta^2 + |G|^2\sigma_\varepsilon^2}{|1 - Aa|^2}\,. \tag{3.7}$$

In the above, the symbol A denotes the function $A(e^{-i\omega})$, with a similar convention for B, C and D. (Equations (3.5), (3.6) and (3.7) were first derived by Akaike (1967).)

4. Conditions on the transfer function

So far, we have assumed that each of the transfer functions A, G, a and Γ represents a 'physically realisable' system, so that, for example, X_t depends only on present and *past* values of U_t, and similarly N_t depends only on present and *past* values of ε_t. However, in what follows we shall need rather more than this, since we will wish to assume: (i) that U_t, Y_t, N_t and n_t are *stationary* processes, and (ii) that the operators A, G, a and Γ are 'invertible', so that, for example, N_t may be written also as an autoregressive process, of the form $G^{-1}N_t = \varepsilon_t$. The first condition requires that each of the functions $A(z)$, $G(z)$, $a(z)$ and (z) has no *singularities* inside or on the unit circle, $|z| \leq 1$, and the second condition requires each of these functions to have no *zeros* in $|z| \leq 1$. (See, for example, Whittle, 1963, Chapter 2.) We therefore impose the following condition:

> Each of the functions $A(z)$, $A^{-1}(z)$, $G(z)$, $G^{-1}(z)$, $a(z)$, $a^{-1}(z)$, $\Gamma(z)$ and $\Gamma^{-1}(z)$ is analytic in the region $|z| \leq 1$.

(It may be noted that the equivalent condition for a continuous time model is that each of the above functions be analytic in the *lower half-plane*.) We could relax the above condition by allowing $G(z)$ and $\Gamma(z)$ to have poles (of finite order) at the point $z = 1$. In this case N_t and η_t are nonstationary processes of the 'accumulated' type (see Whittle, 1963, Chapter 8), but can be reduced to a stationary form by differencing each a sufficient number of times. For example, if $G(z)$ and $\Gamma(z)$ each have a pole of order p at $z = 1$, then $\Delta^p N_t$ and $\Delta^p n_t$ are stationary processes, and the subsequent analysis remains valid if we replace U_t by $\Delta^p U_t$ and Y_t by $\Delta^p Y_t$ (assuming that we know the value of p, a priori). This is, in fact, the type of model for N_t used by Box and Jenkins (1962, 1970). Thus, our conclusions will hold equally well for systems infected by nonstationary noise, provided the nonstationarity is of the above type.

5. Estimation of the system transfer function (A)

Suppose we are given N pairs of observations on the input $\{U_t\}$ and output $\{Y_t\}$ of a general system, and wish to model their relationship in the form,

$$Y_t = AU_t + G\varepsilon_t, \tag{5.1}$$

i.e. we wish to construct the 'best linear form' of the relationship between input and output, allowing for possible correlation between the processes $\{U_t\}$ and $\{\varepsilon_t\}$, but assuming that A and G satisfy the conditions stated in Section 2. Given observations $(U_1, \ldots, U_N)$ and $(Y_1, \ldots, Y_N)$, we wish to compare the two estimates of A obtained (i) by minimising

$$S_{1,N} = \sum_{t=1}^{N} \{G^{-1}(Y_t - AU_t)\}^2 = \sum_{t=1}^{N} \hat{\varepsilon}_t^2$$

with respect to both A and G, and (ii) by minimising

$$S_{2,N} = \sum_{t=1}^{N} (Y_t - AU_t)^2 = \sum_{t=1}^{N} (G\hat{\varepsilon}_t)^2$$

with respect to A. We denote the estimates obtained by methods (i) and (ii) by $\hat{A}_1$ and $\hat{A}_2$ respectively.

We will call $\hat{A}_1$ the 'weighted least-squares' estimate and $\hat{A}_2$ the 'simple least-squares' estimate. If the system is open loop and ε_t is a Gaussian process, the estimates of the parameters obtained by minimising $S_{1,N}$ are equivalent to maximum likelihood estimates. However, it should be noted that we are not claiming optimal properties for either of these estimates in the case of closed-loop systems. In general, both $\hat{A}_1$ and $\hat{A}_2$ will be biased estimates of A, the magnitude of the bias depending on the value of $(1 - aA)$, as discussed in Section 6. Our interest in these two particular estimates arises from the fact that $\hat{A}_1$ corresponds to the original search procedure mentioned in Section 2, whereas $\hat{A}_2$ corresponds to stage 1 of the alternative iterative approach. It is interesting, therefore, to examine the circumstances under which the 'crude' estimate $\hat{A}_2$ will attain an asymptotic form identical with, or close to, the asymptotic form of the more refined estimate $\hat{A}_1$.

For simplicity we consider the asymptotic case where $N \to \infty$, so that we have a semi-infinite realisation of both U_t and Y_t, and, assuming ergodicity we replace $S_{1,N}$ by $S_1 = E(\hat{\varepsilon}_t^2)$, and replace $S_{2,N}$ by $S_2 = E(G\hat{\varepsilon}_t)^2$. The weighted least-squares estimate $\hat{A}_1$ is now obtained by minimising

$$S_1 = E(\hat{\varepsilon}_t^2) = E\{G^{-1}(Y_t - AU_t)\}^2 \tag{5.2}$$

with respect to both A and G, but remembering that A is restricted to the class of physically relisable transfer functions—or 'backward transforms', so that the power series expansion of $A(z)$ contains only positive powers of z. In this formulation neither A nor G are assumed to be finitely parameterised, but, for example, S_2 is considered as a functional of an arbitrary function A which is constrained to satisfy the above conditions. In the sense that one is seeking the 'best linear relationship' between input and output, the asymptotic forms of the problem may be regarded essentially as that of the *linear prediction* of (i) Y_t from $\{U_s, s \leq t\}$, and (ii) $G^{-1}Y_t$ from $\{G^{-1}U_s;\ s \leq t\}$. In this respect, S_1 and S_2 may be interpreted as 'prediction error' criteria.

Before proceeding further, let us consider the unrealistic case when $A(z)$ is not restricted to being a backward transform but may contain positive and negative powers of z. In this case the relationship between Y_t and U_t takes the form

$$Y_t = \sum_{u=-\infty}^{\infty} a_u U_{t-u} + \sum_{v=0}^{\infty} g_v \varepsilon_{t-v} \tag{5.3}$$

and the estimation of A is now trivial. For, suppose first that G is fixed, and write $Y'_t = G^{-1}Y_t$ and $U'_t = G^{-1}U_t$. Then S_1 may be written

$$S_1 = E(Y' - AU'_t)^2, \tag{5.4}$$

and it is well known that the (unrestricted) estimate of A which minimises (5.4) is given by (cf. Priestley, 1981, Chapter 10), with an obvious notation,

$$\hat{A}'_1(\omega) = \frac{f_{Y'U'}(\omega)}{f_{U'U'}(\omega)}. \tag{5.5}$$

Similarly, the (unrestricted) estimate of A which minimises $S_2 = E(Y_t - AU_t)^2$ is given by

$$\hat{A}'_2(\omega) = \frac{f_{YU}(\omega)}{f_{UU}(\omega)}. \tag{5.6}$$

But, we have, of course,

$$f_{Y'U'}(\omega) = |G^{-1}(e^{-i\omega})|^2 f_{YU}(\omega),$$

$$f_{U'U'}(\omega) = |G^{-1}(e^{-i\omega})|^2 f_{UU}(\omega).$$

Hence, $\hat{A}'_1(\omega) \equiv \hat{A}'_2(\omega)$, and $\hat{A}'_1(\omega)$, being obviously independent of the fixed value of G, is the least-squares estimate of A obtained by minimising S_1 with respect to both A and G. Thus, *when A is not restricted to the class of backward transforms, methods (i) and (ii) lead to identical estimates.*

We now return to the original form of the problem and investigate under what conditions the restricted estimates, $\hat{A}_1$ and $\hat{A}_2$, will be identical. The above result suggests that this will certainly be the case if the expression $f_{YU}(\omega)/f_{UU}(\omega)$ is itself a backward transform—as would be true if we were dealing with an *open-loop* system. We will show, however, that the equivalence of $\hat{A}_1$ and $\hat{A}_2$ may hold even in closed-loop systems, provided the transfer function A, G, a and Γ satisfy a certain condition.

Now S_1 is most expeditiously evaluated by considering the spectral density function of the process

$$P_t = G^{-1}(Y_t - AU_t).$$

Using the device described in Section 3, we obtain

$$f_{PP}(\omega) = |G^{-1}|^2[f_{YY}(\omega) - A^* f_{YU} - A f^*_{YU} + |A|^2 f_{UU}]. \tag{5.7}$$

We now assume that $f_{UU}(\omega)$ has the canonical factorisation

$$f_{UU}(\omega) = |\alpha_{UU}(\omega)|^2,$$

where $\alpha_{UU}(z)$ has no zeros or singularities in the region $|z| \leq 1$. Then $f_{PP}(\omega)$ may be written

$$f_{PP}(\omega) = |G^{-1}|^2 \left[\left| \frac{f_{YU}(\omega)}{\alpha^*_{UU}(\omega)} - A\alpha_{UU}(\omega) \right|^2 + \left\{ f_{YY}(\omega) - \frac{|f_{YU}(\omega)|^2}{f_{UU}(\omega)} \right\} \right]. \quad (5.8)$$

But, $S_1 = \int_{-\pi}^{\pi} f_{PP}(\omega)\, d\omega$, so that $\hat{A}_1$ is obtained by minimising

$$\begin{aligned} S_1 = &\int_{-\pi}^{\pi} |G^{-1}|^2 \left| \frac{f_{YU}(\omega)}{\alpha^*_{UU}(\omega)} - A\alpha_{UU}(\omega) \right|^2 d\omega \\ &+ \int_{-\pi}^{\pi} |G^{-1}|^2 \left\{ f_{YY}(\omega) - \frac{|f_{YU}(\omega)|^2}{f_{UU}(\omega)} \right\} d\omega \end{aligned} \quad (5.9)$$

with respect to both A and G. In order to take account of the fact that A must be a backward transform, we adopt the well-known device of decomposing

$$\left\{ \frac{f_{YU}(\omega)}{G\alpha^*_{UU}(\omega)} \right\}$$

into the sum of its backward and forward transforms. Following Whittle (1963, Chapter 2), we use the notation $[B(z)]_+$ and $[B(z)]_-$ to denote the backward and forward transforms respectively of a function $B(z)$. More precisely, if

$$[B(z)] = \sum_{j=-\infty}^{\infty} b_j z^j,$$

then

$$[B(z)]_+ = \sum_{j=0}^{\infty} b_j z^j, \qquad [B(z)]_- = \sum_{j=-\infty}^{-1} b_j z^j.$$

Also, we will write $[B(\omega)]_+$ and $[B(\omega)]_-$ for $[B(e^{-i\omega})]_+$ and $[B(e^{-i\omega})]_-$ respectively.

From (5.9) we now obtain

$$\begin{aligned} S_1 = &\int_{-\pi}^{\pi} \left| \left[\frac{f_{YU}(\omega)}{G\alpha^*_{UU}(\omega)} \right]_+ - \frac{A\alpha_{UU}(\omega)}{G} \right|^2 d\omega + \int_{-\pi}^{\pi} \left| \left[\frac{f_{YU}(\omega)}{G\alpha^*_{UU}(\omega)} \right]_- \right|^2 d\omega \\ &+ \int_{-\pi}^{\pi} |G^{-1}|^2 \left\{ f_{YY}(\omega) - \frac{|f_{YU}(\omega)|^2}{f_{UU}(\omega)} \right\} d\omega . \end{aligned} \quad (5.10)$$

In deriving (5.10) we have used the fact that

$$\left\{ \left[\frac{f_{YU}(\omega)}{G\alpha^*_{UU}(\omega)} \right]_+ - \frac{A\alpha_{UU}(\omega)}{G} \right\}$$

is itself a backward transform (since G^{-1}, A and α_{UU} are all backward transforms) and therefore the integral of the product of this expression with

$$\left[\frac{f_{YU}(\omega)}{G\alpha^*_{UU}(\omega)}\right]^*_-$$

vanishes.

For each fixed G, the form of A which minimises S_1 is readily found by noting that the second and third integrals in (5.10) do not involve A and are both positive, and that the first integral vanishes when

$$A = \hat{A}_1(G) = \frac{G}{\alpha_{UU}(\omega)}\left[\frac{f_{YU}(\omega)}{G\alpha^*_{UU}(\omega)}\right]_+ . \tag{5.11}$$

Similarly, setting $G \equiv 1$, the form of A which minimises S_2 is given by

$$A = \hat{A}_2 = \frac{1}{\alpha_{UU}(\omega)}\left[\frac{f_{YU}(\omega)}{\alpha^*_{UU}(\omega)}\right]_+ \tag{5.12}$$

(cf. Whittle, 1963, p. 42).

Arguing as in the unrestricted case, it is clear that if $\hat{A}_1(G) \equiv \hat{A}_2$ for all G, then $\hat{A}_1(G)$ is equal to the weighted least-squares estimate $\hat{A}_1$ obtained by minimising S_1 with respect to both A and G. More generally if $\hat{A}_1(G) \neq \hat{A}_2$, then assuming that S_1 is a sufficiently 'well-behaved' functional of A and G so that $\min_{A,G} S_1 = \min_G\{\min_A S_1\}$, we find $\hat{A}_1 = \hat{A}_1(\hat{G})$, where $\hat{G}$ is obtained by substituting (5.11) in (5.10) and minimising S_1 with respect to G. (Note that when minimising S_1 with respect to G we must take account of the constraints: (a) that G is a backward transform, and (b) that $g_0 = 1 \Rightarrow \int_{-\pi}^{\pi} G(\omega)\,d\omega = 2\pi$.) *It now follows that a necessary and sufficient condition for the equivalence of $\hat{A}_1$ and $\hat{A}_2$ is that*

$$\left[\hat{G}^{-1}\frac{f_{YU}(\omega)}{\alpha^*_{UU}(\omega)}\right]_+ = \hat{G}^{-1}\left[\frac{f_{YU}(\omega)}{\alpha^*_{UU}(\omega)}\right]_+ . \tag{5.13}$$

However, it follows from (5.11) and (5.12) that $\hat{A}_1(G) \equiv \hat{A}_2$, *for all G*, provided that

$$\left\{\frac{f_{YU}(\omega)}{\alpha^*_{UU}(\omega)}\right\}$$

is a backward transform. (Recall that G^{-1} is certainly a backward transform.)

Also, from (5.1) we have

$$f_{YU}(\omega) = Af_{UU}(\omega) + f_{NU}(\omega), \tag{5.14}$$

so that

$$\frac{f_{YU}(\omega)}{\alpha^*_{UU}(\omega)} = A\alpha_{UU}(\omega) + \frac{f_{NU}(\omega)}{\alpha^*_{UU}(\omega)} .$$

Since A and α_{XX} are both backward transforms,

$$\left[\frac{f_{YU}(\omega)}{\alpha^*_{UU}(\omega)}\right]_- = \left[\frac{f_{NU}(\omega)}{\alpha^*_{UU}(\omega)}\right]_- .$$

Thus $\hat{A}_1 \equiv \hat{A}_2$, provided that

$$\left\{\frac{f_{NU}(\omega)}{\alpha^*_{UU}(\omega)}\right\}$$

is a backward transform, i.e. provided that

$$\left[\frac{f_{NU}(\omega)}{\alpha^*_{UU}(\omega)}\right]_- \equiv 0 . \tag{5.15}$$

We therefore propose the following definition:

> "The process $\{U_t\}$ and $\{N_t\}$ will be termed 'orthogonal' if condition (5.15) holds. In this case, the transfer functions A and G may be estimated 'independently', in the sense described above".

This definition includes, as a special case, the usual interpretation of 'orthogonality' in the sense that (5.15) will certainly be true if $\{U_t\}$ and $\{N_t\}$ are uncorrelated processes.

It is important to note that the 'equivalence' between $\hat{A}_1$ and $\hat{A}_2$ referred to above is a *purely algebraic* one, i.e. we are saying that under the above conditions the expressions S_1 and S_2 are both minimised by the same functional form of A. Since we are considering only the asymptotic forms of S_1 and S_2, this result does not, of course, have any bearing on the relationship between the sampling properties of the corresponding estimates derived from $S_{1,N}$ and $S_{2,N}$ in the finite sample case. In fact, it is fairly clear that, in general, these two estimates would not have the same asymptotic efficiency.

It is, however, interesting to observe the tenuous connection between condition (5.15) and the conditions for the efficiency of least-squares estimates derived by Greander and Rosenblatt (1957), Hannan (1970) and Watson (1967). The problem studied by these writers is somewhat different from ours in that they were dealing with the case of regression on a finite number of deterministic functions with correlated residuals, whereas in our case the 'regression' is on an infinite number of past values of a stationary process. The general form of their conclusions, however, is that the least-squares estimates (obtained by minimising the sum of squares of the *correlated* residuals) are asymptotically efficient provided the spectrum of the residuals is constant over all the elements of 'spectrum' of the regression functions. Whilst condition (5.15) is certainly different from the above, it nevertheless shares the feature of depending purely on the interrelationship between the spectral properties of the 'regression functions' and the spectral properties of the residuals. As it stands, (5.15) is expressed in terms of the relationship between the cross spectrum of $\{U_t\}$ and $\{N_t\}$ and the spectrum of $\{U_t\}$, rather than in terms of the spectrum of $\{N_t\}$ and the spectrum of $\{U_t\}$. But if we now introduce the precise form of the feedback loop joining Y_t with U_t, we may express the cross spectrum $f_{NU}(\omega)$ in terms of the spectrum of $\{N_t\}$, $f_{NN}(\omega)$, and the connection between the two types of conditions is strengthened.

6. The case of linear feedback plus noise

Having obtained expressions for $\hat{A}_1(G)$ and $\hat{A}_2$, we can now study their relationship on the assumption that the observations do, in fact, come from a linear system with superimposed linear feedback. Thus, we now return to the scheme of Fig. 2 and examine the form of condition (5.15) for this system. From (3.3) we have

$$(1-aA)U = aG\varepsilon + \Gamma\eta = aN + \Gamma\eta .$$

Consequently,

$$2\pi f_{NU}(\omega) = 2\pi f^*_{UN}(\omega) = 2\pi\left\{\frac{a}{1-aA}\right\}^* f_{NN}(\omega) = \frac{\sigma_\varepsilon^2 a^*|G|^2}{(1-aA)^*} . \tag{6.1}$$

Using (6.1) in conjunction with (3.6) yields

$$\frac{f_{NU}(\omega)}{\alpha^*_{UU}(\omega)} = \left\{\alpha_{UU}(\omega)\frac{f_{NU}(\omega)}{f_{UU}(\omega)}\right\} = \frac{\alpha_{UU}(\omega)\sigma_\varepsilon^2 a^*|G|^2(1-aA)}{\sigma_\eta^2|\Gamma|^2+\sigma_\varepsilon^2|a|^2|G|^2} .$$

Since $\alpha_{UU}(\omega)$ is necessarily a backward transform, condition (5.15) will be satisfied if the function

$$H = \frac{\sigma_\varepsilon^2 a^*|G|^2(1-aA)}{\sigma_\eta^2|\Gamma|^2+\sigma_\varepsilon^2|a|^2|G|^2} \tag{6.2}$$

is a backward transform. Thus, *the estimates* A_1 *and* A_2 *will be identical if the function* $H(z)$ *is analytic inside the unit circle,* $|z|\leq 1$. When A, G, a and Γ are rational functions (of z) the condition becomes that *H must have no poles inside the unit circle.* Consider the following special cases:

I: $a = A^{-1}$

Here, $H\equiv 0$, but this represents a degenerate system. In fact, we assumed in Section 2 that $aA \neq I$ to prevent degeneracy in the expressions for the spectra and cross spectrum of U_t and Y_t. The degeneracy arises from the fact that $a = A^{-1} \Rightarrow A\Gamma\eta = -G\varepsilon$, which contradicts the assumption that ε and η are uncorrelated processes. Note, however, that when $[H]_- \equiv 0$ the bias of both the estimates, $\hat{A}_1$ and $\hat{A}_2$, is proportional to H. Thus, we would hope that $(1-aA)$ was small at all frequencies.

II: $\Gamma\equiv 0$

This case corresponds to a system where the feedback loop is 'noise-free', and (6.2) now reduces to

$$H=(a^{-1}-A)\,,$$

which is clearly analytic in $|z|\leqslant 1$, by the previously stated conditions on a and A. *Thus, in the case of a noise-free feedback loop the two estimates $\hat{A}_1$ and $\hat{A}_2$ are identical.*

III: $a\equiv 0$

This corresponds to an open-loop system, and, as we have mentioned, we would expect $\hat{A}_1$ and $\hat{A}_2$ to be identical here. This fact is now easily verified by noting that $a\equiv 0\Rightarrow H\equiv 0$.

IV: $\sigma_\varepsilon^2/\sigma_\eta^2$ *small*

It is clear from (6.2) that $|H(\omega)|$ will be small at all points if the ratio $\{\sigma_\varepsilon^2/\sigma_\eta^2\}$ is sufficiently small. Of course, $|H(\omega)|$ may be small without H being a backward transform, but one easily shows from (5.11) and (5.12) that

$$\hat{A}_1(G)-\hat{A}_2=\frac{G}{\alpha_{UU}(\omega)}\left[\frac{1}{G}[\alpha_{UU}(\omega)H(\omega)]_-\right]_+ .$$

Hence, if $|H(\omega)|$ is small for all ω, we may expect $|\hat{A}_1(G)-\hat{A}_2|$ to be small (for all G), and consequently we may expect $|\hat{A}_1-\hat{A}_2|$ to be small (at all frequencies).

The assumption that $\sigma_\varepsilon^2/\sigma_\eta^2$ is small would seem quite reasonable in cases where the system is under manual control; for in these cases, one would expect the deviations from linearity in feedback loop to be much larger than the deviations from linearity in the system itself.

To summarise the results of this section, we have shown that the least-squares estimate of A obtained on the assumption that $G\equiv 1$ is identical with the unconstrained least-squares estimate when either (i) the feedback loop is noise-free, or (ii) the system is open loop. Otherwise, the result holds if H has no poles in the region $|z|\leqslant 1$. If H does not satisfy this condition, the two estimates may not be identical but will be close to each other when $\{\sigma_\varepsilon^2/\sigma_\eta^2\}$ is sufficiently small.

7. Numerical estimation of the system transfer function and noise structure

Suppose now that we are dealing with the practical situation where we have a finite number of observations and have decided to use the estimate $\hat{A}_2$.

Further, we are willing to assume that A is a rational function of the form

$$A(z)=\frac{\sum_{k=0}^{p}\alpha_k z^k}{\sum_{j=0}^{q}\beta_j z^j}, \quad (\beta_0=1),$$

but do *not* assume that we know the values of p and q, a priori. The problem now becomes that of estimating $\{\alpha_k\}$ and $\{\beta_j\}$ by minimising

$$S_{2,N}=\sum_{t=\min(p,q)}^{N}\{Y_t+\beta_1 Y_{t-1}+\cdots+\beta_q Y_{t-q} - \alpha_0 U_t-\alpha_1 U_{t-1}-\cdots-\alpha_p U_{t-p}\}^2 .$$

The numerical procedure is, in effect, equivalent to a multiple regression analysis of Y_t on $Y_{t-1}, Y_{t-2}, \ldots, Y_{t-q}, U_t, U_{t-1}, \ldots, U_{t-p}$, and this is the simplest way of estimating $\{\alpha_k\}$ and $\{\beta_j\}$. Moreover, an initial indication of the values of p and q may be obtained by starting off with fairly high values and using standard regression techniques to decide which variables may be rejected. This procedure, however, should be used with caution since the usual assumptions of classical regression analysis are not satisfied in this type of problem. One technique which has been found to work successfully is to combine multiple regression analysis with cross-spectral analysis in the following manner.

First, compute the 'nonparametric' estimate of A (compare (5.6)), namely

$$\hat{A}(\omega)=\hat{f}_{YU}(\omega)/\hat{f}_{UU}(\omega), \tag{7.1}$$

where $\hat{f}_{YU}(\omega)$ and $\hat{f}_{UU}(\omega)$ denote estimated cross-spectral and spectral density functions. (For methods of estimating f_{YU} and f_{UU}, see, for example, Jenkins and Watts, 1968 and Priestley, 1981.) Then, using the regression analysis described above to determine which Y_t's and U_t's are to be retained in the model and the associated estimates $\{\hat{\alpha}_k\}$ and $\{\hat{\beta}_j\}$, compute the corresponding transfer function,

$$\hat{\hat{A}}(\omega)=\frac{\sum_k \hat{\alpha}_k\, e^{-ik\omega}}{\sum_j \hat{\beta}_j\, e^{-ij\omega}}. \tag{7.2}$$

Now compare the graphs of $|\hat{A}(\omega)|^2$ and $|\hat{\hat{A}}(\omega)|^2$. If there is close agreement, then we may be reasonably confident that the regression analysis has produced a fairly accurate model. If, on the other hand, the fit is poor, then we retrace the steps by which the regression analysis rejected the various Y_t's and U_t's. Suppose that at the final stage of the rejection procedure the variable U_{t-3} (say) was thrown out. Then we would recompute $|\hat{\hat{A}}(\omega)|$, retaining $\hat{\alpha}_3$, and once more compare the fit with $|\hat{A}(\omega)|^2$. This procedure would then be repeated until a satisfactory fit was obtained—assuming, of course, that the initially chosen values of p and q were adequate. If this were not the case, we would eventually be led to increasing the values of p and q until we obtained good agreement between the two functions.

Alternatively, we may compare the residual variance, S_R^2, estimated from the regression analysis with the residual variance, S_0^2, corresponding to the 'non-parametric' estimate (7.1). It is easily shown that S_0^2 is given by (compare the third term in (5.10))

$$S_0^2 = \int_{-\pi}^{\pi} \left\{ \hat{f}_{YY}(\omega) - \frac{|\hat{f}_{YU}(\omega)|^2}{\hat{f}_{UU}(\omega)} \right\} \mathrm{d}\omega$$

$$= \int_{-\pi}^{\pi} \hat{f}_{YY}(\omega)\{1 - \hat{C}_{YU}^2(\omega)\}\, \mathrm{d}\omega ,$$

where

$$\hat{C}_{YU}(\omega) = \frac{|\hat{f}_{YU}(\omega)|}{\{\hat{f}_{UU}(\omega)\hat{f}_{YY}(\omega)\}^{1/2}}$$

is the estimated coherency between Y and U. Equivalently, S_0^2 may be evaluated from

$$S_0^2 = S_Y^2 - \int_{-\pi}^{\pi} \hat{f}_{UU}(\omega)|\hat{A}(\omega)|^2\, \mathrm{d}\omega ,$$

where S_Y^2 is the sample variance of $(Y_1, Y_2, \ldots, Y_n)$. A comparison of S_R^2 and S_0^2 should indicate the adequacy of the regression model, i.e. should indicate how well the rational function (7.2) approximates to the estimated transfer function given by (7.1).

It is important to remember that the estimate (7.1) is of the 'unrestricted' type, i.e. $\hat{A}(\omega)$ may not turn out to be a backward transform—unless, of course, H satisfies the required condition. We should not, therefore, expect to find 'near perfect' agreement between the estimates (7.1) and (7.2). In fact, if the Fourier coefficients of $\hat{A}(\omega)$ exhibit a substantial tail in the 'forward' direction, one would have to resort to an estimate of the form $\hat{A}_2(\omega)$ given by (5.12). This type of estimate requires the factorisation of the spectral density function, $\hat{f}_{XX}(\omega)$, and one numerical procedure for this operation has been given by Whittle (1963, p. 35). Once we have estimated A, we may then compute the estimated values of $\{N_t\}$ using

$$\hat{N}_t = Y_t - \hat{A}U_t , \tag{7.3}$$

and use these values to fit the transfer function G. One usually assumes that G is also a rational function—corresponding to an ARMA (mixed autoregressive/moving average) model for $\hat{N}_t$—and there are now a number of well-established techniques for fitting ARMA models to data.

Summary

We compare two methods of estimating transfer functions from input/output records when the output contains an additive noise component of unknown structure. The two methods correspond to (a) a weighted least-squares estimate ($\hat{A}_1$), and (b) a simple least-squares estimate ($\hat{A}_2$). The estimate $\hat{A}_1$ is more attractive theoretically since it is equivalent to a maximum likelihood approach when the system is open loop and the noise process is Gaussian, but it is difficult to compute numerically. The alternative estimate, $\hat{A}_2$, is much simpler to compute and is the one commonly used in practice, particularly as a first approximation in an iterative approach. Both $\hat{A}_1$ and $\hat{A}_2$, however, will, in general, be biased estimates when the system is closed loop. We determine conditions under which the asymptotic form of A_2 will be identical with, or close to, the asymptotic form of A_1 in the case of a closed-loop system in which the feedback is linear with an additive noise component.

References

Akaike, H. (1967). Some problems in the application of the cross-spectral method. In: C. J. Harris, ed., *Advanced Seminar on Spectral Analysis of Time Series.* Wiley, New York.

Akaike, H. (1968). On the use of a linear model for the identification of feedback systems. *Ann. Inst. Statist. Math.* **20**, 425.

Akaike, H. (1976). Canonical correlations analysis of time series and the use of an information criterion. In: R. Mehra and D. G. Lainiotis, eds., *Advances and Case Studies in System Identification.* Academic Press, New York.

Aoki, M. (1967). *Optimisation of Stochastic Processes.* Academic Press, New York.

Box, G. E. P. and Jenkins, G. M. (1962). Some statistical aspects of adaptive optimisation and control. *J. R. Statist. Soc.* **B24**, 297–343.

Box, G. E. P. and Jenkins, G. M. (1963). Further contributions to adaptive quality control. *Proc. of the 34th Session of the Inter. Statist. Inst. Ottawa, Canada.*

Box, G. E. P. and Jenkins, G. M. (1966). Recent advances in forecasting and control. Paper read at the European Meeting of Statisticians, London.

Box, G. E. P. and Jenkins, G. M. (1970). *Time Series, Forecasting and Control.* Holden-Day, San Francisco.

Box, G. E. P. and MacGregor, J. F. (1974). The analysis of closed loop dynamic stochastic systems. *Technometrics* **16**, 391–398.

Caines, P. E. and Chan, C. W. (1975). Feedback between stationary stochastic processes. *IEEE Trans. on Automatic Control* **AC-20**, 498–508.

Chatfield, C. (1975). *The Analysis of Time Series: Theory and Practice.* Chapman Hall, London.

Cox, D. R. and Snell, E. J. (1968). A general definition of residuals. *J. R. Statist. Soc.* **B30**, 248–275.

Diprose, K. V. (1968). Self-adaptive control system. *Aeronautical Journal of Royal Aeronautical Soc.* **72**, 367–372.

Diprose, K. V. (1978). Identification in the frequency domain of systems with closed loops. Technical Report, School of Electrical Engineering, University of Bath, U.K.

Grenander, U. and Rosenblatt, M. (1957). *Statistical Analysis of Stationary Time Series.* Wiley, New York.

Gustavson, I., Ljung, L. and Soderstrom, T. (1976). Identification of processes in closed loop: identifiability and accuracy aspects. *Proc. 9th IFAC Symposium on Identification, Tbilisi, U.S.S.R.*

Hannan, E. J. (1970). *Multiple Time Series.* Wiley, New York.
Harris, C. J. (1976). Problems in system identification and control. *Bull. Inst. Math. Appl.* **12**, 139–150.
Jenkins, G. M. and Watts, D. G. (1968). *Spectral Analysis and Its Applications.* Holden-Day, San Francisco.
Priestley, M. B. (1968). Discussion on paper by Cox and Snell, *J. R. Statist. Soc.* **B30**, 273–274.
Priestley, M. B. (1969). Estimation of transfer functions in closed loop stochastic systems. *Automatica* **5**, 623–632.
Priestley, M. B. (1981). *Spectral Analysis and Time Series.* Academic Press, London.
Watson, G. S. (1967). Linear least squares regression. *Ann. Math. Stat.* **38**, 1679–1699.
Wellstead, P. E. and Edmunds, J. M. (1975). Least squares identification of closed loop systems. *Int. J. Control.* **21**, 689.
Whittle, P. (1963). *Prediction and Regulation.* English Universities Press, London.

D. R. Brillinger and P. R. Krishnaiah, eds., *Handbook of Statistics, Vol. 3*
© Elsevier Science Publishers B.V. (1983) 293–319

14

The Bispectral Analysis of Nonlinear Stationary Time Series with Reference to Bilinear Time-Series Models

T. Subba Rao

1. Introduction

A considerable amount of work has been reported in the engineering literature on the effects of nonlinearity in various physical systems (see Cunningham, 1958). In this context, the effects of nonlinearity are studied by obtaining the distribution of zero crossings, threshold crossings and peaks of response functions, etc. However, in recent years several research workers, notably Brillinger (1965), Brillinger and Rosenblatt (1967a,b) and Rosenblatt and Van Ness (1965), have pointed out the importance of higher-order spectra in the analysis of nonlinear time series. In this paper we consider the bispectral analysis of bilinear time series, and show the importance of higher-order analysis in distinguishing the linear models and bilinear models. The statistical tests for linearity and Gaussianity are considered. The optimum estimation of the bispectral density is also considered. The bispectral analysis of two well-known time series is included.

2. Stationary time series and second-order spectral analysis

In this section we briefly discuss the second-order spectral analysis of stationary time series, and an extensive discussion of spectral analysis can be found, for example, in Brillinger (1975), Hannan (1970), Jenkins and Watts (1968), Koopmans (1974) and Priestley (1981).

Let $\{X(t)\}$ be a discrete parameter, real-valued time series. We say that the time series $\{X(t)\}$ is second order stationary if

(i) $E(X(t)) = \mu$, independent of t,

(ii) $\operatorname{var}(X(t)) = \sigma_x^2$, independent of t, $\qquad$ (2.1)
$\operatorname{cov}(X(t)X(t+s)) = R(s) =$ a function of the lag s only $(s = 0, \pm1, \pm2, \ldots)$.

The autocovariance function $R(s)$ is symmetric, i.e. $R(s) = R(-s)$. It is well known that there exists a function $F(\omega)$, which is bounded and nondecreasing such that

$$R(s) = \int_{-\pi}^{\pi} e^{is\omega} \, dF(\omega) . \tag{2.2}$$

The function $F(\omega)$ is known as the integrated (nonnormalised) spectrum of the stationary time series $\{X(t)\}$. If $F(\omega)$ is differentiable and $dF(\omega) = f(\omega)\, d\omega$, then we can write (2.2) as

$$R(s) = \int_{-\pi}^{\pi} e^{is\omega} f(\omega) \, d\omega . \tag{2.3}$$

The function $f(\omega)$ is known as the (nonnormalised) spectral density function of the stationary time series $\{X(t)\}$.

From (2.3) we have the inverse Fourier transform

$$f(\omega) = \frac{1}{2\pi} \sum_{-\infty}^{\infty} R(s)\, e^{-is\omega} , \quad |\omega| \leq \pi . \tag{2.4}$$

Since $R(s) = R(-s)$, we obtain $f(\omega) = f(-\omega)$. The spectral density function $f(\omega)$ is always positive. From (2.3) we have

$$R(0) = \sigma_x^2 = \int_{-\pi}^{\pi} f(\omega)\, d\omega . \tag{2.5}$$

In an engineering context, σ_x^2 represents the total power contained in the process $\{X(t)\}$ and the right-hand side integral (2.5) represents a frequency decomposition of the total power. In other words, $f(\omega)\, d\omega$ represents the power contained in the frequency band $(\omega, \omega + d\omega)$ of the process $\{X(t)\}$. Since $f(\omega)$ is the Fourier transform of the autocovariance function of $R(s)$, it is natural to define $f(\omega)$ as the second-order spectral density function.

If the process $\{X(t)\}$ is Gaussian, it is well known that all the information in the process $\{X(t)\}$ is contained in the mean and covariances $\{R(s)\}$ and as such second-order spectral analysis on the process $\{X(t)\}$ is sufficient to draw all the useful information about this process. If the process is not Gaussian, one may have to perform higher-order spectral analysis, and the simplest higher-order spectral analysis one can perform is the bispectral analysis which we shall discuss in the following sections.

3. Second-order covariance and spectral analysis of linear time-series models

The object in time-series model building can be described as follows. Suppose $X(t)$ denotes the given time series, then the object is to seek a function $h(\cdot)$ which is such that

$$h(X(t), X(t-1), \ldots) = e(t), \tag{3.1}$$

where $\{e(t)\}$ is a sequence of independent, identically distributed random variables with $E(e(t)) = 0$, $E(e^2(t)) = \sigma_e^2$. The class of linear models is given by restricting $h(\cdot)$ to be a linear function of $X(t), X(t-1), \ldots$, in which case, (3.1) reduces to

$$\sum_{u=0}^{\infty} h_u X(t-u) = e(t) \tag{3.2}$$

or

$$H(B)X(t) = e(t),$$

where $H(Z) = \Sigma_{u=0}^{\infty} h_u Z^u$. If $H(Z) \neq 0$, $|Z| < 1$, (3.2) may be written as

$$X(t) = H^{-1}(B)e(t) = \Gamma(B)e(t) = \sum_{u=0}^{\infty} g_u e(t-u), \tag{3.3}$$

where $\Gamma(Z) = \Sigma_{u=0}^{\infty} g_u Z^u$. The (nonnormalised) spectral density function of the process $X(t)$ satisfying (3.3) is given by

$$f(\omega) = \frac{\sigma_e^2}{2\pi} |\Gamma(\omega)|^2, \tag{3.4}$$

where $\Gamma(\omega) = \Sigma_{u=0}^{\infty} g_u \mathrm{e}^{-\mathrm{i}u\omega}$ is known as the 'transfer function'. In other words, if the relation between $\{e(t)\}$ and $\{X(t)\}$ is linear, then the system can be completely defined by a *single transfer function* $\Gamma(\omega)$. Recently, Rosenblatt (1980, 1981) has shown that if $X(t)$ is Gaussian and satisfies the relation (3.3), then only the modulus of the transfer function $\Gamma(\omega)$, i.e. $|\Gamma(\omega)|$, is identifiable. However, if $X(t)$ is non-Gaussian, but satisfies (3.3), then it is possible to estimate the transfer function except for a possible linear phase shift using higher-order spectral estimates. If the input/output relationship is nonlinear, one may have to define an infinite number of higher-order transfer functions to characterise the system (Brillinger, 1970).

The two linear representations (3.2) and (3.3), though quite general, are not very useful in practice since the parameters in the representations are infinite. By making further assumptions on $H(Z)$ and $\Gamma(Z)$, we can obtain finite parameter models which are special cases of (3.2) and (3.3).

If we assume that $H(Z) = 1 + a_1 Z + a_2 Z^2 + \cdots + a_k Z^k$, then (3.2) reduces to

$$X(t) + a_1 X(t-1) + \cdots + a_k X(t-k) = e(t) \tag{3.5}$$

which is an autoregressive model of order $k(\mathrm{AR}(k))$.

If we restrict $\Gamma(Z)$ to be of the form $\Gamma(Z) = 1 + b_1 Z + \cdots + b_l Z^l$, then the model (3.3) reduces to the moving average model of order l (MA(l)),

$$X(t) = e(t) + b_1 e(t-1) + \cdots + b_l e(t-l)\,. \tag{3.6}$$

A parsimonious representation can be obtained by assuming $\Gamma(Z)$ to be of the rational form

$$\Gamma(Z) = \frac{1 + b_1 Z + \cdots + b_l Z^l}{1 + a_1 Z + \cdots + a_k Z^k}\,. \tag{3.7}$$

Then (3.3) reduces to the autoregressive moving average (ARMA(k, l)) representation

$$\begin{aligned} X(t) &+ a_1 X(t-1) + \cdots + a_k X(t-k) \\ &= e(t) + b_1 e(t-1) + \cdots + b_l e(t-l)\,. \end{aligned} \tag{3.8}$$

If $X(t)$ satisfies the model (3.8), one can easily show that the covariance function $R(s)$ satisfies the recursive relation

$$R(s) + a_1 R(s-1) + a_2 R(s-2) + \cdots + a_k R(s-k) = 0 \quad \text{if } s \geq l+1\,. \tag{3.9}$$

Equations (3.9) are known as Yule–Walker equations and in the later sections we show that similar equations are satisfied by the bilinear time-series models.

It is important to note that in writing the representations (3.2) and (3.3) we have assumed that we can find a linear function $h(\cdot)$ such that the resulting errors $\{e(t)\}$ are independent. The well-known Wold's theorem in stationary processes clearly shows that this is not always the case. This theorem states that if $X(t)$ is a stationary process (with an absolutely continuous spectral density function), then $X(t)$ can be written in the form

$$X(t) = \sum_{u=0}^{\infty} a_u \eta(t-u)\,, \tag{3.10}$$

where $a_0 = 1$ and $\Sigma\, a_u^2 < \infty$, $\{\eta(t)\}$ is an uncorrelated process. In other words, a stationary process (with an absolutely continuous spectrum) can only be written as a linear combination of an uncorrelated process $\{\eta(t)\}$, and not in general in terms of an independent process $\{e(t)\}$.

As far as second-order properties are concerned, $e(t)$ and $\eta(t)$ have identical properties, but they may differ substantially in many other aspects (see Granger and Andersen, 1978; Priestley, 1978). In order to illustrate this point further, consider the process $X(t)$ satisfying the relation (Priestley, 1978) $X(t) = e(t) + \beta e(t-1)e(t-2)$, where $\{e(t)\}$ are defined as above. Then it can be shown that $E(X(t)) = 0$, $E(X(t)X(t+s)) = 0$, $s \neq 0$ and $E(X(t)/X(s), s \leq t-1) = \beta e(t-1)e(t-2)$. This example illustrates the fact that although the process $X(t)$, which is nonlinear, is identified as a white noise process by the second-order covariance analysis, the past does not contain information on the future, and it is, therefore, possible to forecast its future. For this model, one can show that a higher moment analysis would distinguish this model from a linear model.

4. Bilinear time-series models and second-order spectral analysis

It may not always be possible to represent a non-Gaussian process $X(t)$ in the form (3.4) with $\{e(t)\}$ mutually independent. Motivated by the work of Volterra (1930) in series expansion of continuous functions, Wiener (1958) has made a systematic study of the nonlinear representations of the continuous parameter stochastic process $X(t)$. Wiener's representation, which is in terms of Hermite polynomials, can be considered as an analogue of Wold's decomposition theorem (see Robenblatt, 1979). Volterra (1930) has shown that, under certain conditions, the process $X(t)$ can be written as

$$X(t)=\sum_{i=1}^{\infty}\left[\sum_{u_1}\cdots\sum_{u_i} g(u_1,u_2,\ldots,u_i)\prod_{j=1}^{i} e(t-u_j)\right]. \tag{4.1}$$

Equation (4.1) is known as Volterra expansion and the kernels $\{g_1(u), g_2(u_1, u_2)\ldots\}$ are Volterra kernels. As pointed out earlier, we have to define an infinite sequence of higher-order transfer functions (see Brillinger, 1970) to characterise the above relationship. For example, we can define the kth-order transfer function as

$$\Gamma_k(\omega_1,\omega_2,\ldots,\omega_k)=\sum_{u_1}\sum_{u_2}\cdots\sum_{u_k} g_k(u_1,u_2,\ldots,u_k)\,\mathrm{e}^{-\mathrm{i}(u_1\omega_1+\cdots+u_k\omega_k)}. \tag{4.2}$$

Although the Volterra and Wiener expansions are very general and elegant, in practice, it is difficult to estimate the infinite set of kernels. In view of this, it is natural to seek a finite parameter representation of sufficient generality to describe the nonlinear relationship. Recently, control theorists have introduced the class of bilinear models which have been found useful for describing many nonlinear phenomena. Since the 'bilinear models' are nearly linear, their structural properties are similar to those of linear models. The analysis of bilinear time-series models has been considered by Granger and Andersen (1978) and Subba Rao (1981).

Let $\{X(t)\}$ be a discrete parameter time series, satisfying the difference equation

$$X(t)+\sum_{j=1}^{p} a_jX(t-j)=\sum_{j=0}^{r} c_je(t-j)+\sum_{l=1}^{m}\sum_{l'=1}^{k} b_{ll'}X(t-l)e(t-l'), \tag{4.3}$$

where $c_0=1$ and $\{e(t)\}$ is a sequence of independent random variables. We define the model (4.3) as a bilinear time-series model $\mathrm{BL}(p, r, m, k)$ and the process $\{X(t)\}$ as a bilinear process. In their monograph, Granger and Andersen (1978) have considered the statistical properties of the model $\mathrm{BL}(1, 0, 1, 1)$. Subba Rao (1981) has investigated the properties of the model $\mathrm{BL}(p, 0, p, 1)$ and considered the statistical estimation of the $\mathrm{BL}(p, 0, p, q)$ model. The autoregressive moving average model $\mathrm{ARMA}(p, r)$ can be obtained from (4.3) by setting $b_{ll'} = 0$ for all l and l'. In this paper we briefly describe the properties of the $\mathrm{BL}(p, 0, p, 1)$ model.

Vector form of $\mathrm{BL}(p, 0, p, 1)$ *model*

It is convenient to represent $\mathrm{BL}(p, 0, p, 1)$ in a state space form as follows. Let us define the matrices

$$\boldsymbol{A} = \begin{pmatrix} -a_1 & -a_2 & \cdots & & -a_p \\ 1 & 0 & \cdots & & 0 \\ 0 & 1 & \cdots & & 0 \\ & \vdots & \ddots & & \\ 0 & 0 & \cdots & 1 & 0 \end{pmatrix}, \qquad \boldsymbol{B} = \begin{pmatrix} b_{11} & b_{21} & b_{31} & \cdots & b_{p1} \\ 0 & 0 & 0 & \cdots & 0 \\ & \vdots & \vdots & \ddots & \\ 0 & 0 & 0 & \cdots & 0 \end{pmatrix} \tag{4.4}$$

and $\boldsymbol{C}' = (1, 0, 0, \ldots, 0)$, $\boldsymbol{H}' = (1, 0, \ldots, 0)$, $\boldsymbol{x}'(t) = (X(t), X(t-1), \ldots, X(t-p+1))$. With this notation, we can write the model (4.3) in the form

$$\begin{aligned} \boldsymbol{x}(t) &= \boldsymbol{A}\boldsymbol{x}(t-1) + \boldsymbol{B}\boldsymbol{x}(t-1)e(t-1) + \boldsymbol{C}e(t), \\ X(t) &= \boldsymbol{H}'\boldsymbol{x}(t). \end{aligned} \tag{4.5}$$

We define the model (4.5) as a vector form of the bilinear model $\mathrm{BL}(p, 0, p, 1)$ and denote it by $\mathrm{VBL}(p)$. In the following, we assume that $\{e(t)\}$ are independent and each is distributed as $N(0, 1)$.

Second-order covariance analysis of $\mathrm{VBL}(p)$

We have from (4.5),

$$E(X(t)) = \boldsymbol{H}'E(\boldsymbol{x}(t)), \quad \mathrm{cov}(X(t)X(t+s)) = \boldsymbol{H}' \,\mathrm{cov}\,(\boldsymbol{x}(t)\boldsymbol{x}(t+s))\boldsymbol{H}. \tag{4.6}$$

Let the spectral radius of a matrix $\boldsymbol{A}$, $\rho(\boldsymbol{A})$ be

$$\rho(\boldsymbol{A}) = \max_i\{|\lambda_i(\boldsymbol{A})|\}, \tag{4.7}$$

where $\lambda_i(\boldsymbol{A})$ is the ith eigenvalue of $\boldsymbol{A}$. It has been shown (Subba Rao, 1981) that a sufficient condition for the first-order asymptotic stationarity is that $\rho(\boldsymbol{A}) < 1$. Under this condition, the mean

$$\boldsymbol{\mu} = E(\boldsymbol{x}(t)) = (\boldsymbol{I} - \boldsymbol{A})^{-1}\boldsymbol{BC}. \tag{4.8}$$

The sufficient condition for the second-order stationarity is that

$$\rho(\boldsymbol{A} \otimes \boldsymbol{A} + \boldsymbol{B} \otimes \boldsymbol{B}) < 1\,, \tag{4.9}$$

where $\boldsymbol{A}\otimes\boldsymbol{A}$, $\boldsymbol{B}\otimes\boldsymbol{B}$ are Kronecker product matrices. Under this condition, the expression for the covariances of the process $\boldsymbol{x}(t)$ can be shown to satisfy the following relations.

Let $\boldsymbol{C}(s) = E(\boldsymbol{x}(t+s) - \boldsymbol{\mu})(\boldsymbol{x}(t) - \boldsymbol{\mu})'$, then we have

$$\begin{aligned} \boldsymbol{C}(0) &= \boldsymbol{AC}(0)\boldsymbol{A}' + \boldsymbol{BC}(0)\boldsymbol{B}' + \boldsymbol{\Delta}_2\,, \\ \boldsymbol{C}(1) &= \boldsymbol{AC}(0) + \boldsymbol{\Delta}_3\,, \\ \boldsymbol{C}(s) &= \boldsymbol{AC}(s-1) = \boldsymbol{A}^{s-1}\boldsymbol{C}(1)\,, \quad (s = 2, 3, \ldots)\,, \end{aligned} \tag{4.10}$$

where

$$\begin{aligned} \boldsymbol{\Delta}_2 &= \boldsymbol{B\mu\mu'B'} + \boldsymbol{A\mu\mu'A'} + \boldsymbol{ASB'} + \boldsymbol{BS'A} + 2\boldsymbol{BCC'B'} + \boldsymbol{CC'} - \boldsymbol{\mu\mu'}\,, \\ \boldsymbol{\Delta}_3 &= \boldsymbol{A\mu\mu'} + \boldsymbol{BS} - \boldsymbol{\mu\mu'}\,, \\ \boldsymbol{S} &= \boldsymbol{A\mu C'} + \boldsymbol{BCC'} + \boldsymbol{C\mu'A'} + \boldsymbol{CC'B'}\,. \end{aligned} \tag{4.11}$$

If the matrices $\boldsymbol{A}$ and $\boldsymbol{B}$ are of the form (4.4) and $H' = (1, 0, \ldots, 0)$, the relations (4.10) reduce to

$$R(s) + a_1R(s-1) + \cdots + a_pR(s-p) = 0\,, \quad s \geq 2\,. \tag{4.12}$$

Equations (4.12) *are the same as the Yule–Walker equations* (3.9) *for an* ARMA$(p, 1)$ *and thus show that the bilinear model* BL$(p, 0, p, 1)$ *has the same covariance structure as an* ARMA$(p, 1)$. In deriving (4.12), we have used the fact that $\{e(t)\}$ are independent and each $e(t)$ is distributed as a $N(0, 1)$ variable.

The spectral density function $f(\omega)$ can be obtained from $\boldsymbol{C}(s)$ by taking its Fourier transform. For an illustration, consider the BL(1, 0, 1, 1) model

$$X(t) + aX(t-1) = bX(t-1)e(t-1) + e(t)\,. \tag{4.13}$$

The mean and variance of this model are

$$\mu = \frac{b}{1+a}\,, \qquad R(0) = \sigma_x^2 = \frac{1+2b^2}{1-a^2-b^2} - \frac{4ab^2}{(1+a)(1-a^2-b^2)} - \mu^2\,,$$

$$R(1) = (-a)R(0) + \frac{b^2}{1+a}\,, \qquad R(s) = (-a)^{s-1}R(1) \quad (s = 2, 3, \ldots)\,.$$

The graph of the spectral density functions of the process $X(t)$ generated from

(4.13) when (i) $a = -0.4$, $b = 0.4$, (ii) $a = -0.4$, $b = 0.8$ are plotted in Figs. 1 and 2, respectively.

In both cases the spectral density function decreases as ω increases; and the overall shape is similar to the shape of the spectrum of an AR(1) model.

The above analysis clearly shows that a second-order covariance analysis (spectral analysis) cannot distinguish between a linear ARMA model and a (nonlinear) bilinear model. This suggests the use of higher-order spectral analysis.

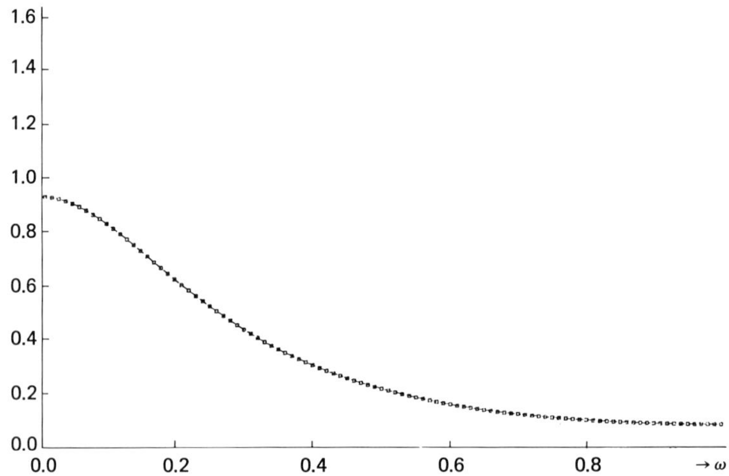

Fig. 1. Spectral density function ($a = -0.4$, $b = 0.4$).

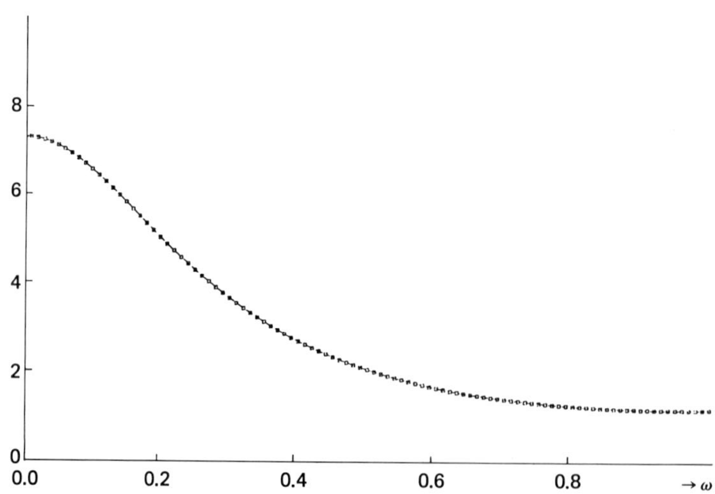

Fig. 2. Spectral density function ($a = -0.4$, $b = 0.8$).

5. Volterra series expansion of VBL(p) model

In this section we obtain the Volterra expansion for VBL(p) model and thus derive the higher-order transfer functions which characterise the model.

We now consider the VBL(p) model given by (4.5) and the solution of the (4.5) is sought in the form of a power series expansion. In this paper we use the 'reversion method'. (For further details, see Cunningham, 1958, p. 133.) We consider the model

$$\boldsymbol{x}(t)=\boldsymbol{A}\boldsymbol{x}(t-1)+\lambda\boldsymbol{B}\boldsymbol{x}(t-1)e(t-1)+\lambda\boldsymbol{C}e(t)\,, \tag{5.1}$$

where λ is a numerical parameter introduced to facilitate the solution, but ultimately λ is allowed to become unity. A solution for (5.1) is sought in the form

$$\boldsymbol{x}(t)=\sum_{j=1}^{\infty}\lambda^{j}\boldsymbol{x}_j(t)\,. \tag{5.2}$$

Substituting (5.2) into (5.1) and then equating powers of λ on both sides, we get

$$\boldsymbol{x}_1(t)=\boldsymbol{A}\boldsymbol{x}_1(t-1)+\boldsymbol{C}e(t)\,, \tag{5.3}$$

$$\boldsymbol{x}_i(t)=\boldsymbol{A}\boldsymbol{x}_i(t-1)+\boldsymbol{B}\boldsymbol{x}_{i-1}(t-1)e(t-1)\quad(i=2,3,\ldots)\,. \tag{5.4}$$

Assuming $\boldsymbol{x}_i(0)=\boldsymbol{0}$ $(i=1,2,\ldots)$ and $e(t)=0$ $(t\leqslant 0)$, we can show that

$$\begin{aligned}\boldsymbol{x}_1(t)&=\sum_{j_1=0}^{t}\boldsymbol{A}^{j_1}\boldsymbol{C}e(t-j_1)\,,\\ \boldsymbol{x}_i(t)&=\sum_{j_1=0}^{t-1}\boldsymbol{A}^{j_1}\boldsymbol{B}\boldsymbol{x}_{i-1}(t-1-j_1)e(t-1-j_1)\\ &=\sum_{u_1=0}^{t}\cdots\sum_{u_i=0}^{t}\boldsymbol{A}^{u_1-1}\boldsymbol{B}\boldsymbol{A}^{u_2-u_1-1}\boldsymbol{B}\cdots\boldsymbol{A}^{u_{i-1}-u_{i-2}-1}\boldsymbol{B}\boldsymbol{A}^{u_i-u_{i-1}}\boldsymbol{C}\delta(u_1-1)\\ &\quad\times\prod_{j=2}^{i-1}\delta(u_j-u_{j-1}-1)\delta(u_i-u_{i-1})\prod_{j=1}^{i}e(t-u_j)\,,\end{aligned} \tag{5.5}$$

where the step function $\delta(u)$ is defined as

$$\delta(u)=\begin{cases}1 & \text{if } u\geqslant 0\,,\\ 0 & \text{otherwise}\,.\end{cases}$$

The Volterra series expansion of $\boldsymbol{x}(t)$ is obtained by putting $\lambda=1$ in (5.1) and (5.2) and the final solution of $X(t)$ (assuming we observe $\boldsymbol{X}(t)=\boldsymbol{H}'x(t)$) can be written in the form

$$X(t) = \sum_{i=1}^{\infty} \boldsymbol{H}'\boldsymbol{x}_i(t) = \sum_{i=1}^{\infty} \left[\sum_{u_1=0}^{t} \cdots \sum_{u_i=0}^{t} V_i(u_1 \cdots u_i) \prod_{j=1}^{i} e(t-u_j) \right], \tag{5.6}$$

where the Volterra kernels $V_i(u_1 \cdots u_i)$ are given by

$$\begin{aligned} V_1(u_1) &= \boldsymbol{H}'\boldsymbol{A}^{u_1}\boldsymbol{C}, \\ V_i(u_1, u_2, \ldots, u_i) & \\ &= \boldsymbol{H}'\boldsymbol{A}^{u_1-1}\boldsymbol{B}\delta(u_1-1) \prod_{j=2}^{i-1} [\boldsymbol{A}^{u_j-u_{j-1}-1}\boldsymbol{B}\delta(u_j-u_{j-1}-1)\boldsymbol{A}^{u_i-u_{i-1}}\boldsymbol{C}] \\ & \qquad (i = 2, 3, \ldots). \end{aligned} \tag{5.7}$$

A comparison of the expansion (5.6) with (4.1) shows that for bilinear processes the coefficients $g_i(u_1 \cdots u_i)$ of the Volterra expansion (4.1) are zero if any of $u_1, u_2, \ldots, u_i$ is zero.

The kernels in the expansion (5.6) are not symmetric, but this expansion can be written in terms of symmetric kernels by defining

$$\begin{aligned} W_1(u_1) &= V_1(u_1), \\ W_1(u_1, u_2, \ldots, u_i) &= \frac{1}{i!} \sum_{\text{per}} V_i(u_1, u_2, \ldots, u_i) \end{aligned} \tag{5.8}$$

when the summation Σ_{per} is taken over all possible permutations of the variables $u_1, u_2, \ldots, u_i$. The Volterra expansion can now be written as

$$X(t) = \sum_{i=1}^{\infty} \left[\sum_{u_1=0}^{t} \cdots \sum_{u_i=0}^{t} W_i(u_1, u_2, \ldots, u_i) \prod_{j=1}^{i} e(t-u_j) \right], \tag{5.9}$$

where, for example, the first two kernels are given by

$$\begin{aligned} W_1(u_1) &= \boldsymbol{H}'\boldsymbol{A}^{u_i}\boldsymbol{C}, \\ W_2(u_1, u_2) &= \tfrac{1}{2}[\boldsymbol{H}'\boldsymbol{A}^{u_1-1}\boldsymbol{B}\boldsymbol{A}^{u_2-u_1}\boldsymbol{C}\delta(u_1-1)\delta(u_2-u_1) \\ &\qquad + \boldsymbol{H}'\boldsymbol{A}^{u_2-1}\boldsymbol{B}\boldsymbol{A}^{u_1-u_2}\boldsymbol{C}\delta(u_2-1)\delta(u_1-u_2)]. \end{aligned} \tag{5.10}$$

We can now define the kth-order transfer function as

$$\Gamma_k(\omega_1, \omega_2, \ldots, \omega_k) = \sum_{u_1=0}^{\infty} \cdots \sum_{u_k=0}^{\infty} W_k(u_1, u_2, \ldots, u_k)\, e^{-i(u_1\omega_1 + \cdots + u_k\omega_k)} \qquad (k = 1, 2, \ldots). \tag{5.11}$$

The transfer functions corresponding to the kernels (5.10) (assuming the spectral radius of $\boldsymbol{A}$ is less than 1) are

$$\Gamma_1(\omega) = \boldsymbol{H}'(\boldsymbol{I} - \boldsymbol{A}\,\mathrm{e}^{-\mathrm{i}\omega})^{-1}\boldsymbol{C}, \tag{5.12}$$

$$\begin{aligned}\Gamma_2(\omega_1, \omega_2) = \tfrac{1}{2}[&\boldsymbol{H}'(\boldsymbol{I} - \boldsymbol{A}\,\mathrm{e}^{-\mathrm{i}(\omega_1+\omega_2)})^{-1}\boldsymbol{B}(\boldsymbol{I} - \boldsymbol{A}\,\mathrm{e}^{-\mathrm{i}\omega_1})^{-1}\boldsymbol{C} \\ &+ \boldsymbol{H}'(\boldsymbol{I} - \boldsymbol{A}\,\mathrm{e}^{-\mathrm{i}(\omega_1+\omega_2)})^{-1}\boldsymbol{B}(\boldsymbol{I} - \boldsymbol{A}\,\mathrm{e}^{-\mathrm{i}\omega_2})^{-1}\boldsymbol{C}]\,\mathrm{e}^{-\mathrm{i}(\omega_1+\omega_2)}.\end{aligned} \tag{5.13}$$

For the BL(1, 0, 1, 1) model (4.13), the first two transfer functions are

$$\begin{aligned}\Gamma_1(\omega) &= \frac{1}{1 + a\,\mathrm{e}^{-\mathrm{i}\omega}}, \\ \Gamma_2(\omega_1, \omega_2) &= \frac{b\,\mathrm{e}^{\mathrm{i}(\omega_1+\omega_2)}}{2(1 + a\,\mathrm{e}^{-\mathrm{i}(\omega_1+\omega_2)})}\left[\frac{1}{1 + a\,\mathrm{e}^{-\mathrm{i}\omega_1}} + \frac{1}{1 + a\,\mathrm{e}^{-\mathrm{i}\omega_2}}\right].\end{aligned} \tag{5.14}$$

These transfer functions agree with the corresponding expressions derived by Priestley (1978) for the BL(1, 0, 1, 1) model.

Although an infinite number of transfer functions have to be defined in the case of nonlinear systems, one can see that in the case of bilinear models, where the norm of $\boldsymbol{B}$ is less than unity, the higher-order transfer functions $\Gamma_k(\omega_1 \cdots \omega_k)$ tend to zero as k tends to infinity. Hence, as such, a finite number of transfer functions would adequately characterise the bilinear model VBL(p).

6. Spectral and bispectral density functions and their estimation

We assume that the time series $\{X(t)\}$ is a real-valued process, has finite sixth-order moments and is weakly stationary up to the sixth order, so that for all t,

$$\mu = E(X(t)), \qquad R(s) = \operatorname{cov}(X(t)X(t+s)), \quad (s = 0, \pm 1, \pm 2, \ldots)$$

and

$$\begin{aligned}C(\tau_1, \tau_2) &= E(X(t) - \mu)(X(t+\tau_1) - \mu)(X(t+\tau_2) - \mu) \\ &= C(\tau_1, \tau_2) = C(-\tau_1, \tau_2 - \tau_1) = C(\tau_1 - \tau_2, -\tau_2).\end{aligned} \tag{6.1}$$

The spectral density, $f(\omega)$, is the Fourier transform of $R(s)$ and the Fourier transform of $C(\tau_1, \tau_2)$ is defined as the bispectral density function. It is given by

$$f(\omega_1, \omega_2) = \frac{1}{(2\pi)^2} \sum_{\tau_1=-\infty}^{\infty} \sum_{\tau_2=-\infty}^{\infty} C(\tau_1, \tau_2)\,\mathrm{e}^{-\mathrm{i}\tau_1\omega_1 - \mathrm{i}\tau_2\omega_2}, \quad -\pi \leqslant \omega_1, \omega_2 \leqslant \pi. \tag{6.2}$$

In view of the symmetric relations (6.1), we have

$$\begin{aligned}f(\omega_1, \omega_2) = f(\omega_2, \omega_1) = f(-\omega_1, -\omega_1 - \omega_2) &= f(-\omega_1 - \omega_2, \omega_2) \\ &= \overline{f(-\omega_1, -\omega_2)},\end{aligned} \tag{6.3}$$

where $\overline{f(-\omega_1, -\omega_2)}$ is the complex conjugate of $f(-\omega_1, -\omega_2)$. The (ω_1, ω_2) plane can be divided into twelve sectors, and because of the symmetric relations the bispectral density is completely specified on any one of the twelve sectors, including the boundaries (see Van Ness, 1966).

There is a considerable literature on the estimation of $f(\omega)$; nevertheless, we briefly describe the estimation of $f(\omega)$ to develop the necessary ideas and terminology for later use.

Estimation of $f(\omega)$

Let $(X(1), X(2), \ldots, X(N))$ be a realisation of $\{X(t)\}$. Then the natural estimates of μ, $R(\tau)$ and $C(\tau_1, \tau_2)$, respectively, are

$$\bar{X} = \frac{1}{N}\sum_{t=1}^{N} X(t), \qquad \hat{R}(\tau) = \frac{1}{N}\sum_{t=1}^{N-|\tau|} (X(t) - \bar{X})(X(t+|\tau|) - \bar{X}) \quad (\tau = 0, \pm 1, \ldots, \pm(N-1))$$

and

$$\hat{C}(\tau_1, \tau_2) = \frac{1}{N}\sum_{t=1}^{N-\gamma} (X(t) - \bar{X})(X(t+\tau_1) - \bar{X})(X(t+\tau_2) - \bar{X}), \tag{6.4}$$

where $\gamma = \max(0, \tau_1, \tau_2)$; $\tau_1 \geqslant 0$, $\tau_2 \geqslant 0$.

A form of spectrum estimate is

$$\hat{f}(\omega) = \frac{1}{2\pi}\sum_{\tau=-(N-1)}^{N-1} \lambda\left(\frac{\tau}{M}\right)\hat{R}(\tau)\cos\omega\tau, \tag{6.5}$$

where $M = M(N)$ and $\lambda(\cdot)$ is a lag window generator. If $\lambda(s) = 0$ for $|s| > 1$, M corresponds to the truncation point. We assume that the function $\lambda(s)$ is a bounded, even and square integrable such that $\lambda(0) = 1$. The integer M (M is a function of N) is chosen such that $M \to \infty$ as $N \to \infty$ but $M/N \to 0$. Let r be the largest integer such that

$$k^{(r)} = \lim_{s \to 0} \frac{1 - \lambda(s)}{|s|^r}.$$

The integer r is known as the 'characteristic exponent' of the function $\lambda(s)$. It is well known (see Parzen, 1957; Priestley, 1981) that $E(\hat{f}(\omega)) \simeq f(\omega)$ and

$$\operatorname{var}(\hat{f}(\omega)) \simeq \frac{M}{N} f^2(\omega) \int_{-\infty}^{\infty} \lambda^2(s)\,\mathrm{d}s \quad (\omega \neq 0, \pi),$$

which shows that $\hat{f}(\omega)$ is a consistent estimate of $f(\omega)$. The basic problem in the estimation of $f(\omega)$ is to find a suitable lag window $\lambda(s)$. During the decade 1955–1965, several authors suggested various windows, some of which depend on unknown parameters of the spectral density function $f(\omega)$. A comparison of

Table 1
Lag window generators

Daniell window	$\lambda_D(s) = \dfrac{\sin(s\pi)}{s\pi}$	
Tukey–Hamming window	$\lambda_T(s) = \begin{cases} 0.54 + 0.46 \cos \pi s \\ 0 \end{cases}$	$\|s\| < 1$ otherwise
Parzen window	$\lambda_P(s) = \begin{cases} 1 - 6s^2 + 6\|s\|^3 \\ 2(1 - \|s\|)^3 \\ 0 \end{cases}$	$\|s\| \leqslant \frac{1}{2}$ $\frac{1}{2} \leqslant s \leqslant 1$ otherwise
Bartlett–Priestley window	$\lambda_{BP}(s) = \dfrac{3}{(\pi s)^2} \left\{ \dfrac{\sin \pi s}{\pi s} - \cos \pi s \right\}$	

these windows has been made by Neave (1972). Some of the windows which are currently used are given in Table 1. All these windows, given in Table 1, have characteristic exponent 2.

Taking the relative mean square error as the optimality criterion, Priestley (1962) has shown that the Bartlett–Priestley window is optimal amongst all nonnegative windows with the characteristic component 2. (Bartlett (1966, p. 316) has suggested the same window independently.) In fact, Priestley (1981) defines an '*efficiency index*' of a window which is proportional to the relative mean square error. He shows that the Bartlett–Priestley window has the smallest index value amongst all the nonnegative windows with the characteristic component 2. We now consider the estimation of the bispectral density function using the spectral window approach, and it is also possible to estimate the bispectrum using the fast Fourier transforms and the method of complex demodulation (for example, see Huber et al. 1971; Godfrey, 1965).

Estimation of the bispectral density function

Let $K_0(\theta_1, \theta_2)$ be a bounded and *nonnegative* function satisfying

(i) $$\int_{-\infty}^{\infty}\int K_0(\theta_1, \theta_2)\, d\theta_1\, d\theta_2 = 1$$

(ii) $$\int_{-\infty}^{\infty}\int K_0^2(\theta_1, \theta_2)\, d\theta_1\, d\theta_2 < \infty\,; \qquad \int_{-\infty}^{\infty}\int \theta_i^2 K_0(\theta_1, \theta_2)\, d\theta_1\, d\theta_2 < \infty\,, \quad i = 1, 2 \tag{6.6}$$

(iii) $$K_0(\theta_1, \theta_2) = K_0(\theta_2, \theta_1) = K_0(\theta_1, -\theta_1 - \theta_2) = K_0(-\theta_1 - \theta_2, \theta_2)\,.$$

Let $\lambda(\tau_1, \tau_2)$ be its inverse Fourier transform,

$$\lambda(\tau_1, \tau_2) = \int_{-\infty}^{\infty}\int e^{i\tau_1\theta_1 + i\tau_2\theta_2} K_0(\theta_1, \theta_2)\, d\theta_1\, d\theta_2\,.$$

$\lambda(\tau_1, \tau_2)$ also satisfies the symmetric conditions

$$\lambda(\tau_1, \tau_2) = \lambda(\tau_2, \tau_1) = \lambda(-\tau_1, \tau_2 - \tau_1) = \lambda(\tau_1 - \tau_2, -\tau_2).$$

Then the bispectral estimate $\hat{f}(\omega_1, \omega_2)$ is given by

$$\hat{f}(\omega_1, \omega_2) = \frac{1}{(2\pi)^2} \sum_{\tau_1=-(N-1)}^{N-1} \sum_{\tau_2=-(N-1)}^{N-1} \lambda\left(\frac{\tau_1}{M}, \frac{\tau_2}{M}\right) \hat{c}(\tau_1, \tau_2)\, \mathrm{e}^{-\mathrm{i}\tau_1\omega_1 - \mathrm{i}\tau_2\omega_2}, \tag{6.7}$$

where M, which is a function of N, is a window parameter chosen such that $M^2/N \to 0$ as $M \to \infty$, $N \to \infty$.

The variance of the bispectral estimate, when (ω_1, ω_2) are defined in one of the twelve sectors is given by (see Brillinger and Rosenblatt, 1967a,b; Rosenblatt and Van Ness, 1965; Van Ness, 1966)

$$\operatorname{var}(\hat{f}(\omega_1, \omega_2) = \frac{M^2}{N} \cdot \frac{V_2}{2\pi} f(\omega_1) f(\omega_2) f(\omega_1 + \omega_2) \quad (0 < \omega_2 < \omega_1),$$

where

$$V_2 = \int_{-\infty}^{\infty}\int \lambda^2(u_1, u_2)\, \mathrm{d}u_1\, \mathrm{d}u_2 = (2\pi)^2 \int_{-\infty}^{\infty}\int K_0^2(\theta_1\theta_2)\, \mathrm{d}\theta_1\, \mathrm{d}\theta_2. \tag{6.8}$$

If the parameter M is chosen such that, as $N \to \infty$, $M \to \infty$, $M/N \to 0$, then the bispectral estimate is a consistent estimate of $f(\omega_1, \omega_2)$. Under certain conditions on the spectral windows, Gabr and Subba Rao (1979) (see also Gabr, 1981) have shown that the mean square error of the bispectral estimate is minimum for the class of estimates of the form given by (6.7) if the bispectral window $K(\theta_1, \theta_2)$ is chosen as

$$K^*(\theta_1, \theta_2) = \begin{cases} \dfrac{\sqrt{3}}{\pi^3}\left[1 - \dfrac{1}{\pi^2}(\theta_1^2 + \theta_2^2 + \theta_1\theta_2)\right] & \text{if } (\theta_1, \theta_2) \in G, \\ 0 & \text{otherwise}, \end{cases} \tag{6.9}$$

where the region G is given by the set $\{(\theta_1, \theta_2);\ \theta_1^2 + \theta_2^2 + \theta_1\theta_2 \leq \pi^2\}$.

In order to calculate an analytic expression for the inverse transform of $K^*(\theta_1, \theta_2)$, we approximate the region G by the set G_1 given by $\{(\theta_1, \theta_2);\ |\theta_1| + |\theta_2| + |\theta_1 + \theta_2| < 2\pi\}$. An approximate expression for the lag window is now given by

$$\begin{aligned}
\lambda^*(s_1, s_2) &= \int_{-\pi}^{\pi}\int \mathrm{e}^{\mathrm{i}s_1\theta_1 + \mathrm{i}s_2\theta_2} K^*(\theta_1, \theta_2)\, \mathrm{d}\theta_1\, \mathrm{d}\theta_2 \\
&\simeq \frac{8}{7\pi^3}\left\{\frac{1}{\pi}\left[\frac{(2s_1^2 + 2s_2^2 + s_1s_2)}{s_1^3s_2^3}\cos(s_2 - s_1)\pi\right.\right. \\
&\quad + \frac{(2s_1^2 + 5s_2^2 - 5s_1s_2)\cos s_1\pi}{s_2^3(s_2 - s_1)^3} + \left.\frac{(5s_1^2 + 2s_2^2 - 5s_1s_2)\cos s_2\pi}{(s_1 - s_2)^3 s_1^3}\right] \\
&\quad - \left[\frac{(s_2 - s_1)}{s_1^2s_1^2}\sin(s_2 - s_1)\pi + \frac{s_1}{s_2^2(s_2 - s_1)^2}\sin(s_1\pi)\right. \\
&\quad + \left.\left.\frac{s_2}{s_1^2(s_2 - s_1)^2}\sin(s_2\pi)\right]\right\}.
\end{aligned} \tag{6.10}$$

$\to \omega$

An alternative to the optimal lag window $\lambda^*(s_1, s_2)$ is to choose the two-dimensional lag window as a product of the one-dimensional windows used in the estimation of $f(\omega)$.

These product windows can be obtained from

$$\lambda_{sG}(s_1, s_2) = \lambda(s_1)\lambda(s_2)\lambda(s_1 - s_2), \tag{6.11}$$

where $\lambda(s)$ are given in Table 1. It is interesting to compare the shape of these windows with respect to the optimal lag window (6.10). These windows are plotted in Figs. 3, 4, 5, 6 and 7. The optimal lag window has a much flatter surface when compared to other windows, and the rate of decay of $\lambda(s_1, s_2)$ as $s_1 \to \infty$, $s_2 \to \infty$ is much slower than for other windows. This means that the

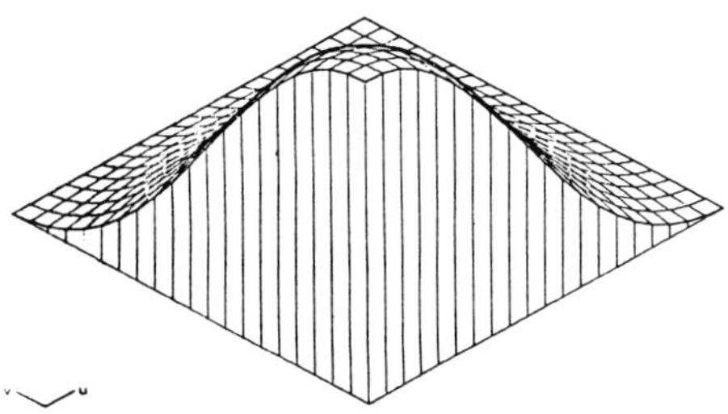

Fig. 3. Daniell window.

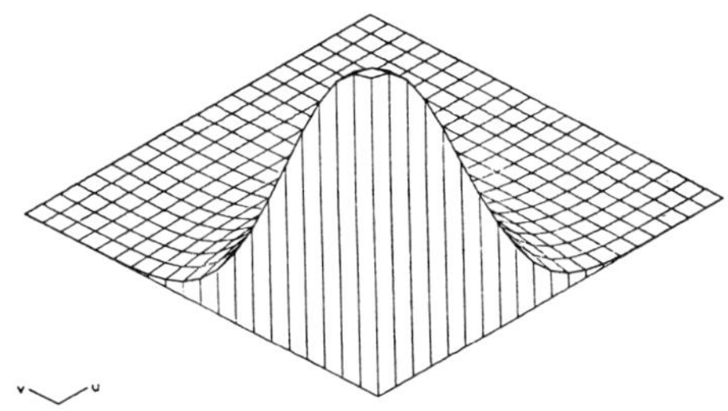

Fig. 4. Parzen window.

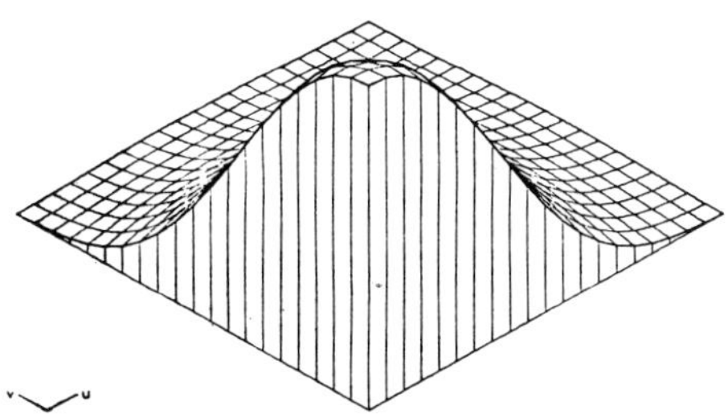

Fig. 5. Tukey window.

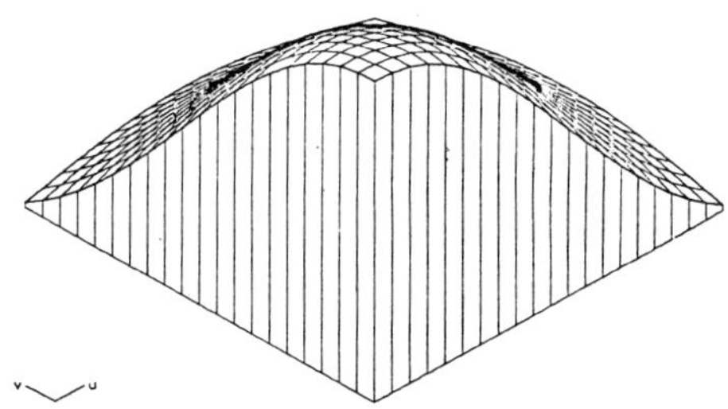

Fig. 6. Barlett–Priestley window.

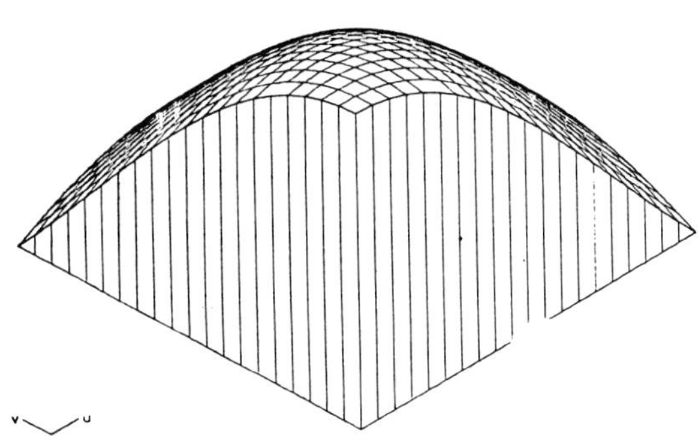

Fig. 7. Optimum window.

Fourier transform of this window will be like a two-dimensional Dirac delta function concentrating all its mass around the origin (0, 0) and this, of course, is a desirable property.

7. Bispectral analysis of the bilinear time-series model BL(1, 0, 1, 1)

In this section we obtain an exact expression for the bispectral density function of BL(1, 0, 1, 1) model. For higher-order bilinear models, the expressions for higher-order spectra are very difficult to obtain. Besides, it must be noted that high-order moments need not always exist.

Let the time series $\{X(t)\}$ satisfy the model BL(1, 0, 1, 1),

$$X(t) + aX(t-1) = bX(t-1)e(t-1) + e(t), \tag{7.1}$$

where $\{e(t)\}$ is a sequence of independent identically distributed $N(0, 1)$ variables. The time series $\{X(t)\}$ generated from the bilinear time-series model (7.1) is asymptotically second-order stationary if $a^2 + b^2 < 1$. Under this condition, the expressions for mean, variance and covariance are given in Section 4. In order to obtain expressions for the third-order moments and hence the bispectral density, we proceed as follows.

From (7.1) we can show

$$\begin{aligned} \mu &= b/(1+a), \\ \mu_2 &= E[X^2(t)] = [1 + 2b^2 - 4ab\mu]/(1 - a^2 - b^2), \qquad (7.2) \\ \mu_3 &= E[X^3(t)] = \frac{1}{1 + 3ab^2 + a^3}[b^3Q_3 + 3a^2bQ_1 + 3\mu(1 - 6ab^2)], \end{aligned}$$

where

$$Q_1 = E[X^3(t-1)e(t-1)] = \frac{3}{1-b^2}(1 + a^2\mu_2 + 2b^2 - 4ab\mu),$$

$$Q_2 = E[X^2(t-1)e^2(t-1)] = \frac{1}{1+3ab^2}[-a^3\mu_3 + b^3Q_3 + 2a^2bQ_1 + 9\mu],$$

$$Q_3 = E[X^3(t-1)e^3(t-1)] = \frac{3}{1-b^2}(5 + 4b^2 + 3a^2\mu_2 - 12ab\mu).$$

Hence, $C(0, 0) = \mu_3 - 2\mu\mu_2 + 2\mu^3$. A sufficient condition for μ_3 to be finite is that $a^2 + 3b^2 < 1$.

All the third-order moments can be obtained by solving a set of difference equations which are too long to describe here. These equations can be solved using the generating functions. Then one can show that the bispectral density function of $\{X(t)\}$ is of the form

$$f(\omega_1, \omega_2) = \frac{1}{(2\pi)^2}\{C(0,0) + F_1(\omega_1+\omega_2) + F_1(-\omega_1) + F_1(-\omega_2) + F_2(\omega_1) \\ + F_2(\omega_2) + F_2(-\omega_1-\omega_2) + g_1(\omega_1+\omega_2)[F_3(\omega_2) + F_3(\omega_1)] \\ + g_1(-\omega_2)F_3(\omega_2) + g_1(-\omega_2)F_3(\omega_1) \\ + [g_1(-\omega_1) + g_1(-\omega_2)]F_3(-\omega_1-\omega_2)\}, \tag{7.3}$$

where

$$g_1(\omega) = \frac{z}{1+az}, \quad z = e^{i\omega},$$

$$g_2(\omega) = [-a\mu_2 + (1+2a)\mu^2]g_1(\omega),$$

$$g_3(\omega) = \frac{1}{1-(a^2+b^2)z}[(\nu_2 - \mu\mu_2)z^2 + 4a^2bz^2g_2(\omega)],$$

$$F_1(\omega) = (-a\mu_3 + bQ_1 - \mu\mu_2)g_1(\omega) - 2\mu g_2(\omega),$$

$$F_2(\omega) = \{\nu_1 - (1-2a)\mu\mu_2 - 2(1+2a)\mu^3\}z + g_3(\omega),$$

$$F_3(\omega) = \{-a\nu_1 - (1+3a)a\mu\mu_2 + (2+7a+6a^2)\mu^3\}z - ag_3(\omega) - \mu g_2(\omega) \\ + \{-2abg_2(\omega) - a^2\mu(-a\mu_2 + (1+2a)\mu^2)g_1(\omega)\}z,$$

$$\nu_1 = a^2\mu_3 + b^2Q_2 - 2abQ_1 + \mu,$$

$$\nu_2 = (a^2+b^2)\nu_1 + 4a^2b\mu_2 + (1+2b^2-8ab^2)\mu.$$

The bispectral density function can now be calculated from (7.3) for any values of a and b. For our illustration we have chosen $a = -0.4$, $b = 0.4$, and the modulus of the bispectral density function is estimated from this sample using the optimum weight function. The modulus of the bispectral density function calculated from (7.3) is shown in Fig. 8. A time series $(X(t); t = 1, 2, \ldots, 1000)$ is generated from (7.1) when $a = -0.4$, $b = 0.4$. The bispectral density function is estimated from this sample using the optimum weight functions. The truncation point M is chosen to be equal to 30. The modulus of the estimated bispectral

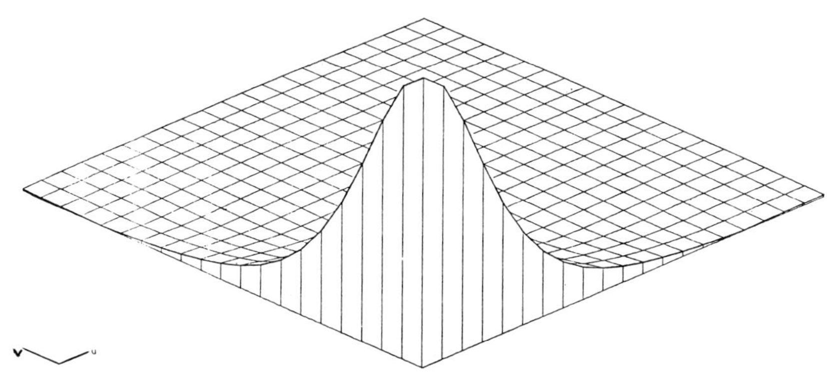

Fig. 8. Parametric bispectrum.

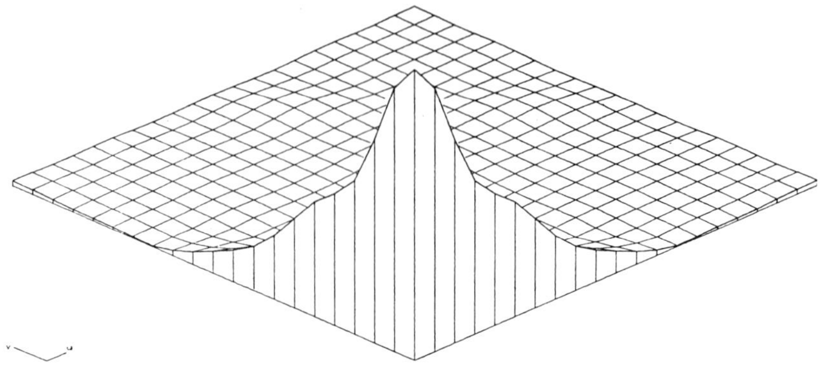

Fig. 9. Estimated bispectrum.

density function is plotted in Fig. 9. The bispectrum obtained by using the optimum weight function compares well with the theoretical bispectrum.

8. Tests for departure from linearity and Gaussianity

The use of cumulant spectra in assessing the departure from linearity (and Gaussianity) has been pointed out by Brillinger and Rosenblatt (1967a) and Brillinger (1965). In this section we briefly discuss the actual construction of a test statistic based on the bispectral density function for testing these hypotheses (for details, see Subba Rao and Gabr, 1980). These tests are illustrated with two well-known examples in time series.

Let us now assume that the time series $\{X(t)\}$ has the linear representation

$$X(t) = \sum_{u=-\infty}^{\infty} a(u)e(t-u), \tag{8.1}$$

where $\{e(t)\}$ are mutually independent with $E(e(t)) = 0$, $E(e^2(t)) = \sigma_e^2$, $E(e^3(t)) = \mu_3$. Then the following relations can be obtained easily (Brillinger, 1965).

$$f(\omega) = \frac{\sigma_e^2}{2\pi} |H(\omega)|^2 ,$$

$$f(\omega_i, \omega_j) = \frac{\mu_3}{(2\pi)^2} H(-\omega_i - \omega_j)H(\omega_i)H(\omega_j) , \tag{8.2}$$

where

$$H(\omega) = \sum_{u=-\infty}^{\infty} a(u)\, \mathrm{e}^{-iu\omega} .$$

Thus, if we write

$$X_{ij} = \frac{|f(\omega_i, \omega_j)|^2}{f(\omega_i)f(\omega_j)f(\omega_i+\omega_j)}, \tag{8.3}$$

then from (8.2) we obtain

$$X_{ij} = \frac{\mu_3^2}{2\pi\sigma_e^6} \quad \text{all } i, j. \tag{8.4}$$

The relation (8.2) shows that if $\mu_3 = 0$, then the bispectral density function is zero for all ω_i and ω_j. Of course, if the random variables $\{e(t)\}$ are Gaussian, then $\mu_3 = 0$ and hence $f(\omega_i, \omega_j) = 0$.

However, it may happen that the process $\{X(t)\}$ is linear but not necessarily Gaussian. In this situation, X_{ij} given by (8.3) is constant for all i and j. In other words, the constancy of X_{ij} is a test for adequacy of linearity of the process (Brillinger, 1965). This leads us to the construction of the tests (i) H_0: $f(\omega_i, \omega_j) = 0$ for all ω_i and ω_j, and (ii) H_1: X_{ij} is constant for all ω_i and ω_j. Acceptance of H_0 implies that the process is linear and $\mu_3 = 0$. Rejection of H_0, but acceptance of H_1, implies that the process $X(t)$ is linear but $\mu_3 \neq 0$.

To test the hypothesis $f(\omega_i, \omega_j) = 0$, all ω_i, ω_j, we proceed in two stages. The first stage consists of testing the hypothesis $f(\omega_i, \omega_j) = 0$ when ω_i and ω_j are restricted to the range $0 < \omega_i < \pi$, $\omega_i < \omega_j < \pi$ (excluding the boundaries and the origin of the region (2) of Fig. 10). Within this region the bispectral

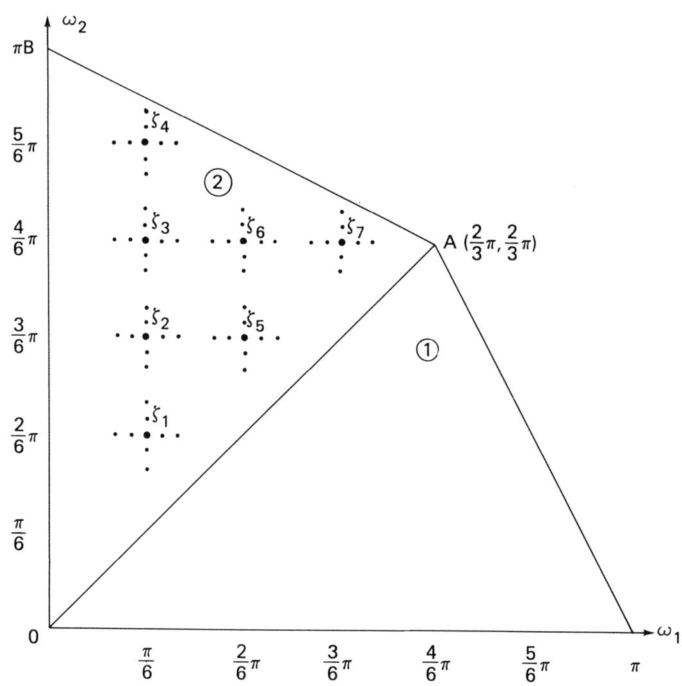

Fig. 10. Bispectrum sample with $K = 6$, $L = 4$, $r = 2$, $P = 7$, $n = 9$.

estimate is approximately a complex normal and the test we describe below depends on the complex analogue of Hotelling T^2 statistic (see Giri, 1965; Khatri, 1965). If we accept the null hypothesis at the first stage, then in order to confirm that the time series $\{X(t)\}$ is linear and $\mu_3 = 0$, we must also test that $f(\omega_i, \omega_j) = 0$ when ω_i and ω_j are defined at the origin and on the boundaries. This is our second stage of the test. In this context, we use Hotelling T^2 for real random variables (Anderson, 1958; Kshirsagar, 1972).

We now consider the first stage of the procedure. We construct a column vector from the set of bispectral densities $f(\omega_i, \omega_j)$ defined on the plane $0 < \omega_i < \pi$, $\omega_i < \omega_j < \pi$. To obtain this column vector we proceed as follows.

Divide the interval $(0, \pi)$ into K equally spaced intervals, where $K \ll N$. In view of the symmetric relations (6.3) satisfied by the bispectral density function, it is sufficient to restrict the frequencies (ω_i, ω_j) to the region OAB of Fig. 10. The equations of the lines OA, AB and OB which define the region OAB are, respectively, $\omega_j - \omega_i = 0$, $\omega_j = -\frac{1}{2}\omega_i + \pi$ and $\omega_i = 0$, where $\omega_i = i\pi/K$, $\omega_j = j\pi/K$ $(i = 1, 2, \ldots, L;\ j = i+1, i+2, \ldots, \gamma(i);\ L = [2K/3])$. The restriction that $L = [2K/3]$ and $\gamma(i) = K - [i/2] - 1$ comes from the fact that the frequencies (ω_i, ω_j) must be within the region OAB.

Let $\eta_{ij} = f(\omega_i, \omega_j)$ and for each i $(i = 1, 2, \ldots, L)$ define the vector

$$\boldsymbol{\eta}'_i = (\eta_{i,i+1}, \eta_{i,i+2}, \ldots, \eta_{i,\gamma(i)}) \tag{8.5}$$

and let the vector $\boldsymbol{\eta}$ be defined by

$$\boldsymbol{\eta}' = (\boldsymbol{\eta}'_1, \boldsymbol{\eta}'_2, \ldots, \boldsymbol{\eta}'_L)\,. \tag{8.6}$$

We now relabel the elements of $\boldsymbol{\eta}$ as

$$\boldsymbol{\eta}' = (\zeta_1, \zeta_2, \ldots, \zeta_P)\,, \tag{8.7}$$

where $P = \sum_{i=1}^{L} (\gamma(i) - i)$, so that for each l $(1 \leqslant l \leqslant P)$, $\zeta_l = \eta_{ij}$ for some i, j, satisfying $1 \leqslant i \leqslant L$; $i + 1 \leqslant j \leqslant \gamma(i)$ (see Fig. 10 for an illustration).

We now form a set of (approximately) uncorrelated estimates of each ζ_i by constructing a 'fine' frequency grid around each (ω_i, ω_j) point. Specifically, for each (ω_i, ω_j), let

$$\omega_{i_p} = \omega_i + \frac{pd\pi}{N}, \quad p = -r, -r+1, \ldots, 0, 1, \ldots, r,$$

$$\omega_{j_q} = \omega_j + \frac{qd\pi}{N}, \quad q = -r, -r+1, \ldots, -1, 1, \ldots, r, \quad (q \neq 0),$$

where the distance 'd' is chosen so that the bispectral estimates at neighbouring points on this fine grid are approximately uncorrelated. (In effect, this means that d must be chosen so that $\pi d/N$ must be greater than the bandwidth of the spectral window corresponding to the lag window $\lambda(s)$.) Now let $\hat{f}(\omega_{i_p}, \omega_{j_q})$

denote the estimated bispectral density function (as given by (6.7)) at the points $(\omega_{i_p}, \omega_{j_p})$. Assuming that the true bispectral density function is sufficiently smooth so as to be effectively constant over the 'fine' frequency grid, we may write

$$E[\hat{f}(\omega_{i_p}, \omega_{j_q})] = f(\omega_i, \omega_j) \quad \text{all } p, q .$$

We may thus regard the set of estimators $\{\hat{f}(\omega_{i_p}, \omega_{j_q})\}$ as $n = 4r + 1$ approximately uncorrelated and unbiased estimates of $f(\omega_i, \omega_j)$. To facilitate the analogy with standard multivariate tests, we now form the bispectral estimates

$$\left\{\hat{f}\left(\omega_i - \frac{rd\pi}{N}, \omega_j\right), \hat{f}\left(\omega_i - \frac{(r-1)d\pi}{N}, \omega_j\right), \ldots \quad \ldots, \right.$$
$$\left. \hat{f}\left(\omega_i + \frac{rd\pi}{N}, \omega_j\right) \ldots \quad \ldots \hat{f}\left(\omega_i, \omega_j + \frac{rd\pi}{N}\right)\right\}$$

into an $n \times 1$ vector, which after relabelling, can be denoted by

$$\boldsymbol{\xi}'_l = (\xi_{l_1}, \xi_{l_2}, \ldots, \xi_{l_n}) ,$$

where

$$\xi_{l_1} = \hat{f}\left(\omega_i - \frac{rd\pi}{N}, \omega_j\right), \qquad \xi_{l_2} = \hat{f}\left(\omega_i - \frac{(r-1)d\pi}{N}, \omega_j\right), \qquad \text{etc.}$$

Using this device, we obtain an $n \times 1$ vector of estimates for each element ζ_l $(1 \leq l \leq n)$ of the vector $\boldsymbol{\eta}$ defined by (8.7). We may thus form the complete set of bispectral estimates into a 'data matrix', $\boldsymbol{D}$,

$$D = \begin{bmatrix} \xi_{11} & \xi_{12} & \cdots & \xi_{1n} \\ \xi_{21} & \xi_{22} & \cdots & \xi_{2n} \\ \xi_{P1} & \xi_{P2} & \cdots & \xi_{Pn} \end{bmatrix} = \begin{bmatrix} \boldsymbol{\xi}'_1 \\ \boldsymbol{\xi}'_2 \\ \boldsymbol{\xi}'_P \end{bmatrix} = [\boldsymbol{\xi}_{(1)}, \boldsymbol{\xi}_{(2)}, \ldots, \boldsymbol{\xi}_{(n)}] ,$$

where

$$\boldsymbol{\xi}'_{(i)} = (\xi_{1j}, \xi_{2i}, \ldots, \xi_{Pi}) \quad (i = 1, 2, \ldots, n) .$$

For large N, $\xi_{(i)}$ $(i = 1, 2, \ldots, n)$ is distributed as complex normal with mean $\boldsymbol{\eta}$ and variance–covariance matrix $\boldsymbol{\Sigma}_\xi$. Under the null hypothesis that the process is linear (and $\mu_3 = 0$), the mean vector $\boldsymbol{\eta} = \boldsymbol{0}$. The maximum likelihood estimates of $\boldsymbol{\eta}$ and $\boldsymbol{\Sigma}_\xi$ are, respectively,

$$\hat{\boldsymbol{\eta}} = \frac{1}{n}\sum_{l=1}^{n} \boldsymbol{\xi}_{(l)} , \qquad \hat{\boldsymbol{\Sigma}}_\xi = \frac{A}{n} , \qquad A = \sum_{l=1}^{n} (\boldsymbol{\xi}_{(l)} - \hat{\boldsymbol{\eta}})(\boldsymbol{\xi}_{(l)} - \hat{\boldsymbol{\eta}})^* .$$

The likelihood ratio test for testing the hypothesis $\boldsymbol{\eta} = \boldsymbol{0}$ against the alternative $\boldsymbol{\eta}^* \boldsymbol{\Sigma}_\xi^{-1} \boldsymbol{\eta} > 0$ leads to the rejection of the hypothesis if the statistic (see Giri, 1965; Kharti, 1965) $T^2 = n\hat{\boldsymbol{\eta}}^* \boldsymbol{A}^{-1} \hat{\boldsymbol{\eta}}$ is greater than a constant λ, where λ is determined by the significance level α. Under the null hypothesis the statistic

$$\mathscr{F}_1 = \frac{2(n-P)}{2P} T^2$$

is distributed as a central F with $(2P, 2(n-P))$ degrees of freedom. If the null hypothesis is accepted, we proceed to the second stage of the testing procedure.

Test for linearity

To test whether $\{X(t)\}$ is linear, but allowing for $\mu_3 \neq 0$, we now use the property that the ratio X_{ij} given by (8.3) is constant for all ω_i and ω_j. As before, we form a column vector of the ratios $\{X_{ij}\}$, $0 < \omega_i < \pi$, $\omega_i < \omega_j < \pi$. It may be noted that in this column vector we may include the elements X_{ij} defined at the origin and all the points on the line $\omega_i = 0$ since X_{ij}'s are always real. Let

$$\hat{X}_{ij} = \frac{|\hat{f}(\omega_i, \omega_j)|^2}{\hat{f}(\omega_i)\hat{f}(\omega_j)\hat{f}(\omega_i + \omega_j)} \tag{8.8}$$

be an estimate of X_{ij}. Then, for large N, $\hat{X}_{ij}$ is approximately normally distributed (Brillinger, 1965, p. 1368).

As before, we can form the $\{\hat{X}_{ij}\}$ into a $P \times 1$ column vector

$$\boldsymbol{Y}' = (Y_1, Y_2, \ldots, Y_P),$$

where, for each l, $Y_l = \hat{X}_{ij}$ for some pair of integers (i, j).

If the null hypothesis is true, then $E(Y_1) = E(Y_2) = \cdots = E(Y_P)$. This corresponds to a classical problem of symmetry in multivariate analysis (Anderson, 1958; Kshirsagar, 1972). The test is well known and the details can be found in the above references. Let $\boldsymbol{Y}_1, \boldsymbol{Y}_2, \ldots, \boldsymbol{Y}_n$ be a random sample of size n, and let

$$\bar{\boldsymbol{Y}} = \frac{1}{n}\sum_{i=1}^{n} Y_i, \qquad \hat{\boldsymbol{S}}_Y = \sum_{i=1}^{n} (\boldsymbol{Y}_i - \bar{\boldsymbol{Y}})(\boldsymbol{Y}_i - \bar{\boldsymbol{Y}})' \quad \text{and } \hat{\boldsymbol{\Sigma}}_Y = \frac{1}{n}\hat{\boldsymbol{S}}_Y.$$

Define a column vector $\boldsymbol{\beta}$ of order $Q \times 1$, where $Q = P - 1$, such that

$$\boldsymbol{\beta} = \boldsymbol{B}\boldsymbol{Y},$$

where $\boldsymbol{B}$ is a matrix of order $Q \times P$ and it is of the form

$$\boldsymbol{B} = \begin{bmatrix} 1 & -1 & 0 & \cdots & 0 \\ 0 & 1 & -1 & \ddots & 0 \\ & & \ddots & \ddots & \\ 0 & 0 & \cdots & 1 & -1 \end{bmatrix}$$

Under the null hypothesis, $\boldsymbol{\beta}$ is asymptotically jointly normally distributed with mean vector $\mathbf{0}$ and variance–covariance matrix $\boldsymbol{B\Sigma}_Y\boldsymbol{B}'$. The likelihood ratio test for testing the null hypothesis leads to the rejection of the hypothesis if the statistic $T^2 = n\bar{\boldsymbol{\beta}}'\hat{\boldsymbol{S}}\bar{\boldsymbol{\beta}}$, where $\bar{\boldsymbol{\beta}} = \boldsymbol{B}\bar{\boldsymbol{Y}}$, $\hat{\boldsymbol{S}} = \boldsymbol{BS}_Y\boldsymbol{B}'$, is greater than a constant λ_0 where λ_0 is determined by the significance level α. The statistic

$$\mathcal{F}_2 = \frac{n-Q}{Q} T^2$$

has, under the null hypothesis, an F distribution with $(Q, n-Q)$ degrees of freedom. The statistic is invariant and it is independent of the choice of $\boldsymbol{B}$ (Anderson, 1958, p. 111).

Numerical illustrations

Sunspot numbers

We consider the Wolfer annual sunspot numbers for the years 1700–1955 (Waldmeirer, 1961) consisting of 256 observations. The bispectral density function is estimated using the optimum weight function and the truncation point is chosen to be equal to $M = 30$. The modulus of the bispectral density function is calculated for several frequencies (ω_1, ω_2) but in Fig. 11, we have given the plot of the modulus of the bispectrum for the frequencies $\omega_1 = \omega_2 = 0.10\pi\,(0.01\pi)\,0.30\pi$. There is a 'big' spike in the neighbourhood of the frequency $\omega_1 = \omega_2 = 0.565$ which corresponds to the periodicity 11 years (approximately).

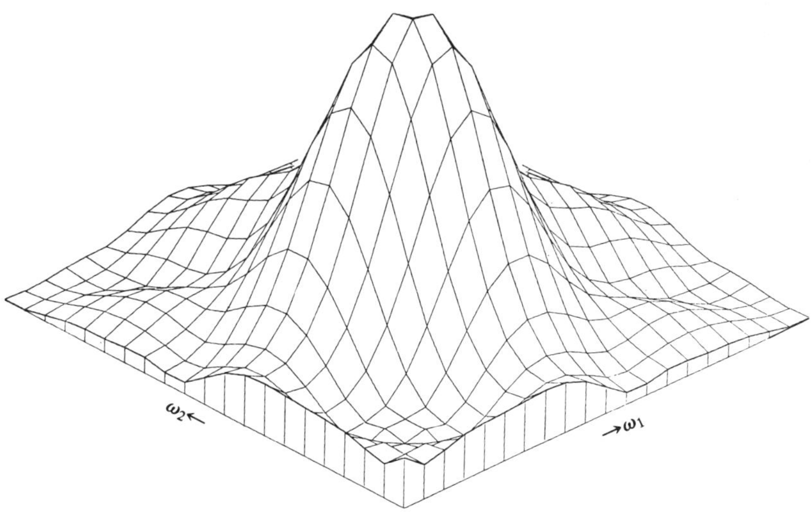

Fig. 11. Sunspot numbers.

Canadian lynx data

The second series we consider is the annual record of the numbers of Canadian lynx trapped in the Mackenzie river district of North West Canada for the years 1821–1934 (inclusive), giving a total of 114 observations. The bispectral density function is estimated using the optimum weight function, and the modulus calculated over the same frequency range as the above example is given in Fig. 12. There is a dominant spike in the neighbourhood of frequency $\omega_1 = \omega_2 = 0.6597$ which corresponds to, approximately, the periodicity 9.5 years.

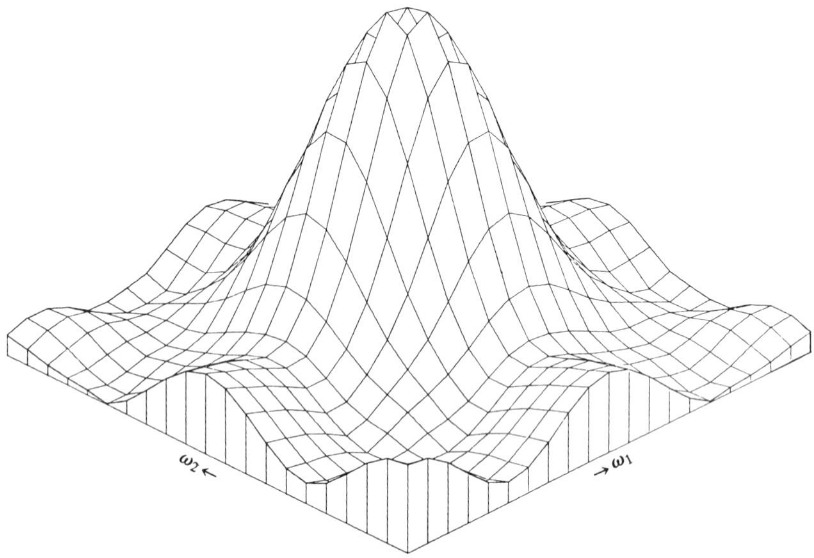

Fig. 12. Canadian lynx data.

Logarithm of the Canadian lynx data

The logarithm transformation has been proposed by several research workers as a means of making the Canadian lynx data nearly Gaussian (see Campbell and Walker, 1977). Hence it is interesting to consider this series as well for test purposes.

The parameters M, K, L, d, r, P and n that are used for constructing the $\mathscr{F}_1$ and $\mathscr{F}_2$ statistics are given in Table 2.

The values of T^2 and $\mathscr{F}_1$ are given in Table 3.

The values of $\mathscr{F}_1$ for the three series are much greater than the percentage point confirming the general belief that the series are non-linear.

We now proceed to test the hypothesis that the series may be linear but $\mu_3 = 0$. To test the hypothesis we use the statistic $\mathscr{F}_2$. The values of T^2 and $\mathscr{F}_2$ are given in Table 4.

From Table 4, it is clear that the sunspot numbers and Canadian lynx data are obviously not linear, but, surprisingly, the logarithm of the Canadian lynx data is linear, though not Gaussian.

Table 2
Values of (M, K, L, d, r, P, n)

	N	M	K	L	d	r	P	n
Sunspot numbers	256	30	6	4	8	2	7	9
Canadian lynx data	114	25	6	4	3.5	2	7	9
Logarithm of Canadian lynx data	114	25	6	4	3.5	2	7	9

Table 3
Values of $(T^2, \mathscr{F}_1)$

	T^2	$\mathscr{F}_1$	5% upper point of $F_{(14,4)}$
Sunspot numbers	11828.063	3379.447	5.89
Canadian lynx data	5038.336	1439.525	5.89
Logarithm of Canadian lynx data	2156.627	616.179	5.89

Table 4
Values of $(T^2, \mathscr{F}_2)$

	T^2	$\mathscr{F}_2$	5% upper point of $F_{(6,3)}$
Sunspot numbers	620.59	310.29	8.94
Canadian lynx data	1174.58	587.29	8.94
Logarithm of Canadian lynx data	8.39	4.20	8.94

Conclusions

In this paper we have considered the second-order and third-order spectral analysis of nonlinear time series, and in particular the analysis of bilinear time-series models which have been introduced recently. We show that the second-order spectral analysis is not sufficient to distinguish between linear time-series models and bilinear time-series models, since, as far as second-order properties are concerned, they are the same. In view of this, one has to resort to higher-order spectral analysis. The bispectral analysis of the time series may suggest any departure from linearity and normality; and statistical test are constructed for dealing with these problems. Of course, these tests are not exhaustive in the sense that for some nonlinear processes bispectrum can be zero. Still, in quite a number of situations, tests based on bispectrum can provide a useful guide to the presence of nonlinearity as the numerical illustrations included in this paper suggest.

References

Anderson, T. W. (1958). *An Introduction to Multivariate Statistical Analysis.* Wiley, New York.

Bartlett, M. S. (1966). *An Introduction to Stochastic Processes*, 2nd edition. Cambridge University Press, Cambridge, Mass.

Brillinger, D. R. (1965). An introduction to polyspectrum. *Ann. Math. Statist.* **36**, 1351–1374.

Brillinger, D. R. (1975). *Time Series, Data Analysis and Theory.* Holt, Rinehart and Winston, New York.

Brillinger, D. R. (1970). The identification of polynomial systems by means of higher order spectra. *J. Sound Vibration* **12**(3), 301–313.

Brillinger, D. R. and Rosenblatt, M. (1967a). Asymptotic theory of estimates of kth order spectra. In: B. Harris, ed., *Spectral Analysis of Time Series*, 153–188. Wiley, New York.

Brillinger, D. R. and Rosenblatt, M. (1967b). Computation and interpretation of kth order spectra. In: B. Harris, ed., *Spectra Analysis of Time Series*, 189–232. Wiley, New York.

Campbell, M. J. and Walker, A. M. (1977). A survey of statistical work on the Mackenzie river series of annual Canadian lynx trappings for the years 1871–1934 and a new analysis. *J. R. Statist. Soc.* **A140**, 411–431.

Cunningham, W. J. (1958). *Introduction to Non-Linear Analysis.* McGraw-Hill, New York.

Gabr, M. M. and Subba Rao, T. (1979). A note on the estimation of the bispectral density function of a stationary time series. Department of Mathematics, UMIST, Technical Report No. 123.

Gabr, M. M. (1981). Bispectral analysis of non-linear time series and the statistical theory of bilinear time series models wth applications. Unpublished Ph.D thesis, submitted to UMIST.

Giri, N. (1965). On the complex analogues of T and R tests. *Ann. Math. Statist.* **36**, 664–670.

Godfrey, M. D. (1965). An exploratory study of bispectrum of economic time series. *Applied Statistics* **14**, 48–69.

Granger, C. W. J. and Anderson, A. P. (1978). *Introduction To Bilinear Time Series Models.* Vandenhoeck & Ruprecht, Göttingen.

Hannan, E. J. (1970). *Multiple Time Series.* Wiley, New York.

Huber, P. J., Kleiner, B. and Gasser, T. (1971). Statistical methods for investigating phase relations in stationary stochastic processes. *I.E.E.E. Trans. Audio and Electroacoustics* **AU19–20**, 78–86.

Jenkins, G. M. and Watts, D. G. (1968). *Spectral Analysis and Its Applications.* Holden-Day, New York.

Khatri, C. G. (1965). Classical statistical analysis bound on a certain multivariate Gaussian distribution. *Ann. Math. Statist.* **36**, 98–107.

Koopmans, L. H. (1974). *The Spectral Analysis of Time Series*. Academic Press, New York.

Kshirsagar, A. M. (1972). *Multivariate Analysis*. Marcel-Dekker, New York.

Neave, H. R. (1972). A comparison of lag window generators. *J. Amer. Statist. Assoc.* **67**, 152–158.

Parzen, E. (1957). On choosing an estimate of the spectral density function of a stationary time series. *Ann. Math. Statist.* **28**, 921–932.

Priestley, M. B. (1962). Basic considerations in the estimation of spectra. *Technometrics* **4**, 551–564.

Priestley, M. B. (1978). Non-linear models in time series analysis. *The Statistician* **27**, 159–176.

Priestley, M. B. (1981). *Spectral Analysis and Time Series*. Academic Press, London.

Rosenblatt, M. and Van Ness, J. W. (1965). Estimation of the bispectrum. *Ann. Math. Statist.* **36**, 1120–1136.

Rosenblatt, M. B. (1979). Linearity and non-linearity in time series: prediction. *Proc. 42nd Session Internat. Statist. Inst.* (Manila), Book 1, pp. 423–434.

Rosenblatt, M. B. (1980). Linear process and bispectra. *J. Appl. Prob.* **17**, 265–270.

Rosenblatt, M. B. (1981). Cumulants and cumulant spectra. Chapter 17 in this volume.

Subba Rao, T. (1981). On the theory of bilinear time series models. *J. R. Statist. Soc.* **B43**, 244–255.

Subba Rao, T. and Gabr, M. M. (1980). A test for linearity of stationary time series. *J. Time Series Analysis* **1**, 145–158.

Subba Rao, T. and Gabr, M. M. (1981). An introduction to bispectral analysis of stationary time series and the bilinear time series models (lecture notes under preparation).

Van Ness, J. W. (1966). Asymptotic normality of bispectral estimates. *Ann. Math. Statist.* **37**, 1257–1275.

Volterra, V. (1930). *Theory of Functionals and an Integral and Integro-Differential Equations*. Dover, New York.

Waldmeirer, M. (1961). *The Sunspot Activity in the Years 1610–1960.* Schulthese, Zürich.

Wiener, N. (1958). *Non-Linear Problems in Random Theory*. Wiley, New York.

D. R. Brillinger and P. R. Krishnaiah, eds., *Handbook of Statistics, Vol. 3*
© Elsevier Science Publishers B.V. (1983) 321–342

15

Frequency-Domain Analysis of Multidimensional Time-Series Data

Enders A. Robinson

1. Introduction

Information occurring in the form of an image is basic in our technological world. A few examples are television, radar, photographic and X-ray images. Many of these images are of fine quality, others are of lesser quality. However, some poor quality images are so unique and are of such importance that it is worthwhile to devise techniques through which the degrading factors can be removed and the image quality enhanced. Such image restoration techniques exist either as digital or optical processing schemes. As we know, images are representations of objects that are indirectly sensed and that various forms of wave motion are the physical mechanisms which perform the sensing. The image, as well as the object, is a spatial entity and so can be described in terms of spatial coordinates. A two-dimensional image, for example, would be measured in terms of length and breadth. However, the wave motion by which the sensing takes place is a time phenomenon, so in this respect the time coordinate becomes an essential feature. At each spatial point we can conceivably measure a time function, or time series, and the ensemble of such time series over a set of spatial points comprises multidimensional time-series data.

Let us give an example. The X-ray imaging of the human body is essential in many medical procedures. The desired object is within the human body, and we can observe the object by sending X-rays (i.e. wave motion) through the body and recording the X-ray image without the body. The problem is to reconstruct either optically or digitally the shape or properties of the object within the body from the measurements (the image) made outside of the body. The utilization of the properties of the wave motion which connect object to image is essential to the solution of this reconstruction problem.

In all practical problems the multidimensional time series represent an enormous amount of data, often enough data to challenge the capacities of our greatest digital computers. How do we cope with the mathematical analyses of such large quantities of data? Any purely statistical approach which involves the computation of correlation functions and spectra usually becomes overwhelmed by the sheer number of such functions and the need for classifying them in some recognizable and coherent scheme. However, in the case of data generated by physical wave motion, we can use the wave equation. The wave equation links wave data together into a unified physical whole. This equation was first formulated three hundred years ago through the work of Isaac Newton, James Bernoulli, Jean Le Rond D'Alembert and Leonhard Euler. The wave equation is a second-order hyperbolic partial differential equation, and so in a mathematical sense cannot fairly be described as a simple equation. However, in a physical sense if we consider how complicated wave motion can appear, it is indeed remarkable that its essential properties can be described by an equation that can be written down in a few mathematical symbols. The fact that so much of our physical world can be described by the second-order equations of mathematical physics is of deep significance, and indicates that beauty and simplicity are not foreign to each other.

In this paper we are concerned with frequency-domain analysis of multi-dimensional time-series data. The underlying physical concept is that the data represent wave motion so the structure is governed by the wave equation. Our approach will make use of Fourier transformations in order to render space–time data into wavenumber–frequency data. Let us note that we usually use the word frequency to mean temporal frequency (i.e. radians per second), whereas we use the word 'wavenumber' to mean spatial frequency (i.e. radians per meter).

2. The wave equation

We live in a three-dimensional world denoted by the x, y and z coordinates. The other vital coordinate is that of time t. Thus we deal with *four coordinates* x, y, z and t. Many important physical systems can be understood as manifestations of wave phenomena. It is therefore appropriate to begin our discussion with wave motion. A significant property of waves is that waves carry energy over time from one spatial point to another. That is, waves represent nature's way of transporting energy. In studying wave motion the independent variables are x, y, z and t. The dependent variable represents the *disturbance*, i.e. the quantity undergoing wave motion. In this paper we assume this quantity is a scalar and denote it by $u(x, y, z, t)$. The basic equation which governs the wave motion is the *scalar wave equation*, whose homogeneous form is

$$\frac{\partial^2 u}{\partial x^2}+\frac{\partial^2 u}{\partial y^2}+\frac{\partial^2 u}{\partial z^2}-\frac{1}{v^2}\frac{\partial^2 u}{\partial t^2}=0\,. \tag{2.1}$$

A derivation of this equation from first principles can be found in most books on mathematical physics. For background materials, the reader is referred to Robinson and Silvia (1981).

Let us look at this wave equation (2.1). On the left is the sum of the three second partial derivatives with respect to each of the space variables. Also on the left is the negative of the second partial derivative with respect to time divided by the quantity v^2. Throughout this paper, we assume that v is a positive constant. The quantity v is the velocity of the traveling wave.

In the two-dimensional case, we assume that the wave motion or disturbance u is a function of only x, z, t and not a function of y. Then the second partial derivative of $u(x, z, t)$ with respect to y is zero, and the wave equation reduces to its two-dimensional form

$$\frac{\partial^2 u}{\partial x^2} + \frac{\partial^2 u}{\partial z^2} - \frac{1}{v^2}\frac{\partial^2 u}{\partial t^2} = 0 . \tag{2.2}$$

In the one-dimensional case, we assume that the disturbance is of the form $u(x, t)$, so the wave equation reduces to its one-dimensional form

$$\frac{\partial^2 u}{\partial x^2} - \frac{1}{v^2}\frac{\partial^2 u}{\partial t^2} = 0 . \tag{2.3}$$

3. Frequency-domain analysis

Both from a conceptual and a computational point of view, the entire multidimensional wave problem becomes considerably simplified by the introduction of frequency-domain techniques. For the time being we shall treat the case of two spatial dimensions x and z, as this case illustrates all the essential properties that we want to develop. First we must agree on the choice of a sign convention of the Fourier transform. For each coordinate x, z, t, there is a choice of sign. Mathematicians, physicists, and electrical engineers often use different sign conventions, and there are good reasons for their choices. Because we are interested in the systems approach we will make use of the electrical engineering convention, and write the *direct Fourier transform* of the space–time function $u(x, z, t)$ as

$$U(k_x, k_z, \omega) = \int_{-\infty}^{\infty}\int_{-\infty}^{\infty}\int_{-\infty}^{\infty} u(x, z, t)\, \mathrm{e}^{-\mathrm{i}(\omega t - k_x x - k_z z)}\, \mathrm{d}x\, \mathrm{d}z\, \mathrm{d}t . \tag{3.1}$$

The *inverse Fourier transformation* is then given by

$$u(x, z, t) = \frac{1}{8\pi^3}\int_{-\infty}^{\infty}\int_{-\infty}^{\infty}\int_{-\infty}^{\infty} U(k_x, k_z, \omega)\, \mathrm{e}^{\mathrm{i}(\omega t - k_x x - k_z z)}\, \mathrm{d}k_x\, \mathrm{d}k_z\, \mathrm{d}\omega . \tag{3.2}$$

The variable ω is the angular frequency, whereas the variables k_x and k_z are angular wavenumbers (or angular spatial frequencies). Generally, the word angular is understood, so it can be dropped. If t is measured in seconds and x and z in meters, then ω is in units of radians per second and k_x and k_z are each in units of radians per meter. The kernel in the inverse Fourier transform is the exponential

$$e^{i(\omega t - k_x x - k_z z)} . \tag{3.3}$$

Notice that the temporal frequency ω has a positive sign, whereas the spatial frequencies k_x and k_z have negative signs. This usage of signs is the one according to the electrical engineering convention. If we take the time derivative of the inverse Fourier transform, we obtain

$$\frac{\partial u}{\partial t} = \frac{1}{8\pi^3} \int_{-\infty}^{\infty} \int_{-\infty}^{\infty} \int_{-\infty}^{\infty} [i\omega U] \, e^{i(\omega t - k_x x - k_z z)} \, dk_x \, dk_z \, d\omega . \tag{3.4}$$

Thus the operation $\partial/\partial t$ in the time domain corresponds to multiplication by $i\omega$ in the frequency domain. In other words, the engineering sign convention associates $+i\omega$ with $\partial/\partial t$, which makes sense because electrical engineers (EE) work more often with time than space. Thus, under the EE convention, we see that $-ik_x$ is associated with $\partial/\partial x$ and likewise $-ik_z$ is associated with $\partial/\partial z$. The physics convention is just the opposite choice of signs. Mathematicians, on the other hand, would associate the same sign, whether it be plus or minus, with all the derivatives. Why then do engineers and physicists have opposite signs associated with time and space? The reason is that by convention in physics as well as in engineering, waves move in the positive direction on the space axis as time increases. As a result, the sign on the spatial frequencies must be opposite to the sign on the temporal frequency. We can illustrate this property by looking at the above exponential kernel. Its real part is

$$\cos(\omega t - k_x x - k_z z) . \tag{3.5}$$

For fixed k_x, k_z, ω, this is a sinusoidal wave (in this case, a cosine wave). A crest of the wave occurs when the quantity within the parentheses is zero, or a multiple of 2π. The rate of change of a crest along the x axis is

$$\frac{\partial x}{\partial t} = \frac{\omega}{k_x} , \tag{3.6}$$

which is called the horizontal phase velocity (where x is the horizontal axis), whereas

$$\frac{\partial z}{\partial t} = \frac{\omega}{k_z} \tag{3.7}$$

is the vertical phase velocity (where z is the vertical axis). The engineering choice of signs (as well as the physics choice) makes these phase velocities positive, so the wave travels in a positive direction.

Let us now define a plane wave. The cosine wave just considered is an example of a plane wave. Fig. 1 shows a typical ray along which the wave moves and also shows typical wavefronts along which a crest (or a trough) lies. The ray and wavefront are at right angles. The angle θ is the angle between the ray and the vertical, and also is the angle between the wavefront and the horizontal. The wave crest travels in the direction of the ray with velocity v. Suppose that it takes one unit of time for the crest to move from A and B. Thus the length of AB is v. The energy of the wave travels along the rays, so the segment AB represents the physical velocity v. If we look at the horizontal axis only, it appears that the crest moves from A to C. If AB is equal to v, then AC is equal to $v/\sin\theta$. The apparent velocity $v/\sin\theta$ is called the horizontal phase velocity. The horizontal phase velocity $v/\sin\theta$ does not represent a physical velocity. Because $\sin\theta$ is less than or equal to one, the horizontal phase velocity is greater than or equal to the velocity v. Similarly, the velocity of the crest (or wavefront) along the vertical axis is the vertical phase velocity $v/\cos\theta$. The expression for the wavefront is the line given by

$$z = \frac{vt}{\cos\theta} - x\tan\theta\,. \tag{3.8}$$

If we solve for t, we obtain

$$t(x, z) = \frac{x}{v}\sin\theta + \frac{z}{v}\cos\theta\,, \tag{3.9}$$

which is the time at which the wavefront passes any point (x, z). If we use $t(x, z)$ to define the time delay τ, then the expression for an arbitrary wave shape $f(t)$ delayed by the travel time τ is

$$f(t-\tau) = f\left(t - \frac{x}{v}\sin\theta - \frac{z}{v}\cos\theta\right). \tag{3.10}$$

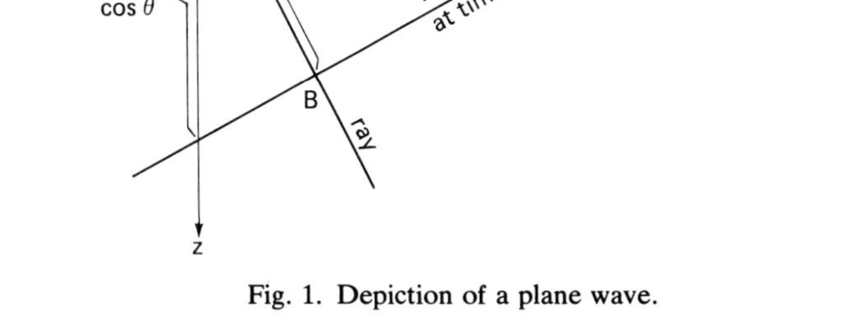

Fig. 1. Depiction of a plane wave.

This gives the expression for a traveling plane wave. In the case of the sinusoidal wave that we treated previously, the function f was a cosine, i.e. we had

$$\cos(\omega t - k_x x - k_z z) = \cos \omega\left(t - \frac{k_x}{\omega} x - \frac{k_z}{\omega} z\right). \tag{3.11}$$

On comparing the above two expressions, we obtain

$$\frac{\sin \theta}{v} = \frac{k_x}{\omega}, \qquad \frac{\cos \theta}{v} = \frac{k_z}{\omega}. \tag{3.12}$$

Thus we obtain the following relationships

$$\sin \theta = \frac{v k_x}{\omega}, \qquad \cos \theta = \frac{v k_z}{\omega}, \tag{3.13}$$

which relate physical angles to velocity and Fourier components. If we now make use of $\sin^2 \theta + \cos^2 \theta = 1$, we obtain

$$k_x^2 + k_z^2 = \frac{\omega^2}{v^2}, \tag{3.14}$$

which is called the *dispersion relation of the scalar wave equation.*

This dispersion relation is extremely important, so now let us derive it in another way. Let us take the complete Fourier transform of the scalar wave equation

$$\frac{\partial^2 u}{\partial x^2} + \frac{\partial^2 u}{\partial z^2} = \frac{1}{v^2}\frac{\partial^2 u}{\partial t^2}. \tag{3.15}$$

Because $\partial/\partial x$ corresponds to $-ik_x$, it follows that $\partial^2/\partial x^2$ corresponds to $(-ik_x)^2 = -k_x^2$. Likewise, $\partial^2/\partial z^2$ corresponds to $-k_z^2$ and $\partial^2/\partial t^2$ corresponds to $-\omega^2$. Thus, in the wavenumber–frequency domain, the scalar wave equation becomes

$$-k_x^2 U - k_z^2 U = -(\omega^2/v^2) U. \tag{3.16}$$

Thus the scalar wave equation is satisfied whenever the dispersion relation

$$k_x^2 + k_z^2 = \omega^2/v^2 \tag{3.17}$$

is satisfied.

We are now in a position to explain how multidimensional time-series data generated by physical wave motion is different from ordinary data. Take the data $u(x, z, t)$ and perform a frequency-domain analysis to obtain $U(k_x, k_z, \omega)$. Let us now look at an arbitrary value of $U(k_x, k_z, \omega)$ that is not zero. We have a wavefield if and only if the coordinates k_x, k_z, ω of this value of U satisfy the dispersion relation. Thus the dispersion relation represents the interrelationships existing in wavefield data. These interrelations make it possible to extrapolate wavefield data either over space or time. The solution to many important problems rests in the success of such extrapolations.

4. The distributed source model in seismic exploration

As an application of the frequency-domain analysis of multidimensional time-series data, we want to consider a specific physical situation, namely the problem of the seismic exploration for oil and gas in the sedimentary rock layers in the earth. The nature of the problem can be visualized by regarding the earth's crust as a stack of approximately horizontal sheets of rock, all with different properties and thicknesses, some of the layers being tilted, bent, or deformed, and some containing traps of varying shapes and sizes. Some of these traps may contain accumulations of oil and natural gas. The geophysicist sets off sources of energy at the surface and he analyzes the resulting responses recorded at the surface. From this information he wants to determine the thickness and physical properties of each sedimentary layer, details of the interfaces between the layers, and the locations and shapes of the traps in the layers. With this physical picture we model the earth's crust as an acoustically translucent, multilayered medium with the layering exposed along the vertical z axis, with depth z measured positively downward from the surface $z = 0$. The traps for petroleum may have an areal extent of only a square kilometer, and be as much as seven kilometers deep within the earth, with no geologic evidence at the surface that there is a trap below.

As is well known, the contact or interface between two different rock materials can cause reflections of seismic waves. Compared to the wavelength of the seismic waves, the magnitude of the roughness of these interfaces is not great, and as a result the quality of seismic reflections can be good. In many cases, the strata interfaces are quite smooth, with surface roughness of approximately one meter. To seismic waves with wavelengths up to one hundred meters, these surfaces appear to have a smoothness equivalent to that experienced by light striking a high-grade optical reflecting surface. Thus, by and large, the subsurface interfaces act as partially reflecting mirrors, and the seismic reflection process represents specular partial reflection rather than diffuse scattering.

Suppose now that we have the seismic source and receiver at the same point on the surface of the earth. According to ray theory, most of the energy that is reflected from a certain interface will travel down and up on the same ray, namely the ray that strikes the interface at right angles. The situation in seismic field recording corresponds to many such shot-receiver points which are activated in a sequential manner. A time series is obtained for each such point. However, once we have collected all these data, we may imagine instead a thought experiment whereby all the sources were actuated simultaneously instead of sequentially. We, therefore, have one overall wavefield which satisfies the wave equation. However, a further important simplification is possible. Because the energy goes down and up on the same ray path, we can consider that the sources were distributed along the interfaces and the seismic recordings at the surface just represent energy traveling up along the ray paths. In the field situation the waves go down and then return upward along the same ray, whereas in the hypothetical thought experiment they only go up along the ray. Thus in the field situation we record two-way time, whereas in the hypothetical situation we record one-way time. One-way time t is just one half of two-way time. It is a great conceptual advantage to consider the wavefield as generated by this hypothetical model, which is called the *distributed source model.* In seismic prospecting it is often the case that all the physical sources are explosions, i.e. the sources are impulses (or spikes) at time zero. By lining up the time origin on each time series, we obtain the so-called seismic record section. This record section may be considered to be the wavefield $s(x, y, t)$ observed at the surface (x, y) of the earth due to impulsive sources distributed along the subsurface interfaces where all the sources are set off at the same instant $t = 0$.

The distributed source model of seismic prospecting provides the following physical process. The interfaces are defined within the earth as a function of three-dimensional space (x, y, z), where z is depth. The surface of the ground is $z = 0$ and depth z is measured positively into the ground. Let the function which gives the *reflectivity* strength of the interfaces be denoted by $r(x, y, z)$.

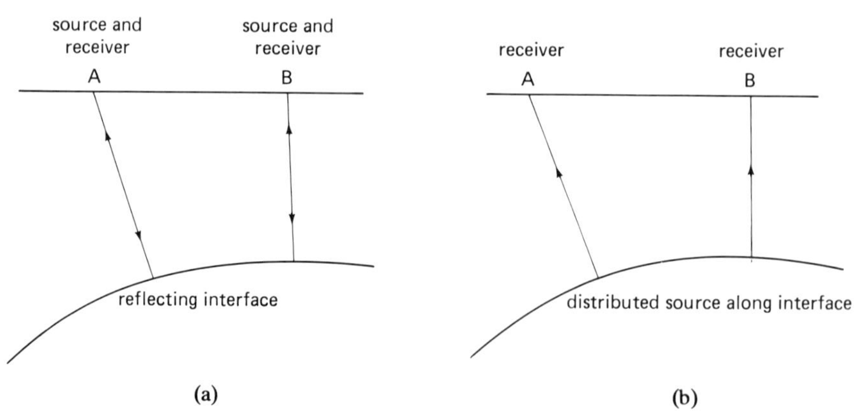

Fig. 2. (a) In seismic data acquisition, the surface points A and B are activated sequentially. The source and receiver are at the same point for each time series obtained. (b) Under the distributed source model, the sources are regarded as being distributed along the interface, and they are all fired simultaneously and recorded at the surface points such as A and B.

For short, we can call r the reflectivity function. According to the distributive source model, the impulses at time zero are made equal to the reflection strengths, so $r(x, y, z)$ also may be interpreted as the wavefield at time $t = 0$. On the other hand, the time-series data recorded at the surface $z = 0$, which we have denoted by the function $s(x, y, t)$, may be interpreted as the surface wavefield, i.e. the wavefield at the surface $z = 0$. Let us now denote the entire wavefield by $u(x, y, z, t)$. Then, as we have just seen, the *initial value* of the wavefield is given by the *reflectivity*, i.e.

$$u(x, y, z, t = 0) = r(x, y, z),$$

whereas the *surface value* of the wavefield is given by the *time-series data*, i.e.

$$u(x, y, z = 0, t) = s(x, y, t).$$

This physical process can be simulated within a computer. The computer memory can be used as a representation of physical space (x, y, z), and time t evolves from initial time 0. Thus we would start with the initial value $r(x, y, z)$ and let the computer simulate the wavefield $u(x, y, z, t)$ as time t increases. This type of simulation corresponds to what occurs in nature. A nice feature of such a simulation is that the wavefield evolves in a unique way depending upon the initial condition, and generally small errors in the initial conditions as well as in model specification do not propagate in an ever-increasing way. Thus given the initial conditions and the structure of the earth, we can find the wavefield everywhere, and in particular the surface wavefield $s(x, y, t)$. Such a simulation represents the solution of a time-evolution problem, which is also called a forward problem.

In geophysical exploration, on the other hand, we are faced with what is called an inverse problem. The time-series data that we have have been measured at the surface of the ground and we would like to determine the structure of the underground interfaces which delineate the sedimentary rock layers. In other words, we are given the surface time-series data $s(x, y, t)$ and we would like to extrapolate the wavefield downward in depth z. That is, in the inverse problem we take data at the earth's surface and extrapolate the data into the earth. Under our distributed impulsive source model, we can thus obtain the structure of the interfaces by letting $t = 0$ in our computed wavefield $u(x, y, z, t)$. It is for this reason that this model is so valuable. Thus the solution to the inverse problem gives the interface reflectivity function $r(x, y, z)$ from knowledge of the surface data $s(x, y, t)$. As is well known, such inverse problems do not have the nice stability and uniqueness properties as do direct problems. However, with the ever-increasing use of computers, much more confidence in the solution of inverse problems has been gained by means of diligent research and careful analysis in recent years, and the outlook for even better solutions is bright.

5. Migration of seismic data

As we have seen in the previous section, a seismic exploration program collects data on the surface of the ground. These data show seismic events due to reflections from subsurface interfaces. The purpose of the process known as *seismic migration* is to construct a geometric picture of these geologic interfaces from the time-series data measured at the surface. In terms of the distributed source model, the problem of migration is to construct the reflectivity function $r(x, z)$ of the subsurface interfaces from the surface data $s(x, t)$. Accurate migration requires a high-quality estimate of the velocity which is usually obtained independently by means of other seismic processing techniques. In this paper we assume that velocity v is a known positive constant.

Reflection seismic prospecting was first used commercially for oil exploration in the 1930s. During the 1940s various mechanical migration methods were being used by seismic interpreters. Seismic migration was put on a sound mathematical basis in terms of wave motion in the pioneering work of Hagedoorn (1954). His work was in the space–time domain so the various computer implementations which followed Hagedoorn's work were in space and time variables. These migration programs made use of wavefront superposition methods, diffraction (hyperbolic) summation methods, and finite-difference approximations. All of these methods represent digital approximations to the solution of the wave equation in the space–time domain. Migration in the frequency domain had been known for some time, but it was the work of Stolt (1978), Gazdag (1978), and Chen and Jacewitz (1978) that gave it the practical edge over the space–time methods. The frequency-domain approach is based upon the Cauchy–Poisson solution of the wave equation by use of Fourier transforms. For fuller treatment of the Cauchy–Poisson solution of the wave equation, see Webster (1950) and Tychonov and Samarski (1967). The advantages of frequency-domain migration include rapid computing time because of the fast Fourier transform, good performance under low signal-to-noise ratio, and good results in the case of complicated structures. One disadvantage is that the method as presented here does not handle variations in the velocity v as a function of depth. Stolt (1978) devised an approximate correction for velocity by an axis stretching procedure which made the method practical for field data. Gazdag (1978) extended the frequency-domain method so that it can perform in situations where the velocity varies as a function of depth. Further improvements were made by Bardan (1980). For a survey of the whole field of digital seismic processing, see Robinson (1980).

In this paper we want to give an exposition of frequency-domain migration. Fourier, Cauchy, Poisson and the other great nineteenth-century mathematicians were able to solve many of the important problems of wave motion, and their work is now found in books on partial differential equations. However, with the tremendous increase in computer capacity during the last decade, and with the excellent quality and large amount of time-series data available, their work takes on new meaning. Their work can now be used to solve practical data-processing problems, and the purpose of this paper is to illustrate one such case and so indicate how such methods can be applied to other time-series problems.

In the next section, we consider the case of one spatial dimension as an introductory treatment. Then we consider the case of two spatial dimensions. We do not treat the case of the three spatial dimensions as it follows the same pattern as the two-dimensional case.

6. One-dimensional migration

In this section we want to consider the case of only one spatial dimension, namely, the depth z. Such a situation can be visualized by putting a distributed explosive charge $r(z)$ down an oil well, where the function $r(z)$ is determined by the reflectivity of the interfaces cut by the well. At time $t = 0$, the impulsive charges are detonated, and the time series $s(t)$ is measured at the top of the well. If the wavefield is denoted by $u(z, t)$, then we have the initial condition $u(z, 0) = r(z)$ and the surface condition $u(0, t) = s(t)$.

Let us now introduce frequency methods. We wish to consider the two-dimensional Fourier transform of the wavefield $u(z, t)$. A two-dimensional Fourier transform is computed as a sequence of one-dimensional Fourier transforms. Thus we can represent the steps as

$$\begin{array}{ccc} u(z, t) & \rightarrow & U(k_z, t) \\ \downarrow & & \downarrow \\ U(z, \omega) & \rightarrow & U(k_z, \omega) \end{array} . \tag{6.1}$$

The two paths indicate that there is a choice of which variable is transformed first. Note that the capital U is actually used for three different transforms, namely the complete transform and the two partial transforms. The arguments of U indicate which transform is meant.

Now we must make use of the wave equation, which is

$$\frac{\partial^2 u}{\partial z^2} = \frac{1}{v^2}\frac{\partial^2 u}{\partial t^2} . \tag{6.2}$$

We perform a partial Fourier transform with respect to z, and obtain

$$\frac{\mathrm{d}^2 U(k_z, t)}{\mathrm{d}t^2} = -v^2 k_z^2 U(k_z, t) . \tag{6.3}$$

Note that the partial derivative with respect to t has become a full derivative because we have transformed out the space variable z. Equation (6.3) is thus a second-order ordinary differential equation. By direction substitution, we can verify that each of the functions

$$e^{-ivk_zt}, \qquad e^{ivk_zt} \tag{6.4}$$

are solutions. Let us now look at the dispersion relation for the wave equation. In this one-dimensional case, it is

$$k_z^2 = \omega^2/v^2\,, \tag{6.5}$$

which has two solutions, namely

$$k_z = -\omega/v, \qquad k_z = \omega/v\,. \tag{6.6}$$

The wavefield, expressed as an inverse Fourier transform, is

$$u(z, t) = \frac{1}{4\pi^2}\int_{-\infty}^{\infty}\int_{-\infty}^{\infty} U(k_z, \omega)\, e^{i(\omega t - k_z z)}\, dk_z\, d\omega\,. \tag{6.7}$$

We see that the kernel

$$e^{i(\omega t - k_z z)} \tag{6.8}$$

represents an upgoing wave (i.e. traveling in the direction of negative z) if ω and k_z are of opposite signs, whereas it represents a downgoing wave (i.e. traveling in the direction of positive z) if ω and k_z are of the same sign. Because our sources are at depth and our receiver is on the surface of the ground, we are only interested in upgoing waves, so we can discard the case $k_z = \omega/v$. Thus we choose the case $k_z = -\omega/v$ and use the solution

$$e^{i\omega t} = e^{-ivk_zt} \tag{6.9}$$

of the differential equation so the kernel

$$e^{i(\omega t - k_z z)} = e^{i(-vk_zt - k_z z)} = e^{-ik_z(vt+z)} \tag{6.10}$$

represents an upgoing wave. If we let $U(k_z, t = 0)$ represent the initial condition, then the required full solution of the differential equation is

$$U(k_z, t) = U(k_z, 0)\, e^{-ivk_zt}\,. \tag{6.11}$$

Because the initial condition is

$$u(z, 0) = r(z)\,, \tag{6.12}$$

where $r(z)$ is the reflectivity function, we have

$$U(k_z, 0) = R(k_z)\,, \tag{6.13}$$

where $R(k_z)$ is the Fourier transform of the reflectivity $r(z)$. Thus we have

$$U(k_z, t) = R(k_z)\,\mathrm{e}^{-\mathrm{i}vk_z t}. \tag{6.14}$$

The inverse Fourier transform of this function gives the wavefield; that is,

$$u(z, t) = \frac{1}{2\pi}\int_{-\infty}^{\infty} [R(k_z)\,\mathrm{e}^{-\mathrm{i}vk_z t}]\,\mathrm{e}^{-\mathrm{i}k_z z}\,\mathrm{d}k_z. \tag{6.15}$$

The surface data are the value of the wavefield $u(z, t)$ on the surface $z = 0$; that is,

$$s(t) = u(z = 0, t) = \frac{1}{2\pi}\int_{-\infty}^{\infty} R(k_z)\,\mathrm{e}^{-\mathrm{i}vk_z t}\,\mathrm{d}k_z. \tag{6.16}$$

This equation represents the solution of the forward (or time-evolution) problem; that is, given $r(z)$, find $s(t)$. Equation (6.16) can be simplified by the change of variable given by $\omega = -vk_z$. Thus we have $\mathrm{d}\omega = -v\,\mathrm{d}k_z$, so

$$R(k_z)\,\mathrm{d}k_z = R(-\omega/v)(-1/v)\,\mathrm{d}\omega. \tag{6.17}$$

When we change variables, we note that the integral with $\mathrm{d}k_z$ over $-\infty$ to ∞ becomes an integral with $(-1/v)\,\mathrm{d}\omega$ over ∞ to $-\infty$. If the minus sign is used to change the limits to $-\infty$ to ∞, we obtain

$$s(t) = \frac{1}{2\pi}\int_{-\infty}^{\infty} [(1/v)R(-\omega/v)]\,\mathrm{e}^{\mathrm{i}\omega t}\,\mathrm{d}\omega. \tag{6.18}$$

This equation says that the surface data are the inverse Fourier transform of the function $(1/v)R(-\omega/v)$. We thus have the following algorithm for the time-evolution (or forward) solution

$$r(z) \rightarrow R(k_z) \rightarrow (1/v)R(-\omega/v) \rightarrow s(t). \tag{6.19}$$

That is, we compute the direct space Fourier transform of the reflectivity $r(z)$, multiply it by $-(1/v)$ and change its argument k_z to $-\omega/v$, and finally take the inverse time Fourier transform to obtain the surface time series $s(t)$.

Next let us do the inverse problem, that is, given the surface time series $s(t)$, find the reflectivity $r(z)$. We can go backwards. We thus have the algorithm

$$s(t) \rightarrow S(\omega) \rightarrow vS(-vk_z) \rightarrow r(z). \tag{6.20}$$

That is, we compute the direct time Fourier transform $S(\omega)$ of the data $s(t)$, multiply it by v and change its argument ω to $-vk_z$, and finally take the inverse space Fourier transform to obtain the reflectivity $r(z)$.

The derivation of this algorithm can also be found along the same lines as the time-evolution algorithm. The wave equation becomes the ordinary differential equation

$$\frac{d^2U(z, \omega)}{dz^2} = -\frac{\omega^2}{v^2} U(z, \omega), \tag{6.21}$$

which has two solutions, namely

$$e^{-i(\omega/v)z}, \qquad e^{i(\omega/v)z}. \tag{6.22}$$

Because the kernel

$$e^{i(\omega t - k_z z)} \tag{6.23}$$

represents an upgoing wave (i.e. traveling in the direction of negative z) if ω and k_z are of opposite signs, and because we are only interested in upgoing waves, we use the solution

$$k_z = -\omega/v \tag{6.24}$$

of the dispersion relation. In other words, we choose the same sign as before, as we are dealing with the same physical situation. Thus we pick the solution

$$e^{-ik_z z} = e^{i(\omega/v)z} \tag{6.25}$$

of the differential equation, so the kernel

$$e^{i(\omega t - k_z z)} = e^{i[\omega t + (\omega/v)z]} = e^{i\omega(t + z/v)} \tag{6.26}$$

represents an upgoing wave. If we let

$$S(\omega) = U(z = 0, \omega) \tag{6.27}$$

represent the surface boundary condition, the required full solution of the differential equation is

$$U(z, \omega) = S(\omega)\, e^{i(\omega/v)z} \tag{6.28}$$

so the wavefield is

$$u(z, t) = \frac{1}{2\pi} \int_{-\infty}^{\infty} [S(\omega)\, e^{i(\omega/v)z}]\, e^{i\omega t}\, d\omega. \tag{6.29}$$

Finally, the reflectivity is the value of the wavefield at $t = 0$; that is,

$$r(z) = u(z, t=0) = \frac{1}{2\pi}\int_{-\infty}^{\infty} S(\omega)\, e^{i(\omega/v)z}\, d\omega\,. \tag{6.30}$$

This equation represents the solution of the inverse (or depth extrapolation) problem. We can simplify this expression by the change of variables given by $k_z = -\omega/v$. Then we have

$$S(\omega)\, d\omega = S(-vk_z)(-v)\, dk_z \tag{6.31}$$

so

$$r(z) = \frac{1}{2\pi}\int_{-\infty}^{\infty} [vS(-vk_z)]\, e^{-ik_z z}\, dk_z\,. \tag{6.32}$$

This is the required result to justify the algorithm we gave previously.

7. Two-dimensional migration

Let us now consider the case of two spatial dimensions, namely, a horizontal coordinate x and a vertical coordinate z (with depth positive downward). The solutions of the time-evolution problem and the depth-extrapolation problem are similar to the one-dimensional case, so we will only treat the depth-extrapolation problem. The given surface time-series data are $s(x, t)$ and the required depth reflectivity data are $r(x, z)$. The wave-equation dispersion relation is

$$k_x^2 + k_z^2 = \frac{\omega^2}{v^2}\,. \tag{7.1}$$

As before, we want to specify upgoing waves only. This means that the vertical phase velocity

$$\frac{\partial z}{\partial t} = \frac{\omega}{k_z} \tag{7.2}$$

must be negative (i.e. the wave must travel in the negative z direction). Thus we can describe an upgoing wave as one for which the signs of ω and k_z are opposite. Thus, when we solve the dispersion relation for ω in terms of k_z, we shall pick the sign of the square root so ω and k_z have opposite signs. That is, in the general solution

$$\omega = \pm v[k_x^2 + k_z^2]^{1/2} \tag{7.3}$$

we pick the sign so that ω and k_z have the opposite sign. We can write this chosen solution as

$$\omega = -vk_z[1 + (k_x/k_z)^2]^{1/2} . \tag{7.4}$$

Solving the dispersion relation for k_z, we have

$$k_z = \pm[(\omega/v)^2 - k_x^2]^{1/2} . \tag{7.5}$$

If we pick the sign so that ω and k_z have opposite signs, we obtain

$$k_z = -(\omega/v)[1 - (vk_x/\omega)^2]^{1/2} . \tag{7.6}$$

Let us now derive the algorithm. On the two-dimensional wave equation we perform a Fourier transformation with respect to x and t. We obtain the ordinary differential equation

$$\frac{\mathrm{d}^2U(k_x, z, \omega)}{\mathrm{d}z^2} = -\left[\frac{\omega^2}{v^2} - k_x^2\right]U(k_x, z, \omega) , \tag{7.7}$$

which has the two solutions

$$\mathrm{e}^{-\mathrm{i}[(\omega/v)^2 - k_x^2]^{1/2}z} , \qquad \mathrm{e}^{\mathrm{i}[(\omega/v)^2 - k_x^2]^{1/2}z} . \tag{7.8}$$

For upgoing waves it is necessary that k_z and ω have opposite signs. Thus we write k_z for upgoing waves as

$$k_z = -(\omega/v)[1 - (k_xv/\omega)^2]^{1/2} \tag{7.9}$$

and pick the solution

$$\mathrm{e}^{-\mathrm{i}k_zz} = \mathrm{e}^{\mathrm{i}(\omega/v)[1-(k_xv/\omega)^2]^{1/2}z} \tag{7.10}$$

of the differential equation. Thus the kernel

$$\mathrm{e}^{\mathrm{i}(\omega t - k_xx - k_zz)} = \mathrm{e}^{-\mathrm{i}k_xx}\,\mathrm{e}^{\mathrm{i}\omega\{t+(z/v)[1-(k_xv/\omega)^2]^{1/2}\}} \tag{7.11}$$

represents an upgoing wave. If we let

$$S(k_x, \omega) = U(k_x, z = 0, \omega) \tag{7.12}$$

represent the surface boundary condition, then the required full solution of the differential equation is

$$U(k_x, z, \omega) = S(k_x, \omega)\,\mathrm{e}^{\mathrm{i}(\omega/v)[1-(k_xv/\omega)^2]^{1/2}} \tag{7.13}$$

so the wavefield is

$$u(x, z, t) = \frac{1}{4\pi^2} \int_{-\infty}^{\infty} \int_{-\infty}^{\infty} \{S(k_x, \omega)\, e^{i(\omega/v)[1-(k_x v/\omega)^2]^{1/2}}\}\, e^{i(\omega t - k_x x)}\, dk_x\, d\omega\,. \tag{7.14}$$

Finally, the reflectivity is the value of the wavefield at $t = 0$; that is,

$$\begin{aligned} r(x, z) &= u(x, z, t = 0) \\ &= \frac{1}{4\pi^2} \int_{-\infty}^{\infty} \int_{-\infty}^{\infty} S(k_x, \omega)\, e^{i\{(\omega/v)[1-(k_x v/\omega)^2]^{1/2} - k_x x\}}\, dk_x\, d\omega\,. \end{aligned} \tag{7.15}$$

This equation represents the solution of the inverse (or depth-extrapolation) problem. We can simplify (7.15) by the change of variables given by

$$\omega = -vk_z[1 + (k_x/k_z)^2]^{1/2}\,, \tag{7.16}$$

which is found from the dispersion relation by requiring that ω and k_z have opposite signs. The derivative

$$\frac{\partial\omega}{\partial k_z} = -\frac{v}{[1 + (k_x/k_z)^2]^{1/2}} \tag{7.17}$$

is known as the vertical group velocity. The vertical group velocity, like the vertical phase velocity, is negative for upgoing waves. We have

$$\begin{aligned} S(k_x, \omega)\, dk_x\, d\omega &= S(k_x, \omega)\, dk_x \frac{\partial\omega}{\partial k_z} dk_z \\ &= S(k_x, -vk_z[1 + (k_x/k_z)^2]^{1/2}) \frac{-v}{[1 + (k_x/k_z)^2]^{1/2}}\, dk_x\, dk_z \end{aligned} \tag{7.18}$$

so

$$r(x, z) = \frac{1}{4\pi^2} \int_{-\infty}^{\infty} \int_{-\infty}^{\infty} \frac{vS(k_x, -vk_z[1 + (k_x/k_z)^2]^{1/2})}{[1 + (k_x/k_z)^2]^{1/2}}\, e^{-i(k_x x + k_z z)}\, dk_x\, dk_z\,. \tag{7.19}$$

This algorithm may be indicated by

$$s(x, t) \rightarrow S(k_x, \omega) \rightarrow \frac{vS(k_x, -vk_z[1 + (k_x/k_z)^2]^{1/2})}{[1 + (k_x/k_z)^2]^{1/2}} \rightarrow r(x, z)\,.$$

That is, the migration algorithm consists of the three steps:

(1) Fourier transform the surface data with respect to x and t;

(2) Multiply the Fourier transform by $v/[1 + (k_x/k_z)^2]^{1/2}$ and change its argument ω to $-vk_z[1 + (k_x/k_z)^2]^{1/2}$;

(3) Take the inverse Fourier transform with respect to k_x and k_z. The final result is the reflectivity $r(x, z)$.

8. Two-dimensional example

Let us now look at a two-dimensional example. Suppose that the surface data show only one event. This event lies upon the straight line

$$t = a + mx\,, \tag{8.1}$$

where the constant a is the t intercept and the constant m is the slope. For definiteness, let us assume that the angle α that the line makes with the x axis is an acute angle so the slope $m = \tan\alpha$ is positive. The surface data are thus the line spike model given by

$$s(x, t) = \delta(t - mx - a)\,. \tag{8.2}$$

We now want to calculate the algorithm given in Section 7 in order to find the reflectivity $r(x, z)$.

In step (1) we compute the Fourier transform. We have

$$S(k_x, \omega) = \int_{-\infty}^{\infty}\int_{-\infty}^{\infty} \delta(t - mx - a)\, \mathrm{e}^{-\mathrm{i}(\omega t - k_x x)}\, \mathrm{d}x\, \mathrm{d}t\,. \tag{8.3}$$

Using the sifting property of the delta function with respect to $t = mx + a$, we have

$$S(k_x, \omega) = \int_{-\infty}^{\infty} \mathrm{e}^{-\mathrm{i}[\omega(mx+a) - k_x x]}\, \mathrm{d}x\,, \tag{8.4}$$

which is

$$S(k_x, \omega) = \mathrm{e}^{-\mathrm{i}\omega a} \int_{-\infty}^{\infty} \mathrm{e}^{-\mathrm{i}x(m\omega - k_x)}\, \mathrm{d}x\,. \tag{8.5}$$

We now make use of the integral expression for the delta function given by

$$\delta(k) = \frac{1}{2\pi}\int_{-\infty}^{\infty} \mathrm{e}^{-\mathrm{i}kx}\, \mathrm{d}x\,. \tag{8.6}$$

We thus have

$$S(k_x, \omega) = 2\pi\, \mathrm{e}^{-\mathrm{i}\omega a}\delta(m\omega - k_x)\,. \tag{8.7}$$

Thus the Fourier transform is a line spike along the line $\omega = (1/m)k_x$. See Fig.

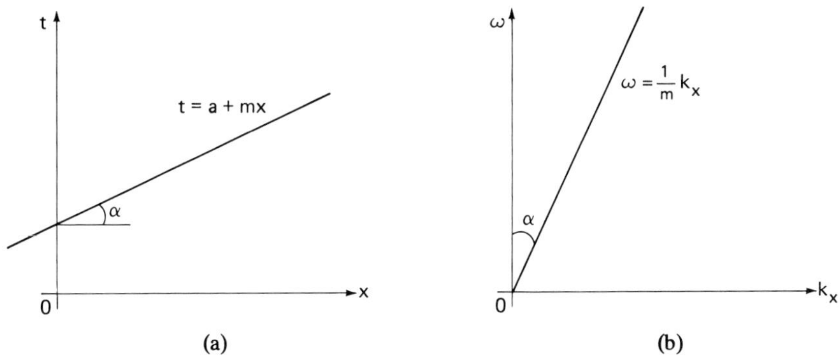

Fig. 3. (a) The event line in the space–time domain. (b) The event line in the wavenumber–frequency domain.

3. This line goes through the origin. The information as to the intercept a of the original line is contained in the phase factor $e^{-i\omega a}$. We have now obtained the Fourier transform of the surface data, which is the first step in the algorithm.

For convenience, let us choose our time and distance scales so that the constant velocity v is equal to one (i.e. v is equal to one distance unit per time unit). This convention means that an interpreter can plot the geologic section $r(x, z)$ on a transparency over the surface section $s(x, t)$ in order to graphically see a wave arrival (i.e. a surface event) move in order to become the interface that produced the event. In the 1940s and 1950s, before the production use of computers, seismic interpreters by geometrical constructions transformed seismic events on the (x, t) plane into the corresponding interfaces in the (x, z) plane. This movement was called 'migration', and this usage represented the origin of the term.

Let us now perform the second step of the algorithm, namely the change of variable and the scale factor. With $v = 1$, the change of variable is given by

$$\omega = -k_z[1 + (k_x/k_z)^2]^{1/2} . \tag{8.8}$$

Again we note that ω and k_z have opposite algebraic sign, because the positive square root is always implied. Let us now define the angle β by the equation

$$\omega = -k_z/\cos\beta . \tag{8.9}$$

In Fig. 4 we see how β is defined geometrically. On the horizontal axis we measure k_x, and on the vertical axis we measure both ω and $-k_z$ (where, as we know, ω and $-k_z$ have the same algebraic sign, which for illustrative purposes we take to be positive). The transformation is as follows. For a given pair (k_x, ω), swing a circle with center at origin and radius ω. The point of intersection of the vertical line through k_x defines $-k_z$. As we know from the

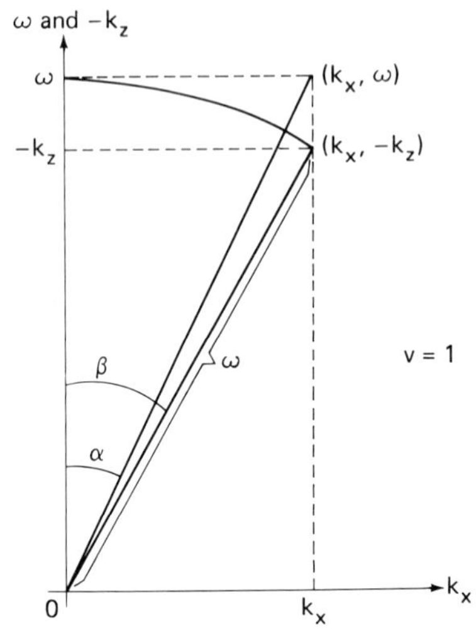

Fig. 4. Construction of the angle β.

equation $k_x = m\omega$ (where $m = \tan\alpha$), the line from the origin to the point (k_x, ω) defines the angle α with the vertical. Similarly, the line from the origin to the constructed point $(k_x, -k_z)$ defines the angle β with the vertical. Thus we have a graphical way of constructing the angle β.

Let us now evaluate the quantities ωa and $m\omega$ which occur in $S(k_x, \omega)$. We have

$$\omega a = -(k_z/\cos\beta)a\,. \tag{8.10}$$

If we define the constant b as

$$b = \frac{a}{\cos\beta}\,, \tag{8.11}$$

then

$$\omega a = -k_z b\,. \tag{8.12}$$

From Fig. 4 we have (because the circle has radius ω)

$$m = \tan\alpha = \frac{k_x}{\omega} = \sin\beta\,. \tag{8.13}$$

Equation (8.13), namely $\tan\alpha = \sin\beta$, is known as the *migrator's equation.* Now let us find $m\omega$. We have

$$\omega m = -(k_z/\cos\beta)\tan\alpha = -(k_z/\cos\beta)\sin\beta = -k_z\tan\beta\,. \tag{8.14}$$

The scale factor in the algorithm is

$$\frac{1}{[1+(k_x/k_z)^2]^{1/2}} = \cos\beta\,. \tag{8.15}$$

We are now ready to perform step (2) of the algorithm. The function $S(k_x, \omega)$ is transformed into

$$\cos\beta S(k_x, -k_z/\cos\beta)\,, \tag{8.16}$$

which is

$$(\cos\beta)2\pi\, e^{-i\omega a}\delta(m\omega - k_x) = 2\pi(\cos\beta)\, e^{ik_z b}\delta(-k_z\tan\beta - k_x)\,. \tag{8.17}$$

We are now ready to perform step (3) of the algorithm. This step says that we must perform the inverse Fourier transform on the above expression. We have

$$\begin{aligned} r(x, z) &= \frac{\cos\beta}{2\pi}\int_{-\infty}^{\infty}\int_{-\infty}^{\infty} e^{ik_z b}\delta(-k_z\tan\beta - k_x)\, e^{-i(k_x x + k_z z)}\, dk_x\, dk_z \\ &= \frac{\cos\beta}{2\pi}\int_{-\infty}^{\infty} e^{ik_z b}\, e^{-i(-xk_z\tan\beta + k_z z)}\, dk_z \\ &= \frac{\cos\beta}{2\pi}\int_{-\infty}^{\infty} e^{-ik_z(-x\tan\beta - b + z)}\, dk_z \\ &= \cos\beta\delta(z - b - x\tan\beta)\,. \end{aligned} \tag{8.18}$$

The reflectivity function is therefore a line spike model with the *interface* given by the line

$$z = b + x\tan\beta \qquad \text{(interface line)}\,. \tag{8.19}$$

This is the interface which gives rise to the line

$$t = a + x\tan\alpha \qquad \text{(event line)}\,. \tag{8.20}$$

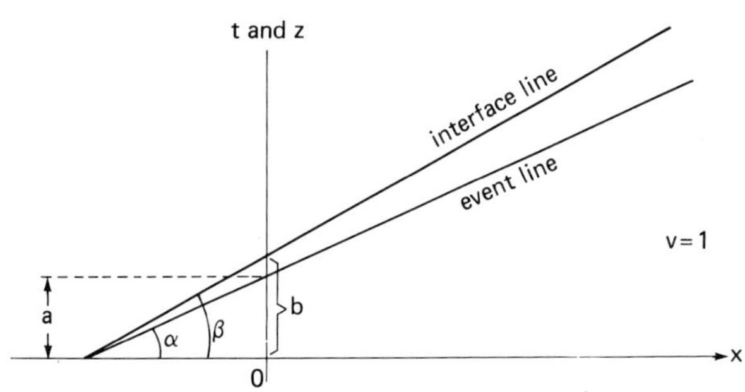

Fig. 5. Migration as the movement of the event line to the interface line.

of the received *event* on the surface section. These two lines are related by the migrator's equation (8.13) and the intercept equation (8.11). Because we have chosen our scales so $v = 1$, we can plot these two lines on the same set of axes, as in Fig. 5. The movement of the lines from event to interface constitutes migration. One can verify that both of these lines have a common x intercept.

9. Concluding remarks

To the interpreter of seismic data from geologically complex regions, the process of seismic migration is a necessity in order to transform the observed surface events to their proper spatial position in the subsurface. In this paper we have developed the theory of the wavenumber–frequency method of migration which is the most popular and successful method in use by the geophysical exploration industry. The reader will note that the method is based on physical principles but as it turns out, it also has good statistical properties. The reason is that the wave equation organizes the time-series data in a systematic fashion, so that methods based on the wave equation work powerfully against the unsystematic elements of random noise. Of course, such methods can only be applied to processes characterized by wave equation, and for this reason these methods are limited to physical situations where wave motion plays a central role.

References

Bardan, V. (1980). Migration of seismic sections using the scalar wave equation. *St. Cerc. Geol., Geofiz., Georg., Geofizica* **18**, 41–59.

Chen, J. H. and Jacewitz, C. A. (1978). *Fundamentals of Frequency Domain Migration.* Seismograph Service Corp., Tulsa, OK.

Gazdag, J. (1978). Wave equation migration with the phase-shift method. *Geophysics* **43**, 1342–1351.

Hagedoorn, J. G. (1954). A process of seismic reflection interpretation. *Geophysical Prospecting* **2**, 85–127.

Robinson, E. A. (1980). *University Course in Digital Seismic Methods Used in Petroleum Exploration.* International Human Resources Development Corp., Boston.

Robinson, E. A. and Silvia, M. T. (1981). *Digital Foundations of Time Series Analysis, Vol.* 2, *Wave-Equation Space-Time Processing.* Holden-Day, Oakland, CA.

Stolt, R. H. (1978). Migration by Fourier transform. *Geophysics* **43**, 23–48.

Tychonov, A. N. and Samarski, A. A. (1967). *Partial Differential Equations of Mathematical Physics* (2 volumes). Holden-Day, Oakland, CA.

Webster, A. G. (1950). *Partial Differential Equations of Mathematical Physics.* Hafner, New York.

D. R. Brillinger and P. R. Krishnaiah, eds., *Handbook of Statistics, Vol. 3*

16

Review of Various Approaches to Power Spectrum Estimation

P. M. Robinson

1. Introduction

1.1. Introductory definitions and theory

A sequence of real-valued random variables $\{x(t)\}$ is defined on all the integers $t = 0, \pm 1, \ldots$. No generality is lost by assigning the time interval the value 1. We assume $x(t)$ has finite second moments and moreover that the expectations of $x(t)$ and $x(t)x(t+u)$ do not depend on t, for all t, u. Then $x(t)$ is said to be wide-sense stationary and we define

$$\mu = Ex(t), \qquad \gamma(u) = E(x(t) - \mu)(x(t+u) - \mu).$$

We call μ the *mean* of $x(t)$ and $\gamma(u)$ the uth *autocovariance* of $x(t)$.

The $\gamma(u)$ must decay to zero at a suitably rapid rate as $|u| \to \infty$ for their Fourier inverse

$$f(\lambda) = \frac{1}{2\pi} \sum_{u=-\infty}^{\infty} e^{-iu\lambda} \gamma(u) \tag{1}$$

to exist for all real λ. A sufficient condition for such existence is that the Fourier series converge absolutely,

$$\sum_{u=-\infty}^{\infty} |\gamma(u)| < \infty. \tag{2}$$

Condition (2) implies in fact that $f(\lambda)$ obeys the Lipschitz condition $\sup_\lambda |f(\lambda + \delta) - f(\lambda)| = O(|\delta|^\alpha)$, for all real δ and some $\alpha > \frac{1}{2}$. Sometimes stronger assumptions are necessary such as $f(\lambda)$ is differentiable. For our purposes, the smoother $f(\lambda)$ (or equivalently the more rapidly the $\gamma(u)$ decay), the better it will be.

The function $f(\lambda)$ is called the *power spectrum* (or *spectral density*, or *spectrum*) and its argument λ, the *frequency*. The *period* is inversely proportional to frequency. We are concerned with the problem of estimating $f(\lambda)$ at one or more values of λ, given at most a *single* realization of finitely many consecutive $x(t)$, i.e. of $x(t)$, $t = 1, 2, \ldots, T$, the time origin being arbitrary. The availability of two or more such realizations would alter the problem, but this case has been little studied, perhaps being of relatively minor practical importance, and we shall not discuss it. Our terms of reference include data sets in which observations on $x(t)$ are unavailable for some values of t between 1 and T (*missing* data), or in which one knows only that the value of $x(t)$ lies within a given real interval, for some or all such t (*censored* data).

Finitely many observations can be expected to yield good estimates of $f(\lambda)$ at only a finite number of λ values, and indeed this number will typically be small relative to T. A common practice is to select a representative set of λ values, perhaps an equally spaced one. In this connection the following properties of $f(\lambda)$ are important. Because $e^{-iu\lambda}$ is periodic of period 2π, so also is $f(\lambda)$. Wide-sense stationary implies that $\gamma(-u) = \gamma(u)$, so $f(\lambda)$ is an even, real function. Therefore, it suffices to estimate $f(\lambda)$ over the interval $[0, \pi]$. The frequency π is called the *Nyquist* or *sampling* frequency. Because the sequence $\{\gamma(u)\}$ is nonnegative definite, i.e. $\Sigma_u \Sigma_v a_u a_v \gamma(u - v) \geqslant 0$ for all real sequences $\{a_u\}$, it follows that $f(\lambda)$ is a nonnegative function. An analogy is suggested between $f(\lambda)$ and $\gamma(u)$ on the one hand, and the probability density function and characteristic function of a real symmetric random variable on the other.

Equation (1) can be inverted,

$$\gamma(u) = \int_{-\pi}^{\pi} e^{iu\lambda} f(\lambda)\, d\lambda = \int_{-\pi}^{\pi} \cos u\lambda f(\lambda)\, d\lambda ,$$

in particular,

$$\gamma(0) = \int_{-\pi}^{\pi} f(\lambda)\, d\lambda .$$

The variance of $x(t)$, $\gamma(0)$, is decomposed by frequency, $f(\lambda)\,\Delta\lambda$ being the approximate contribution to $\gamma(0)$ from the narrow frequency band $(\lambda, \lambda + \Delta\lambda)$, and Tukey (1961), Jackson and Lawton (1969) have investigated the analogy with random effects models of the analysis of variance. The magnitude of $f(\lambda)$ is thus a measure of importance of the frequency λ. When the $x(t)$ are uncorrelated (so $\gamma(u) = 0$ for all $u \neq 0$), all frequencies are equally important, $f(\lambda)$ being constant for all λ. As a general rule, a power spectrum which is large for small values of λ, and decreases as $\lambda \to \pi$, reflects an $x(t)$ with smooth, slowly changing, realizations, whereas a rapidly oscillating process is indicated by the reverse spectral shape. Spectral peaks can also occur between 0 and π, suggesting important cycles or resonances, such as seasonal effects.

1.2. *Purposes of power spectrum estimation*

(i) *Description*

Like other statistical samples, a time series of values of $x(t)$ for $t = 1, 2, \ldots, T$ requires a descriptive summary statistic, particularly when T is large. Because it need be computed and inspected over only the range of frequencies $[0, \pi]$, the spectrum estimate is convenient and conveys meaningful information. It has the desirable property, furthermore, that estimates at distinct frequencies tend to be nearly statistically independent when T is large. Estimates of $\gamma(u)$ contain identical information but they do not share the latter property.

(ii) *Detecting hidden periodicities*

Many time series in the natural sciences and economics contain very strong periodic effects, and their detection was the objective of some of the earliest investigations of time series (Schuster, 1898). An important periodic effect will manifest itself in a readily identifiable spectral peak at the corresponding frequency, its influence on the process being measured by the magnitude of the peak. The presence of spectral peaks leads, however, to serious difficulties in power spectrum estimation.

(iii) *Hypothesis testing*

Hannan (1961) proposed a test for a jump in the spectral distribution at a given frequency in terms of power spectral estimates. A question frequently asked is whether $x(t)$ is white noise, i.e. $\gamma(u) = 0$ for all $u \neq 0$: because, for example, least-squares estimators of time-series regressions are efficient if the residuals are white noise, or because a strictly stationary point process is Poisson if the intervals are independent. The white noise hypothesis corresponds to a flat spectrum, and indeed it is very easy to obtain good estimates of a flat spectrum. Zaremba (1960) gives a test for a more general spectral shape.

(iv) *Discrimination and classification*

In some applications, such as in seismography, the object is to distinguish between two stationary time series or to classify a series, and the power spectrum is a convenient discriminator (Grenander, 1974; Shumway and Unger, 1974; Dargahi-Noubary and Laycock, 1979).

(v) *Model identification*

Box and Jenkins (1970) have proposed that the integers p and q in the stationary autoregressive moving average model

$$x(t) + \sum_{j=1}^{p} a_j x(t-j) = e(t) + \sum_{j=1}^{q} b_j e(t-j) \tag{3}$$

($e(t)$ unobservable white noise), be determined by examination of time-domain statistics. Because the values of p and q correspond, loosely speaking, to the numbers of peaks and troughs in $f(\lambda)$, power spectrum estimates might play a useful role in model identification.

(vi) *Parameter estimation*

To carry things a stage further, the coefficients a_j, b_j in the model (3) and parameters in more general models can be estimated by means of power spectral estimates. Hannan (1963) proposed the efficient estimation of time-series regressions by inverse weighting by nonparametric estimates of the spectrum of the residual process, an approach which has the advantage of avoiding precise assumptions about the correlation structure of the residuals.

(vii) *Prediction and smoothing*

The Wiener–Kolmogorov theory of prediction and smoothing leads to frequency-domain formulas which require power spectrum estimates for their practical implementation (Kolmogorov, 1941; Wiener, 1949; Bhansali, 1974).

(viii) *Seasonal adjustment*

Nerlove (1964) and Hannan (1970a) consider spectral methods of seasonally adjusting economic time series.

1.3. Limitations of spectrum estimation

(i) *Stationarity assumption*

Many stochastic processes are intrinsically nonstationary and the power spectrum ceases to be a meaningful concept although Parzen (1961b) and Herbst (1964) apply it to processes that are only 'asymptotically stationary' and Priestley (1965b) provides an estension to nonstationary processes. Usually some detrending is necessary and the way in which this is done can crucially affect spectral estimation of the stationary component.

(ii) *Gaussianity*

The statement that $x(t)$ is *Gaussian* means that the joint distribution of $x(t_1), \ldots, x(t_k)$ is k-variate normal for all integers $t_1, \ldots, t_k, k$. A Gaussian process is entirely characterized by its first two moments and cross-moments, so the Gaussian case is ideal for spectrum analysis. Non-Gaussian processes are not always adequately described by the spectrum, a striking case being a stationary discrete-valued process which can take only the values 0 and 1. Spectra of such processes are often estimated but other forms of analysis are more informative. Sometimes simple nonlinear instantaneous transformations, such as Box–Cox transformations, produce a more Gaussian character, but they may also lead to difficulties of interpretation.

(iii) *Series length*

Because of their nonparametric nature, spectrum estimates are unlikely to be accurate or reliable unless based on a substantial amount of data. In many applications, particularly economics, T is not large and practitioners prefer to invest in a finite parameter model such as (3).

(iv) *Aliasing problem*

The interpretation of spectral estimates is complicated by the phenomenon of *aliasing*. As a rule, the sampling interval is not intrinsic to the underlying process, which is defined over a continuum of time points. If $x(t)$, $-\infty < t < \infty$ is a wide-sense continuous stationary process with mean $\mu = Ex(t)$ and autocovariance function $\gamma(u) = E(x(t) - \mu)(x(t+u) - \mu)$, $-\infty < t,\ u < \infty$, its power spectrum, when it exists, is defined by

$$g(\lambda) = \frac{1}{2\pi}\int_{-\infty}^{\infty} e^{-iu\lambda}\gamma(u)\,du\,, \quad -\infty < \lambda < \infty\,.$$

By comparison with (1),

$$f(\lambda) = g(\lambda) + \sum_{|j|=1}^{\infty} g(\lambda + 2\pi j)\,. \tag{4}$$

The spectrum of the sampled discrete process at a frequency λ in $(-\pi, \pi)$ is thus the sum of spectra of the underlying continuous process at the 'aliased' frequencies $\lambda + 2\pi j$, $j = 0, \pm 1, \ldots$, and it is impossible to disentangle these effects. Equation (4) is merely a manifestation of the obvious fact that uncountably many continuous curves can be drawn through the discrete $x(t)$, $t = 0, \pm 1, \ldots$. In some applications, it is possible to choose the sampling interval sufficiently small that $g(\lambda)$ is negligible for $\lambda > \pi$, and thus $f(\lambda) = g(\lambda)$, $0 \leq \lambda \leq \pi$. The bias in estimating $g(\lambda)$ by $f(\lambda)$ is investigated by Robinson (1976) and Splettstosser (1980).

1.4. Data preparation

Unless the series is too long for this to be feasible, it should always be graphed. Graphing is helpful in deciding whether any preliminary transformation of the data is required, to deal with outliers or remove deterministic and nonstationary components.

The presence of *outliers*—extreme or surprising observations, or gross recording errors—is a considerable cause for concern, and Kleiner, Martin and Thomson (1979) have demonstrated the extent to which they can contaminate spectral estimates. These authors suggest systematic procedures for robust estimation, as an alternative to ad hoc replacement of outliers by more reasonable-looking interpolated values.

Verification of the *stationarity* assumption is usually desirable, followed by a detrending of the data if necessary. We can split the series into consecutive sections, containing roughly equal numbers of observations. Simple summary statistics, such as the sample means and variances of the $x(t)$ are computed for each section, and the results compared.

Two methods of detrending time series have proved popular. First-differencing—replacing $x(t)$ by $x(t) - x(t-1)$—and if necessary further differencing, is advocated by Box and Jenkins (1970) and others for the removal of certain deterministic and stochastic trends. The principal alternative is regression methods, usually ordinary least squares. Polynomial regressors are traditional, but over a long time span the recently developed spline functions are more effective. Strong cyclic trends such as seasonal effects do not necessarily imply nonstationarity but are also often removed in advance, by trigonometric regression or linear filtering, to avoid the difficulties inherent in spectral analysis of series with large spectral peaks. A further problem that can be handled by regression methods is the presence of brief transient signals (Brillinger, 1973).

Even if $x(t)$ can be assumed stationary with smooth spectrum, its mean μ will be unknown. If spectral estimates are formed from the raw data, asymptotically the effect of a nonzero μ will manifest itself only at the zero frequency, but in finite series it will produce bias at other frequencies. Usually, therefore, *mean-correction* will be necessary, replacing $x(t)$ by $x(t)-\bar{x}$, where $\bar{x}=T^{-1}(x(1)+\cdots+x(T))$.

The use of regression methods, whether it be to remove polynomial or trigonometric trends, or to mean-correct the data, itself introduces a bias problem, because one is using the estimated residuals from the regression, rather than the true ones. Hannan (1958) suggests a correction which reduces bias, analogous to the factor $T/(T-1)$ classically applied to $T^{-1}\Sigma_1^T(x(t)-\bar{x})^2$ to estimate the population variance. (See also Priestley, 1964 and Nicholls, 1967.) To avoid repeated references to this problem *we shall assume in the sequel, with the exception of part of Subsection* 4.2, *that* $x(t)$ *has known, zero mean.*

1.5. Basic statistics

Earliest attempts at spectral analysis were based on the *discrete Fourier transform* of the series

$$w(\lambda)=\frac{1}{\sqrt{2\pi T}}\sum_{t=1}^{T}x(t)\,\mathrm{e}^{\mathrm{i}t\lambda}\,.$$

As a preparation for estimation of $f(\lambda)$ over the whole frequency range $[0,\pi]$, $w(\lambda)$ is typically calculated at the $[T/2]$ *fundamental frequencies* $\lambda=\lambda_j=2\pi j/T$, $j=1,2,\ldots,[T/2]$, the $w(\lambda_j)$ representing a unitary transformation of the series. Naive computation of all these statistics requires the order of T^2 operations, a prohibitively expensive task even on today's electronic computers when T is very large. The fast Fourier transform (FFT) algorithm (Good, 1958; Cooley and Tukey, 1965) requires only $\mathrm{O}(T\log_2 T)$ operations when T is a power of 2, and considerable savings are possible even when T is not highly composite. One can append T' zeros to the series so that $T+T'$ is highly composite and obtain $w(\lambda)$ at the frequencies $2\pi j/(T+T')$. As well as cutting computational time, the FFT may incur less round-off error than the naive method.

We can write $w(\lambda)$ as

$$\tilde{w}(\lambda)=\frac{1}{\sqrt{2\pi T}}\sum_{t=-\infty}^{\infty}x(t)h_T(t)\,\mathrm{e}^{\mathrm{i}t\lambda}\,,$$

where $h_T(t) = 1$, $1 \leq t \leq T$, $= 0$, otherwise. Because of its lack of smoothness at 1 and T, this $h_T(t)$ has a Fourier transform which does not decay rapidly to 0 away from the origin, leading to the possibility of contamination from other frequencies. We can reduce this contamination by letting $h_T(t)$ fall away smoothly to 0 at both ends. A popular $h_T(t)$ which accomplishes this while leaving most of the data intact, is the *cosine bell* (Koopmans, 1974, p. 302). An alternative, which has a minimum-bias property, is the class of prolate spheroidal functions (Thomson, 1977). Such a function $h_T(t)$ is called a *taper*, *fader* or *data window*.

The *periodogram*, or *sample spectral density* is defined as

$$I(\lambda) = |w(\lambda)|^2 .$$

If $f(\lambda)$ is continuous, $I(\lambda)$ is an *asymptotically unbiased* estimator of $f(\lambda)$. It is not, however, a *consistent* estimator of $f(\lambda)$, the mean square error $E(I(\lambda) - f(\lambda))^2$ having nonzero limit as $T \to \infty$. In fact for T large, the $I(\lambda_j)/f(\lambda_j)$ behave like independent χ_2^2 random variables under general conditions. (For a discussion of the asymptotic properties of $I(\lambda)$, see Olshen, 1967.) When graphed, the periodogram thus exhibits a highly erratic behaviour, and it will be necessary to smooth it in order to obtain reasonable spectral estimates.

Before the advent of the FFT, the prevalent approach to spectral analysis was based, not on the discrete Fourier transform or periodogram, but on the *sample autocovariances*

$$c(u) = \frac{1}{T} \sum_{t=1}^{T-u} x(t)x(t+u) , \quad u \geq 0 ,$$

and $c(-u) = c(u)$. Under mild conditions, $c(u)$ is a consistent estimator of $\gamma(u)$. When T is very large, the computation of the $c(u)$ for many values of u will be expensive, but Sande (1966) shows that again the FFT can improve on the naive method. Augment the series $x(1), \ldots, x(T)$ by $T - 1$ zeros, $x(T+1) = \cdots = x(2T-1) = 0$ and compute, by the FFT

$$\tilde{w}(\lambda_j') = \frac{1}{\sqrt{2\pi(2T-1)}} \sum_{t=1}^{2T-1} x(t)\, e^{it\lambda_j'} , \quad |j| \leq T - 1 ,$$

where $\lambda_j' = 2\pi j/(2T-1)$ and $\tilde{w}(\lambda_j')$ differs from $w(\lambda_j')$ only by a scale factor. Compute the $I(\lambda_j') = |w(\lambda_j')|^2 = ((2T-1)/T)|\tilde{w}(\lambda_j')|^2$. Finally, use the formula

$$c(u) = \frac{2\pi}{2T-1} \sum_{j=-T+1}^{T-1} I(\lambda_j')\, e^{i\lambda_j' u}$$

to compute the $c(u)$ by the FFT. We have thus used the FFT twice, which requires $O(T \log_2 T)$ operations if T is a power of 2. Many of the spectral estimates based on the $c(u)$ do not use the $c(u)$ for all $|u| \leq T - 1$, but if we want M of them, the naive method requires $O(TM)$ operations.

2. Quadratic spectrum estimators

2.1. Two classes of estimator

The quadratic estimators can all be expressed as

$$\hat{f}(\lambda) = \sum_{t,u=1}^{T}\sum a_{tu}(\lambda)x(t)x(u),$$

where $\{a_{tu}(\lambda)\}$ is a given array of functions of λ.

The periodogram $I(\lambda)$ is of this form, but although it is asymptotically unbiased, it is not consistent. Our aim, then, is to find alternative choices of $a_{tu}(\lambda)$ which will reduce variation without introducing unacceptable bias.

The periodogram may be written

$$I(\lambda) = \frac{1}{2\pi T}\sum_{t,u=1}^{T}\sum x(t)x(u)\,\mathrm{e}^{\mathrm{i}(t-u)\lambda} = \frac{1}{2\pi}\sum_{u=-T+1}^{T-1} c(u)\,\mathrm{e}^{-\mathrm{i}u\lambda}\,.$$

We can compare the rightmost expression with the Fourier series expansion of $f(\lambda)$ (1). As $T\to\infty$, we may have $c(u)\to\gamma(u)$ in probability but the numbers of summands also increases as T, resulting in the lack of consistency of $I(\lambda)$. It is suggested, then, that the influence of the $c(u)$ be damped for large u by the introduction of a *lag window* $k_M(u)$, a sequence of weights such that $k_M(u)\to 0$ as $|u|\to\infty$. We have the *weighted covariance* estimators

$$\hat{f}_C(\lambda) = \frac{1}{2\pi}\sum_{u=-T+1}^{T-1} k_M(u)c(u)\,\mathrm{e}^{-\mathrm{i}\lambda u} = \frac{1}{2\pi}\sum_{u=-T+1}^{T-1} k_M(u)c(u)\cos u\lambda\,, \qquad (5)$$

where the $k_M(u)$ satisfy

$$k_M(-u) = k_M(u) \leq k_M(0) = 1\,.$$

The integer M plays a very important role in the asymptotic properties of spectrum estimators. In many cases, $k_M(u)=0$, $|u|>M$, so M could be called the *lag number*. By choosing M small relative to T, the variability of the estimator is brought under control; by regarding M as increasing to infinity with T, asymptotic unbiasedness will be possible.

The second class of estimators is motivated by recalling that the $I(\lambda_j)$ are approximately independent for large T, and because $f(\lambda)$ is assumed smooth, their bias as estimates of $f(\lambda)$ will diminish the closer λ_j is to λ. We consider the *smoothed periodogram* estimators

$$\hat{f}_P(\lambda) = \frac{2\pi}{T}\sum_{j=1}^{T} K_M(\lambda-\lambda_j)I(\lambda_j)\,,$$

where the *spectral window* $K_M(\lambda)$ is periodic of period 2π and becomes more and more heavily concentrated around $\lambda = 0$,

$$\lim_{M\to\infty} K_M(\lambda) = \begin{cases} \infty, & \lambda = 0\,, \\ 0, & \lambda \neq 0\,. \end{cases} \tag{6}$$

The data may have been tapered in which case $\tilde{w}(\lambda)$ is computed instead of $w(\lambda)$, and then $I(\lambda) = |\tilde{w}(\lambda)|^2$.

To compare the estimators $\hat{f}_C$ and $\hat{f}_P$, assume that $k_M(u)$ and $K_M(\lambda)$ are related by

$$k_M(u) = \int_{-\pi}^{\pi} K_M(\lambda)\, \mathrm{e}^{\mathrm{i}u\lambda}\, \mathrm{d}\lambda\,.$$

Then

$$\hat{f}_C(\lambda) = \int_{-\pi}^{\pi} K_M(\lambda - \theta) I(\theta)\, \mathrm{d}\theta \tag{7}$$

and $\hat{f}_P(\lambda)$ can be thought of as an approximating sum to $\hat{f}_C(\lambda)$. Because the λ_j cover the interval $[-\pi, \pi]$ more and more densely as $T \to \infty$, the two estimators can be regarded as equivalent asymptotically.

Most lag windows take the form

$$k_M(u) = k(u/M)\,.$$

Parseval's equality leads to

$$\int_{-\pi}^{\pi} K_M^2(\theta)\, \mathrm{d}\theta \doteqdot \frac{M}{2\pi} \int_{-\infty}^{\infty} k^2(u)\, \mathrm{d}u\,.$$

A parameter whose value is often quoted is the *equivalent degrees of freedom*

$$r = \frac{T}{\pi} \left\{ \int_{-\pi}^{\pi} K_M^2(\theta)\, \mathrm{d}\theta \right\}^{-1} \doteqdot \frac{2T}{M} \left\{ \int_{-\infty}^{\infty} k^2(u)\, \mathrm{d}u \right\}^{-1}.$$

We can control r by choice of M and $k(u)$. For given $k(u)$, the *variance* of $\hat{f}(\lambda)$ will vary directly with M, so it seems desirable for M to be small. However, a small M corresponds to a large *bandwidth*, that $K_M(\theta)$ is not heavily concentrated around 0. Thus the estimator will have poor *resolution*, and may be severely *biased* because of undue influence of frequencies neighbouring λ. For given M, we face a similar dilemma in the choice of $k(u)$. As in many other statistical problems, a trade-off between bias and variance is involved. Earlier, considerable discussion took place about what constitutes a useful definition of bandwidth, and about the relationship between bandwidth, bias, resolution and variance of spectral estimators (see Grenander, 1951, 1958; Bartlett and Medhi, 1955; Lomnicki and Zaremba, 1957, 1959; Akaike, 1968; Priestley, 1965a).

2.2. *Historical note*

It is some interest to reflect upon the historical development of spectrum estimators. Daniell (1946) proposed a form of (7), averaging I over frequencies neighbouring λ. Bartlett (1948, 1950), Tukey (1949) and a number of subsequent authors looked at the form (5) of $\hat{f}_C(\lambda)$, proposing various candidates for $k_M(u)$. An important reference is the book of Blackman and Tukey (1959). Tukey has indeed contributed much to the terminology of the subject. The smoothed periodogram estimators were thought to be computationally impractical until Jones (1965) reappraised the periodogram and Cooley and Tukey (1965) published the FFT algorithm. Versions of $\hat{f}_P$, particularly those in which $K_M(\lambda)$ vanishes outside a small band around 0, became very popular. More recently, renewed interest has been shown in the weighted covariance estimators. At the same time, a variety of alternative, nonquadratic, spectrum estimators has become available (see Section 3) and these are now favoured by many practitioners.

2.3. *Statistical properties of estimators*

To save space we discuss only the weighted covariance estimators $\hat{f}_C(\lambda)$, the smoothed periodogram estimators $\hat{f}_P(\lambda)$ having similar asymptotic properties, as indicated above. For fixed M,

$$\lim_{T\to\infty} E(\hat{f}_C(\lambda)) = \int_{-\pi}^{\pi} K_M(\lambda-\theta)f(\theta)\,\mathrm{d}\theta = \int_{-\pi}^{\pi} K_M(\theta)f(\lambda-\theta)\,\mathrm{d}\theta\,. \tag{8}$$

By (6), we can then increase M such that (8)$\to f(\lambda)$, so $\hat{f}_C(\lambda)$ is asymptotically unbiased. For large T, we have under fairly general conditions, for $\lambda \neq 0, \pi$,

$$\operatorname{var}(\hat{f}_C(\lambda)) \doteqdot \frac{2\pi}{T}\int_{-\pi}^{\pi} K_M^2(\lambda-\theta)f^2(\theta)\,\mathrm{d}\theta\,, \tag{9}$$

$$\operatorname{cov}(\hat{f}_C(\lambda), \hat{f}_C(\nu)) \doteqdot \frac{2\pi}{T}\int_{-\pi}^{\pi} K_M(\lambda-\theta)K_M(\nu-\theta)f^2(\theta)\,\mathrm{d}\theta\,.$$

Notice that if $K_M(\theta)$ is *band-limited*, being zero outside of a narrow band around the origin (as in the case of Daniell's (1946) estimator), the covariance will actually be 0 for λ and ν sufficiently apart. This property will hold approximately for many other choices of $K_M(\theta)$. As a rule, spectrum estimators are approximately uncorrelated if $|\lambda-\nu| \geqslant 2\pi/M$, so a common practice is to estimate $f(\lambda)$ at frequencies $\lambda = 2\pi j/M$, $0 \leqslant j \leqslant M/2$.

For large T and M, (9) may be further approximated,

$$\operatorname{var}(\hat{f}_C(\lambda)) \doteqdot (2/r)f^2(\lambda)\,. \tag{10}$$

The approximation (10) is widely used in practice but it may be seriously in error if T and M are not sufficiently large, or f is far from constant around λ. De Jongh (1980) studies closer approximations. An exact expression for the variance is given by Neave (1971). To extend (10), asymptotically $\hat{f}_C(\lambda)$ can be approximated by a $(f(\lambda)/r)\chi_r^2$ random variable, a result that is important in statistical inference for spectral estimates. It can be easily used to set confidence intervals for $f(\lambda)$ (or, more often, for $\log f(\lambda)$ which usually has a much smaller range). Some finite-sample distribution theory of spectrum estimators is given by Grenander, Pollak and Slepian (1959) and Poskitt (1978); for empirical examinations of small-sample properties, see Hatanaka (1972) and Granger and Hughes (1968). Exact finite-sample results are rather complicated to use and must be based on precise distributional assumptions, usually Gaussianity. Moreover, the exact distribution will depend on the true $f(\lambda)$; for practical use this must be replaced by an estimate, so again only approximate inference procedures are available, and it is not clear that they will be much more accurate than the asymptotic ones described above.

A different asymptotic theory is relevant when M and T are both large but M is small relative to T, i.e. $r \to \infty$. From (10), it follows that then $\text{var}(\hat{f}_C(\lambda)) \to 0$, so $\hat{f}_C(\lambda)$ is a *mean square consistent* estimator of $f(\lambda)$. Furthermore, $r^{1/2}(\hat{f}_C(\lambda) - f(\lambda))$ converges to a normal distribution (for details, see Hannan (1970b, p. 289)). Further aspects of the asymptotic theory of spectrum estimators are developed by Grenander and Rosenblatt (1957), Parzen (1957, 1958), Hannan (1960, 1970b) and Brillinger (1969, 1975).

The bias of spectrum estimators deserves further comment. For convenience, let $T \to \infty$ and keep M fixed, and consider (8). It was observed that $K_M(\theta)$ is heavily concentrated around $\theta = 0$, but if it is nonzero for some $\theta \neq 0$, a large value of $f(\lambda - \theta)$ will lead to bias. This phenomenon is known as *leakage*, and leakage can be transmitted from any of the frequencies in $[-\pi, \pi]$. A second type of bias is of local origin. Assuming we can expand f in a Taylor series,

$$E(\hat{f}_C(\lambda)) \doteq f(\lambda) \int_{-\pi}^{\pi} K_M(\theta)\, \mathrm{d}\theta + f'(\lambda) \int_{-\pi}^{\pi} \theta K_M(\theta)\, \mathrm{d}\theta + \frac{f''(\lambda)^2}{2} \int_{-\pi}^{\pi} \theta^2 K_M(\theta)\, \mathrm{d}\theta$$

and $k_M(0) = 1$, $k_M(u) = k_M(-u)$ imply

$$\int_{-\pi}^{\pi} K_M(\theta) = 1, \qquad \int_{-\pi}^{\pi} \theta K_M(\theta) = 0,$$

thence

$$E(\hat{f}_C(\lambda)) \doteq f(\lambda) + \frac{f''(\lambda)^2}{2} \int_{-\pi}^{\pi} \theta^2 K_M(\theta)\, \mathrm{d}\theta .$$

The second term on the right is a measure of bias, and it has two components.

The first component $f''(\lambda)^2/2$ will be small if f is flat around λ, or is changing linearly but will be large positive (negative) if f has a trough (peak) at λ. This indicates that the shape of the underlying spectrum can have a profound influence on our ability to obtain good estimates. Generally the bias will vary over frequency, and $\hat{f}_C(\lambda)$ will appear smoother than $f(\lambda)$. The other bias component, $\int \theta^2 K_M(\theta)\,\mathrm{d}\theta$, will be small if $K_M(\theta)$ decreases rapidly to 0 away from $\theta = 0$.

A variety of additional criteria has been proposed for evaluating the goodness of spectrum estimators:

$$E(\hat{f}(\lambda) - f(\lambda))^2: \quad \text{Grenander and Rosenblatt (1957);}$$

$$\int_{-\pi}^{\pi} E(\hat{f}(\lambda) - f(\lambda))^2\,\mathrm{d}\lambda: \quad \text{Lomnicki and Zaremba (1957);}$$

$$E(\max_{\lambda} |\hat{f}(\lambda) - f(\lambda)|^2): \quad \text{Parzen (1961a);}$$

$$\int_{-\pi}^{\pi} \left\{\frac{E(\hat{f}(\lambda) - f(\lambda))^2}{f(\lambda)^2}\right\} \mathrm{d}\lambda: \quad \text{Jenkins and Watts (1968).}$$

These measures are of only limited practical use; each is somewhat arbitrary and emphasizes different properties, and no single spectrum estimator will be optimal with respect to all.

2.4. Suggested windows

Over the years a very large number of lag/spectral windows have been proposed. Space does not permit a listing of all, or a comprehensive discussion of the properties of any.

The most common form of smoothed periodogram estimator in use involves only the n neighbouring λ_j frequencies, with equal weights

$$\hat{f}_P(\lambda) = \frac{1}{n} \sum{}' I(\lambda_j),$$

summing over the n λ_j closest to λ. This approximates a suggestion of Daniell (1946). The equivalent degrees of freedom are $2n = T/M$. The resolution of the above estimator is not good, but there is little problem with leakage, particularly if the $I(\lambda_j)$ are computed after applying a data window.

We shall now concentrate on the weighted covariance estimators, bearing in mind that these have an approximate smoothed periodogram representation. We restrict ourselves to cases where $k_M(u) = k(u/M)$.

One class of spectral windows is of the form

$$K_M(\lambda) \propto \left\{\frac{\sin(\lambda M/q)}{\lambda M/q}\right\}^q, \quad \lambda \neq 0, \tag{11}$$

for integers q. Members of class (11) are concentrated around $\lambda = 0$, and are zero at $\lambda = \pi jq/M$, for all integers j. In between, *side lobes* appear, and these produce leakage. (The exact Daniell window has no side lobes.) For $q = 1$, $K_M(\lambda)$ corresponds to the lag window $k(u) = 1$, $|u| \leq 1$, $= 0$, $|u| > 1$, called the *truncated* window. The magnitudes of the side lobes in this $K_M(\lambda)$ are such that it is rarely used; indeed, because this $K_M(\lambda)$ is sometimes negative, a negative spectrum estimate can result. On increasing q, the side lobes of (11) are damped and for q even $K_M(\lambda)$, and thence $\hat{f}(\lambda)$ are always nonnegative. A modification of the *Bartlett* (1950) window is the case $q = 2$, in which $K_M(\lambda)$ is essentially Féjèr's kernel, and the technique of Cesaro summation is being employed. Other windows also borrow from ideas for the summation of Fourier series. For example, one of the most widely used windows is that of Parzen (1961a); it is (11) with $q = 4$ and is closely related to the Jackson–de la Vallée Poussin kernel. As $q \to \infty$, $K_M(\lambda)$ given by (11) takes on the shape of a *normal probability density function* (and so does $k(u)$). This form is recommended by Daniels (1962). An alternative window, also considered by Daniels (1962), is based on the *Laplace probability density function.*

A popular rival to the Parzen window is the window of *Tukey–Hanning*,

$$k(u) = \begin{cases} \frac{1}{2}(1 + \cos u), & |u| \leq 1, \\ 0, & |u| > 1. \end{cases}$$

It has less bias, but larger variance than the Parzen window. A closely related window is the *Tukey–Hamming*

$$k(u) = \begin{cases} 0.54 + 0.46 \cos u, & |u| \leq 1, \\ 0, & |u| > 1, \end{cases}$$

whose spectral window has smaller first side lobe than the Tukey–Hanning.

A further class of windows is

$$k(u) \propto (1 + (u/a)^{2j}), \quad j = 1, 2, \ldots,$$

for some constant a. Notice that this window requires use of the $c(u)$ for all $|u| \leq T - 1$, whereas the truncated, Bartlett, Parzen, Tukey–Hanning and Hamming use only the first M $c(u)$. This class was shown by Cogburn and Davis (1974) to arise from smoothing spline considerations. A similar approach is taken by Byrd, Tapia and Thompson (1978). In the spectral analysis of very long series, particularly, there are often substantial computational savings to be made by constructing a large number of (possibly overlapping) subseries of contiguous observations, and then combining the autocovariance estimates or periodograms of the subseries, see Bartlett (1950). One such approach (suggested by A. Kolmogorov), which employs a polynomial taper and appears to possess some computational advantages, has been shown by Zurbenko (1978) to possess low mean squared error and weak dependence on distant frequencies. For other suggestions concerning window design, see Jenkins (1961), Wonnacott (1961), Akaike (1962), Priestley (1962), Welch (1967), Papoulis (1973), Eberhard (1973), Yuen (1978), Geckinli (1980) and Pelkowitz and Havelock (1980), for example. The range of possible windows is unlimited and often it will make little difference which is used. Ideally, because different windows are suited to different situations, the choice of window should be based on what we know about the data and the underlying model. However, many practitioners will use the most readily available computer library package, and then it is the producers of such packages who really make the choice. Some of these packages, for example MIDAS, do however allow the user to choose between a small number of windows. One can at least experiment with different values of M, equivalently different bandwidths. As M increases, the bandwidth is narrowed, so to speak, and more and more detail emerges, at the cost of some instability. This technique is known as *window closing*.

3. Nonquadratic spectrum estimators

It has been seen that peaks or troughs in the true spectrum are liable to cause serious bias in quadratic spectrum estimators. Researchers were led to search for alternative estimators which can better explain such behaviour.

3.1. Prewhitening and recolouring

This is actually a technique to be used in conjunction with window estimation, rather than an alternative. In Subsection 2.3, it was seen that variation in the spectrum is liable to lead to biased spectral estimates, as a result of either poor resolution or leakage. Press and Tukey (1956) proposed a prefiltering technique that will often reduce this bias. Suppose that $f(\lambda)$ can be expressed as

$$f(\lambda) = \frac{b(\lambda)}{\left|\sum_{j=0}^{p} a_j \, e^{ij\lambda}\right|^2} \tag{12}$$

where $b(\lambda)$ is relatively flat, whereas the denominator can produce any number of sharp peaks in $f(\lambda)$ by appropriate choice of p and a_j. (If $b(\lambda)$ were *flat* over $[-\pi, \pi]$, (12) would be a pth-order autoregressive spectrum for $x(t)$.) To prewhiten the data we fit a pth-order autoregression by least squares or by the Yule–Walker relations, and then apply windowed estimation procedures to the residuals, to estimate $b(\lambda)$. Then we 'recolour', estimating $f(\lambda)$ by

$$\hat{f}(\lambda) = \left|\sum_{j=0}^{p} \hat{a}_j \, e^{ij\lambda}\right|^{-2} \hat{b}(\lambda) .$$

3.2. Autoregressive/maximum entropy estimators

If we are prepared to choose the autoregressive order sufficiently large, we can effectively make $b(\lambda)$ flat and dispense altogether with windowed estimation. More precisely, any continuous spectrum that is bounded away from zero can be arbitrary well approximated, uniformly over frequency, by an autoregressive spectrum, for large enough p. The autoregressive spectral estimator and the closely related maximum entropy estimator (see Parzen, 1969; Akaike, 1969; Burg, 1975) have been widely used, and objective ways of choosing p have been proposed. The pth-order autoregressive spectrum has the form

$$f_A(\lambda) \propto \left\{ \sum_{u=-p}^{p} \alpha_u \, \mathrm{e}^{-\mathrm{i}\lambda u} \right\}^{-1} . \tag{13}$$

Many of the quadratic windowed estimators have $k_M(u) = 0$, $|u| > M$, in which case the probability limit is

$$f_M(\lambda) \propto \sum_{u=-M}^{M} \beta_u \, \mathrm{e}^{-\mathrm{i}\lambda u} ,$$

when $T \to \infty$ but M stays fixed. The parameters p and M are regarded as increasing, albeit slowly, to ∞ as $T \to \infty$, and for large p and M it is difficult to choose between the two types of estimator. However, in practice, p and M are finite and $f_A(\lambda)$ is a finite autoregressive spectrum whereas $f_M(\lambda)$ is a finite moving average spectrum. Because pure autoregressive models tend to provide better fits to time series than pure moving average models of comparable order, this may help to explain the popularity of the autoregressive spectral estimator. We shall not discuss the autoregressive estimator further because it is the subject of a whole chapter of the present volume (Chapter 11).

3.3. Estimators based on other parametric models

Again our treatment is abbreviated. To generalize the autoregressive model, we have the autoregressive moving average model (3), with spectrum

$$f(\lambda) = \left| \sum_{j=0}^{q} b_j \, \mathrm{e}^{\mathrm{i}j\lambda} \right|^2 \Bigg/ \left| \sum_{j=0}^{p} a_j \, \mathrm{e}^{\mathrm{i}j\lambda} \right|^2 . \tag{14}$$

It is a better approximation to spectra with narrow troughs, as well as peaks. The parameters a_j and b_j are estimated by statistically efficient procedures after initial determination of p and q. More generally, we can model the spectrum as $f(\lambda; \theta)$, a given function of λ and an unknown but estimable parameter vector θ. Ideally, a conceptual model will be the source. The problem then becomes one of classical parametric statistics. A Gaussian likelihood can be maximized to obtain estimates of θ. Several approximate forms lead to computationally simpler estimates, with equivalent asymptotic efficiency. A good deal of effort has gone into proving that such estimates are strongly consistent and asymptotically normal under mild conditions. These matters are discussed by Bloomfield (1973), Davies (1973), Dzhaparidze (1974), Hannan (1973), Ibragimov (1967), Robinson (1978) and Whittle (1961).

3.4. Pisarenko estimator

Denote by C the $M\times M$ matrix with uvth element $c(u-v)$. Let ϕ_u, $u=1,\ldots,M$ be the eigenvalues of C, with corresponding $M\times 1$ eigenvectors ψ_u, $u=1,\ldots,M$. Let $H(\phi)$, $0<\phi<\infty$, be a strictly monotonic function, with inverse function $h(\cdot)$. Pisarenko (1972) suggested the estimator

$$\hat{f}(\lambda)=h\left(\frac{1}{2\pi M}\sum_{j=1}^{M}H(\phi_j)\left|\sum_{j=1}^{M}\psi_{uj}\,\mathrm{e}^{\mathrm{i}j\lambda}\right|^2\right),\tag{15}$$

where ψ_{uj} is the jth element of ψ_u. When $H(\phi)=\phi$, $\hat{f}(\lambda)$ is simply the Bartlett windowed estimator (and modifications of (15) would reduce to alternative quadratic windowed estimators). When $H(\phi)$ is nonlinear, nonquadratic estimators result. For $H(\phi)=1/\phi$, we have the 'high-resolution' estimator of Capon (1969)

$$\hat{f}(\lambda)\propto\left\{\sum_{u=-M+1}^{M-1}\left(1-\frac{|u|}{M}\right)d(u)\,\mathrm{e}^{-\mathrm{i}u\lambda}\right\}^{-1},$$

where $d(u-v)$ is the uvth element of a Toeplitz approximation to C^{-1}. (The high-resolution estimator is closely related to the maximum entropy and autoregressive spectral estimators, cf. (13).) Other choices for H considered by Pisarenko (1972) are $H(\phi)=\phi^r$ and $H(\phi)=\log\phi$. The Pisarenko estimators have similar asymptotic properties to the quadratic ones, but they aim to provide less biased estimators in the region of spectral peaks.

3.5. G-estimator

Let $m\geq 1$, and $A_m=a_0+a_1+\cdots+a_m$. For a given positive integer n, the e_n-transform of A_m is defined as

$$e_n[A_m;l]=\frac{\begin{vmatrix}A_m & A_{m+l} & \cdots & A_{m+nl}\\ a_m & a_{m+l} & & a_{m+nl}\\ \vdots & \vdots & & \vdots\\ a_{m+(n-1)l} & a_{m+nl} & \cdots & a_{m+(2n-1)l}\end{vmatrix}}{\begin{vmatrix}1 & 1 & \cdots & 1\\ a_m & a_{m+l} & & a_{m+nl}\\ \vdots & \vdots & & \vdots\\ a_{m+(n-1)l} & a_{m+nl} & \cdots & a_{m+(2n-1)l}\end{vmatrix}}$$

We have

$$\lim_{n\to\infty} e_n[A_m; l] = \lim_{m\to\infty} A_m = A_\infty ,$$

and the convergence of e_n is known to be much more rapid than that of A_m (Gray, Houston and Morgan, 1978). Because the power spectrum has an infinite series representation (1), we can apply the e_n-transformation as a means of summing it; the windows considered in Subsection 2.4 indeed being a means to the same end. The G-estimator is defined by

$$\hat{f}(\lambda) = e_n\left[c(0) + 2\sum_{u=1}^{M} c(u)\cos u\lambda ; l\right].$$

It is shown in Gray, Houston and Morgan (1978) that if $f(\lambda)$ is in fact the autoregressive moving average spectrum (14), then

$$f(\lambda) = e_n\left[\gamma(0) + 2\sum_{u=1}^{M} \gamma(u)\cos u\lambda ; l\right]$$

for all sufficiently large n. Thus we can expect the G-estimator to be most useful if the spectrum closely approximates that of a low-order autoregressive moving average process. Gray, Houston and Morgan (1978) discuss the computation of the G-estimator.

4. Spectrum estimation from incomplete data

4.1. Missing and unequally spaced data

We say that $x(t)$ is *missing* for some t if there is no information about the value of $x(t)$. Missing data can arise from a number of causes, such as failure of recording equipment, clerical errors, rejection of outliers, or because of an inability to observe the phenomenon at certain times. The time series will contain gaps where observations have been missed, but if the gaps are very small and infrequent, satisfactory results will probably be obtained by simple methods of interpolating the missing values, followed by application of the standard technqiues described previously. Such ad hoc procedures seem less acceptable when a significant proportion of the T possible observations has been missed, and certainly their statistical properties are likely to be rather messy. Over the past twenty years, several methods have been proposed for obtaining consistent spectrum estimators from a variety of patterns of missing data.

The usual approach has been in terms of windowed quadratic estimators. A commonsense modification to deal with missing observations replaces $c(u)$ by

$$\tilde{c}(u) = \frac{T-u}{TT_u} \sum_t{}' x(t)x(t+u), \quad u \geq 0,$$

where the sum is over all t such that both $x(t)$ and $x(t+u)$ are observed, and T_u is the number of such summands. Then we form

$$\hat{f}(\lambda) = \frac{1}{2\pi} \sum_{u=-T+1}^{T-1} k_M(u)\tilde{c}(u) \cos u\lambda \tag{16}$$

as in Subsection 2.1. Parzen (1963) formalized and somewhat generalized this approach by introducing the notion of an *amplitude modulating sequence* $v(t)$

$$v(t) = \begin{cases} 1 & \text{if } x(t) \text{ is observed}, \\ 0 & \text{if } x(t) \text{ is missing}. \end{cases}$$

It is not essential that $v(t)$ be stationary, but the asymptotics require *asymptotic stationarity*:

$$\lim_{T\to\infty} \bar{v}, \qquad \lim_{T\to\infty} c_v(u)$$

exist for all u, where

$$\bar{v} = \frac{1}{T}\sum_{t=1}^{T} v(t), \qquad c_v(u) = \frac{1}{T}\sum_{t=1}^{T-u} v(t)v(t+u),$$

for $u \geq 0$. The $v(t)$ may be generated by an underlying random process in which case the convergence is stochastic. One forms the *amplitude modulated process*

$$y(t) = v(t)x(t),$$

which is $x(t)$ when $x(t)$ is observed and 0 otherwise. Defining

$$c_y(u) = \frac{1}{T}\sum_{t=1}^{T-u} y(t)y(t+u), \quad u \geq 0,$$

we have $\tilde{c}(u) = c_y(u)/c_v(u)$ and thence (16) is

$$\hat{f}(\lambda) = \frac{1}{2\pi} \sum_{u=-T+1}^{T-1} k_M(u) \frac{c_y(u)}{c_v(u)} \cos u\lambda. \tag{17}$$

Jones (1971) discusses the computation of (17). He shows how the FFT can be used when T is large: the complex series $y(t) + iv(t)$ is Fourier transformed, the periodograms of $y(t)$ and $v(t)$ are derived, and then Fourier transformed to get $c_y(u)$ and $c_v(u)$.

Denoting the limit of $c_v(u)$ by $\gamma_v(u)$, it follows that $c_y(u)$ has limit $\gamma_v(u)E(x(t)x(t+u)) = \gamma_v(u)\gamma(u)$, so long as $x(t)$ and $v(t)$ are independent, and this assumption could serve to define missing data: $x(t)$ is missing due to extraneous causes and not as a result of the value it would have taken. Assuming $\gamma_v(u) \neq 0$, we have $c_y(u)/c_v(u) \to \gamma(u)$, and the consistency of $\hat{f}(\lambda)$ then follows like that of the windowed estimators from complete data in Subsection 2.3. Note that $v(t)$ need not be a 0:1 process, although it is difficult to think of other useful examples of amplitude modulation.

A disadvantage of (17) is that the sequence $c_y(u)/c_v(u)$ is not nonnegative definite and so use of a $k_M(u)$ corresponding to a spectral window $K_M(\lambda)$ which is everywhere nonnegative does not guarantee that $\hat{f}(\lambda)$ will always be nonnegative. It may be possible to design alternative windows which will ensure nonnegativity.

The variance of (17) is studied by Parzen (1963) and Jones (1971). Substantial simplifications result when the data are systematically missing: one unequally spaced pattern is periodically repeated. The simplest such example is the case of α observed values being followed by β missing values, followed by α observed values and so on (see Jones, 1962b; Parzen, 1963; Alekseev and Savitskii, 1973). It is necessary that $\alpha > \beta$. Jones (1962b) finds that the harmonic frequencies brought in by the periodic method of sampling adds to the variance, and calls this *variance leakage*.

When $v(t)$ is generated by a random process for which $Ev(t)v(t+u) = \gamma_v(u)$ is known, then an obvious modification of (17) is to replace $c_v(u)$ by $\gamma_v(u)$. Scheinok (1965) follows this approach in the case that $v(t)$ is a sequence of Bernoulli trials, so $\gamma_v(0) = \theta$, $\gamma_v(u) = \theta^2$, $u \neq 0$, where θ is the probability that $x(t)$ is not missed. Bloomfield (1970) considers a more general class of $v(t)$ processes. In practice, $\gamma_v(u)$ will not be perfectly known but at best will depend on finitely many parameters. These parameters can be estimated from the observed sequence $v(t)$, $t = 1, \ldots, T$, if possible by a statistically efficient method such as maximum likelihood. For example, for Bernoulli trials we have $\hat{\gamma}_v(0) = \bar{v}$, $\hat{\gamma}_v(u) = \bar{v}^2$, $u \neq 0$.

Because of improved sampling capabilities, it is sometimes possible to increase the frequency at which economic variables are observed. For example, an economic time series may consist of quarterly observations initially, followed by monthly observations. The first time segment will then, in effect, contain missing observations. Neave (1970a) applies Parzen's (1963) amplitude modulation technique to this problem to obtain estimates of the form (17). An important modification is needed to produce a sensible asymptotic theory, however. In order to prevent the early, infrequent, observations from being swamped by the later, frequent, ones, we define the time span over which $x(t)$ is observed as $[-b(T)+1, a(T)]$, where $a(T)$ and $b(T)$ are positive integers that increase with T, such that $a(T)+b(T) = T$ and $a(T)/b(T)$ converges to a finite, nonzero constant. We observe $x(t)$ for all integers $t \in [0, a(T)]$, but only at intervals of $r > 1$ for $t \in [-b(T)+1, -1]$, so $r = 3$ for the quarterly/monthly example mentioned above. As indicated previously, (17) may produce negatively biased, even negative, spectrum estimators, so Neave (1970a) proposes an alternative estimator

$$\bar{f}(\lambda) = f^*(\lambda)\bar{f}_2(\lambda)/\bar{f}_1(\lambda),$$

where $f^*(\lambda)$ is the windowed estimator obtained from the unit spaced second segment $x(1), \ldots, x(a(T))$; $\bar{f}_1(\lambda)$ is the windowed estimator obtained from the skip-sampled sequence $x(0), x(r), \ldots, x([a(T)/r])$; $\bar{f}_2(\lambda)$ is the windowed estimator obtained from $x(-[(b(T)-1)/r])$, $x(-[(b(T)-1)/r]+r), \ldots, x(0)$, $x(r), \ldots, x([a(T)/r])$. Both $\bar{f}_2(\lambda)$ and $\bar{f}_2(\lambda)$ are periodic of period $2\pi/r$. Because f^*, $\bar{f}_1$ and $\bar{f}_2$ are also computed from equally spaced data, the use of a nonnegative spectral window will guarantee that each is nonnegative, in which case so is $\bar{f}$.

Missing data are really only a special case, albeit a very important one, *of unequally spaced data.* Actually time intervals are quite often unequal, but if the deviations are small they are ignored. For example, calendar monthly economic data are unequally spaced because of differences in the number of days, or working days, per month. Granger (1962) analyses the effect of such deviations on spectral estimates and finds them to be negligible in the case of instantaneously measured variables, but possibly significant for aggregated flow variables.

Time intervals may be chosen to be unequally spaced in order to avoid the aliasing problem described in Section 1. Suppose $x(t)$ is defined on the whole real line, and consider the estimation of its spectrum $g(\lambda)$, $-\infty < \lambda < \infty$, from the sequence $\{x(t_n), n = 1, \ldots, T\}$ where $t_n > t_{n-1}$ all n. Shapiro and Silverman (1960) (see also Beutler, 1970) show that if the increments $t_n - t_{n-1}$ are independent Poisson variables, alias-free estimation of $g(\lambda)$ is possible. For most other examples that come to mind, some aliasing creeps in, however, such as for 'jittered sampling', $t_n = n + \varepsilon_n$ where the ε_n are independent random variables with zero means and variances much less than 1. Akaike (1960) examines the effect of the timing errors ε_n on the spectral estimators of the discrete sequence $x(t)$, $t = 0, \pm 1, \ldots$. For details of how to construct spectrum estimates from unequally spaced data, see Jones (1962a) and Brillinger (1972).

Poisson sampling is not always technically feasible. Often the frequency of observation is bounded from below, and it is desired to keep it constant for reasons of convenience or economy. If two or more recorders are available however, extension of the estimation frequency range may still be possible, as shown by Neave (1970b). One recorder is calibrated to read at the minimum interval, 1; the other at $1+\delta$, $0<\delta<1$. If $\delta = 1/n$ for some integer $n > 1$, then the combined data from the two recorders enable frequencies up to $n\pi$ to be detected.

Finite parameter models have been used to estimate spectra of continuous processes from discrete observations. Robinson (1977) and Jones (1979) estimate rational spectral densities, $g(\lambda) \propto |\Sigma_0^p \alpha_j(\mathrm{i}\lambda)^j|^2/|\Sigma_0^q \beta_j(\mathrm{i}\lambda)^j|^2$ from arbitrary unequally spaced observations. Robinson (1980a) shows how the parameters can be identified in the case of Poisson sampling, providing a parametric version of results of Shapiro and Silverman (1960).

4.2. Censored data

If $x(t)$ is *censored*, we can narrow down its value to a proper subset of the possible values. Thus $x(t)$ is not observed because of the value it would have taken. For a comparison of missed and censored data in time-series analysis, see Robinson (1980b). The usual approach to spectral estimation of censored data has been via windowed quadratic estimators, following direct estimation of sample autocovariances. Finite parameter modelling does not seem promising because of the computational difficulties of maximum likelihood estimation (Robinson, 1980b).

Limitations in storage space sometimes require *clipping* of time series. If $x(t)$ is *hard-clipped* or *hard-limited*, we store only the sign of $x(t)$,

$$y(t) = \begin{cases} 1 & \text{if } x(t) \geqslant 0\,, \\ -1 & \text{if } x(t) < 0\,. \end{cases}$$

If $x(t)$ is Gaussian with zero mean, we have the well-known relation

$$\gamma(u) = \gamma(0)\sin\left(\frac{\pi}{2}\,Ey(t)y(t+u)\right), \qquad u \neq 0\,.$$

We can thus compute

$$c_y(u) = \frac{1}{T}\sum_{t=1}^{T-u} y(t)y(t+u)\,, \quad u > 0\,,$$

and estimate $\gamma(u)$ by

$$\hat{\gamma}(u) = \gamma(0)\sin\left(\frac{\pi}{2}\,c_y(u)\right) \tag{18}$$

if $\gamma(0)$ is known. Finally, $f(\lambda)$ is estimated by

$$\hat{f}(\lambda) = \frac{1}{2\pi}\sum_{-T+1}^{T-1} k_M(u)\hat{\gamma}(u)\,\mathrm{e}^{-\mathrm{i}\lambda u}\,. \tag{19}$$

This procedure is studied by Brillinger (1968), Hinich (1967) and Schlittgen (1978). The estimated autocorrelations $\hat{\gamma}(u)/\gamma(0)$ given by (18) are all between -1 and 1, but it is not clear that the $\hat{\gamma}(u)$ will necessarily form a nonnegative definite sequence, in which case negative spectral estimates may result.

A different setup is assumed in Robinson (1980b, 1980c); we observe $x(t)$ if and only if $x(t) > b_t$, where the b_t are known numbers. (The case that $x(t)$ is observed when $x(t) < b_t$ is handled by changing signs.) The problem of mean-correction of censored data is not covered by our remarks in Subsection 1.4. If $Ex(t) = \mu$, μ unknown, and $b_t < \mu$ for all t, the sample median m of $x(t)$ can still be obtained, and this will consistently estimate μ if $x(t)$ is a symmetric random variable. Putting $y(t) = x(t) - m$,

$$\hat{\gamma}(0) = \frac{1}{T'} \sum' y(t)^2$$

is a consistent estimator of $\gamma(0)$, where the sum is over all t for which $x(t) \geqslant m$. The autocovariances $\gamma(u)$, $u \neq 0$, can be estimated by assuming Gaussianity and using relations for incomplete moments in terms of the autocorrelation function. One possible estimator of $\gamma(u)$ is

$$\hat{\gamma}(u) = \hat{\gamma}(0)\left\{\left[\frac{2\pi}{\hat{\gamma}(0)}\right]^{1/2} \frac{2}{T''_u} \sum'' y(t) - 1\right\}, \tag{20}$$

where the sum is over t such that $x(t) \geqslant m$, $x(t+u) \geqslant m$, and T''_u is the number of such summands. Then $\hat{f}(\lambda)$ is formed, as in (19). Unfortunately, the implied correlation estimates $\hat{\gamma}(u)/\hat{\gamma}(0)$ are not necessarily between -1 and 1, and the $\hat{\gamma}(u)$ are not a nonnegative definite sequence. The estimator (20) can be used only when $b_t < \mu$, all t. This requirement is relaxed by Robinson (1980c), where more complicated moment estimators, and nonlinear least-squares estimators of $\gamma(u)$ are proposed, from which weighted covariance spectral estimators can be formed.

Another type of censored data is discussed by McNeil (1967). A less drastic form of clipping is imposed than that in Brillinger (1968) and Hinich (1967): $x(t)$ is *digitalized*. We observe

$$y(t) = nd + \varepsilon\,, \quad \text{if } (n - \tfrac{1}{2})d - \varepsilon \leqslant x(t) \leqslant (n + \tfrac{1}{2})d + \varepsilon\,.$$

Then the autocovariance functions of $x(t)$ and $y(t)$ are related by

$$\gamma_y(u) = g(\gamma(u))\,, \tag{21}$$

if $x(t)$ is Gaussian, where g has a unique inverse. Once g is found, which may be a complicated business, we can invert (21) to deduce an estimate of $\gamma(u)$ from one of $\gamma_y(u)$, and thence compute (19). A situation that is in most respects more general is covered by Rodemich (1966). We observe $y(t) = h(x(t))$, where h is an odd, bounded, nondecreasing function. An example is $y(t) = x(t)$ if $|x(t)| < K$; $= K$, if $x(t) > K$; $= -K$ if $x(t) < -K$, so that the values that are large in magnitude are censored. Again a relationship of the form (21) is obtained when $x(t)$ is Gaussian.

References

Akaike, H. (1960). Effect of timing-error on the power-spectrum of sampled data. *Ann. Inst. Statist. Math.* **11**, 145–165.

Akaike, H. (1962). On the design of lag windows for the estimation of spectra. *Ann. Inst. Statist. Math.* **14**, 1–21.

Akaike, H. (1968). On the use of an index of bias in the estimation of power spectra. *Ann. Inst. Statist. Math.* **20**, 55–69.

Akaike, H. (1969). Fitting autoregressive models for prediction. *Ann. Inst. Statist. Math.* **21**, 243–247.

Alekseev, V. G. and Savitskii, Yu. A. (1973). Estimation of the spectrum of a Gaussian stochastic process on the basis of a realization of the process with omissions. *Prob. Inf. Transm.* **9**, 50–54.

Bartlett, M. S. (1948). Smoothing periodograms from time series with continuous spectra. *Nature* **161**, 686–687.

Bartlett, M. S. (1950). Periodogram analysis and continuous spectra. *Biometrika* **37**, 1–16.

Bartlett, M. S. and Medhi, J. (1955). On the efficiency of procedures for smoothing periodograms from time series with continuous spectra. *Biometrika* **42**, 143–150.

Beutler, F. J. (1970). Alias-free randomly-timed sampling of stochastic processes. *IEEE Trans. Information Theory* **16**, 147–152.

Bhansali, R. J. (1974). Asymptotic properties of the Wiener–Kolmogorov predictor. *J. R. Statist. Soc. Ser. B* **36**, 61–73.

Blackman, R. B. and Tukey, J. W. (1959). *The Measurement of Power Spectra.* Dover, New York.

Bloomfield, P. (1970). Spectral analysis with randomly missing observations. *J. R. Statist. Soc. Ser. B* **32**, 369–380.

Bloomfield, P. (1973). An exponential model for the spectrum of a scalar time series. *Biometrika* **60**, 217–226.

Box, G. E. P. and Jenkins, G. M. (1970). *Time Series Analysis, Forecasting and Control.* Holden-Day, San Francisco.

Brillinger, D. R. (1968). Estimation of the cross-spectrum of a stationary bivariate Gaussian process from its zeros. *J. R. Statist. Soc. Ser. B* **30**, 145–159.

Brillinger, D. R. (1969). Asymptotic properties of spectral estimates of second order. *Biometrika* **56**, 375–390.

Brillinger, D. R. (1972). The spectral analysis of stationary interval functions. In: L. Lecam, J. Neyman and E. L. Scott, eds., *Proc. Seventh Berkeley Symp. Math. Statist. Prob.*, 483–513. University of California Press, Berkeley, California.

Brillinger, D. R. (1973). A power spectral estimate which is insensitive to transients. *Technometrics* **15**, 559–562.

Brillinger, D. R. (1975). *Time Series, Data Analysis and Theory.* Holden-Day, San Francisco.

Burg, J. P. (1975). Maximum entropy spectral analysis. Ph.D. dissertation, Stanford University, Stanford, California.

Byrd, R. H., Tapia, R. A. and Thompson, J. R. (1978). Optimal smoothing of direct estimates of the power spectrum. *Comm. Statist.* **87**, 335–344.

Capon, J. (1969). High resolution frequency wavenumber spectral analysis. *Proc. IEEE* **57**, 1408–1418.

Cogburn, R. and Davis, H. T. (1974). Periodic splines and spectral estimation. *Ann. Statist.* **2**, 1108–1126.

Cooley, J. W. and Tukey, J. W. (1965). An algorithm for the machine calculation of complex Fourier series. *Math. Comp.* **19**, 297–301.

Daniell, P. J. (1946). Discussion of paper by M. S. Bartlett. *J. R. Statist. Soc. Suppl.* **8**, 27.

Daniels, H. E. (1962). The estimation of spectral densities. *J. R. Statist. Soc. Ser. B* **24**, 185–198.

Dargahi-Noubary, G. R. and Laycock, P. J. (1979). Spectral ratio discriminants and information theory. Technical Report No. 116, University of Manchester Institute of Science and Technology, Manchester, U.K.

Davies, R. B. (1973). Asymptotic inference in stationary Gaussian series. *Adv. Appl. Prob.* **5**, 469–497.

De Jongh, P. (1980). Personal communication.

Dzhaparidze (1974). A new method for estimating spectral parameters of a stationary regular time series. *Th. Prob. Appl.* **19**, 122–132.

Eberhard, A. (1973). An optimal discrete window for the calculation of power spectra. *IEEE Trans. Audio Electroacoust.* **21**, 37–43.

Geckinli, N. C. (1980). Power spectra estimation using discrete Fourier transform. *Proc. First European Signal Processing Conf.*, 567–576. North-Holland, Amsterdam.

Good, I. J. (1958). The iteration algorithm and practical Fourier series. *J. R. Statist. Soc. Ser. B* **20**, 361–372.

Granger, C. W. J. (1962). The effect of varying month-length on the analysis of economic time series. Research Memo. No. 41, Princeton University Economic Research Program, Princeton, N.J.

Granger, C. W. J. and Hughes, A. O. (1968). Spectral analysis of short series—a simulation study. *J. R. Statist. Soc. Ser. A* **131**, 83–99.

Gray, H. L., Houston, A. G. and Morgan, F. W. (1978). On *G*-spectral estimation. In: D. F. Findley, ed., *Applied Time Series Analysis*, 39–138. Academic Press, New York.

Grenander, U. (1951). On empirical spectral analysis of stochastic processes. *Ark. Mat.* **1**, 502–531.

Grenander, U. (1958). Bandwidth and variance in estimation of the spectrum. *J. R. Statist. Soc. Ser. B* **20**, 152–157.

Grenander, U. (1974). Large sample discrimination between two Gaussian processes with different spectra. *Ann. Statist.* **2**, 347–352.

Grenander, U., Pollak, H. O. and Slepian, D. (1959). The distribution of quadratic forms in normal variates: a small sample theory with applications to spectral analysis. *J. SIAM* **7**, 374–401.

Grenander, U. and Rosenblatt, M. (1957). *Statistical Analysis of Stationary Time Series.* Wiley, New York.

Hannan, E. J. (1958). The estimation of the spectral density after trend removal. *J. R. Statist. Soc. Ser. B.* **20**, 323–333.

Hannan, E. J. (1960). *Time Series Analysis.* Methuen, London.

Hannan, E. J. (1961). Testing for a jump in the spectral function. *J. R. Statist. Soc. Ser. B.* **23**, 394–404.

Hannan, E. J. (1963). Regression for time series. In: M. Rosenblatt, ed., *Time Series Analysis*, 17–37. Wiley, New York.

Hannan, E. J. (1970a). The seasonal adjustment of economic time series. *Int. Econ. Rev.* **11**, 1–29.

Hannan, E. J. (1970b). *Multiple Time Series.* Wiley, New York.

Hannan, E. J. (1973). The asymptotic theory of linear time series models. *J. Appl. Prob.* **10**, 130–145.

Hatanaka, M. (1972). The estimation of spectra and cross-spectra on short time series data. *Int. Econ. Rev.* **13**, 679–704.

Herbst, L. J. (1964). Spectral analysis in the presence of variance fluctuations. *J. R. Statist. Soc. Ser. B.* **26**, 354–360.

Hinich, M. (1967). Estimation of spectra after hard clipping of Gaussian processes. *Technometrics* **9**, 391–400.

Ibragimov, I. A. (1967). On maximum likelihood estimation of parameters of the spectral density of stationary time series. *Th. Prob. Appl.* **12**, 115–119.

Jackson, J. E. and Lawton, W. H. (1969). Comparison of ANOVA and harmonic components of variance. *Technometrics* **11**, 75–90.

Jenkins, G. M. (1961). General considerations in the analysis of spectra. *Technometrics* **3**, 98–131.

Jenkins, G. M. and Watts, D. G. (1968). *Spectrum Analysis and its Applications.* Holden-Day, San Francisco.

Jones, R. H. (1962a). Spectral estimates and their distributions II. *Skand. Aktuar.* **45**, 135–153.

Jones, R. H. (1962b). Spectral analysis with regularly missed observations. *Ann. Math. Statist.* **33**, 455–461.

Jones, R. H. (1965). A reappraisal of the periodogram in spectral analysis. *Technometrics* **7**, 531–542.

Jones, R. H. (1971). Spectrum estimation with missing observations. *Ann. Inst. Statist. Math.* **23**, 387–398.

Jones, R. H. (1979). Spectrum estimation from unequally spaced data. Preprint.

Kleiner, B., Martin, R. D. and Thomson, D. J. (1979). Robust estimation of power spectra. (With discussion) *J. R. Statist. Soc. Ser. B.* **41**, 313–351.

Kolmogorov, A. N. (1941). Interpolation und Extrapolation von stationaren zufalligen Folgen. *Bull. Acad. Sic. De L'U.R.S.S.* **5**, 3–14.

Koopmans, L. H. (1974). *The Spectral Analysis of Time Series.* Academic Press, New York.

Lomnicki, Z. A. and Zaremba, S. K. (1957). On estimating the spectral density of a stochastic process. *J. R. Statist. Soc. Ser. B.* **19**, 13–37.

Lomnicki, Z. A. and Zaremba, S. K. (1959). Bandwidth and resolvability in statistical spectral analyis. *J. R. Statist. Soc. Ser. B.* **21**, 169–171.

McNeil, D. R. (1967). Estimating the covariance and spectral density functions from a clipped stationary time series. *J. R. Statist. Soc. Ser. B.* **29**, 180–185.

Neave, H. A. (1970a). Spectral analysis of a stationary time series using initially scarce data. *Biometrika* **57**, 111–122.

Neave, H. A. (1970b). Extending the frequency range of spectrum estimates by the use of two data recorders. *Technometrics* **12**, 877–890.

Neave, H. A. (1971). The exact error in spectrum estimates. *Ann. Math. Statist.* **42**, 961–975.

Nerlove, M. (1964). Spectral analysis of seasonal adjustment procedures. *Econometrica* **32**, 241–286.

Nicholls, D. F. (1967). Estimation of the spectral density function when testing for a jump in the spectrum. *Aust. J. Statist.* **9**, 103–108.

Olshen, R. A. (1967). Asymptotic properties of the periodogram of a discrete stationary process. *J. Appl. Prob.* **4**, 508–528.

Papoulis, A. (1973). Minimum-bias window for high resolution spectral estimation. *IEEE Trans. Information Theory.* **19**, 9–12.

Parzen, E. (1957). On consistent estimates of the spectrum of a stationary time series. *Ann. Math. Statist.* **28**, 329–348.

Parzen, E. (1958). On asymptotically efficient consistent estimates of the spectral density function of a stationary time series. *J. R. Statist. Soc. Ser. B.* **20**, 303–322.

Parzen, E. (1961a). Mathematical considerations in the analysis of spectra. *Technometrics* **3**, 167–190.

Parzen, E. (1961b). Spectral analysis of asymptotically stationary time series. *Bull. Int. Statist. Inst.* 33rd Session, 162–179.

Parzen, E. (1963). On spectral analysis with missing observations and amplitude modulation. *Sankhya Ser. A.* **25**, 383–392.

Parzen, E. (1969). Multiple time series modelling. In: P. R. Krishnaiah, ed., *Multivariate Analysis—II*, 383–409. Academic Press, New York.

Pelkowitz, L. and Havelock, D. (1980). Applications of time-varying filter theory to high resolution spectrum estimation. *Proc. First European Signal Processing Conf.*, 705–710. North-Holland, Amsterdam.

Pisarenko, V. F. (1972). On the estimation of spectra by means of non-linear functions of the covariance matrix. *Geophys. J. R. Astron. Soc.* **28**, 511–531.

Poskitt, D. S. (1978). Approximating the exact finite sample distribution of a spectral estimator. *Econometrica* **46**, 21–32.

Press, H. and Tukey, J. W. (1956). Power spectral methods of analysis and their application to problems in airplane dynamics. *Bell Systems Monograph* 2606.

Priestley, M. B. (1962). Basic considerations in the estimation of spectra. *Technometrics* **4**, 551–564.

Priestley, M. B. (1964). Estimation of the spectral density in the presence of harmonic components. *J. R. Statist. Soc. Ser. B.* **26**, 123–132.

Priestley, M. B. (1965a). The role of bandwidth in spectral analysis. *Appl. Statist.* **14**, 33–47.

Priestley, M. B. (1965b). Evolutionary spectra and non-stationary processes. *J. R. Statist. Soc. Ser. B.* **27**, 204–237.

Robinson, P. M. (1976). Some problems in the identification and estimation of continuous time systems from discrete time data. In: R. K. Mehra and D. G. Lainiotis, eds., *System identification: Advances and Case Studies*, 407–439. Academic Press, New York.

Robinson, P. M. (1977). Estimation of a time series model from unequally spaced data. *Stoch. Proc. Appl.* **6**, 9–24.

Robinson, P. M. (1978). Alternative models for stationary stochastic processes. *Stoch. Proc. Appl.* **8**, 141–152.

Robinson, P. M. (1980a). Continuous model fitting from discrete data. In: D. R. Brillinger and G. C. Tiao, eds., *Directions in Time Series*, 263–278. Institute of Mathematical Statistics.

Robinson, P. M. (1980b). Estimation and forecasting for time series containing censored or missing observations. In: *Time Series*, 167–182. North-Holland, Amsterdam.

Robinson, P. M. (1982). Analysis of time series from mixed distributions. *Ann. Statist.* **10**, 915–925.

Rodemich, E. R. (1966). Spectral estimates using nonlinear functions. *Ann. Math. Statist.* **37**, 1237–1256.

Sande, G. (1966). On an alternative method of calculating autocovariance functions. Unpublished manuscript.

Scheinok, P. A. (1965). Spectral analysis with randomly missed observations: the binomial case. *Ann. Math. Statist.* **36**, 971–977.

Schlittgen, R. (1978). Ein niehlparameterischer Ansatz in der Zeitreihenanalyse. Diskussionsarbeit No. 1, Institut fur Quantitatave Okonomik und Statistik, Freie Universitat Berlin, Berlin.

Schuster, A. (1898). On the investigation of hidden periodicities with applications to a supposed 26 day period of the Greenwich Observatory during the years 1871–1895. *Camb. Phil. Trans.* **18**, 107–135.

Shapiro, H. S. and Silverman, R. A. (1960). Alias-free sampling of random noise. *J. SIAM.* **8**, 225–249.

Shumway, R. H. and Unger, A. N. (1974). Linear discriminant functions for stationary time series. *J. Amer. Statist. Assoc.* **69**, 948–956.

Splettstosser, W. (1980). On the aliasing error in the sampling representation of random signals. *Proc. First European Signal Processing Conf., Short Communications*, 17–18.

Thomson, D. J. (1977). Spectrum estimation techniques for characterization and development of WT4 Waveguide—I. *Bell System Tech. J.* **56**, 1769–1815.

Tukey, J. W. (1949). The sampling theory of power spectrum estimates. *Proc. on Applications of Autocovariance Analysis to Physical Problems*, 47–67. Office of Naval Research, Washington, D.C.

Tukey, J. W. (1961). Discussion emphasizing the connection between analysis of variance and spectrum analysis. *Technometrics* **3**, 1–29.

Welch, P. D. (1967). The use of the fast Fourier transform for estimation of spectra: a method based on time averaging over short, modified periodograms. *IEEE Trans. Audio Electroacoust.* **15**, 70–73.

Whittle, P. (1961). Gaussian estimation in stationary time series. *Bull. Int. Statist. Inst.* **39**, 105–130.

Wiener, N. (1949). *The Extrapolation, Interpolation and Smoothing of Stationary Time Series with Engineering Applications.* Wiley, New York.

Wonnacott, T. H. (1961). Spectral analysis combining a Bartlett window with an associated inner window. *Technometrics* **3**, 235–243.

Yuen, C. K. (1978). Quadratic windowing in the segment averaging method for power spectrum computation. *Technometrics* **20**, 195–200.

Zaremba, S. K. (1960). A test of fit for the spectral density function of a stochastic process. *Monat. Math.* **64**, 68–79.

Zurbenko, I. G. (1978). On a statistic for the spectral density of a stationary sequence. *Soviet. Math. Dokl.* **19**, 203–266.

D. R. Brillinger and P. R. Krishnaiah, eds., *Handbook of Statistics, Vol. 3*

17

Cumulants and Cumulant Spectra*

M. Rosenblatt

Introduction

At first, some of the basic properties of cumulants are discussed. Some of the reasons for interest in cumulants are presented. Spectra and cumulant spectra for stationary processes are introduced. The relations between continuous and discrete time parameter processes are dealt with in some detail. A class of estimates of cumulant spectra is introduced and asymptotic properties of these estimates are noted. There are some remarks on nonlinear systems. The use of cumulant spectra in resolving the structure of non-Gaussian linear schemes and corresponding phase information is indicated. Finally there are a few comments on applications.

1. Definition

Consider the random variables $X_1, \ldots, X_k$. Let $\varphi(t_1, \ldots, t_k)$ be the joint characteristic function of the random variables

$$\varphi(t_1, \ldots, t_k) = E \exp\left\{i \sum_{j=1}^{k} t_j X_j\right\} = \varphi(t) . \tag{1.1}$$

If mixed moments $EX^\nu = E(X_1^{\nu_1} \cdots X_k^{\nu_k}) = m_\nu$,

$$\nu = (\nu_1, \ldots, \nu_k) , \qquad \nu_j \geq 0 , \qquad |\nu| = \sum_{j=1}^{k} \nu_j , \qquad \nu! = \prod_{j=1}^{k} \nu_j !$$

exist up to a certain order $|\nu| \leq k$, they can be read off as coefficients in the Taylor expansion of φ about zero

*Research supported in part by Office of Naval Research Contract N00014-81-K-0003.

$$\varphi(t) = \int \Big\{ \sum_{|\nu| \le k} (\mathrm{i}t)^\nu x^\nu / \nu! + \mathrm{o}(|t|^k) \Big\} \, \mathrm{d}G(x)$$

$$= \sum_{|\nu| \le k} (\mathrm{i}t)^\nu m_\nu / \nu! + \mathrm{o}(|t|^k) . \tag{1.2}$$

The joint cumulants

$$c_\nu = \mathrm{cum}(X_1^{\nu_1}, \ldots, X_k^{\nu_k}) \tag{1.3}$$

can similarly be read off as coefficients in the Taylor expansion of log φ about zero

$$\log \varphi(t) = \sum_{|\nu| \le k} (\mathrm{i}t)^\nu c_\nu / \nu! + \mathrm{o}(|t|^k) . \tag{1.4}$$

There are formulas relating the cumulants of order k or less to the moments of order k or less and vice versa. It is enough to write out these relations for $E(X_1 \cdots X_k)$ and $\mathrm{cum}(X_1, \ldots, X_k)$ since those involving higher powers of the random variables are obtained just by identifying random variables. Consider the set of integers $\{1, 2, \ldots, k\}$. Let $x = (\nu_1, \ldots, \nu_p)$ be a partition of $\{1, 2, \ldots, k\}$. Then

$$E(X_1 \cdots X_k) = \sum_\nu C_{\nu_1} \cdots C_{\nu_p} \tag{1.5}$$

where we sum over all such partitions and C_ν is the joint cumulant of the X's with subscripts in ν. Let μ_ν be the mean of the product of the X's with subscripts in ν. The inverse relation is then given by

$$\mathrm{cum}(X_1, \ldots, X_k) = \sum_\nu (-1)^{p-1} (p-1)! \mu_{\nu_1} \cdots \mu_{\nu_p} . \tag{1.6}$$

This implies that existence of all moments up to order k is equivalent to the existence of all cumulants up to order k.

At this point one can suggest a possible reason for interest in cumulants. One classical way of proving a limit theorem for the distribution of a sequence of random k-vectors $X^{(n)}$, $n = 1, 2, \ldots,$ is to show that the νth moment $\nu = (\nu_1, \ldots, \nu_k)$, $\nu_i = 0, 1, 2, \ldots,$ $i = 1, \ldots, k$, of $X^{(n)}$ as $n \to \infty$ tends to the νth moment of the conjectured limiting distribution F as $n \to \infty$ for all ν when the distribution F is within the domain of the moment problem (that is, the distribution F is determined by its moments). From our earlier remarks it is clear that we could equivalently obtain such a limit theorem if we show that the νth cumulant of $X^{(n)}$ tends to the νth cumulant of F for all ν and F is within the domain of the moment problem. Later a context within which such an approach is useful will be described.

Of course, the probability structure of Gaussian random vectors (or processes) is determined by first and second order moments. The interest in higher order moments can then be motivated by a need to detect departures from a Gaussian structure or interest in resolving nonlinearity.

Here we note a few basic properties of cumulants:

(a) $\operatorname{cum}(a_1X_1, \ldots, a_kX_k) = a_1 \cdots a_k \operatorname{cum}(X_1, \ldots, X_k)$ where the a_j's are constants.

(b) If $X_1, \ldots, X_k$ can be partitioned into two disjoint sets which are independent of each other, then $\operatorname{cum}(X_1, \ldots, X_k) = 0$.

(c) The cumulant is a symmetric and multilinear function of its arguments.

(d) If $X_1, \ldots, X_k$ and $Y_1, \ldots, Y_k$ are independent, then $\operatorname{cum}(X_1 + Y_1, \ldots, X_k + Y_k) = \operatorname{cum}(X_1, \ldots, X_k) + \operatorname{cum}(Y_1, \ldots, Y_k)$.

The relations given above hold for complex-valued random variables. It is clear from the definition that

$$\operatorname{cum}(X) = EX, \qquad \operatorname{cum}(X_j, \bar{X}_k) = \operatorname{cov}(X_j, X_k)\,.$$

If the means of the random variables are zero

$$\operatorname{cum}(X_j, X_k, X_l) = EX_jX_kX_l\,.$$

A discussion of cumulants can be found in [11]. There an algorithm for the computation of cumulants of polynomial forms is given.

2. Spectra

Let $x(t)$, $-\infty < t < \infty$, be a weakly stationary process

$$\begin{aligned} &Ex(t) \equiv m\,, \qquad E|x(t)|^2 < \infty\,, \\ &\operatorname{cov}[x(t), x(\tau)] = r(t-\tau) \end{aligned} \tag{2.1}$$

that is continuous in mean square. For convenience assume $m = 0$. Then there are corresponding spectral representations of the covariance function

$$r(t) = \int_{-\infty}^{\infty} e^{it\lambda}\, dF(\lambda)\,, \tag{2.2}$$

with F a bounded nondecreasing function, and of the process $x(t)$ itself

$$x(t) = \int_{-\infty}^{\infty} e^{it\lambda}\, dz(\lambda) \tag{2.3}$$

with $z(\lambda)$ a process of orthogonal increments, that is,

$$Ez(\lambda) \equiv 0 , \qquad E\,\mathrm{d}z(\lambda)\overline{\mathrm{d}z(\mu)} = \delta(\lambda - \mu)\,\mathrm{d}F(\lambda) . \tag{2.4}$$

Here $\delta(\lambda)$ is the Kronecker delta, i.e.

$$\delta(\lambda) = \begin{cases} 1 & \text{if } \lambda = 0 , \\ 0 & \text{otherwise.} \end{cases}$$

If $x(t)$ is real-valued, then

$$\mathrm{d}z(\lambda) = \overline{\mathrm{d}z(-\lambda)} , \tag{2.5}$$

$$\mathrm{d}F(\lambda) = \mathrm{d}F(-\lambda) . \tag{2.5'}$$

A case of common interest is that in which F is absolutely continuous with $f(\lambda) = F'(\lambda)$.

If $x(t)$ is strictly stationary and moments of order $k > 2$ exist, then

$$m_k(t_1, \ldots, t_k) = E[x(t_1) \cdots x(t_k)] \tag{2.6}$$

will depend only on the time differences $t_2 - t_1, \ldots, t_k - t_1$:

$$m_k(t_1, \ldots, t_k) = r_k(t_2 - t_1, \ldots, t_k - t_1) . \tag{2.7}$$

A representation for r_k analogous to (2.2) with

$$r_k(\tau_1, \tau_2, \ldots, \tau_{k-1}) = \int \exp\left\{ \mathrm{i} \sum_{\alpha=1}^{k-1} \tau_\alpha \lambda_\alpha \right\} \mathrm{d}F_k(\lambda_1, \ldots, \lambda_{k-1}) \tag{2.8}$$

and F_k of bounded variation is not valid generally. However, if $x(t)$ satisfies a sufficiently strong mixing condition, such a relation will be valid. Even under such circumstances the spectral mass $\mathrm{d}F_k$ will be singular on certain submanifolds of the $(k-1)$-dimensional frequency space if $k > 3$. But if cumulants are considered instead of moments, under these conditions the corresponding spectral mass will no longer be singular. Let

$$c_k(t_2 - t_1, \ldots, t_k - t_1) = \mathrm{cum}(x(t_1), x(t_2), \ldots, x(t_k)) . \tag{2.9}$$

The existence of a representation (2.8) for $k \leq j$ is equivalent to that of a corresponding representation for $k \leq j$

$$c_k(\tau_1, \ldots, \tau_{k-1}) = \int \exp\left\{ \mathrm{i} \sum_{\alpha=1}^{k-1} \tau_\alpha \lambda_\alpha \right\} \mathrm{d}G_k(\lambda_1, \ldots, \lambda_{k-1}) \tag{2.10}$$

with G_k of bounded variation. However, with a sufficiently strong mixing condition assumed for the process $x(\cdot)$, the spectral cumulant mass G_k will be absolutely continuous with density g_k. Also

$$\begin{aligned}\operatorname{cum}(\mathrm{d}z(\lambda_1),\ldots,\mathrm{d}z(\lambda_k)) &= \delta(\lambda_1+\cdots+\lambda_k)\,\mathrm{d}G_k(\lambda_1,\ldots,\lambda_{k-1})\\ &= \delta(\lambda_1+\cdots+\lambda_k)g_k(\lambda_1,\ldots,\lambda_{k-1})\,\mathrm{d}\lambda_1\cdots\mathrm{d}\lambda_{k-1}\,. \end{aligned} \tag{2.11}$$

Since the cumulants or cumulant spectra of order $k>2$ are zero if the process is Gaussian, they will usually be of primary interest for a non-Gaussian or nonlinear problem.

If we deal with a column vector-valued stationary process $x(t)=(x_a(t);\ a=1,\ldots,m)$, there will still be a vector-valued representation of the form (2.3) with

$$E\,\mathrm{d}z(\lambda)\,\mathrm{d}z(\mu)'=\delta(\lambda+\mu)\,\mathrm{d}F(\lambda)\,. \tag{2.12}$$

Here the matrix A' denotes the conjugated transpose of the matrix A. The function $F(\lambda)$ is now an $m\times m$ Hermitian matrix-valued function that is bounded and nondecreasing. The covariance function $r(t)=Ex(\tau+t)x(\tau)'$ is still given by (2.2) but is an $m\times m$ matrix-valued function. We still assume that $x(t)$ has real-valued components so that (2.5) still holds but (2.5′) is replaced by

$$\mathrm{d}F(-\lambda)=\mathrm{d}F(\lambda)'\,. \tag{2.13}$$

At a small risk of confusion the following notation is introduced for moments, cumulants, and spectra.

$$\begin{aligned} m_{a_1,\ldots,a_k}(t_1,\ldots,t_k) &= Ex_{a_1}(t_1)x_{a_2}(t_2)\cdots x_{a_k}(t_k) \\ &= r_{a_1,\ldots,a_k}(t_2-t_1,\ldots,t_k-t_1)\,,\\ c_{a_1,\ldots,a_k}(t_2-t_1,\ldots,t_k-t_1) &= \operatorname{cum}(x_{a_1}(t_1),\ldots,x_{a_k}(t_k))\,,\\ \operatorname{cum}(\mathrm{d}z_{a_1}(\lambda_1),\ldots,\mathrm{d}z_{a_k}(\lambda_k)) &= \delta(\lambda_1+\cdots+\lambda_k)\,\mathrm{d}G_{a_1,\ldots,a_k}(\lambda_1,\ldots,\lambda_{k-1})\\ &= \delta(\lambda_1+\cdots+\lambda_k)g_{a_1,\ldots,a_k}(\lambda_1,\ldots,\lambda_{k-1})\,\mathrm{d}\lambda_1\cdots\mathrm{d}\lambda_{k-1}\,. \end{aligned} \tag{2.14}$$

The assumption that the process $x(t)$ has real-valued components implies that

$$g_{a_1,\ldots,a_k}(\lambda_1,\ldots,\lambda_{k-1})=\bar g_{a_1,\ldots,a_k}(-\lambda_1,\ldots,-\lambda_{k-1})\,. \tag{2.15}$$

It should be noted that if the labels $a_1,\ldots,a_k$ are permuted by a permutation P and a_i is taken into a_j by P, then λ_i should be replaced by λ_j. We have given the g's as functions of $k-1$ variables $\lambda_1,\ldots,\lambda_{k-1}$ but it should be understood that there is a hidden variable $\lambda_k=-\lambda_1-\cdots-\lambda_{k-1}$. This is an unsymmetric but conventional notation. If a number of the indices a_i are the same (as is the case when we deal with a univariate time series since then $a_1=\cdots=a_k$), there are then a corresponding number of symmetries satisfied by $g_{a_1,\ldots,a_k}(\lambda_1,\ldots,\lambda_{k-1})$. A number of these symmetries are discussed in [3] for the case $k=3$ and $k=4$ when one has a discrete time parameter.

If the process $x(t)$ is discretely sampled with sampling interval h, we have for $t=0,\pm 1,\pm 2,\ldots$

$$x(th)=\int_{-\infty}^{\infty} e^{ith\lambda}\,dz(\lambda)=\int_{-\pi/h}^{\pi/h} e^{ith\lambda}\,d_hz(\lambda)\,, \tag{2.16}$$

$$d_hz(\lambda)=\sum_{j=-\infty}^{\infty} dz\left(\lambda+\frac{2\pi j}{h}\right),\quad |\lambda|<\frac{\pi}{h},$$

and the second order covariance function r is given in terms of the second order spectral density function f by

$$r(th)=\int_{-\infty}^{\infty} e^{ith\lambda} f(\lambda)\,d\lambda=\int_{-\pi/h}^{\pi/h} e^{ith\lambda}\,{}_hf(\lambda)\,d\lambda\,, \tag{2.17}$$

$${}_hf(\lambda)=\sum_{j=-\infty}^{\infty} f\left(\lambda+\frac{2\pi j}{h}\right),$$

$$E\{d_hz(\lambda)\,d_hz(\mu)'\}=\delta(\lambda+\mu)\,{}_hf(\lambda)\,d\lambda\,,\quad |\lambda|<\frac{\pi}{h},$$

if r is integrable. The effect described here in which the original representation in terms of f is replaced by one in terms of ${}_hf$ because of the discrete sampling is referred to as 'aliasing'. The relation (2.17) can be rewritten in terms of a periodic extension of the δ function

$$\eta(\lambda)=\sum_{j=-\infty}^{\infty}\delta(\lambda+2\pi j) \tag{2.18}$$

as

$$E\{d_hz(\lambda)\,d_hz(\mu)'\}=\eta\left(\frac{\lambda+\mu}{h}\right){}_hf(\lambda)\,d\lambda\,. \tag{2.19}$$

Of course, there are higher order effects of aliasing in that for $t_1,\ldots,t_{k-1}=0,\pm 1,\pm 2,\ldots$

$$c_{a_1,\ldots,a_k}(t_1h,\ldots,t_{k-1}h)$$

$$=\int_{-\pi/h}^{\pi/h}\cdots\int_{-\pi/h}^{\pi/h} {}_hg_{a_1,\ldots,a_k}(\lambda_1,\ldots,\lambda_{k-1})\exp\left\{i\sum_{\alpha=1}^{k-1}\lambda_\alpha t_\alpha h\right\}d\lambda_1\cdots d\lambda_{k-1}\,, \tag{2.20}$$

with

$$
\begin{aligned}
&{}_h g_{a_1,\ldots,a_k}(\lambda_1,\ldots,\lambda_{k-1}) \\
&= \sum_{j_1,\ldots,j_{k-1}=-\infty}^{\infty} g_{a_1,\ldots,a_k}\left(\lambda_1+\frac{2\pi j_1}{h},\ldots,\lambda_{k-1}+\frac{2\pi j_{k-1}}{h}\right), \qquad (2.21)\\
&|\lambda_1|,\ldots,|\lambda_{k-1}|<\frac{\pi}{h}.
\end{aligned}
$$

At times we will write out ${}_h g_{a_1,\ldots,a_k}$ and $g_{a_1,\ldots,a_k}$ in a more symmetric form as functions of k variables $\lambda_1,\ldots,\lambda_k$ where it is to be understood that $\lambda_k = -\lambda_1-\lambda_2-\cdots-\lambda_{k-1}$. Thus

$$
g_{a_1,\ldots,a_k}(\lambda_1,\ldots,\lambda_{k-1}) = g_{a_1,\ldots,a_k}(\lambda_1,\ldots,\lambda_{k-1},\lambda_k). \qquad (2.22)
$$

3. Spectral estimates

At first spectral estimates will be described in the univariate second order case. After this initial discussion, remarks will be made about the general multivariate kth order context.

Let $x(\tau)$ be a univariate stationary process sampled at $\tau = th$, $t = 0, 1, \ldots, N-1$. The periodogram ${}_hI^{(T)}(\lambda)$ is given by

$$
{}_hI^{(T)}(\lambda) = \frac{h}{2\pi T}\left|\sum_{t=0}^{T-1} x(th)\,\mathrm{e}^{-\mathrm{i}th\lambda}\right|^2.
$$

If the spectral density f is continuous, one can show that

$$
E_hI^{(T)}(\lambda) \to {}_hf(\lambda), \quad |\lambda| < \frac{\pi}{h},
$$

as $T\to\infty$. A further summability condition on fourth order cumulants implies that

$$
\operatorname{cov}[{}_hI^{(T)}(\lambda), {}_hI^{(T)}(\mu)] \to \eta((\lambda-\mu)h)[1+\eta(2\lambda h)]{}_hf^2(\lambda)
$$

as $T\to\infty$ if $0 \leqslant \lambda, \mu$. Though the periodogram is not a reasonable estimate of the aliased spectral density ${}_hf$, decent estimates can be manufactured out of it in the following manner. Let $W(u)$ be a bounded weight function symmetric about zero with bounded support such that $\int W(u)\,\mathrm{d}u = 1$. Set

$$
W_T(u) = B_T^{-1} W(B_T^{-1}u) \qquad (3.1)
$$

with $B_T \to 0$ as $T\to\infty$. The conditions on W can be relaxed appreciably and we have chosen the conditions cited above for convenience. The bandwidth B_T is chosen so that $TB_T \to \infty$ as $T \to \infty$. An estimate of ${}_hf(\lambda)$ is given by

$$f^{(T)}(\lambda) = \int_{-\pi/h}^{\pi/h} W_T(\lambda - \alpha)\, {}_hI^{(T)}(\alpha)\, d\alpha \tag{3.2}$$

$$= \frac{h}{2\pi} \sum_{t=-(T-1)}^{T-1} w_T(th)\, {}_hm^{(T)}(th) \exp(-ith\lambda)$$

with

$${}_hm^{(T)}(th) = \frac{1}{T} \sum_{0 \leq \tau,\, \tau+t \leq T-1} x(\tau h) x((\tau + t)h) \tag{3.3}$$

and

$${}_hw_T(th) = \int_{-\pi/h}^{\pi/h} W_T(u) \exp(-ithu)\, du\,. \tag{3.4}$$

We have a representation of $f^{(T)}(\lambda)$ as a smoothed version of a periodogram or as a Fourier transform of weighted estimates of second order moments. This aspect of such a dual representation also holds for higher order spectral estimates. For reasons of economy we shall only consider the representation of higher order spectral estimates obtained by smoothing analogues of a periodogram. Given the continuity of f and summability of fourth order cumulants, it follows that

$$\lim_{T\to\infty} B_T T \operatorname{cov}[f^{(T)}(\lambda), f^{(T)}(\mu)]$$

$$= \frac{2\pi}{h} \eta((\lambda - \mu)h)[1 + \eta(2\lambda h)]\, {}_hf^2(\lambda) \int W^2(u)\, du \tag{3.5}$$

for $0 \leq \lambda, \mu$, if $B_T \to 0$, $B_T T \to \infty$. Also it is clear then that

$$\lim_{T\to\infty} Ef^{(T)}(\lambda) = {}_hf(\lambda)\,. \tag{3.6}$$

Under further cumulant mixing conditions, the estimates $f^{(T)}(\lambda)$ at a finite number of λ values are asymptotically normal with the mean and covariance properties indicated above. The cumulant mixing condition (see [2]) will be given in the context of higher order multivariate cumulant spectral estimates. The mixing condition can be thought of as a form of asymptotic independence of the process.

Let $x(t) = (x_a(t);\ a = 1, \ldots, m)$ now be an m-variate stationary series. We shall say a cumulant mixing condition is satisfied if

$$\sum_{t_1, \ldots, t_{k-1}=-\infty}^{\infty} |t_j c_{a_1, \ldots, a_k}(t_1 h, \ldots, t_{k-1} h)| < \infty \tag{3.7}$$

for $j = 1, \ldots, k-1$, any k-tuple $a_1, \ldots, a_k$ $(k = 2, 3, \ldots)$ and any $h > 0$ with $k = 2, 3, \ldots$. Assume that this mixing condition holds. Consider the finite Fourier transform

$$d_a^{(T)}(\lambda) = \sum_{t=0}^{T-1} x_a(th) \exp(-\mathrm{i}th\lambda)\,. \tag{3.8}$$

kth order analogues of the periodogram can be defined in terms of these finite Fourier transforms

$$I_{a_1,\ldots,a_k}^{(T)}(\lambda_1, \ldots, \lambda_k) = \left(\frac{h}{2\pi}\right)^{k-1} \frac{1}{T}\, d_{a_1}^{(T)}(\lambda_1) \cdots d_{a_k}^{(T)}(\lambda_k)\,. \tag{3.9}$$

Let $W(u_1, \ldots, u_k)$ be a bounded piecewise continuous weight function on $\Sigma_1^k u_j = 0$ with bounded support such that

$$\int \cdots \int W(u_1, \ldots, u_k)\delta\Big(\sum_1^k u_j\Big)\, \mathrm{d}u_1 \cdots \mathrm{d}u_k = 1 \tag{3.10}$$

and

$$W(-u_1, \ldots, -u_k) = W(u_1, \ldots, u_k)\,. \tag{3.11}$$

Set

$$W_T(u_1, \ldots, u_k) = B_T^{-k+1} W(B_T^{-1}u_1, \ldots, B_T^{-1}u_k)\,. \tag{3.12}$$

Assume as before that $x(t) = (x_a(t);\ a = 1, \ldots, m)$ is m-variate stationary and that one observes $x(th)$, $t = 0, 1, \ldots, T-1$, with $h > 0$ fixed. An estimate $g_{a_1,\ldots,a_k}^{(T)}(\lambda_1, \ldots, \lambda_k)$, $\Sigma_1^k \lambda_j = 0$, of ${}_h g_{a_1,\ldots,a_k}(\lambda_1, \ldots, \lambda_{k-1})$, $|\lambda_i| \leq \pi/h$, is given by

$$\begin{aligned} g_{a_1,\ldots,a_k}^{(T)}&(\lambda_1, \ldots, \lambda_k) \\ &= \left(\frac{2\pi}{h}\right)^{k-1} T^{-k+1} \sum_{s_1=-\infty}^{\infty} \cdots \sum_{s_k=-\infty}^{\infty} W_T\Big(\lambda_1 - \frac{2\pi s_1}{Th}, \ldots, \lambda_k - \frac{2\pi s_k}{Th}\Big) \\ &\quad \cdot \Phi\Big(\frac{2\pi s_1}{T}, \ldots, \frac{2\pi s_k}{T}\Big) I_{a_1,\ldots,a_k}^{(T)}\Big(\frac{2\pi s_1}{Th}, \ldots, \frac{2\pi s_k}{Th}\Big) \end{aligned} \tag{3.13}$$

with the kth order periodogram specified by (2.31). Here the function $\Phi(u_1, \ldots, u_k) = 1$ if $\Sigma_1^k u_j \equiv 0 \bmod 2\pi$ but $\Sigma_{j\in J} u_j \not\equiv 0 \bmod 2\pi$ where J is any nonvacuous proper subject of $1, 2, \ldots, k$ and $\Phi(u_1, \ldots, u_k) = 0$ at all other $(u_1, \ldots, u_k)$. The set $\Sigma_1^k u_j \equiv 0$ is sometimes called the principal manifold and a set of the form $\Sigma_{j\in J} u_j = 0$ with J a proper subset of $1, 2, \ldots, k$ a proper submanifold. The use of the factor Φ in the estimate is essential so as to ensure that the periodogram values on proper submanifolds are deleted. If left in, they would have a bad effect on the estimate. If $B_T \to 0$ as $T \to \infty$, then

$$\lim_{T\to\infty} E g_{a_1,\ldots,a_k}^{(T)}(\lambda_1, \ldots, \lambda_k) \to {}_h g_{a_1,\ldots,a_k}(\lambda_1, \ldots, \lambda_{k-1})\,. \tag{3.14}$$

Also if $B_T^{k-1}T \to \infty$ as $T \to \infty$, then

$$\lim_{T\to\infty} B_T^{k-1} T \operatorname{cov}[g^{(T)}_{a_1,\ldots,a_k}(\lambda_1,\ldots,\lambda_k), g^{(T)}_{a'_1,\ldots,a'_k}(\mu_1,\ldots,\mu_k)]$$

$$= \frac{2\pi}{h} \sum_P \eta((\lambda_1 - \mu_{P(1)})h) \cdots \eta((\lambda_k - \mu_{P(k)})h) f_{a_1,a'_{P(1)}}(\lambda_1) \cdots f_{a_k,a'_{P(k)}}(\lambda_k)$$

$$\times \int \cdots \int W(\tau_1,\ldots,\tau_k) W(\tau_{P(1)},\ldots,\tau_{P(k)}) \delta\Big(\sum_1^k \tau_j\Big) \, d\tau_1 \cdots d\tau_k . \tag{3.15}$$

Here the sum is over all permutations P of the set $1, 2, \ldots, k$. Let all estimates of order k have the same bandwidth B_T. Under the assumptions made it follows that all estimates of order k are asymptotically normal with the means and covariance indicated above. Estimates of different order k will typically have different bandwidths satisfying the appropriate condition $B_T^{k-1}T \to \infty$, $B_T \to 0$ as $T \to \infty$. It then follows that estimates of different order are asymptotically independent. A sketch of the details of the arguments leading to these results can be found in [2]. An early discussion of bispectral estimates can be found in [16]. An exposition in which similar techniques are used to deal with point processes can be found in Brillinger's 1972 paper in the 7th Berkeley Symposium.

4. Nonlinear systems

Often one wishes to determine the response of a time invariant nonlinear system. There are a number of ways in which this might be done. We shall briefly comment on one procedure. Suppose a Gaussian stationary process $x(t)$ is considered as input to the system. Assume $x(t)$ has spectral density function $f(\lambda)$. The process $x(t)$ has the Fourier representation

$$x(t) = \int e^{it\lambda} \, dz(\lambda)$$

with $z(\lambda)$ a complex-valued Gaussian process of orthogonal increments

$$E \, dz(\lambda) \, dz(\mu) = \delta(\lambda + \mu) f(\lambda) \, d\lambda , \qquad dz(\lambda) = \overline{dz(-\lambda)} .$$

Under broad conditions, the output $Y(t)$ of a time invariant system when the Gaussian process $X(t)$ is fed into it can be represented as a sum of Wiener–Ito integrals

$$Y(t)=\sum_{j=0}^{\infty}\frac{1}{j!}q\int\cdots\int_{\lambda_k\pm\lambda_l\neq 0}\exp\{i(\lambda_1+\cdots+\lambda_j)t\}$$
$$\times R_j(\lambda_1,\ldots,\lambda_j)\,dz(\lambda_1)\cdots dz(\lambda_j) \tag{4.1}$$

with the functions $R_j(\lambda_1,\ldots,\lambda_j)$ complex-valued symmetric functions satisfying the conditions

$$\begin{aligned}&R_j(\lambda_1,\ldots,\lambda_j)=\overline{R_j(-\lambda_1,\ldots,-\lambda_j)},\\&\sum_{j=0}^{\infty}\frac{1}{j!}\int\cdots\int|R_j(\lambda_1,\ldots,\lambda_j)|^2f(\lambda_1)\cdots f(\lambda_j)\,d\lambda_1\cdots d\lambda_j<\infty.\end{aligned} \tag{4.2}$$

The cross-cumulant spectral density $f_{x\cdots xy}(\lambda_1,\ldots,\lambda_j)$ is then given by

$$R_j(-\lambda_1,\ldots,-\lambda_j)f(\lambda_1)\cdots f(\lambda_j). \tag{4.3}$$

This suggests that a way of determining the jth order term $R_j(\lambda_1,\ldots,\lambda_j)$ is to estimate the cumulant spectral density $f_{x\cdots xy}(\lambda_1,\ldots,\lambda_j)$ and the spectral density of the x process. The Fourier representation (4.1) in terms of Wiener–Ito integrals was chosen for convenience. Alternative Fourier representations or representations directly in the time domain can also be given. Ways of going from one representation to another are discussed in [9] and [5]. A more extensive treatment of related questions can be found in [1].

5. Non-Gaussian linear processes and phase information

Linear processes have been discussed probabilistically. A brief description follows. Consider independent, identically distributed random variables v_t, $t=\ldots,-1,0,1,\ldots$ with mean zero $Ev_t\equiv 0$ and variance one $Ev_t^2=1$. A sequence of real constants $\{\alpha_j\}$ is given with

$$\sum_{j=-\infty}^{\infty}\alpha_j^2<\infty. \tag{5.1}$$

Now

$$x_t=\sum_{j=-\infty}^{\infty}\alpha_j v_{t-j} \tag{5.2}$$

is the linear process determined by $\{\alpha_j\}$ and $\{v_t\}$. The spectral density of $\{x_t\}$

$$f(\lambda)=\frac{1}{2\pi}|\alpha(e^{-i\lambda})|^2 \tag{5.3}$$

is determined by the frequency response function

$$\alpha(e^{-i\lambda}) = \sum_j \alpha_j e^{-ij\lambda} . \tag{5.4}$$

In the case of a Gaussian linear process $\{x_t\}$, only the modulus of $\alpha(e^{-i\lambda})$, $|\alpha(e^{-i\lambda})|$ is identifiable since the probability structure of the process is determined by second order moments, or equivalently by the spectral density. It is curious that here the Gaussian case is the atypical rather than the typical case as it often is. For if $\{x_t\}$ is non-Gaussian and generated by $\{v_t\}$ having a marginal distribution (non-Gaussian) with all its moments finite, then under fairly broad conditions the phase information in $\arg\{\alpha(e^{-i\lambda})\}$ is identifiable up to an additive linear term. For the kth order cumulant spectral density of $\{x_t\}$ is

$$b_k(\lambda_1, \ldots, \lambda_{k-1}) = \frac{1}{(2\pi)^{k-1}} \alpha(e^{-i\lambda_1}) \cdots \alpha(e^{-i\lambda_{k-1}})\alpha(e^{i(\lambda_1+\cdots+\lambda_{k-1})})\gamma_k \tag{5.5}$$

with γ_k the kth cumulant of the independent random variables v_t. Since v_t is non-Gaussian with all its moments finite, it follows that for some k, $\gamma_k \neq 0$. Of course, one need not assume all moments finite. It is enough to assume that γ_k exists and is not zero for some $k \neq 0$. Also, let

$$\sum_j |j|\,|\alpha_j| < \infty \tag{5.6}$$

and

$$\alpha(e^{-i\lambda}) \neq 0 \quad \text{for all } \lambda .$$

Let

$$h(\lambda) = \arg\left\{\alpha(e^{-i\lambda}) \frac{\alpha(1)}{|\alpha(1)|}\right\} . \tag{5.7}$$

The equation giving $h(\lambda)$ above formula (8) in [15] should read as (5.7) does. Under the conditions cited above

$$h(\lambda_1) + \cdots + h(\lambda_{k-1}) - h(\lambda_1 + \cdots + \lambda_{k-1})$$

can be determined. This implies that $\arg\{\alpha(e^{-i\lambda})\}$ is identifiable except possibly for an additive linear term. A possible estimate based on bispectral estimates is suggested in [14] for the case $k = 3$. Similar questions are considered in [15].

It should be noted that most finite parameter statistical investigations center on Gaussian stationary autoregressive moving average processes, that is, solutions of a system

$$\sum_{j=0}^{p} b_j x_{t-j} = \sum_{k=0}^{q} a_k v_{t-k} \tag{5.8}$$

(b_0, $a_0 \neq 0$). Here it is usual to assume that the roots of the polynomials

$$A(z)=\sum_{k=0}^{q} a_k z^k, \qquad B(z)=\sum_{j=0}^{p} b_j z^k \tag{5.9}$$

all have modulus greater than one. A stationary solution of (5.8) has

$$\alpha(e^{-i\lambda})=\frac{A(e^{-i\lambda})}{B(e^{-i\lambda})}.$$

If the process is Gaussian, any real zero $z_j \neq 0$ can be replaced by its inverse and pairs of nonzero conjugate roots by their paired conjugated inverses $\bar{z}_j^{-1}$ without changing the probability structure of the process if the process is appropriately rescaled. Given distinct real roots there are 2^{p+q} ways of specifying the zeros without changing the structure of a Gaussian $\{x_t\}$. In the case of a non-Gaussian structure satisfying (5.8), the 2^{p+q} specifications of zeros mentioned above will usually correspond to different processes. The actual location of the zeros can in principle be estimated in the case of a non-Gaussian solution of (5.8) if $\gamma_k \neq 0$ for some $k > 2$.

A continuous analogue of a first order autoregressive scheme with Poisson noise is discussed in [19]. This model has been used in the analysis of streamflow data.

6. Applications

Perhaps the earliest application is that of Hasselman et al. [6] in resolving nonlinear aspects of ocean wave propagation. Sunspot records are analyzed in [3]. Cartwright [4] carries out a fourth order analysis of tides and surges. Huber et al. [8] apply bispectral analysis to the resolution of electroenece-phalograph records. These higher order techniques are used in the analysis of oceanographic temperature and salinity readings among others in [13]. A series of papers consider bispectral methods in the analysis of turbulence and the nonlinear spectral transfer of energy (see [7, 12, 18]). Kim and Powers [10] employ bispectral estimates and problems in plasma physics. The paper of Sato et al. [17] uses bispectral methods to deal with non-Gaussian noise. The National Bureau of Standards Technical Note 1036 of P.V. Tryon titled "The bispectrum and higher-order spectra: a bibliography" is a useful bibliography of related literature from 1953 to 1980.

References

[1] Brillinger, D.R. (1970). The identification of polynomial systems by means of higher order spectra. *J. Sound Vib.* **12**(3), 301–313.

[2] Brillinger, D. R. and Rosenblatt, M. (1967). Asymptotic theory of estimates of kth order spectra. In: B. Harris, ed., *Spectral Analysis of Time Series*, 153–188. Wiley, New York.

[3] Brillinger, D. R. and Rosenblatt, M. (1967). Computation and interpretation of kth order spectra. In: B. Harris, ed., *Spectral Analysis of Time Series*, 189–232. Wiley, New York.

[4] Cartwright, D. E. (1968). A unified analysis of tides and surges round North and East Britain. *Philos. Trans. Roy. Soc. London, Ser. A* **263**, 1–55.

[5] Dobrushin, R. L. (1979). Gaussian and their subordinated self-similar random generalized fields. *Ann. Prob.* **7**, 1–29.

[6] Hasselman, K., Munk, W. and MacDonald, G. (1963). Bispectra of ocean waves. In: M. Rosenblatt, ed., *Time Series Analysis*. Wiley, New York.

[7] Helland, K. N., Lii, K. S. and Rosenblatt, M. (1979). Bispectra and energy transfer in grid-generated turbulence. In: P. R. Krishnaiah, ed., *Developments in Statistics*, Vol. 2, 123–155.

[8] Huber, P. J., Kleimer, B., Gasser, Th. and Dummermuth, G. (1971). Statistical methods for investigating phase relations in stationary stochastic processes. *IEEE Trans. Audio Electroacoust.* **19**, 78–86.

[9] Ito, K. (1952). Complex multiple Wiener integral. *J. Math. Soc. Japan* **22**, 63.

[10] Kim, Y. C. and Powers, E. J. (1978). Digital bispectral analysis of self-excited fluctuation spectra. *Phys. Fluids* **21**(8), 1452–1453.

[11] Leonov, V. P. and Shiryaev, A. N. (1959). On a method of calculation of semi-invariants. *Theor. Prob. Appl.* **4**, 319–329.

[12] Lii, K. S., Rosenblatt, M. and Van Atta, C. W. (1976). Bispectral measurements in turbulence. *J. Fluid Mech.* **77**(1), 45–62.

[13] Roden, G. I. and Bendiner, D. J. (1973). Bispectra and cross-bispectra of temperature, salinity, sound velocity and density fluctuations with depth off Northeastern Japan. *J. Phys. Ocean*, **3**(3), 308–317.

[14] Rosenblatt, M. (1980). Linear processes and bispectra. *J. Appl. Prob.* **17**, 265–270.

[15] Rosenblatt, M. (1979). Linearity and nonlinearity in times series: prediction. *Proc. 42nd Session Internat. Statist. Inst.* (Manila), Book 1, pp. 423–434.

[16] Rosenblatt, M. and Van Ness, J. W. (1965). Estimation of the bispectrum. *Ann. Math. Statist.* **36**, 1120–1136.

[17] Sato, T., Sasaki, K. and Nakamura, Y. (1977). Real-time bispectral analysis of gear noise and its application to contactless diagnosis. *J. Acoust. Soc. Amer.* **62**(2), 382–387.

[18] Van Atta, C. W. (1979). Inertial range bispectra in turbulence. *Phys. Fluids*, **22**(8), 1440–1442.

[19] Weiss, G. Properties of the first order shot noise process. Manuscript.

D. R. Brillinger and P. R. Krishnaiah, eds., *Handbook of Statistics, Vol. 3*
© Elsevier Science Publishers B.V. (1983) 383–408 18

Replicated Time-Series Regression: An Approach to Signal Estimation and Detection

R. H. Shumway

1. Introduction

There are a number of practical problems arising in the analysis of time-correlated data which require extended or generalized versions of classical multiple regression. A particular kind of model which is of interest in signal estimation and detection problems relates an observed output series to a collection of lagged input signals using a deterministic time-varying design matrix. The analysis of signals propagating across an array of sensors monitoring some physical phenomenon requires multiple signal models of this kind. For example, seismic applications (cf. Capon et al., 1967, 1969); Blandford et al., 1976; Shumway and Husted, 1970; Shumway, 1972) pertain to estimating the waveforms and directional properties of propagating plane waves generated by earthquakes or explosions. Other examples can be given relating to estimating the source of an acoustic signal in the ocean as in Clay (1966), MacDonald and Schultheiss (1969), or to detecting an infrasonic signal in the atmosphere as in Mack and Smart (1972) and Smart and Flinn (1971). In radar, it is conventional to use a phased antenna array to recover a return from a number of possible interfering signals (cf. Ksienski and McGhee, 1968; McGarty, 1974; McAulay and McGarty, 1974).

The multiple signal problems mentioned in the previous discussion can all be regarded as special cases of the general time-series regression model

$$y_j(t) = \sum_{u=-\infty}^{\infty} \boldsymbol{\beta}'(u)\boldsymbol{x}_j(t-u) + e_j(t)\,, \tag{1.1}$$

$t = 0, \pm 1, \pm 2, \ldots, j = 1, 2, \ldots, n$, with $\boldsymbol{\beta}(u) = (\beta_1(u), \beta_2(u), \ldots, \beta_q(u))'$ a $q \times 1$ vector of unknown deterministic signals and $\boldsymbol{x}_j(t) = (x_{j1}(t), x_{j2}(t), \ldots, x_{jq}(t))'$ a $q \times 1$ design vector describing the way in which the signals are related to the value $y_j(t)$ recorded by the jth sensor. The error or noise series $e_1(t)$, $e_2(t), \ldots, e_N(t)$ are assumed to be stationary independent identically distributed (i.i.d.) processes whose values measured across time are only weakly dependent as the separation increases. Brillinger (1980) formulates this 'mixing' condition in terms of the cumulants of the process as

$$\sum_{u_1} \cdots \sum_{u_k} |\text{cum}\{e_j(t+u_1), \ldots, e_j(t+u_k), e_j(t)\}| < \infty, \tag{1.2}$$

$k = 1, 2, \ldots, t = 0, \pm 1, \pm 2, \ldots, j = 1, \ldots, N$, and notes that it justifies a number of large sample approximations involving the discrete Fourier transform. It is convenient to summarize the model (1.1) in the vector form

$$\boldsymbol{y}_t = \sum_{u=-\infty}^{\infty} x_{t-u}\boldsymbol{\beta}_u + \boldsymbol{e}_t, \tag{1.3}$$

$t = 0, \pm 1, \pm 2, \ldots,$ with $x_t = (\boldsymbol{x}_1(t), \ldots, \boldsymbol{x}_N(t))$ the $N \times q$ time-varying design matrices and $\boldsymbol{e}_t = (e_1(t), \ldots, e_N(t))$ and $\boldsymbol{y}_t = (y_1(t), \ldots, y_N(t))'$ the vector noise and observation processes. These design matrices are assumed to be well behaved in the sense that

$$\sum_{t=-\infty}^{\infty} (1 + |t|)\|x_t\| < \infty, \tag{1.4}$$

where $\|x_t\| = [\text{tr}(x_t x_t')]^{1/2}$, whereas the regression functions need only satisfy $\|\beta_t\| = c < \infty$ for all t. The autocorrelation matrix of the vector noise process $\boldsymbol{e}_t$ is of the form

$$E[\boldsymbol{e}_{t+u}\boldsymbol{e}_t'] = R_e(u)I_N, \tag{1.5}$$

where I_N denotes the $N \times N$ identity matrix with $\Sigma|R_e(u)| < \infty$;

$$f_e(\lambda) = \sum_{t=-\infty}^{\infty} R_e(u)\exp\{-iu\lambda\}, \qquad -\pi \leq \lambda \leq \pi \tag{1.6}$$

defines the common power spectrum of the identically distributed noise processes. The overall model defined by (1.2) and the attendant conditions will be called the replicated time-series regression model.

Before proceeding any further, it is essential to note that in many practical cases, the design matrices will be functionally related to some $p \times 1$ parameter vector $\boldsymbol{\Theta} = (\theta_1, \ldots, \theta_p)'$, so that an overall nonlinear regression model of the form

$$\boldsymbol{y}_t = \sum_{u=-\infty}^{\infty} x_{t-u}(\boldsymbol{\Theta})\boldsymbol{\beta}_u + \boldsymbol{e}_t \tag{1.3'}$$

may be more sensible. As a simple example of this model, we note that in the case of a plane wave signal $\beta(t)$ propagating across an array with a fixed velocity, the jth sensor observes

$$y_j(t) = \beta(t - T_j(\boldsymbol{\theta})) + e_j(t)\,, \tag{1.7}$$

where the time delay $T_j(\boldsymbol{\theta})$ at the jth channel depends on the parameter $\boldsymbol{\theta} = (\theta_1, \theta_2)'$; θ_1 and θ_2 in this example are related to the velocity and azimuth of the incident wave. The signal model (1.7) is seen to be a special case of (1.3)′ with $p = 2$, $q = 1$, and

$$x_t(\boldsymbol{\theta}) = (\delta(t - T_1(\boldsymbol{\theta})), \ldots, \delta(t - T_N(\boldsymbol{\theta})))'\,, \tag{1.8}$$

where

$$\delta(t) = \begin{cases} 1, & t = 0, \\ 0, & t = \pm 1, \pm 2, \ldots \end{cases} \tag{1.9}$$

denotes the usual Kronecker delta function. This model is used often in the previously mentioned references, where one is typically interested first in detecting a signal and then in estimating the velocity, azimuth, and waveform of the detected signal. A generalization of interest would be the case where multiple interfering signal sources are present, generated perhaps by storms or multipath reflections. The simultaneous resolution of these multiple signal models can be approached by stepwise regression techniques.

Other versions of (1.3) arise in experimental designs where one might observe time series originating from different types of phenomena or as responses to different treatment combinations. Brillinger (1980) has given an exhaustive treatment of the frequency-domain approach to analysis of variance; an example involving a frequency-dependent test of whether underground nuclear explosions and earthquakes have significantly different mean waveforms is given in Shumway (1980). Other regression models which have the same general form as (1.1) can be developed by assuming that no replication is present ($N = 1$) and that the q input series $x_{11}(t), x_{12}(t), \ldots, x_{1q}(t)$ are stationary processes. For example, Akaike (1964), Brillinger (1969, 1974), Hannan (1963, 1970), Parzen (1967), and Wahba (1968, 1969) have given treatments from this point of view. Computational details and some examples can be found in Otnes and Enochson (1978). If one assumes that the relation (1.1) is not a lagged regression and that $x_{j1}(t), \ldots, x_{jq}(t)$ are fixed known functions, typically polynomials or sines and cosines, satisfying

$$y_j(t) = \sum_{k=1}^{q} \beta_k x_{jk}(t) + e_j(t)\,, \tag{1.10}$$

a trend regression model results. Fairly complete asymptotic results for trend functions satisfying regularity conditions due to Grenander (1954) are given in Hannan (1970) or Anderson (1971, 1972).

In this discussion, however, we will concentrate on the versions of (1.3) and (1.3)′ which lead to multiple signal models of interest in various practical situations. The problems considered will divide roughly into those involving tests of hypotheses for various subsets of the q regression signals in (1.3) or

(1.3)′ and those primarily concerned with estimating waveforms for those same signals. The techniques for attacking both of these problems will depend on the properties of the discrete Fourier transform (DFT) of a finite sample from the model (1.3). We begin by discussing the problem of constructing finite time estimators for the q regression functions.

2. Estimation of the regression functions

It is convenient to consider first the problem of constructing the best linear unbiased estimators (BLUE) for the regression vector $\boldsymbol{\beta}_t$ in the general linear model (1.3) over $t \equiv 0, \pm 1, \pm 2, \ldots$. We note that applying an extended version of the Gauss–Markoff theorem (cf. Shumway and Dean, 1968) leads to

$$\hat{\boldsymbol{\beta}}_t = \sum_{u=-\infty}^{\infty} h_u \boldsymbol{y}_{t-u}, \tag{2.1}$$

where h_t is the $q \times N$ matrix valued function $(\Sigma \|h_t\| < \infty)$

$$h_t = (2\pi)^{-1} \int_{-\pi}^{\pi} [\overline{X(\lambda)'} X(\lambda)]^{-1} \overline{X(\lambda)'} \exp\{\mathrm{i}t\lambda\}\, \mathrm{d}\lambda , \tag{2.2}$$

with

$$X(\lambda) = \sum_{t=-\infty}^{\infty} x_t \exp\{-\mathrm{i}\lambda t\} \tag{2.3}$$

the Fourier transform of the $N \times q$ time-varying design matrix x_t. The variance–covariance matrix of the estimated regression functions is given by

$$E(\hat{\boldsymbol{\beta}}_{t+u} - \boldsymbol{\beta}_{t+u})(\hat{\boldsymbol{\beta}}_t - \boldsymbol{\beta}_t)' = (2\pi)^{-1} \int_{-\pi}^{\pi} f_e(\lambda)[\overline{X(\lambda)'} X(\lambda)]^{-1} \exp\{\mathrm{i}u\lambda\}\, \mathrm{d}\lambda, \tag{2.4}$$

with the variance terms obtained by substituting $u = 0$.

As a simple example, consider the model (1.7) where $T_j(\boldsymbol{\theta})$, $j = 1, 2, \ldots, N$ are assumed, for the moment, to be a collection of known time delays which determine where the signal appears on each recording channel. Then, applying (2.2) with the definition (2.3) yields

$$X(\lambda) = (\exp\{-\mathrm{i}\lambda T_1(\boldsymbol{\theta})\}, \ldots, \exp\{-\mathrm{i}\lambda T_N(\boldsymbol{\theta})\})' \tag{2.5}$$

and

$$h_t = (\delta(t + T_1(\boldsymbol{\theta})), \ldots, \delta(t + T_N(\boldsymbol{\theta}))) , \tag{2.6}$$

so that the BLUE estimator (2.1) takes the form

$$\hat{\beta}(t) = N^{-1} \sum_{j=1}^{N} y_j(t + T_j(\boldsymbol{\theta})) , \tag{2.7}$$

which is just the ordinary 'beamformed' estimator for the signal $\beta(t)$. The variance of this estimator is given by (2.4) as $N^{-1}R_e(0)$, where $R_e(\cdot)$ is the common autocorrelation function of the error or noise series in (1.7).

Although the transform (2.2) could be evaluated conveniently in the case of a single regression function, the multiple signal case frequently leads to an intractable expression, and it is simpler to develop a frequency-domain approximation of the form

$$h_t^M = M^{-1} \sum_{m=0}^{M-1} [\overline{X(\omega_m)'}X(\omega_m)]^{-1}\overline{X(\omega_m)'} \exp\{-i\omega_m t\}, \tag{2.8}$$

where $\omega_m = 2\pi m M^{-1}$ for some $M < T$. The approximation for the filtered output (2.1), assuming that y_t is observed at $t = 0, 1, \ldots, T-1$, can be written in the form (M even)

$$\hat{\beta}_t^M = \sum_{u=-M/2+1}^{M/2-1} h_u^M y_{t-u}, \tag{2.9}$$

for the points $t = M/2 - 1, \ldots, -1, 0, 1, \ldots, T - M/2$. The choice for M depends on balancing a desire to reduce the bias and mean square error by increasing M, against the effect of losing $M/2$ points off the beginning and end of the filtered output (2.9). Kirkendall (1974) has shown that the bias in the finite estimator (2.6) is bounded by $4\gamma_M K$ where

$$\gamma_M = \sum_{|t| \geq M/2} \|h_t\| \tag{2.10}$$

and

$$K = \sup_t \|\beta_t\| \sum_{t=-\infty}^{\infty} \|X_t\|. \tag{2.11}$$

The mean square difference between the finite approximation (2.6) and the infinite two-sided BLUE estimator (2.1) satisfies

$$E^{1/2}\|\hat{\beta}_t^M - \hat{\beta}_t\|^2 \leq 4\gamma_M(K + N^{1/2}R_e^{1/2}(0)). \tag{2.12}$$

The quantity γ_M plays a major role in controlling both the bias and the mean square error of the finite approximation; for example, choosing M large enough so that $h_t = 0$ for $|t| \geq M$ can eliminate the error completely, and this should be an objective of the design procedure. On the other hand, the filtered output (2.6) is free of other end effects only for $t = (M/2) - 1, \ldots, T - M/2$, so that M cannot be increased arbitrarily without losing potentially important portions of the signal.

As a nontrivial example of the preceding analysis, we consider the problem of estimating a waveform in a contrived mixture of two signals $\beta_1(t)$ and $\beta_2(t)$ following the model

$$y_j(t) = \beta_1(t) + 0.5(\beta_2(t - T_j) + \beta_2(t + T_j)) + e_j(t)\,, \tag{2.13}$$

$j = 1, 2, \ldots, 10$. The model is applicable to a vertical array of recording instruments where $\beta_1(t)$ represents a generalized mean present at the same time on all series and $\beta_2(t)$ represents a signal coming in at an angle and then being reflected off the surface so that it appears twice on each channel except for the surface. Fig. 1 shows such a mixture on the last of $N = 10$ channels where the regression function of interest is an exponentially decaying sine wave. The arrows show the respective entry points on the tenth series ($T_{10} = 23$) and we note that $\beta_2(t)$ has been completely obscured by the noise. The model described by (2.13) can be written in the form (1.3) by noting that $q = 2$ and the elements in the jth column of the $N \times 2$ matrix $\boldsymbol{x}_t$ are $x_{j1}(t) = \delta(t)$ and $x_{j2}(t) = 0.5(\delta(t - T_j) + \delta(t + T_j))$ respectively. If we define the $2 \times N$ matrix

$$H(\lambda) = [\overline{X(\lambda)'}X(\lambda)]^{-1}\overline{X(\lambda)'}\,, \tag{2.14}$$

the optimum filters (2.2) will be the Fourier transform of this matrix, with elements of the form

$$H_{1j}(\lambda) = \Delta^{-1}(\lambda)[A(\lambda) - B(\lambda)\cos(\lambda T_j) \tag{2.15}$$

and

$$H_{2j}(\lambda) = \Delta^{-1}(\lambda)[N\cos(\lambda T_j) - B(\lambda)]\,, \tag{2.16}$$

$j = 1, 2, \ldots, N$, with

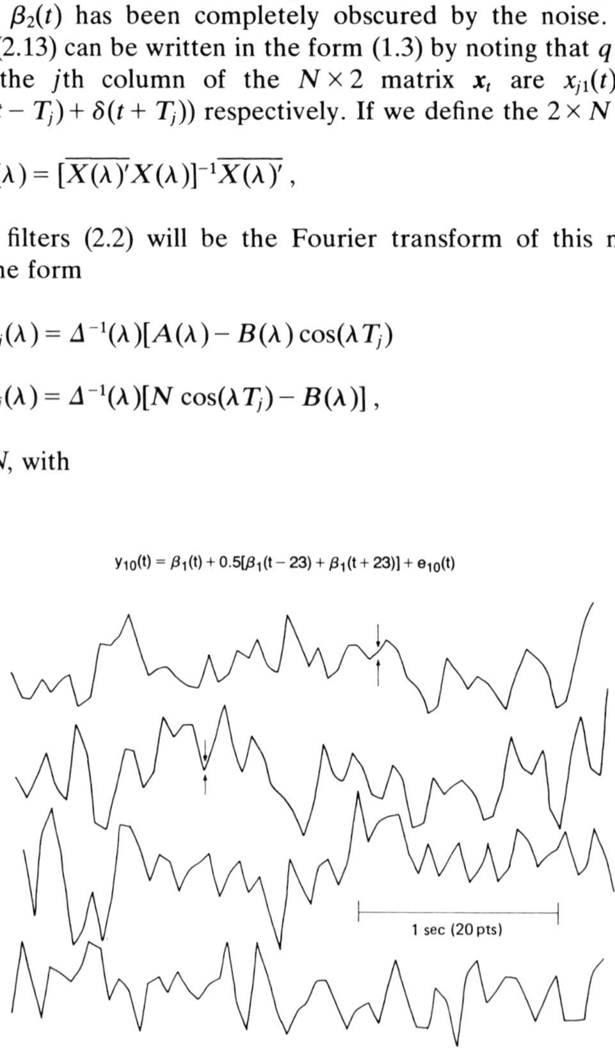

Fig. 1. A portion of the contrived mixture of $\beta_1(t)$ and $\beta_2(t)$ with the arrows denoting the entry points for the second regression function $\beta_2(t)$.

$$A(\lambda) = \sum_{j=1}^{N} \cos^2(\lambda T_j)\,,$$
$$B(\lambda) = \sum_{j=1}^{N} \cos(\lambda T_j)\,, \tag{2.17}$$
$$\Delta(\lambda) = NA(\lambda) - B^2(\lambda)\,.$$

The singularity ($\Delta(0) = 0$) at $\lambda = 0$ can be adjusted by using a ridge type correction or by taking $H(0) = 0$ so that the filters do not pass the zero frequency. In this case, the finite filters, computed using (2.8) with $M = 128$, are shown in Fig. 2, with the second filter preserving the signal by essentially averaging the primary waveform and its reflection, with the negative center peak cancelling the first regression function. The estimated regression function $\beta_2^M(t)$, calculated using (2.9), is shown in Fig. 3 and we see that it is an excellent estimator for the known true signal. The interpretation of these results in terms of significance tests is given in the next section.

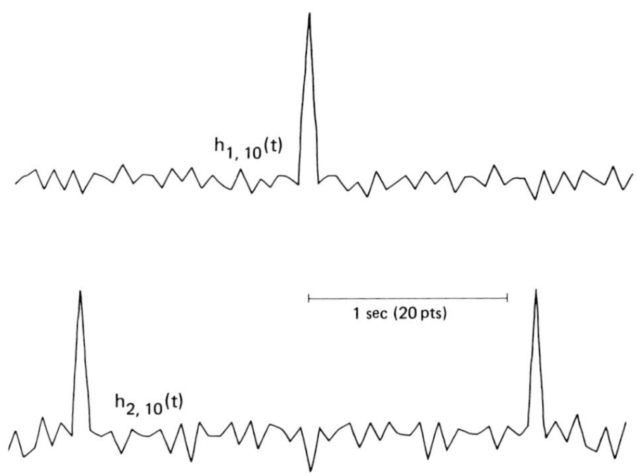

Fig. 2. Portions of the BLUE filters corresponding to eq. (2.8) with $H_{1j}(\cdot)$ and $H_{2j}(\cdot)$ given by eqs. (2.15)–(2.17).

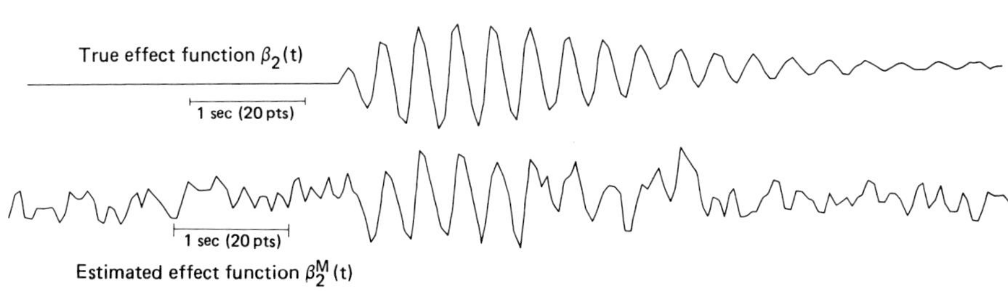

Fig. 3. True regression function compared with the approximate BLUE estimator (2.9).

3. Tests of hypotheses in the frequency domain

In many cases, one may be more interested in determining how many regression functions are present than in estimating the waveforms of the separate functions in time. In this case, a partitioned form of (1.3) may be of more interest, say

$$y_t = \sum_u x^{(1)}_{t-u}\boldsymbol{\beta}^{(1)}_u + \sum_u x^{(2)}_{t-u}\boldsymbol{\beta}^{(2)}_u + e_t\,, \tag{3.1}$$

where $\boldsymbol{\beta}^{(1)}_t = (\beta_1(t), \ldots, \beta q_1(t))'$ and $\boldsymbol{\beta}^{(2)}_t = (\beta_{q_1+1}(t), \ldots, \beta_{q_1+q_2}(t))'$ are $q_1 \times 1$ and $q_2 \times 1$ subvectors of the $q \times 1$ $(q = q_1 + q_2)$ vector $\boldsymbol{\beta}_t = (\boldsymbol{\beta}^{(1)\prime}_t, \boldsymbol{\beta}^{(2)\prime}_t)'$, and $x^{(1)}_t$ and $x^{(2)}_t$ are $N \times q_1$ and $N \times q_2$ time-varying design matrices. The question of interest becomes whether a specified subset, say the q_2 regression signals in $\boldsymbol{\beta}^{(2)}_t$, are present or absent. This generally occurs within some restricted frequency range or band, and it is useful to express the basic model (3.1) in the frequency domain. The discrete Fourier transform (DFT) of a finite sample involving the vector $\boldsymbol{y}_t$, $t = 0, 1, 2, \ldots, T-1$, is

$$\boldsymbol{Y}\hat{}(k) = T^{-1/2} \sum_{t-0}^{T-1} \boldsymbol{y}_t \exp\{-\mathrm{i}\lambda_k t\}\,, \tag{3.2}$$

which is generally evaluated at the frequencies $\lambda_k = 2\pi k/T$, $k = 0, 1, \ldots, T-1$. Now, applying the DFT to both sides of (3.1) for $\lambda_k \doteq \lambda$, leads to

$$\boldsymbol{Y}\hat{}(k) = X\hat{_1}(k)\boldsymbol{B}\hat{_1}(k) + X\hat{_2}(k)\boldsymbol{B}\hat{_2}(k) + \zeta_k + \mathrm{o}_{\mathrm{a.s.}}(1)\,, \tag{3.3}$$

where $\boldsymbol{B}\hat{_1}(k)$, $\boldsymbol{B}\hat{_2}(k)$, $X\hat{_1}(k)$, and $X\hat{_2}(k)$ are the vector DFT's of $\boldsymbol{\beta}^{(1)}_t$, $\boldsymbol{\beta}^{(2)}_t$, $x^{(1)}_t$, and $x^{(2)}_t$; $\boldsymbol{\zeta}_k$ denotes a zero-mean complex normal with covariance matrix $f_e(\lambda)\boldsymbol{I}_N$. The expression $\mathrm{o}_{\mathrm{a.s.}}(1)$ (see Brillinger, 1980) denotes a term tending to zero almost surely as $T \to \infty$.

In this partitioned model, we may be interested first in testing whether the first q_1 time series are related to $\boldsymbol{Y}\hat{}(k)$, that is H_0: $\boldsymbol{B}\hat{_1}(k) = \boldsymbol{0}$, so that the comparison is between

$$\boldsymbol{Y}\hat{}(k) = X\hat{_1}(k)\boldsymbol{B}\hat{_1}(k) + \boldsymbol{\zeta}_k + \mathrm{o}_{\mathrm{a.s.}}(1) \tag{3.4}$$

and

$$\boldsymbol{Y}\hat{}(k) = \boldsymbol{\zeta}_k + \mathrm{o}_{\mathrm{a.s.}}(1)\,, \tag{3.5}$$

or simply, noise alone. This is basically the problem of detecting the $q_1 \times 1$ vector signal $\boldsymbol{B}\hat{_1}(k)$ which, under (3.4), would be estimated as

$$\hat{\boldsymbol{B}}\hat{_1}(k) = [\overline{X\hat{_1}(k)'}X\hat{_1}(k)]^{-1}\overline{X\hat{_1}(k)'}\boldsymbol{Y}\hat{}(k)\,. \tag{3.6}$$

The analogue of the classical F statistic for testing the hypothesis $\boldsymbol{B}\hat{_1}(k) = 0$, (cf.

Shumway, 1970; Brillinger, 1973, 1980) can be written as the ratio of the regression mean square to the error mean square, say

$$F_{2q_1, 2(N-q_1)} = \frac{\mathrm{SSR}_1(k)}{\mathrm{SSE}_1(k)} \frac{(N-q_1)}{q_1}. \tag{3.7}$$

where

$$\mathrm{SSR}_1(k) = \|X_1^{\wedge}(k)\hat{B}_1^{\wedge}(k)\|^2 \tag{3.8}$$

and

$$\begin{aligned} \mathrm{SSE}_1(k) &= \|\boldsymbol{Y}^{\wedge}(k) - X_1^{\wedge}(k)\hat{\boldsymbol{B}}_1^{\wedge}(k)\|^2 \\ &= \|\boldsymbol{Y}^{\wedge}(k)\|^2 - \|X_1^{\wedge}(k)\hat{\boldsymbol{B}}_1^{\wedge}(k)\|^2 \end{aligned} \tag{3.9}$$

denote the regression and error sums of squares. ($\|a\|^2 = \Sigma_m |a_m|^2$ is the usual squared norm of the complex vector $\boldsymbol{a}$.) The notation F_{n_1,n_2} denotes an F statistic with n_1 numerator degrees of freedom and n_2 denominator degrees of freedom. It should be noted that if the numerator (3.8) and denominator (3.9) are summed over K frequencies in the neighborhood of λ, then $2q_1$ and $2(N-q_1)$ are replaced by $2Kq_1$ and $2K(N-q_1)$ respectively in (3.7)–(3.9). The statistic (3.7) gives a test function for detecting the presence of the $q_1 \times 1$ vector signal $\boldsymbol{B}_1^{\wedge}(k)$ at frequency λ. The sums of squares can be arranged in an analysis of power table as shown in Shumway (1970) and Brillinger (1973, 1980).

Another test which is often useful develops from a requirement to test for the presence of the $q_2 \times 1$ signal vector $\boldsymbol{B}_2^{\wedge}(k)$ given that the first q_1 signals are present. This compares the residual sums of squares computed under the models (3.3) and (3.4). We note, for example, for $\boldsymbol{B}^{\wedge}(k) = (\boldsymbol{B}_1^{\wedge}(k)', \boldsymbol{B}_2^{\wedge}(k)')'$ and $X^{\wedge}(k) = (X_1^{\wedge}(k), X_2^{\wedge}(k))$, the usual least-squares estimator for $B^{\wedge}(k)$ in (3.3) is

$$\hat{\boldsymbol{B}}^{\wedge}(k) = [\overline{X^{\wedge}(k)'}X^{\wedge}(k)]^{-1}\overline{X^{\wedge}(k)'}\,\boldsymbol{Y}^{\wedge}(k), \tag{3.10}$$

leading to an F statistic of the form

$$F_{2q_2, 2(N-q_1-q_2)} = \frac{\mathrm{SSR}_2(k)}{\mathrm{SSE}_2(k)} \frac{(N-q_1-q_2)}{q_2}, \tag{3.11}$$

where

$$\begin{aligned} \mathrm{SSR}_2(k) &= [\|\boldsymbol{Y}^{\wedge}(k) - X_1^{\wedge}(k)\hat{\boldsymbol{B}}_1^{\wedge}(k)\|^2 - \|\boldsymbol{Y}^{\wedge}(k) - X^{\wedge}(k)\hat{\boldsymbol{B}}^{\wedge}(k)\|^2] \\ &= [\|X^{\wedge}(k)\hat{\boldsymbol{B}}^{\wedge}(k)\|^2 - \|X_1^{\wedge}(k)\hat{\boldsymbol{B}}_1^{\wedge}(k)\|^2] \end{aligned} \tag{3.12}$$

and

$$\begin{aligned} \mathrm{SSE}_2(k) &= \|\boldsymbol{Y}^{\wedge}(k) - X^{\wedge}(k)\hat{\boldsymbol{B}}^{\wedge}(k)\|^2 \\ &= [\|\boldsymbol{Y}^{\wedge}(k)\|^2 - \|X^{\wedge}(k)\hat{\boldsymbol{B}}^{\wedge}(k)\|^2]. \end{aligned} \tag{3.13}$$

Again, smoothing over K frequencies replaces $2q_2$ and $2(N-q_1-q_2)$ by $2Kq_2$ and $2K(N-q_1-q_2)$ in (3.11)–(3.13). The resulting sums of squares can again be summarized in an analysis of power table.

As an example, consider the mixture of $q=2$ regression functions discussed in the estimation section where we choose $q_1 = q_2 = 1$ and the series are recorded for $T = 1024$ points assumed to be sampled at 20 points per second. The frequencies are of the form $\lambda_k = 2\pi k/1024$, $k = 0, 1, \ldots, 512$, and the power components are averaged over $K = 17$ frequencies. The resulting F statistic (cf. (3.7)) for detecting the presence of $\beta_1(t)$ will have $2(17) = 34$ and $2(17)(10-1) = 306$ degrees of freedom. It is convenient to plot the test statistic as a function of frequency $f = 20\lambda/2\pi$ in cycles per second (Hz). We note from Fig. 4 that the evidence for a single regression function appearing on all channels is very strong over a frequency band ranging from 0 to 5 Hz. The test statistic (3.11) is also shown in Fig. 4, where we have again averaged the sums of squares over $K = 17$ frequencies, leading to F statistics with $2(17) = 34$ and

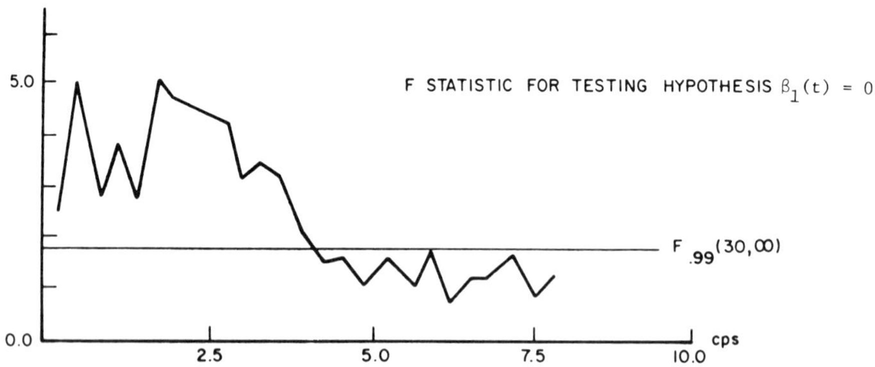

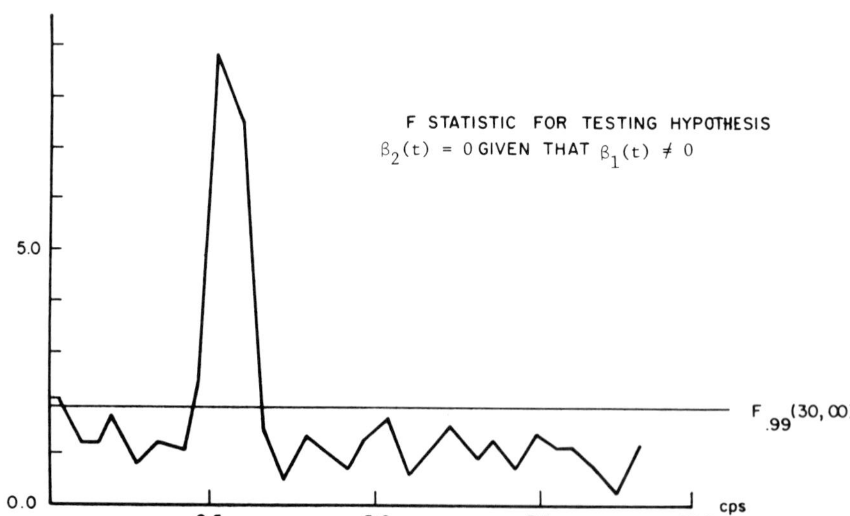

Fig. 4. F statistics for sequentially testing for the presence of $\beta_1(t)$ and $\beta_2(t)$.

$2(17)(10-2)=272$ degrees of freedom. Here, one may note the very significant detection at 3 Hz, corresponding to the frequency of the regression function $\beta_2(t)$.

A problem of interest in many applications involves the case where the matrix of design functions depends nonlinearly on some $p \times 1$ vector $\boldsymbol{\Theta}$, as in (1.3)′. The frequency-domain version of (1.3)′ (see also (3.3)) can be written in the form

$$\boldsymbol{Y}^{\wedge}(k) = X^{\wedge}(k; \boldsymbol{\Theta})\boldsymbol{B}^{\wedge}(k) + \boldsymbol{\zeta}_k + o_{\text{a.s.}}(1)\,. \tag{3.14}$$

Jennrich (1969) has derived the asymptotic properties of general nonlinear least-squares estimators. His results in this case apply to the values $\hat{\boldsymbol{\Theta}}$, $\hat{\boldsymbol{B}}^{\wedge}(k; \hat{\boldsymbol{\Theta}})$, obtained by minimizing

$$\text{SSE}_2(k; \Theta) = \|\boldsymbol{Y}^{\wedge}(k) - X^{\wedge}(k; \boldsymbol{\Theta})\boldsymbol{B}^{\wedge}(k)\|^2\,, \tag{3.15}$$

where

$$\hat{\boldsymbol{B}}^{\wedge}(k; \hat{\Theta}) = S^{-1}(k; \hat{\Theta})\overline{X^{\wedge}(k; \hat{\boldsymbol{\Theta}})'}\,\boldsymbol{Y}^{\wedge}(k)\,. \tag{3.16}$$

with

$$S(k; \hat{\boldsymbol{\Theta}}) = \overline{X^{\wedge}(k; \hat{\boldsymbol{\Theta}})'}X(k; \hat{\boldsymbol{\Theta}})\,. \tag{3.17}$$

Wu (1982) has given regularity conditions involving the derivatives of the matrix $X^{\wedge}(k; \boldsymbol{\Theta})$ which imply the strong consistency $(N \to \infty)$ of $\hat{\boldsymbol{\Theta}}$, $\hat{B}^{\wedge}(k; \hat{\boldsymbol{\Theta}})$, and of

$$\hat{f}_e(\lambda; \hat{\boldsymbol{\Theta}}) = N^{-1}\text{SSE}_2(k; \hat{\boldsymbol{\Theta}}) \tag{3.18}$$

as an estimator of $f_e(\lambda; \boldsymbol{\Theta})$, the spectrum of the error series. He also showed that the estimators were asymptotically normal and gave the limiting covariance structure. These results will be shown later for the important single-signal case $(q = 1)$.

An approach to testing hypotheses for models depending on the parameter $\boldsymbol{\Theta}$ is suggested by noting that if Θ is known, one may make the partition $\boldsymbol{\Theta} = (\boldsymbol{\Theta}_1', \boldsymbol{\Theta}_2')'$, where $\boldsymbol{\Theta}_1$ and $\boldsymbol{\Theta}_2$ are $p_1 \times 1$ and $p_2 \times 1$ parameter vectors and $X^{\wedge}(k; \boldsymbol{\Theta}) = (X_1^{\wedge}(k; \boldsymbol{\Theta}_1), X_2^{\wedge}(k; \boldsymbol{\Theta}_2))$. Thus, we may compare

$$\boldsymbol{Y}^{\wedge}(k) = X_1^{\wedge}(k; \boldsymbol{\Theta}_1)\boldsymbol{B}_1^{\wedge}(k) + \boldsymbol{\zeta}_k + o_{\text{a.s.}}(1) \tag{3.19}$$

against (3.5), using the approximate F statistic

$$F_1(k; \boldsymbol{\Theta}_1) = \frac{\text{SSR}_1(k; \boldsymbol{\Theta}_1)}{\text{SSE}_1(k; \boldsymbol{\Theta}_1)}\frac{(2N - 2q_1)}{2q_1} \tag{3.20}$$

with

$$\hat{\boldsymbol{B}}_1^{\wedge}(k; \boldsymbol{\Theta}_1) = S_{11}^{-1}(\boldsymbol{\Theta}_1)\overline{X_1^{\wedge}(k; \boldsymbol{\Theta}_1)'}\,\boldsymbol{Y}^{\wedge}(k) \tag{3.21}$$

maximizing

$$\begin{aligned}\mathrm{SSE}_1(k;\boldsymbol{\Theta}_1) &= \|\boldsymbol{Y}^\wedge(k) - X_1^\wedge(k;\boldsymbol{\Theta}_1)\hat{B}_1^\wedge(k;\boldsymbol{\Theta}_1)\|^2 \\ &= \|\boldsymbol{Y}^\wedge(k)\|^2 - \|X_1^\wedge(k;\boldsymbol{\Theta}_1)\hat{\boldsymbol{B}}_1^\wedge(k;\boldsymbol{\Theta}_1)\|^2,\end{aligned} \tag{3.22}$$

where we have defined the matrix

$$S_{11}(\boldsymbol{\Theta}_1) = \overline{X_1^\wedge(k;\boldsymbol{\Theta}_1)'X_1^\wedge(k;\boldsymbol{\Theta}_1)}, \tag{3.23}$$

and the regression power is

$$\mathrm{SSR}_1(k;\boldsymbol{\Theta}_1) = \|X_1^\wedge(k;\boldsymbol{\Theta}_1)\hat{\boldsymbol{B}}_1^\wedge(k;\boldsymbol{\Theta}_1)\|^2. \tag{3.24}$$

For a given value $\boldsymbol{\Theta}_1$, the statistic $F_1(k;\boldsymbol{\Theta}_1)$ will be distributed as a doubly noncentral F distribution with $2q_1$ and $2(N-q_1)$ degrees of freedom. The noncentrality parameters are

$$\gamma_1^2(\boldsymbol{\Theta}_1) = 2\overline{\boldsymbol{B}_1^\wedge(k)'}\,\overline{X_1^\wedge(k;\boldsymbol{\Theta}_1^0)'}D_1(\boldsymbol{\Theta}_1)X_1^\wedge(k;\boldsymbol{\Theta}_1^0)\boldsymbol{B}_1^\wedge(k)/f_e(\lambda) \tag{3.25}$$

for the numerator and

$$\gamma_2^2(\boldsymbol{\Theta}_1) = 2\overline{\boldsymbol{B}_1^\wedge(k)'}\,\overline{X_1^\wedge(k;\boldsymbol{\Theta}_1^0)'}(I - D_1(\boldsymbol{\Theta}_1))X_1^\wedge(k;\boldsymbol{\Theta}_1^0)\boldsymbol{B}_1^\wedge(k)/f_e(\lambda) \tag{3.26}$$

for the denominator, with $\boldsymbol{\Theta}_1^0$ the true value for $\boldsymbol{\Theta}_1$ and $D(\boldsymbol{\Theta}_1)$ the matrix product

$$D_1(\boldsymbol{\Theta}_1) = \overline{X_1^\wedge(k;\boldsymbol{\Theta}_1)'}S_{11}^{-1}(k;\boldsymbol{\Theta}_1)X_1^\wedge(k;\boldsymbol{\Theta}_1). \tag{3.27}$$

If $B_1^\wedge(k) = 0$, the signal is not present at any $\boldsymbol{\Theta}_1$ and the test statistic (3.19) becomes a central F distribution for any fixed $\boldsymbol{\Theta}_1$. If $\boldsymbol{B}_1^\wedge(k) \neq 0$ but $\theta_1 = \theta_1^0$, then

$$\gamma_1^2(\boldsymbol{\theta}_1^0) = \overline{2\boldsymbol{B}_1^\wedge(k)'}S_{11}(\boldsymbol{\Theta}_1^0)\boldsymbol{B}_1^\wedge(k)/f_e(\lambda) \tag{3.28}$$

and $\gamma_2^2(\theta_1^0) = 0$, so that the distribution reduces to a singly noncentral F. Generally, the procedures is to plot $F_1(k;\boldsymbol{\Theta}_1)$ as a function of $\boldsymbol{\Theta}_1$, if that is convenient, with the value $\hat{\boldsymbol{\Theta}}_1$ which maximizes $F_1(k;\boldsymbol{\Theta}_1)$ yielding the least-squares estimator. One may then apply the asymptotic arguments to note that the strong consistency of the least-squares estimator $\hat{\boldsymbol{\Theta}}_1$ would imply that

$$F_1(k;\hat{\boldsymbol{\Theta}}_1) = \max_{\theta_1} F_1(k;\boldsymbol{\Theta}_1) \tag{3.29}$$

is distributed proportionally to a singly noncentral F distribution with $2q_1$ and $2(N-q_1)$ degrees of freedom. This approach is similar to that advocated by Doksum and Wong (1981) in another context.

The doubly noncentral F distribution which obtains when $\boldsymbol{\Theta}_1 \neq \boldsymbol{\Theta}_1^0$ can be approximated by a constant multiplied by a central F distribution as in Johnson and Kotz (1970, p. 197) or by more exact methods as in Mudholkar et al. (1976). For the cases considered here, the ratio $\gamma_1^2(\boldsymbol{\Theta}_1)/\gamma_2^2(\boldsymbol{\Theta}_1)$ decreases rapidly as $\boldsymbol{\Theta}_1$ moves away from $\boldsymbol{\Theta}_1^0$ so that using the critical value implied the central F distribution seems to give a conservative approximation.

It should be noted that a modification of the above procedure relating to the likelihood ratio test $\boldsymbol{\Theta}_1 = \boldsymbol{\Theta}_1^0$, $B_1^{\frown}(k) = B_1^{\frown 0}(k)$ has been given by Gallant (1975). The asymptotic results imply that the statistic (3.19) should be distributed as an F with $2q_1 + p_1$ and $2N - 2q_1 - p_1$ degrees of freedom and a slightly different noncentrality parameter. The difficulty in applying those results stems from the fact that $\boldsymbol{B}_1^{\frown}(k)$ enters linearly so that, for $\boldsymbol{B}_1^{\frown 0}(k) = \mathbf{0}$, the parameter $\boldsymbol{\Theta}_1$ can take on any arbitrary value.

In order to test for the presence of the second regression function assuming that the first has been identified, we may use the generalization of (3.11), namely

$$F_2(k;\boldsymbol{\Theta}) = F_2(k;\boldsymbol{\Theta}_1,\boldsymbol{\Theta}_2) = \frac{\mathrm{SSR}_2(k;\boldsymbol{\Theta})}{\mathrm{SSE}_2(k;\boldsymbol{\Theta})}\frac{(N-q_1-q_2)}{q_2} \tag{3.30}$$

with

$$\hat{\boldsymbol{B}}(k;\boldsymbol{\Theta}) = S^{-1}(k;\boldsymbol{\Theta})\overline{\boldsymbol{X}^{\frown}(k;\boldsymbol{\Theta})'}\boldsymbol{Y}^{\frown}(k)$$

minimizing

$$\begin{aligned}\mathrm{SSE}_2(k;\boldsymbol{\Theta}) &= \|\boldsymbol{Y}^{\frown}(k) - \boldsymbol{X}^{\frown}(k;\boldsymbol{\Theta})\boldsymbol{B}^{\frown}(k;\boldsymbol{\Theta})\|^2 \\ &= \|\boldsymbol{Y}^{\frown}(k)\|^2 - \|\boldsymbol{X}^{\frown}(k;\boldsymbol{\Theta})\boldsymbol{B}^{\frown}(k;\boldsymbol{\Theta})\|^2\end{aligned} \tag{3.31}$$

and the regression sum of squares defined by

$$\mathrm{SSR}_2(k;\boldsymbol{\Theta}) = \|\boldsymbol{X}^{\frown}(k;\boldsymbol{\Theta})\boldsymbol{B}^{\frown}(k;\boldsymbol{\Theta})\|^2 - \|\boldsymbol{X}_1^{\frown}(k;\boldsymbol{\Theta}_1)\hat{\boldsymbol{B}}_1^{\frown}(k;\boldsymbol{\Theta}_1)\|^2, \tag{3.32}$$

with $\hat{\boldsymbol{B}}_1^{\frown}(k;\boldsymbol{\Theta}_1)$ given, as before, in (3.21).

For a fixed $\boldsymbol{\Theta}_1$, the distribution of $F(k;\boldsymbol{\Theta})$ may be approximated by a doubly noncentral F distribution for $\boldsymbol{\Theta}_2 \neq \boldsymbol{\Theta}_2^0$ and by a singly noncentral F distribution for $\boldsymbol{\Theta}_2 = \boldsymbol{\Theta}_2^0$. For the applications given here, it is adequate to compare

$$F(k;\boldsymbol{\Theta}_1,\hat{\boldsymbol{\Theta}}_2) = \max_{\boldsymbol{\Theta}_2} F(k;\boldsymbol{\Theta}_1,\boldsymbol{\Theta}_2) \tag{3.33}$$

with the critical point on the central F distribution with $2q_2$ and $2(N-q_1-q_2)$ degrees of freedom. In this case, $\hat{\boldsymbol{\Theta}}_2$ also minimizes $\mathrm{SSE}_2(k;\boldsymbol{\Theta}_1,\boldsymbol{\Theta}_2)$ for a fixed $\boldsymbol{\Theta}_1$. Again, we note that the alternate approach given by Gallant (1975) would indicate that the numerator and denominator degrees of freedom are $2q_2 + p_2$ and $2N - 2q - p_2$ respectively. The modifications (3.20) and (3.30) are important in applications to signal detection where the directional characteristics of the signal depend on the parameter $\boldsymbol{\Theta}$ and must be estimated. The following section considers in more detail a particular problem arising in signal detection.

4. Applications to signal detection

In the processing of time-series data from arrays monitoring data from such diverse subject areas as seismology, radar, underwater acoustics, and infrasonics, one may be interested in the possibility that one or more propagating signals are present. In general, the model expressing the fact that q propagating signals are arriving at an array can be written as

$$y_j(t) = \sum_{m=1}^{q} \beta_m(t - T_j(\boldsymbol{\theta}_m)) + e_j(t)\,, \tag{4.1}$$

$j = 1, 2, \ldots, N$ where $\beta_m(t)$ is the mth signal, $e_j(t)$ is a zero-mean stationary noise process as in (1.1), and $T_j(\boldsymbol{\theta}_m)$ is the time delay experienced by the nth signal of the jth sensor relative to its arrival at the origin. The time delays depend nonlinearly on the $p_m \times 1$ parameter vectors $\boldsymbol{\theta}_m$ which model the particular mode of propagation experienced by the mth signal. We consider here the particular example corresponding to a planar array so that $p_m = 2$ for each of the $m = 1, 2, \ldots, q$ signals. The parametrization will be in terms of the wavenumber vector $\boldsymbol{\theta}_m = (\theta_{m1}, \theta_{m2})'$ which can be related to the velocity c_m and azimuth α_m of the mth plane wave signal. For example, if the coordinates of the jth sensor are $\boldsymbol{r}_j = (r_{j1}, r_{j2})'$, the time delay in (4.1) can be expressed as

$$T_{jm}(\boldsymbol{\theta}_m) = 2\pi\boldsymbol{\theta}_m'\boldsymbol{r}_j/\lambda \tag{4.2}$$

at some frequency of interest λ, and the velocity and azimuth are given by

$$c_m = \frac{\lambda}{2\pi\|\boldsymbol{\theta}_m\|} \tag{4.3}$$

and

$$\alpha_m = \tan^{-1}\left(\frac{\theta_{m1}}{\theta_{m2}}\right). \tag{4.4}$$

If the signals $\beta_m(t)$, $m = 1, \ldots, q$ are assumed to be fixed waveforms, the model (4.1) is just the regression model considered in the first section and in (2.5)–(2.7) for $q = 2$. In the general case, the $N \times q$ matrix of design functions, $x_t(\boldsymbol{\Theta})$, $\boldsymbol{\Theta} = (\boldsymbol{\theta}_1', \boldsymbol{\theta}_2', \ldots, \boldsymbol{\theta}_q')'$, has elements of the form

$$x_{jm}(t) = \delta(t - T_j(\boldsymbol{\theta}_m))\,. \tag{4.5}$$

It follows that the frequency-domain version of the above model can be written in the form

$$\boldsymbol{Y}^{\hat{}}(k) = \boldsymbol{X}^{\hat{}}(k; \boldsymbol{\Theta})\boldsymbol{B}^{\hat{}}(k) + \zeta_k + o_{\text{a.s.}}(1) \tag{4.6}$$

for some frequency $\lambda_k = 2\pi k/T \doteq \lambda$, where $\boldsymbol{B}^{\hat{}}(k)$ is the DFT of the signal vector and the elements of $\boldsymbol{X}^{\hat{}}(k; \boldsymbol{\Theta})$ are given by

$$X_{jm}^{\wedge}(k\,;\,\boldsymbol{\Theta}) = \exp\{-\mathrm{i}\lambda_k T_j(\boldsymbol{\theta}_m)\}$$
$$= \exp\{-2\pi\mathrm{i}\boldsymbol{\theta}_m'\boldsymbol{r}_j\}\,, \tag{4.7}$$

$j = 1, \ldots, N$, $m = 1, \ldots, q$, using (4.2).

The regression model now can be used to approach the problem of identifying and resolving a collection of signals propagating across a planar array. We will consider separately the case of a single signal $q = 1$ and the more general case where a collection of q signals may be present. The identification of a single signal from a specified source can be modeled in terms of detecting a signal in a collection of N stationarily correlated time series as in Shumway (1972). The incorporation of a procedure for estimating the parameter vector $\boldsymbol{\theta}$ into the analysis was treated by Hinich and Shaman (1972) for regularly spaced sensors using maximum likelihood techniques. We consider here a least-squares approach which applies to sensors with coordinates chosen from a general bivariate distribution. In addition, we consider the identification problem for a multiple signal model using stepwise multiple regression techniques.

4.1. Detection of a single propagating signal

In order to derive the regression detector for a single signal $\beta_1(t)$ propagating with some general wavenumber coordinate $\boldsymbol{\theta}_1$ at frequency λ as in (1.7) we may write the frequency-domain version as in (3.19), giving

$$\boldsymbol{Y}^{\wedge}(k) = \boldsymbol{x}_1^{\wedge}(\boldsymbol{\theta}_1)B_1^{\wedge}(k) + \boldsymbol{\zeta}_k + \mathrm{o}_{\mathrm{a.s.}}(1)\,, \tag{4.8}$$

where

$$\boldsymbol{x}_1^{\wedge}(\boldsymbol{\theta}_1) = (\exp\{-2\pi\mathrm{i}\boldsymbol{\theta}_1'\boldsymbol{r}_1\}, \ldots, \exp\{-2\pi\mathrm{i}\boldsymbol{\theta}_1'\boldsymbol{r}_N\})' \tag{4.9}$$

is the $N \times 1$ vector describing the propagation characteristics of the single signal. The approximate least-squares estimator for the signal with θ_1 specified is (cf. (3.67) and (3.17))

$$\hat{B}_1^{\wedge}(k\,;\,\boldsymbol{\theta}_1) = N^{-1}\overline{\boldsymbol{x}_1^{\wedge}(\boldsymbol{\theta}_1)'}\,\boldsymbol{Y}^{\wedge}(k)$$
$$= N^{-1}\sum_{j=1}^{N} \exp\{2\pi\mathrm{i}\boldsymbol{\theta}_1'\boldsymbol{r}_j\}Y_j^{\wedge}(k)\,, \tag{4.10}$$

which is just the ordinary beamformed estimator expressed in the frequency domain. The estimators $\hat{\boldsymbol{\theta}}$ and $\hat{B}_1^{\wedge}(k\,;\,\hat{\boldsymbol{\theta}})$ are defined as the minimizing values for

$$\mathrm{SSE}_1(k\,;\,\boldsymbol{\theta}_1) = \|\boldsymbol{Y}^{\wedge}(k) - \boldsymbol{x}_1^{\wedge}(\theta_1)B_1^{\wedge}(k)\|^2\,. \tag{4.11}$$

In this case, using (3.24),

$$\mathrm{SSR}_1(k\,;\,\boldsymbol{\theta}_1) = N|\hat{B}_1^{\wedge}(k\,;\,\boldsymbol{\theta}_1)|^2 \tag{4.12}$$

is just the ordinary beampower, evaluated at $\boldsymbol{\theta}_1$. Since the range of $\boldsymbol{\theta}_1$ can be restricted in applications, it is convenient to plot the beampower as a function of $\boldsymbol{\theta}_1$ for a fixed k. Then because of (4.11), the value of $\boldsymbol{\theta}_1$ maximizing the beampower will be the approximate least-squares estimator.

The test statistic (3.20) may also be plotted as a function of $\boldsymbol{\theta}_1$, say

$$F_1(k, \boldsymbol{\theta}_1) = \frac{\mathrm{SSR}_1(k; \boldsymbol{\theta}_1)}{\mathrm{SSE}_1(k; \boldsymbol{\theta}_1)} \frac{(2N-2)}{2}, \tag{4.13}$$

which is also maximized at the least-squares estimator $\hat{\boldsymbol{\theta}}_1$, so that $F(k; \hat{\theta}_1)$ can be compared with the critical point on an F distribution with 2 and $2N-2$ degrees of freedom. This is generally more informative than the beampower, $\mathrm{SSR}_1(k; \boldsymbol{\theta}_1)$, since one has an approximate significance test available for any observed peak. The distribution of $F_1(k; \boldsymbol{\theta}_1)$ in (4.13) will be again doubly noncentral F with noncentrality parameters (3.25) and (3.26) given by

$$\gamma_1^2(\boldsymbol{\theta}_1) = \frac{2N|\hat{B_1}(k)|^2}{f_e(\lambda)} \varepsilon^2(\boldsymbol{\theta}_1) \tag{4.14}$$

and

$$\gamma_2^2(\boldsymbol{\theta}_1) = \frac{2N|\hat{B_1}(k)|^2}{f_e(\lambda)} (1 - \varepsilon^2(\boldsymbol{\theta}_1)), \tag{4.15}$$

where

$$\begin{aligned} \varepsilon^2(\boldsymbol{\theta}_1) &= \frac{|\overline{x_1(k; \boldsymbol{\theta}_1)'} x_1(k; \boldsymbol{\theta}_1^0)|^2}{N^2} \\ &= \left| \sum_{j=1}^{N} \exp\{2\pi \mathrm{i}(\boldsymbol{\theta}_1 - \boldsymbol{\theta}_1^0)' \boldsymbol{r}_j\} \right|^2 / N^2 \end{aligned} \tag{4.16}$$

lies between 0 and 1 and takes the value 1 when $\boldsymbol{\theta}_1 = \boldsymbol{\theta}_1^0$, the true value of $\boldsymbol{\theta}_1$. The behavior of the noncentrality parameters will depend on the rate at which $\varepsilon^2(\boldsymbol{\theta}_1)$ decreases as $\boldsymbol{\theta}_1$ moves away from $\boldsymbol{\theta}_1^0$.

The regularity conditions of Jennrich (1969) have been verified by Wu (1982) under the conditions that (1) the measure of bivariate coordinates $\boldsymbol{r}_j$ converges to a bivariate probability distribution with finite fourth moments, and (2) there is positive probability that the location coordinates are linearly independent. Hinich and Shaman (1972) have given conditions for uniformly spaced sensor locations with the locations scaled by N. Hinich (1981) has studied the array response on a geometry with coordinates limited to rational numbers.

Under the above conditions, one can say that $\hat{\boldsymbol{\theta}}_1$ is approximately bivariate normal with mean equal to the true value $\boldsymbol{\theta}_1^0$ and with the covariance matrix estimated consistently by

$$\widehat{\mathrm{cov}}(\hat{\boldsymbol{\theta}}_1) = \tfrac{1}{2}(2\pi)^{-2} |\hat{B_1}(k; \hat{\boldsymbol{\theta}}_1)|^{-2} N^{-1} f_e(\lambda; \hat{\boldsymbol{\theta}}_1) A^{-1}, \tag{4.17}$$

where A is the sample covariance matrix of the bivariate coordinate locations. The elements of A are defined as

$$a_{mn} = N^{-1} \sum_{j=1}^{N} (r_{jm} - \bar{r}_m)(r_{jn} - \bar{r}_n) \tag{4.18}$$

for $m, n = 1, 2$, where $\bar{r}_m$ is the sample mean

$$\bar{r}_m = N^{-} \sum_{j=1}^{N} r_{jm} . \tag{4.19}$$

The error spectral estimator is given by

$$f_e(\lambda ; \hat{\boldsymbol{\theta}}_1) = N^{-1}\, \mathrm{SSE}_1(k ; \hat{\boldsymbol{\theta}}_1) \tag{4.20}$$

where $\mathrm{SSE}_1(k ; \hat{\boldsymbol{\theta}}_1)$ is the error sum of squares defined in (4.11) evaluated at $\hat{\boldsymbol{\theta}}_1$, $\hat{B}_1^{\prime}(k ; \hat{\boldsymbol{\theta}}_1)$. A joint $(1-\alpha)$ confidence ellipse for θ_{11} and θ_{12} can be computed using the inequality

$$(\hat{\boldsymbol{\theta}}_1 - \boldsymbol{\theta}_1)' A (\hat{\boldsymbol{\theta}}_1 - \boldsymbol{\theta}_1) \leq (2\pi)^{-2} |\hat{B}_1^{\prime}(k ; \hat{\boldsymbol{\theta}}_1)|^{-2} \frac{f_e(\lambda ; \hat{\boldsymbol{\theta}}_1)}{(N-1)} F_{2,2N-2;\alpha} , \tag{4.21}$$

where $F_{2,2N-2;\alpha}$ denotes the upper α critical point on an F distribution with 2 and $2N-2$ degrees of freedom (see also Gallant (1975) who recommends 2 and $2N-4$ degrees of freedom). The confidence region (4.21) can be displayed in the usual way as the region enclosed by an ellipse in $(\theta_{11}, \theta_{12})$ coordinates. Confidence intervals for the related velocity and azimuth, c_1 and α_1, defined by (4.3) and (4.4) can be read directly off the plot of (4.21) by noting that c_1 is the length of a radius line centered at $\boldsymbol{\theta}_1 = (0, 0)$ and passing through $\hat{\boldsymbol{\theta}}_1 = (\hat{\theta}_{11}, \hat{\theta}_{12})'$, whereas α is the angle measured clockwise from north. Alternately, the delta method may be applied as in Hinich and Shaman (1972) or Wu (1982).

In order to give an example of the detection procedure for a small ($N = 5$) seismic array, consider the two long-period signals, taken from Lambert and Der (1973), shown in Fig. 5. Those signals were generated by events from Baja, California and Guatamala and were measured at a small subarray within the Large Aperture Seismic Array (LASA) in Montana consisting of a center element and four outside elements spaced approximately uniformly on a circle with radius 30 km. The 'signals' of interest are assumed to be long-period Rayleigh waves with periods of 21 seconds and wavenumber coordinate satisfying $(-0.02 \leq \theta_{11} \leq 0.02,\ -0.02 \leq \theta_{12} \leq 0.02)$ which are roughly compatible with signals traveling faster than 3 km/sec. The sampling rate was 1 point per second, yielding a folding frequency of 0.5 Hz, whereas the frequency interest was $2\pi\lambda = 0.047$ cycles per second. In Fig. 6, it is obvious that the beampower and F statistics have the same maximum, which corresponds to least-squares estimators of $\hat{\boldsymbol{\theta}}_1 = (-0.002769, -0.01146)'$ for the first signal and $\hat{\boldsymbol{\theta}}_2 = (0.008308, 0.009167)'$ for the second signal. Converting these values to velocities v and azimuths (measured clockwise from 12:00 o'clock), using (4.3) and (4.4), we obtain $c_1 = 3.98$ km/sec and $\alpha_1 = 194°$ for the first signal and $c_2 = 3.79$ km/sec and

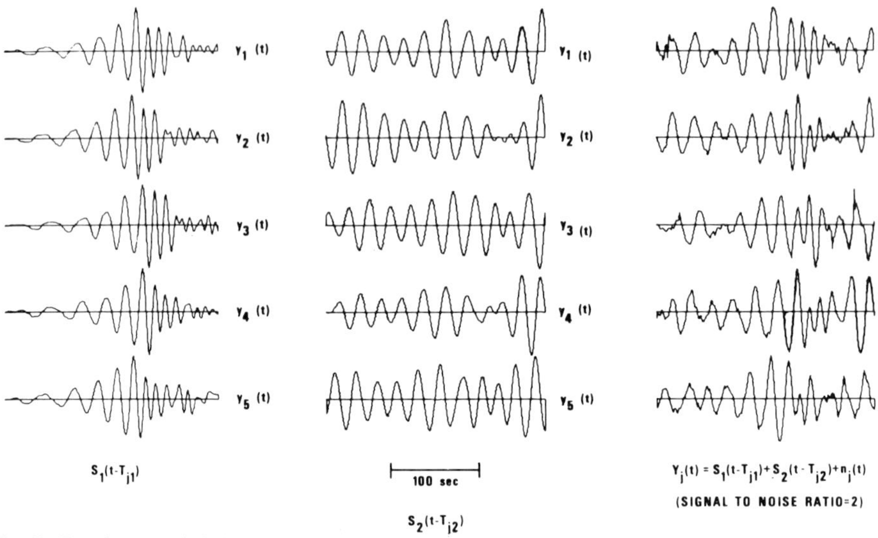

Fig. 5. Two long-period signals separately and combined with noise (taken from Lambert and Der, 1973).

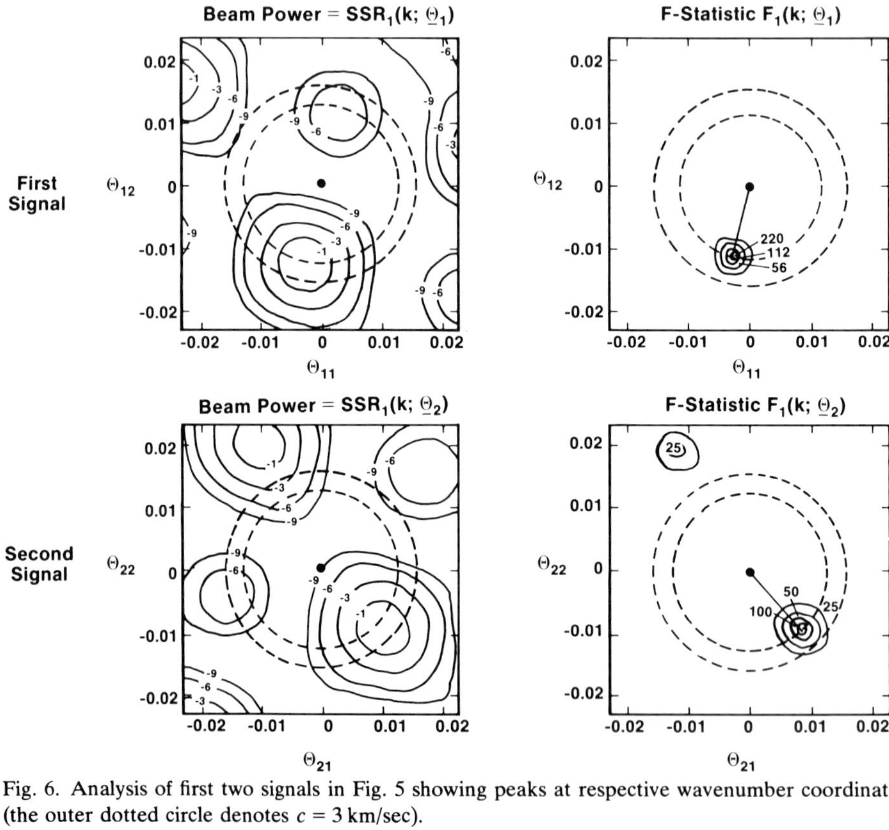

Fig. 6. Analysis of first two signals in Fig. 5 showing peaks at respective wavenumber coordinates (the outer dotted circle denotes $c = 3$ km/sec).

$\alpha_2 = 132°$ for the second signal. The critical value for the F statistic in this case, say $F_{2,8,;0.01} = 8.65$, can be compared with the maximum values 220 and 100 which are observed on the two wavenumber F plots. It can be seen that the values decay rapidly down to insignificant levels once we move away from the estimated center coordinates.

Thus, the problem of detecting a single signal using the wavenumber F detector seems to be reasonably straightforward when there are no interfering signals. However, the third set of observed series in Fig. 5 shows a contrived mixture of the two signals with an additive noise component. The wavenumber plots resulting from analyzing this mixture are shown in Fig. 7, and we note that the observed value of 16 exceeds $F_{2,8;0.05} = 4.46$ so that the peak is significant at the $\alpha = 0.05$ level. The estimated wavenumber coordinates from this mixture correspond to a velocity of $c_2 = 4.19$ km/sec and an azimuth of $\alpha_2 = 125°$. However, we note also that there is little indication in either of the plots that there might be a second signal present at another set of wavenumber coordinates. Since, in this case, the other signal is known to be present, we would like to be able to design further test statistics which could detect additional signals. This problem is considered in the following section.

4.2. Detection in the multiple signal case

We assume now that an array may have more than a single signal present and consider a sequential method for determining the number of unknown signals q as well as their associated wavenumber parameters, say $\boldsymbol{\theta}_1, \boldsymbol{\theta}_2, \ldots, \boldsymbol{\theta}_q$.

More specifically, we assume the presence of q_1 propagating signals at some stage with wavenumber parameters $\boldsymbol{\theta}_1, \boldsymbol{\theta}_2, \ldots, \boldsymbol{\theta}_{q_1}$ and test for the presence of an additional signal. That is, for a model of the form

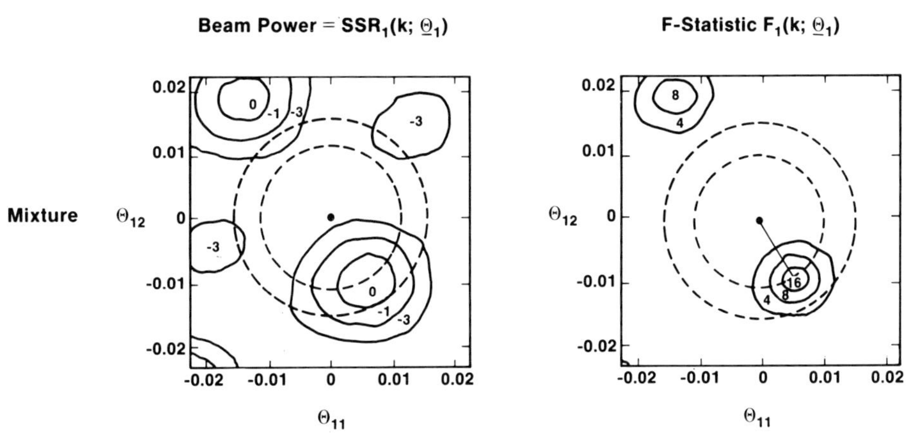

Fig. 7. Analysis of the mixture of two signals showing a weak preliminary detection corresponding to the approximate coordinates of the second signal. (The outer dotted circle denotes $c = 3$ km/sec.)

$$Y\hat{}(k) = \sum_{m=1}^{q_1} \boldsymbol{x}\hat{}_m(\boldsymbol{\theta}_m) B\hat{}_m(k) + \boldsymbol{x}\hat{}_{q_1+1}(\boldsymbol{\theta}_{q_1+1}) B_{q_1+1}(k) + \boldsymbol{\zeta}_k + \mathrm{o}_{\mathrm{a.s.}}(1)\,, \qquad (4.22)$$

we wish to examine the effect of the (q_1+1)st regression signal. As a simple approach, the following stepwise procedure has been used with some success. Assume that a single has been detected by the method given in the previous section with the estimator for $\boldsymbol{\theta}_1$ given by $\hat{\boldsymbol{\theta}}_1$. Then, regard $\boldsymbol{\theta}_1$ as being fixed at $\hat{\boldsymbol{\theta}}_1$ in (4.22) and consider testing the hypothesis that a second signal $B\hat{}_2(k)$ is present at some unspecified wavenumber coordinates $\boldsymbol{\theta}_2$. This would involve fitting the reduced model estimating $B\hat{}_1(k)$ only and then fitting the full model estimating $B\hat{}_1(k)$, $B\hat{}_2(k)$ and $\boldsymbol{\theta}_2$ as in (3.30) to (3.33). If the test statistic $F_2(k; \hat{\boldsymbol{\theta}}_1, \boldsymbol{\theta}_2)$ with 2 and $2N-2$ degrees of freedom indicates that $B\hat{}_2(k)$ is present, then consider calculating $F_2(k; \boldsymbol{\theta}_1, \hat{\boldsymbol{\theta}}_2)$ fixed at $\hat{\boldsymbol{\theta}}_2$, the estimated coordinates of the second signal. The procedure can be repeated until $\hat{\boldsymbol{\theta}}_1$ and $\hat{\boldsymbol{\theta}}_2$ change by negligible amounts. Blandford et al. (1976) have shown that this procedure converges when applied in the time domain with known values for $\boldsymbol{\theta}_1$ and $\boldsymbol{\theta}_2$. In general, at the stage when q_1 signals are present at wavenumber coordinates $\boldsymbol{\theta}_1, \boldsymbol{\theta}_2, \ldots, \boldsymbol{\theta}_{q_1}$ as in (4.22), setting the wavenumber coordinates at the current estimated values leads to a version of (3.28) with 2 and $2(N-q_1)$ degrees of freedom, where $X\hat{}_1(k; \boldsymbol{\Theta}_1) = (\boldsymbol{x}\hat{}_1(\boldsymbol{\theta}_1), \ldots, \boldsymbol{x}\hat{}_{q_1}(\boldsymbol{\theta}_{q_1}))$ and $\hat{\boldsymbol{B}}\hat{}_1(k; \boldsymbol{\Theta}_1) = (B\hat{}_1(k; \boldsymbol{\theta}_1), \ldots, B\hat{}_{q_1}(k; \boldsymbol{\theta}_{q_1}))'$. Furthermore, $\hat{B}\hat{}(k; \boldsymbol{\Theta}_1, \hat{\boldsymbol{\theta}}_{q_1+1})$ and $\hat{\boldsymbol{\Theta}} = (\boldsymbol{\Theta}_1, \hat{\boldsymbol{\theta}}_{q_1+1})'$ jointly minimize

$$\mathrm{SSE}_2(k; \boldsymbol{\Theta}_1, \hat{\boldsymbol{\theta}}_{q_1+1}) = \|Y\hat{}(k)\|^2 - \|X\hat{}(\boldsymbol{\Theta}_1, \hat{\boldsymbol{\theta}}_{q_1+1}) \boldsymbol{B}\hat{}(k; \boldsymbol{\Theta}_1, \hat{\boldsymbol{\theta}}_{q_1+1})\|^2\,, \quad (4.23)$$

regarded as a function of the regression coefficients and the last wavenumber coordinates $\boldsymbol{\theta}_{q_1+1}$.

In order to apply this method to the mixture of two signals in Fig. 5, consider testing the hypothesis that another signal is present against the alternative single signal model indicated by the wavenumber plots in Fig. 7. In this case, since the signal added first was the second signal in Fig. 5, we consider testing $B\hat{}_1(k) = 0$.

Fig. 8 shows that the other signal ($B\hat{}_1(k)$ in this case) is present with an observed F of 29, well over the 0.01 significance value $F_{2,6;0.01} = 10.9$. The estimated wavenumber coordinates for this signal correspond to a velocity of 3.83 km/sec and an azimuth of 184°. After $B\hat{}_1(k)$ is in the model a test was made to consider deleting $B\hat{}_1(k)$ and the result is shown as a contour plot of $F_1(k; \boldsymbol{\theta}_1, \hat{\boldsymbol{\theta}}_2)$ on the right-hand side of Fig. 8. In this case, the maximum value of 81 for the F statistic gives a much stronger detection than the initial value indicated by the wavenumber plot in Fig. 7. No change was indicated for the estimated wavenumber coordinates in this second determination. A search for a possible third signal using the model (4.23) in the special case $q_1 = 2$ yielded a plot with no significant peaks. The two plots of the F statistic in Fig. 8 then, may be regarded as the final diagnostics relating to the number and origin of signals in this particular mixture. Table 1 below compares the known velocity and azimuth computed for the pure signal case with those obtained by the

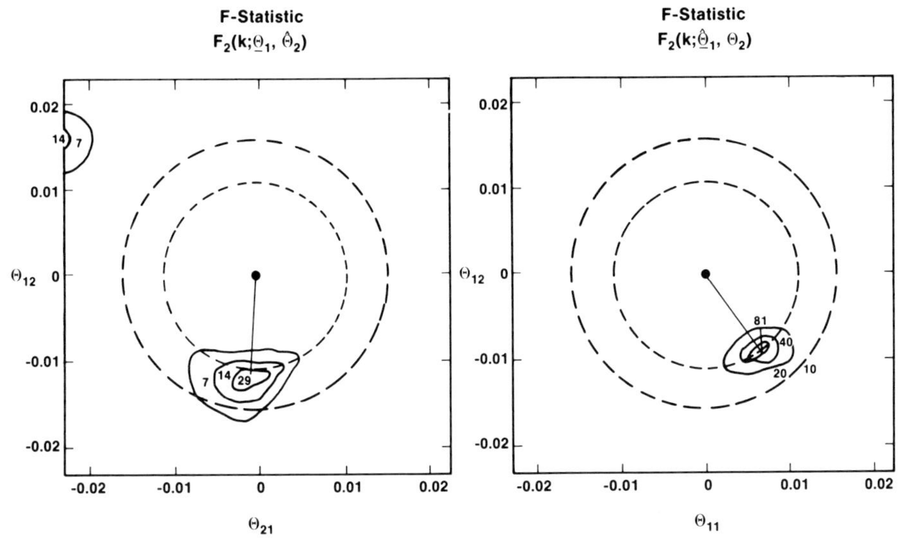

Fig. 8. Stepwise regression detectors for the mixture of two signals shown in the third column of Fig. 5.

Table 1
Estimated velocities (c_i) and azimuths (α_i) for the mixed signal case compared with the known values

	Known	Estimated
c_1 (km/sec)	3.98	3.83
α_1 (°)	194	184
c_2 (km/sec)	3.79	4.19
α_2 (°)	132	125

stepwise resolution of the mixed signal data. It is evident that the estimators from the mixed signal sample are slightly biased. In this case, one might iterate further using Gauss–Newton corrections for the complete parameter vector $(B_1^{\wedge}(k), \theta_{11}, \theta_{12}, B_2^{\wedge}(k), \theta_{21}, \theta_{22})$, but we have not investigated this procedure.

4.3. Other procedures derived from stochastic signal models

While the approach advocated here has been formulated using a deterministic signal model, there may be certain situations where a random signal model might be more appropriate. For example, suppose that in the multiple signal model

$$Y^{\wedge}(k) = \sum_{m=1}^{q} x_m^{\wedge}(\boldsymbol{\theta}_m) B_m^{\wedge}(k) + \zeta_k + o_{a.s.}(1), \tag{4.24}$$

we regard the regression coefficients as being DFT's of normal zero-mean stationary processes. If the signals are uncorrelated with $E|B_m^{\wedge}(k)|^2 = f_m(\lambda)$, then the complex covariance matrix of the vector $\boldsymbol{Y}^{\wedge}(k)$ will be

$$\Sigma(\lambda) = E(\boldsymbol{Y}^{\wedge}(k)\overline{\boldsymbol{Y}^{\wedge}(k)'}) = \sum_{m=1}^{q} f_m(\lambda)\boldsymbol{x}_m^{\wedge}(\boldsymbol{\theta}_m)\overline{\boldsymbol{x}_m^{\wedge}(\boldsymbol{\theta}_m)'} + f_e(\lambda)\boldsymbol{I}_N .$$

If one obtains a sample of $\boldsymbol{Y}^{\wedge}(k)$ over a set of K frequencies in the neighborhood of λ, a nonsingular estimator for the spectral matrix in (4.25) can be constructed as

$$\hat{\Sigma}(\lambda) = K^{-1}\sum_k \boldsymbol{Y}^{\wedge}(k)\overline{\boldsymbol{Y}^{\wedge}(k)'} \tag{4.26}$$

if $K > N$.

The ordinary beampower (4.12) in the single signal case, say (4.12) can be written in the form

$$\mathrm{SSR}_1(k; \boldsymbol{\theta}_1) = N^{-1}\overline{\boldsymbol{x}_1^{\wedge}(\boldsymbol{\theta}_1)'}\boldsymbol{Y}^{\wedge}(k)\overline{\boldsymbol{Y}^{\wedge}(k)'}\boldsymbol{x}_1^{\wedge}(\boldsymbol{\theta}_1)$$

so that smoothing over K frequencies yields

$$\begin{aligned}\mathrm{SSR}_{11} &= K^{-1}\sum_k \mathrm{SSR}_1(k; \boldsymbol{\theta}_1)\\ &= N^{-1}\overline{\boldsymbol{x}_1^{\wedge}(\theta_1)'}\hat{\Sigma}(\lambda)\boldsymbol{x}_1^{\wedge}(\boldsymbol{\theta}_1)\end{aligned} \tag{4.27}$$

which is a Hermitian form in the sample spectral matrix. Capon (1969) considered an alternate detector, defined in terms of the spectral matrix which takes the form

$$\hat{C}(\boldsymbol{\theta}_1) = [\overline{\boldsymbol{x}_1^{\wedge}(\boldsymbol{\theta}_1)'}\hat{\Sigma}^{-1}(\lambda)\boldsymbol{x}_1^{\wedge}(\boldsymbol{\theta}_1)]^{-1} \tag{4.28}$$

and was shown by Capon and Goodman (1970) to be distributed proportionally to a chi-square random variable with $2(K - N + 1)$ degrees of freedom. The proportionality constant depends on $\overline{\boldsymbol{x}_1^{\wedge}(\boldsymbol{\theta}_1)'}\Sigma^{-1}(\lambda)\boldsymbol{x}_1^{\wedge}(\boldsymbol{\theta}_1)$ where $\Sigma(\lambda)$ is the true covariance matrix (4.25), so that a rejection region with a specified significance cannot be defined unless one assumes values for $\boldsymbol{\theta}_1$, $f_1(\lambda)$ and $f_e(\lambda)$. Furthermore, the matrix $\hat{\Sigma}(\lambda)$ will be singular unless one smooths over a broad band or makes a ridge type modification suggested by Capon (1969) which amounts to replacing $\hat{\Sigma}(\lambda)$ by $\hat{\Sigma}(\lambda) + \delta^2 I_N$, where δ^2 is a small positive constant. Some examples are given in Lambert and Der (1973), Woods and Lintz (1973), and Capon et al. (1967, 1969).

Another possible estimator is suggested by the principal component representation (cf. Booker and Ong, 1972)

$$\hat{\Sigma}(\lambda) = \sum_{m=1}^{N} \lambda_m \boldsymbol{a}_m \bar{\boldsymbol{a}}'_m , \tag{4.29}$$

where $\lambda_1, \ldots, \lambda_N$ are the eigenvalues of the spectral matrix $\hat{\Sigma}(\lambda)$ and the eigenfunctions $\boldsymbol{a}_1, \ldots, \boldsymbol{a}_N$ presumed to correspond approximately to the complex vectors $\boldsymbol{x}_m^{\hat{}}(\boldsymbol{\theta}_1)$ appearing in (4.24). In general, plane wave vectors (unlike the eigenfunctions) are not necessarily orthogonal, and Der and Flinn (1975) have shown simulated examples where the principal component resolution gives the incorrect components.

5. Discussion

The approach presented in this paper has concentrated on a lagged regression model which adapts well to applications involving the resolution of propagating signals. Hence, the main thrust of the presentation involved a known design matrix, depending possibly on an unknown parameter vector $\boldsymbol{\Theta}$. The special assumptions made in this particular version of the lagged regression model were supported by noting the great number and variety of applications existing in the physical sciences. In general, the nature of the various physical phenomena under consideration was such that the waveforms of the various signals generated tended to be confined within relatively narrow frequency bands. This suggests that the frequency domain would provide the natural setting for estimation and testing problems.

If the number of signals and their propagation characteristics, as measured by the parameter $\boldsymbol{\Theta}$, are well known, one may develop approximations to BLUE estimators for their waveforms, using equations expressed in the frequency domain. This problem is an important one when the primary object of the processing is to provide an undistorted version of the signal for use in a possible identification procedure. For example, the problem of distinguishing a waveform originating as an underground nuclear test from that generated by an earthquake (cf. Shumway, 1980) requires that one make certain measurements directly from a waveform. The problem of using an array of sensors to process physical data expressed in terms of propagating plane waves then, is exactly the problem of estimating the regression functions in the model that we have considered in the second section.

The case where neither the number nor the general propagation characteristics of the signal sources are known poses two additional problems. Even if we assume that the plausible sources are known exactly, determining the number of signals present in the mixture depends on simultaneous evaluation of a number of different models. If the source characteristics as well as the number of signals are unknown, some of the approximate theory available for nonlinear least-squares methods can be combined with stepwise methods to produce simultaneously, estimators for the number of signals q and their associated wavenumber parameters $\boldsymbol{\Theta} = (\boldsymbol{\theta}'_1, \boldsymbol{\theta}'_2, \ldots, \boldsymbol{\theta}'_q)'$. This approach is given for the general case in Section 3 and investigated briefly with some examples involving the detection of isolated or multiple propagating signals in Section 4. The reader is cautioned that the approximate distribution theory for this case depends both on the sampling properties of the DFT and on regularity conditions for the derivatives with respect to $\boldsymbol{\Theta}$ of the mean regression function $X^{\cdot}(k;\boldsymbol{\theta})\boldsymbol{B}^{\cdot}(k)$. The small sample validity of the F statistic approximation under the null and alternative hypotheses should be verified by simulations like those performed by Gallant (1975).

References

Akaike, H. (1964). On the statistical estimation of a frequency response function of a system having multiple input. *Ann. Inst. Statist. Math.* **20**, 271–298.

Anderson, T. W. (1971). *The Statistical Analysis of Time Series.* Wiley, New York.

Anderson, T. W. (1972). Efficient estimation of regression coefficients in time series. In: L. LeCam, J. Neyman and E. L. Scott, eds., *Proc. Sixth Berkeley Symp. Math. Statist. Prob.*, University of California Press, Berkeley.

Barker, B. W., Der, Z. A. and Shumway, R. H. (1980). Phase velocities of regional phases observed at CPO and LASA. In: Studies of Seismic Wave Characteristics at Regional Distances. Final Report AL-80-1, Teledyne Geotech, Alexandria, VA 22314.

Blandford, R., Cohen, T. and Woods, J. (1976). An iterative approximation to the mixed-signal processes. *Geophys. J. R. Astron. Soc.* **45**, 677–687.

Booker, A. H. and Ong, C. (1972). Resolution and stability of wave number spectral estimates. *Extended Array Evaluation Program* **2**, Texas Instruments, Inc.

Brillinger, D. R. (1969). A search for a relationship between monthly sunspot numbers and certain climatic series. *Bull. Internat. Statist. Inst.* **43**, 293–306.

Brillinger, D. R. (1973). The analysis of time series collected in an experimental design. In: P. R. Krishnaiah, ed., *Multivariate Analysis I.*, 241–256. Academic Press, New York.

Brillinger, D. R. (1974). Fourier analysis of stationary processes. *Proc. IEEE* **62**, 1623–1643.

Brillinger, D. R. (1975). *Time Series: Data Analysis and Theory.* Holt, Rinehart and Winston, New York.

Brillinger, D. R. (1980). Analysis of variance and problems under time series models. In: P. R. Krishnaiah, ed., *Handbook of Statistics*, Vol. 1, 237–278. North-Holland, New York.

Capon, J., Greenfield, R. J. and Kolker, R. J. (1967). Multidimensional maximum-likelihood processing of a large aperture seismic array. *Proc. IEEE* **55**, 192–211.

Capon, J., Greenfield, R. J. and Lacoss, R. T. (1969). Long-period signal processing results for the large aperture seismic array. *Geophysics* **34**, 305–329.

Capon, J. (1969). High resolution frequency-wavenumber spectrum analysis. *Proc. IEEE* **57**, 1408–1418.

Capon, J. and Goodman, N. R. (1970). Probability distributions for estimators of the frequency wavenumber spectrum. *Proc. IEEE* (*letters*) **58**, 1785–1786.

Clay, C. S. (1966). Use of arrays for acoustic transmission in a noisy ocean. *Rev. Geophys.* **4**, 475–507.

Dean, W. C. (1966). Rayleigh wave rejection by optimum filtering of vertical arrays. Seismic Data Laboratory, SDL 166. Teledyne-Goetech, Alexandria, VA 22314.

Der, Z. A. and Flinn, E. A. (1975). The applicability of principal component analysis to the separation of multiple plane wave signals. *Bull. Seism. Soc. of Amer.* **65**, 627–635.

Doksum, K. A. and Wong, C. (1981). Statistical tests after transformation. Submitted.

Ducan, D. B. and Jones, R. H. (1966). Multiple regression with stationary errors. *J. Amer. Statist. Assoc.* **61**, 917–928.

Dunsmuir, W. and Hannan, E. J. (1976). Vector linear time series models. *J. Appl. Prob.* **10**, 130–145.

Gallant, A. R., Gerig, T. M. and Evans, J. W. (1974). Time series realizations obtained according to an experimental design. *J. Amer. Statist. Assoc.* **69**, 639–645.

Gallant, A. R. (1975). The power of the likelihood ratio test of location in nonlinear regression models. *J. Amer. Statist. Assoc.* **70**, 198–203.

Gallant, A. R. (1975). Testing a subset of the parameters of a nonlinear regression model. *J. Amer. Statist. Assoc.* **70**, 927–932.

Goodman, N. R. (1963). Statistical analysis based on a certain multivariate complex Gaussian distribution. *Ann. Math. Statist.* **36**, 152–176.

Grenander, U. (1954). On the estimation of regression coefficients in the case of an autocorrelated disturbance. *Ann. Math. Statist.* **25**, 252–272.

Hannan, E. J. (1963). Regression for time series. In: M. Rosenblatt, ed., *Proc. Symp. Time Series Anal. Brown Univ.*, 17–37 (M. Rosenblatt ed.). Wiley, New York.

Hannan, E. J. (1970). *Multiple Time Series.* Wiley, New York.

Hinich, M. J. (1981). Frequency–wavenumber array processing. *J. Acoust. Soc. Amer.* **69**, 732–737.

Hinich, M. J. and Shaman, P. (1972). Parameter estimation for an r-dimensional plane wave observed with additive Gaussian errors. *Ann. Math. Statist.* **43**, 153–169.

Jennrich, R. I. (1969). Asymptotic properties of non-linear least squares estimators. *Ann. Math. Statist.* **40**, 633–643.

Johnson, N. L. and Kotz, S. (1970). *Continuous Univariate Distributions*—2. Houghton-Mifflin, Boston.

Katzoff, M. J. and Shumway, R. H. (1978). Distributed lag regression with an almost periodic design matrix. *J. Appl. Prob.* **15**, 759–773.

Kirkendall, N. J. I. (1974). Large sample finite approximations in an infinite dimensional distributed-lag regression model. Dissertation, The George Washington University, Washington, D.C.

Ksienski, A. A. and McGhee, R. B. (1968). A decision theoretic approach to the angular resolution and parameter estimation problem for multiple targets. *IEEE Trans. on Aerospace and Electronic Systems* **AES-13**, 620–623.

Lambert, J. W. and Der, Z. A. (1973). Comparison of two-segment maximum likelihood frequency wavenumber spectra with the fast beamed frequency wavenumber spectra (FKPLOT). SDAC Report, TR-74-6, Teledyne Geotech, Alexandria, VA 22314.

MacDonald, V. H. and Schultheiss, P. M. (1969). Optimum passive bearing estimation in a spatially incoherent noise environment. *J. Acoust. Soc. Amer.* **46**, 37043.

Mack, H. and Smart, E. (1972). Frequency domain processing of digital microbarograph data. *J. Geophys. Res.* **77**, 488–490.

McAulay, R. J. and McGarty, T. P. (1974). Maximum likelihood detection of unresolved targets and multipath. *IEEE Trans. on Aerospace and Electronic Systems* **AES-10**, 821–829.

McGarty, T. P. (1974). The effect of interfering signals on the performance of angle of arrival estimates. *IEEE Trans. on Aerospace and Electronic Systems* **AES-10**, 70–77.

Mudholkar, G. S., Chaubey, Y. B. and Lin, C. (1976). Approximations for the doubly noncentral-F distribution. *Comm. Statist. Theor. Meth.* **AS(1)**, 49–63.

Otnes, R. K. and Enochson, L. D. (1978). *Applied Time Series Analysis.* Wiley, New York.

Parzen, E. (1967). On empirical multiple time series analysis. In: L. LeCam, ed., *Proceedings of the Fifth Berkeley Symposium*, Vol. 1, 305–340. University of California Press, Berkeley.

Pisarenko, V. F. (1972). On the estimation of spectra by means of non-linear functions of the covariance matrix. *Geophys. J. R. Astron. Soc.* **28**, 511–531.

Rosenblatt, M. (1956). Some regression problems in time series analysis. In: J. Neyman, ed., *Proc. Third Berkeley Symposium on Math. Statist. and Probability*, Vol. 1, 165–186. University of California Press, Berkeley.

Schweppe, F. C. (1968). Sensor array data processing for multiple signal sources. *IEEE Trans. on Information Theory* **IT-4**, 294–305.

Shumway, R. H. and Dean, W. C. (1968). Best linear unbiased estimation for multivariate stationary processes. *Technometrics* **10**, 523–534.

Shumway, R. H. and Husted, H. (1970). Frequency dependent estimation and detection for seismic arrays. Technical Report No. 242, Seismic Data Laboratory, Teledyne Geotech, Alexandria, VA 22314.

Shumway, R. H. (1970). Applied regression and analysis of variance for stationary time series. *J. Amer. Statist. Assoc.* **65**, 1527–1546.

Shumway, R. H. (1971). On detecting a signal in *N* stationarily correlated noise series. *Technometrics* **13**, 499–519.

Shumway, R. H. (1972). Some applications of a mixed signal processor. Seismic Data Lab Report No. 280, Teledyne Geotech, Alexandria, VA 22314.

Shumway, R. H. (1980). Discriminant analysis for time series (Chapter 1). In: P. R. Krishniah, ed., *Handbook of Statistics*, Vol. II, *Classification, Pattern Recognition and Reduction of Dimensionality*. North-Holland, Amsterdam.

Smart, E. and Flinn, E. (1971). Fast frequency–wavenumber analysis and Fisher signal detection in real-time infrasonic array data processing. *Geophys. J.* **26**, 279–284.

Wahba, Grace (1968). On the distribution of some statistics useful in the analysis of jointly stationary time series. *Annals of Math. Statis.* **39**, 1849–1862.

Wahba, Grace (1969). Estimation of the coefficients in a multidimensional distributed lag model. *Econometrica* **37**, 398–407.

Wirth, M. H., Blandford, R. R. and Shumway, R. H. (1976). Automatic seismic array and network detection. *Bull. Seismolog. Soc. Am.* **66**, 1375–1380.

Woods, J. W. and Lintz, P. R. (1973). Plane waves at small arrays. *Geophysics* **53**, 1023–1041.

Wu, J. S. (1982). Asymptotic properties of nonlinear least squares estimators in a replicated time series model. Ph.D. dissertation, The George Washington University, Washington, D.C.

D. R. Brillinger and P. R. Krishnaiah, eds. *Handbook of Statistics, Vol. 3*

19

Computer Programming of Spectrum Estimation*

Tony Thrall

1. Introduction

In this chapter we present an overview of computational approaches to nonparametric univariate spectrum estimation. Our goal is to familiarize the statistical software user with the terminology and methods embodied in currently available programs, and to mention some recent developments that should soon be available.

The basis of our discussion is a real-valued time series $x(t)$, t an integer, which we observe for $t = 0, 1, 2, \ldots, T-1$. We suppose that $x(t)$ has finite first and second moments

$$E|x(t)|, \qquad E|x(t)|^2 < \infty \quad \text{for all } t \tag{1.1}$$

and that it satisfies certain stationarity and mixing assumptions of the form

$$\begin{aligned} c_x &= Ex(t) \\ c_{xx}(u) &= \operatorname{cov}(x(t+u), x(t)), \quad \text{independent of } t \end{aligned} \tag{1.2}$$

and

$$\sum_{u=-\infty}^{\infty} |c_{xx}(u)| < \infty . \tag{1.3}$$

The stationarity assumption (1.2) implies that the mean value and correlation structure of $x(t)$ do not change over time. The mixing assumption (1.3) implies that any pair of observations far removed in time are essentially uncorrelated. Assumptions (1.1)–(1.3) are sometimes extended to higher than second moments, and (1.3) is sometimes strengthened to be of the form

$$\sum_{u=-\infty}^{\infty} |u|^p |c_{xx}(u)| < \infty \tag{1.4}$$

for some positive integer p.

*This work was supported by the VA Brentwood Medical Center, the Alcohol Research Center, UCLA, and by NSF grant no. CDP80-30837.

We call $c_{xx}(u)$, regarded as a function of integer lag u, the autocovariance function of $x(t)$. Assumption (1.3) enables us to define the Fourier transform

$$s_{xx}(\alpha) = \sum_{u=-\infty}^{\infty} c_{xx}(u) \exp(-2\pi i \alpha u) \tag{1.5}$$

called the spectrum (spectral density, power spectrum) of $x(t)$ at frequency α. From (1.2) and (1.5), it may be shown that

$$s_{xx}(\alpha) \geqslant 0 \,, \tag{1.6}$$

$$s_{xx}(-\alpha) = s_{xx}(\alpha) \,, \tag{1.7}$$

$$s_{xx}(\alpha + n) = s_{xx}(\alpha) \quad \text{for integer } n \,. \tag{1.8}$$

Also from (1.5), we have the inverse relation

$$c_{xx}(u) = \int_{-1/2}^{1/2} s_{xx}(\alpha) \exp(2\pi i \alpha u) \, d\alpha \,. \tag{1.9}$$

Setting $u = 0$ gives

$$\operatorname{var} x(t) = c_{xx}(0) = \int_{-1/2}^{1/2} s_{xx}(\alpha) \, d\alpha \,. \tag{1.10}$$

From (1.10), we see that the variance of $x(t)$ is distributed over a continuous range of frequencies, and that $s_{xx}(\alpha)$ is the density of this distribution. Using the Cramer representation (see, for example, Brillinger, 1975, Section 4.6), we may further interpret $s_{xx}(\alpha)$ as the relative contribution of sinusoidal variation at frequency α to the overall variation exhibited by the time series $x(t)$.

In this paper, we address the problem of computing an estimate of $s_{xx}(\alpha)$ from a single, partial realization $x(0), x(1), \ldots, x(T-1)$ of the time series $x(t)$. Historically, the approach to this problem has been to estimate the autocovariance function

$$\hat{c}_{xx}(u) = \frac{1}{T} \sum_{0 \leq t,\, t+u < T} (x(t+u) - \bar{x})(x(t) - \bar{x}) \tag{1.11}$$

with

$$\bar{x} = \frac{1}{T} \sum_{t=0}^{T-1} x(t) \tag{1.12}$$

for a number of lags u and then to calculate the estimated spectrum $\hat{s}_{xx}(\alpha)$ as a weighted transform of $\hat{c}_{xx}(u)$:

$$\hat{s}_{xx}(\alpha) = \sum_{u=-U}^{U} w_u \hat{c}_{xx}(u) \exp(-2\pi i \alpha u) \,. \tag{1.13}$$

With the popularization of the fast Fourier transform (FFT) by Cooley and Tukey (1965), another approach to spectrum estimation has been widely adopted. This entails Fourier transforming the data

$$J_x(\omega) = \sum_{t=0}^{T-1} x(t) \exp(-2\pi i \omega t) \tag{1.14}$$

at a discrete set of frequencies, e.g.

$$\omega_j = j/T, \quad j = 0, 1, 2, \ldots, T-1, \tag{1.15}$$

forming the periodogram

$$I_{xx}(\omega) = \frac{1}{T} |J_x(\omega)|^2 \tag{1.16}$$

as a crude estimate of the spectrum, and then smoothing periodograms to obtain the final estimate, i.e.

$$\hat{s}_{xx}(\alpha) = \text{weighted average } (I_{xx}(\omega_j)), \quad \omega_j \text{ near } \alpha\,. \tag{1.17}$$

Both covariance-based estimates (1.13) and periodogram-based estimates (1.17) are used in present-day statistical packages (e.g. BMD, BMDP, SAS, SPSS) and subroutine libraries (e.g. IMSL, STATLIB). Both approaches are highly flexible due to a number of refinements which we shall discuss.

General references for this chapter are Bloomfield (1976), a well-written introduction to frequency-domain analysis containing pertinent Fortran subroutine listings, Brillinger (1975), containing precise formulations and proofs for a wealth of useful results, and Thomson (1977), treating in depth the more sophisticated techniques mentioned here.

2. Two methods of estimation

We now turn to a more detailed discussion of the sequence of computations for covariance-based and periodogram-based spectrum estimates.

The weights w_u used for the covariance-based spectrum estimate (1.13) are conveniently generated by an autocovariance weighting function, or lag window, $w(x)$ with Fourier transform

$$W(\alpha) = \int_{-\infty}^{\infty} w(x) \exp(-2\pi i \alpha x)\, dx\,. \tag{2.1}$$

$W(\alpha)$ is often called the spectral window. From (2.1) follows the reciprocal relation

$$w(x) = \int_{-\infty}^{\infty} W(\alpha) \exp(2\pi i \alpha x)\, d\alpha . \tag{2.2}$$

We usually require $w(x)$, $W(\alpha)$ to be real-valued, even functions satisfying

$$w(0) = \int_{-\infty}^{\infty} W(\alpha)\, d\alpha = 1 , \tag{2.3}$$

$$\int_{-\infty}^{\infty} w(x)^2\, dx = \int_{-\infty}^{\infty} W(\alpha)^2\, d\alpha < \infty , \tag{2.4}$$

and

$$\int_{-\infty}^{\infty} |w(x)| dx, \qquad \int_{-\infty}^{\infty} |x|\, |w(x)|\, dx < \infty , \tag{2.5}$$

the last requirement implying the existence, boundedness, and continuity of $W(\alpha)$, $W'(\alpha)$. The lag windows commonly used for covariance-based estimates also vanish outside some interval, i.e.

$$w(x) = 0 \quad \text{for } |x| > x_0 , \tag{2.6}$$

typical examples being

$$w_P(x) = \begin{cases} 1 - 6x^2(1 - |x|), & |x| \le \frac{1}{2} , \\ 2(1 - |x|)^2, & \frac{1}{2} \le |x| \le 1 , \\ 0, & |x| \ge 1 , \end{cases} \tag{2.7}$$

and

$$w_H(x) = \begin{cases} (1 + \cos \pi x)/2, & |x| \le 1 , \\ 0, & |x| > 1 . \end{cases} \tag{2.8}$$

The cubic spline function (2.7) was proposed by de la Valle-Poussin and by E. Parzen, while the cosine-shaped function (2.8) is associated with R. W. Hamming and J. W. Tukey. Both (2.7) and (2.8) have nonnegative Fourier transforms $W(\cdot)$. A concise survey of window functions is given by Geçkinli and Yavuz (1978).

A general class of covariance-based spectrum estimates may be written as

$$\hat{s}_{xx}(\alpha) = \sum_{u=-T+1}^{T-1} w(\beta_T u) \hat{c}_{xx}(u) \exp(-2\pi i \alpha u) \tag{2.9}$$

with expected value

$$\begin{aligned} E\hat{s}_{xx}(\alpha) &= \int_{-\infty}^{\infty} \frac{1}{\beta_T} W\left(\frac{\alpha - \omega}{\beta_T}\right) s_{xx}(\omega)\, d\omega + O\left(\frac{1}{\beta_T T}\right) \\ &= \int_{-\infty}^{\infty} W(\omega) s_{xx}(\alpha - \beta_T \omega)\, d\omega + O\left(\frac{1}{\beta_T T}\right) . \end{aligned} \tag{2.10}$$

See Sections 5.9 and 7.7 of Brillinger (1975) for further details.

The integrand in (2.10) becomes concentrated about frequency α as bandwidth parameter β_T approaches zero, while the error term is reduced as $\beta_T T$ becomes large. Hence the estimate (2.9) is asymptotically unbiased if $\beta_T \to 0$ and $\beta_T T \to \infty$ as $T \to \infty$, e.g. if $\beta_T = O(T^{-\gamma})$ with $0 < \gamma < 1$. If, as in (1.13), lags u with $|u| > U$ are to receive zero weight and if (2.6) holds, we may set

$$\beta_T = x_0/(U + \tfrac{1}{2}) . \tag{2.11}$$

Selection of bandwidth parameter β_T is further discussed in Section 6.

The sequence of computations for the covariance-based spectrum estimate (2.9) is thus:

(a) form $\hat{c}_{xx}(u)$,

(b) multiply by $w(\beta_T u)$,

(c) Fourier transform.

In the next section we describe how steps (a) and (c) may be efficiently carried out using a fast Fourier transform (FFT) algorithm.

Employing the above notation and taking particular note of (2.10), a general class of periodogram-based spectrum estimates may be written as

$$\hat{s}_{xx}(\alpha) = \frac{1}{T} \sum_{j \neq 0 \bmod T} \frac{1}{\beta_T} W\left(\frac{\alpha - \omega_j}{\beta_T}\right) I_{xx}(\omega_j) . \tag{2.12}$$

The expected value of (2.12) is again given by (2.10). By excluding integer multiples, $j = nT$, from the sum in (2.12), the weight given to $I_{xx}(n)$ is zero. This is done because $I_{xx}(n)$ $(= I_{xx}(0))$ is not even a crude estimate of $s_{xx}(n)$ $(= s_{xx}(0))$, but rather reflects the sample average $\bar{x}$, i.e.

$$I_{xx}(n) = \frac{1}{T} |J_x(n)|^2 = T\bar{x}^2 . \tag{2.13}$$

In principle, the summation in (2.12) may extend over all noninteger Fourier frequencies $\omega_j = j/T$. In practice, of course, the summation is finite, for using the periodicity of the periodogram

$$I_{xx}(\omega + n) = I_{xx}(\omega), \quad n \text{ an integer} , \tag{2.14}$$

we may weight the periodogram $I_{xx}(\omega_k)$, $0 < \omega_k < 1$, by

$$W_k(\alpha) = \sum_{n=-\infty}^{\infty} W\left(\frac{\alpha - \omega_k + n}{\beta_T}\right) \Big/ \sum_{0<\omega_l<1} \sum_{n=-\infty}^{\infty} W\left(\frac{\alpha - \omega_l + n}{\beta_T}\right) . \tag{2.15}$$

We have thus 'folded' the weights $(1/\beta_T) W((\alpha - \omega_j)/\beta_T)$, $-\infty < \omega_j < \infty$, ω_j noninteger, into the frequency range $0 < \omega_k < 1$ and normalized these weights so that they sum to unity. By taking advantage of symmetry about $\frac{1}{2}$, i.e.

$$J_x(1 - \omega) = \overline{J_x(\omega)} ,$$

implying,

$$I_{xx}(1-\omega)=I_{xx}(\omega)\,, \tag{2.16}$$

we may further fold weights $W_k(\alpha)$ into the interval $0<\omega_k\leq\frac{1}{2}$, viz.

$$W_k^*(\alpha)=\begin{cases}W_k(\alpha)+W_{T-k}(\alpha), & 0<\omega_k<\frac{1}{2}\,,\\ W_k(\alpha), & \omega_k=\frac{1}{2}\,,\end{cases} \tag{2.17}$$

so that

$$\sum_{0<\omega_k\leq 1/2} W_k^*(\alpha)=\sum_{0<\omega_k<1} W_k(\alpha)=1\,. \tag{2.18}$$

Under these conventions, the spectrum estimate (2.12) becomes

$$\begin{aligned}\hat{s}_{xx}(\alpha)&=\sum_{0<\omega_k\leq 1/2} W_k^*(\alpha)I_{xx}(\omega_k)\\ &=\sum_{0<\omega_k<1} W_k(\alpha)I_{xx}(\omega_k)\,.\end{aligned} \tag{2.19}$$

The periodogram-weighting procedure we have outlined works for any spectral window $W(\alpha)$ (assuming the divisor in (2.15) is nonzero), but we usually simplify the procedure by requiring $W(\alpha)$ to vanish outside some interval, i.e.

$$W(\alpha)=0 \quad \text{for } |\alpha|>\alpha_0\,. \tag{2.20}$$

(In general, (2.6) and (2.20) do not hold simultaneously.) Then for $0<\alpha<1$ and for sufficiently small β_T, $W_k(\alpha)$ is proportional to $W((\alpha-\omega_k)/\beta_T)$ and therefore at most $(1+2\beta_T T\alpha_0)$ out of T periodogram values receive nonzero weight. Common choices for $W(\alpha)$ are the rectangular window

$$W_R(\alpha)=\begin{cases}1, & |\alpha|\leq\frac{1}{2}\,,\\ 0, & |\alpha|>\frac{1}{2}\,,\end{cases} \tag{2.21}$$

and a cosine-shaped window

$$W_C(\alpha)=\begin{cases}\frac{2}{3}\left(1+\cos\dfrac{4\pi\alpha}{3}\right), & |\alpha|\leq\frac{3}{4}\,,\\ 0\,, & |\alpha|>\frac{3}{4}\,,\end{cases} \tag{2.22}$$

(the coefficients of the latter having been chosen so that $W_C(\alpha)$ and $W_C(\alpha)^2$ integrate to unity as discussed in Section 5).

The spectrum estimate (2.9) is the discrete Fourier transform of a weighted sample autocovariance function, while (2.19) is a weighted average of periodogram ordinates. In both cases the weights are generated by sampling a rescaled lag or spectral window function, $w(x)$ or $W(\alpha)$, respectively. This has been done as a matter of convenience, allowing us to obtain asymptotic expressions of the form (2.10). In many spectrum estimation computer programs, the user is offered a limited selection of weighting functions (i.e. windows), but even when the user desires or is required to construct his own set of weights, it is usually convenient to think in terms of a window, or weight-shaping function.

A point to be noted in this framework is that the choice of bandwidth parameter β_T may not in reality by taken as arbitrarily large or small. If (2.6) and (2.9) hold, taking $\beta_T > x_0$ produces a constant spectrum estimate, i.e. $\hat{s}_{xx}(\alpha) = \hat{c}_{xx}(0)$ for all α. If (2.19) and (2.20) hold, taking $\beta_T < (2T\alpha_0)^{-1}$ produces estimates that are either zero or equal to the nearest periodogram value.

Another practical point is that we usually estimate the spectrum at a grid of frequencies, i.e. we compute and plot

$$\hat{s}_{xx}(\alpha_j), \quad \alpha_j = j/n_\alpha,\ j = 0, 1, 2, \ldots, [n_\alpha/2]\,. \tag{2.23}$$

Computational efficiency is achieved for covariance-based estimates by employing the fast Fourier transform (see Section 3). For periodogram-based estimates, computing time will be reduced if the periodogram weights $W_k(\alpha_j)$ need not be recomputed for each α_j. This is the case when a simple average of periodogram values is formed ($W(\alpha) = W_R(\alpha)$) or when the band centers α_j are restricted to a subset of the Fourier frequencies ω_k, and $\hat{s}_{xx}(\alpha_j)$ is a moving average of $I_{xx}(\omega_k)$ with fixed coefficients. Bloomfield (1976) and others have suggested this last approach.

The choices of method, window, and bandwidth parameter are no longer hotly contested as they were during the period of intense development of spectrum estimation techniques following the Cooley–Tukey (1965) paper. These choices, nevertheless, have no simple resolution and must be left to the user's discretion.

On computational grounds, covariance-based spectrum estimates require up to three Fourier transformations, whereas periodogram-based estimates require only one. The covariance-based estimate may be more efficient, nonetheless, when the number of nonzero weights $W_k(\alpha)$ for the corresponding periodogram-based estimate is large. See Cooley, Lewis and Welch (1977, p. 406).

Qualitatively, the two methods produce similar results. To better compare the methods, we may rewrite the covariance-based estimate as (1) a cosine polynomial and (2) a continuously weighted periodogram average. The first version is obtained by supposing that lags u, $|u| > U$, receive zero weight, so that

$$\begin{aligned}\hat{s}_{xx}(\alpha) &= \sum_{u=-U}^{U} W(\beta_T u)\hat{c}_{xx}(u)\exp(-2\pi i\alpha u)\\ &= \hat{c}_{xx}(0) + 2\sum_{u=1}^{U} W(\beta_T u)\hat{c}_{xx}(u)\cos(2\pi\alpha u)\\ &= \hat{c}_{xx}(0) + 2\sum_{u=1}^{U} W(\beta_T u)\hat{c}_{xx}(u)P_u(\cos(2\pi\alpha))\,,\end{aligned} \tag{2.24}$$

where P_u is the Chebyshev polynomial of degree u. Thus (2.24) is a polynomial in $\cos 2\pi\alpha$ of degree U. By (2.11), U and β_T are approximately inversely proportional, so that a smaller bandwidth parameter β_T implies a higher degree polynomial in $\cos 2\pi\alpha$. Conversely, covariance-based smoothing via selection of a larger bandwidth parameter β_T amounts to the selection of a lower degree polynomial.

Expression (2.24) may be used to highlight the difference between the two methods, for the density estimate (2.24) is a polynomial approximation to the underlying spectrum, whereas periodogram-based estimates bear a closer resemblance to histograms.

The next expression for covariance-based estimates emphasizes the similarity between the two methods. Let

$$\tilde{x}(t) = x(t) - \bar{x} \tag{2.25}$$

so that

$$\begin{aligned} I_{\tilde{x}\tilde{x}}(\omega) &= \frac{1}{T} |J_{\tilde{x}}(\omega)|^2 \\ &= \frac{1}{T} \sum_{s=0}^{T-1} \sum_{t=0}^{T-1} \tilde{x}(s)\tilde{x}(t) \exp(-2\pi \mathrm{i}(s-t)\omega) \\ &= \sum_{u=-T+1}^{T-1} \hat{c}_{xx}(u) \exp(-2\pi \mathrm{i}\omega u)\,. \end{aligned} \tag{2.26}$$

We may thus obtain

$$\begin{aligned} \hat{s}_{xx}(\alpha) &= \sum_{|u|<T} W(\beta_T u)\hat{c}_{xx}(u) \exp(-2\pi \mathrm{i}\alpha u) \\ &= \sum_{|u|<T} \left(\int_{-\infty}^{\infty} \frac{1}{\beta_T} W\left(\frac{\alpha-\omega}{\beta_T}\right) \exp(2\pi \mathrm{i}(\alpha-\omega)u)\, \mathrm{d}\omega \right) \\ &\qquad \times \hat{c}_{xx}(u) \exp(-2\pi \mathrm{i}\alpha u) \\ &= \int_{-\infty}^{\infty} \frac{1}{\beta_T} W\left(\frac{\alpha-\omega}{\beta_T}\right) \left(\sum_{|u|<T} \hat{c}_{xx}(u) \exp(-2\pi \mathrm{i}\omega u) \right) \mathrm{d}\omega \\ &= \int_{-\infty}^{\infty} \frac{1}{\beta_T} W\left(\frac{\alpha-\omega}{\beta_T}\right) I_{\tilde{x}\tilde{x}}(\omega)\, \mathrm{d}\omega\,. \end{aligned} \tag{2.27}$$

The covariance-based estimate (2.27) therefore closely resembles the periodogram-based estimate (2.12), especially in light of the fact that

$$I_{\tilde{x}\tilde{x}}(\omega_j) = I_{xx}(\omega_j) \quad \text{for noninteger } \omega_j = j/T\,. \tag{2.28}$$

See Theorem 5.9.1 in Brillinger (1975) for a bound on the difference between the two estimates (2.12) and (2.27).

Summarizing the issue of which method to use, we may say that for moderate values of T, e.g. less than 500, the covariance-based estimate, viewed as a cosine polynomial (2.24), tends to produce a smoother result than the periodogram-based estimate, particularly when the latter employs the rectangular window $W_R(\alpha)$. On the other hand, periodogram-based estimates are often computationally simpler and faster than their covariance-based counterparts.

As for the selection of lag or spectral window and the selection of bandwidth parameter β_T, both choices effect the smoothness of the estimate, i.e. the apparent continuity of $\hat{s}_{xx}(\alpha_j)$. The bandwidth parameter β_T, however, is probably more deserving of our attention for it usually has a more pronounced effect upon the estimated spectrum. Also, the question of exactly what value of β_T to choose may be (and typically should be) circumvented by estimating the spectrum at several bandwidths. From (2.10), we see that what is being estimated is not precisely the spectrum but a smoothed version of the spectrum. Thus the estimand changes with our choice of β_T. The selection of β_T must be a compromise between a small value that better portrays the spectrum at one frequency and a large value which increases the number of periodograms averaged and thus reduces the variance of the estimate. See Section 6 for further remarks.

3. Fast Fourier transform (FFT) algorithms

The spectrum estimate $\hat{s}_{xx}(\alpha)$ requires the Fourier transform of either the estimated autocovariances $\hat{c}_{xx}(u)$ or the data sequence $x(t)$. A direct application of the formula for $J_x(\omega)$, the Fourier transform of $x(t)$, entails T operations (complex multiplications and additions) per frequency. In 1965, Cooley and Tukey published an algorithm for calculating the Fourier transform at T equispaced frequencies involving significantly fewer operations than the number T^2 used in directly applying the formula. It turns out that similar algorithms had been published previously by other authors, but under the impact of the Cooley–Tukey paper, what had once been an impractically long calculation became a widely used tool and a model of computing efficiency. The algorithms for the rapid calculation of the Fourier transform at a discrete set of frequencies are collectively known as the 'fast Fourier transform' or 'FFT'. See Cooley, Lewis and Welch (1977) for a detailed discussion of these algorithms and their history.

A brief description of one version of the algorithm is as follows. Suppose T can be factored into two integers $T = T_1T_2$ and that frequencies ω of interest are of the form $\omega = (j_1T_2 + j_2)/T$, with j_1, j_2 integers. Then

$$
\begin{aligned}
J_x(\omega) &= \sum_{t=0}^{T-1} \exp(-2\pi \mathrm{i}\omega t)x(t) \\
&= \sum_{t_1=0}^{T_1-1} \sum_{t_2=0}^{T_2-1} \exp\left\{-\frac{2\pi \mathrm{i}}{T}(j_1 T_2 + j_2)(t_1 + t_2 T_1)\right\} x(t_1 + t_2 T_1) \\
&= \sum_{t_1=0}^{T_1-1} \exp\left\{-\frac{2\pi \mathrm{i}}{T_1} j_1 t_i\right\} \exp\left\{-\frac{2\pi \mathrm{i}}{T} j_2 t_1\right\} \\
&\quad \times \sum_{t_2=0}^{T_2-1} \exp\left\{-\frac{2\pi \mathrm{i}}{T_2} j_2 t_2\right\} x(t_1 + t_2 T_1) .
\end{aligned} \tag{3.1}
$$

Now consider the following steps:

(a) arrange $x(t_1 + t_2 T_1)$ in a rectangular array, with row-index t_1 and column-index t_2;

(b) Fourier-transform each row at frequencies j_2/T_2, $j_2 = 0, 1, 2, \ldots, T_2 - 1$ so that j_2 becomes the new column index;

(c) for each element (t_1, j_2) of the array multiply by $\exp\{-(2\pi \mathrm{i}/T) j_2 t_1\}$;

(d) Fourier-transform each column of the array at frequencies j_1/T_1, $j_1 = 0, 1, \ldots, T_1 - 1$ so that j_1 becomes the new row index.

The number of required operations is $T_1 T_2^2$ in step (b), T in step (c), and $T_2 T_1^2$ in step (d) for a total of $(1 + T_1 + T_2)T$ operations (we should also mention that a fair amount of computational effort may be required in step (a), where the original sequence $x(t)$ is permuted). If T_2 may be factored as $T_2 = T_2' T_3$, then the $T_1 T_2^2$ operations in step (b) may be replaced by $T_1(1 + T_2' + T_3)T_2' T_3$. In general, the algorithm may be refined and extended so that for $T = T_1, \ldots, T_k$ the required number of operations is of the order $(T_1 + \cdots + T_k)T$. As an example, if $T = 1024 = 2^{10}$, the algorithm requires approximately twenty thousand operations compared to one million operations, approximately, required by direct application of the formula.

Cooley, Lewis and Welch (1977) provide further modifications of the basic algorithm that allow one to take advantage of the sequence $x(t)$ being real rather than complex, that produce an inverse Fourier transform, or that simultaneously calculate the Fourier transform of two real sequences $x(t)$ and $y(t)$. We will concentrate on using the FFT to calculate estimated autocovariances $\hat{c}_{xx}(u)$.

Let U be the largest integer lag at which $\hat{c}_{xx}(u)$ is to be calculated, and take an integer $S \geqslant T + U$. Note that U need not exceed $T - 1$ for our purpose, since $\hat{c}_{xx}(u) = 0$ for $|u| > T - 1$. Now form the sequence

$$
\tilde{x}(t) = \begin{cases} x(t) - \bar{x}, & 0 \leqslant t < t, \\ 0 & T \leqslant t < S . \end{cases} \tag{3.2}
$$

Using the fast Fourier transform to compute

$$
J_{\tilde{x}}(\omega_k) = \sum_{t=0}^{S-1} \tilde{x}(t) \exp(-2\pi \mathrm{i}\omega_k t) = \sum_{t=0}^{T-1} \tilde{x}(t) \exp(-2\pi \mathrm{i}\omega_k t) \tag{3.3}
$$

for $\omega_k = k/S$, $k = 0, 1, 2, \ldots, S-1$, we form the periodogram

$$I_{\tilde{x}\tilde{x}}(\omega_k) = \frac{1}{T}|J_{\tilde{x}}(\omega_k)|^2$$

$$= \sum_{|v|<T} \hat{c}_{xx}(v)\exp(-2\pi i\omega_k v) \tag{3.4}$$

as in (2.26). Using the FFT a second time, we may compute

$$\frac{1}{S}\sum_{k=0}^{S-1} I_{\tilde{x}\tilde{x}}(\omega_k)\exp(2\pi i\omega_k u) = \sum_{|v|<T} \hat{c}_{xx}(v)\frac{1}{S}\sum_{k=0}^{S-1}\exp(2\pi i\omega_k(u-v)). \tag{3.5}$$

The second factor in (3.5) is unity when $v-u$ is a multiple of S and is zero otherwise. For $|u|<T$, (3.5) thus reduces to

$$\begin{cases}\hat{c}_{xx}(u)+\hat{c}_{xx}(u+S), & -T\leqslant u\leqslant 0,\\ \hat{c}_{xx}(u)+\hat{c}_{xx}(u-S), & 0\leqslant u<T,\end{cases} \tag{3.6}$$

and we therefore obtain

$$\hat{c}_{xx}(u) = \frac{1}{S}\sum_{k=0}^{S-1} I_{\tilde{x}\tilde{x}}(\omega_k)\exp(2\pi i\omega_k u), \quad |u|<U, \tag{3.7}$$

as desired. In order to conveniently compute the spectrum estimates

$$\hat{s}_{xx}(\alpha_j) = \sum_{|u|<U} W(\beta_T u)\hat{c}_{xx}(u)\exp(-2\pi i\alpha_j u),$$

$$\alpha_i = j/n_\alpha, \quad j = 0, 1, 2, \ldots, [n_\alpha/2], \tag{3.8}$$

we may take $n_\alpha \geqslant U+1$ and use the FFT for a third time. When $n_\alpha > U+1$, this can be done by padding the calculated covariances with zeros:

$$\hat{c}_{xx}(u) = \begin{cases}\hat{c}_{xx}(u), & |u|\leqslant U,\\ 0, & U<|u|<n_\alpha.\end{cases} \tag{3.9}$$

Cooley, Lewis and Welch (1977) recommend special variants of the FFT for cases, such as (3.8), where the sequence to be Fourier transformed is real and even.

4. Tapering and padding the data

The periodogram can be regarded as a crude estimate of the spectrum. Its primary defect as an estimate is its large variance, but it also suffers from bias.

If

$$J_x(\omega) = \sum_{t=0}^{T-1} x(t) \exp(-2\pi i \omega t)$$

and

$$I_{xx}(\omega) = \frac{1}{T} |J_x(\omega)|^2 ,$$

then

$$EI_{xx}(\omega) = \frac{1}{T} \int_{-1/2}^{1/2} \left[\frac{\sin \pi T(\omega - \alpha)}{\sin \pi(\omega - \alpha)} \right]^2 s_{xx}(\alpha)\, d\alpha + \frac{1}{T} \left[\frac{\sin \pi T\omega}{\sin \pi\omega} \right]^2 c_x^2 , \tag{4.1}$$

where $c_x = Ex(t)$.

When ω is a Fourier frequency, i.e. of the form $\omega = k/T$, the second term will vanish. Nevertheless, bias will remain in the estimate because the function multiplying the spectrum $s_{xx}(\alpha)$ in the first term of (4.1) has not only a major lobe at frequency ω but also has sidelobes of significant magnitude. Thus values of the spectrum $s_{xx}(\alpha)$ at frequencies α removed from ω can affect the expected value. This phenomenon is sometimes called leakage across frequencies.

One way of reducing the effect of the sidelobes is to multiply the data sequence $x(t)$ by a function, or data taper (data window), $h^T(t)$ which is typically taken to be symmetric about $(T-1)/2$ where it assumes a maximum. Tapering thus consists of the operation

$$\hat{x}(t) = h^T(t)x(t) , \tag{4.2}$$

where

$$h^T(t) = h((t + \tfrac{1}{2})/T) \tag{4.3}$$

for some tapering function $h(x)$. As a matter of convenience, we often define $h(x) = 0$ for $x < 0$ or $x > 1$.

We define the Fourier transform of the tapered data to be

$$J_{\hat{x}}(\omega) = \sum_{t=0}^{T-1} \hat{x}(t) \exp(-2\pi i \omega t) . \tag{4.4}$$

The periodogram of the tapered data (sometimes called the modified periodogram) is scaled to have the same order of magnitude as the ordinary periodogram:

$$I_{\hat{x}\hat{x}}(\omega) = \left(\sum_{t=0}^{T-1} h^T(t)^2 \right)^{-1} |J_{\hat{x}}(\omega)|^2 \tag{4.5}$$

with expected value

$$EI_{\tilde{x}\tilde{x}}(\omega) = \left(\int_{-1/2}^{1/2} |H^T(\alpha)|^2 \, d\alpha\right)^{-1} \left\{\int_{-1/2}^{1/2} |H^T(\omega-\alpha)|^2 s_{xx}(\alpha)\, d\alpha + |H^T(\omega)|^2 c_x^2\right\}, \tag{4.6}$$

where

$$H^T(\alpha) = \sum_{t=0}^{T-1} h^T(t) \exp(-2\pi i \alpha t)\,. \tag{4.7}$$

In practice, we typically center the data, i.e. subtract some estimate $\hat{c}_x$ of the mean c_x, prior to tapering. Using the tapered mean

$$\hat{c}_x = \sum_{t=0}^{T-1} h^T(t) x(t) / \sum_{t=0}^{T-1} h^T(t) \tag{4.8}$$

and setting

$$\tilde{x}(t) = h^T(t)(x(t) - \hat{c}_x)\,, \tag{4.9}$$

we have

$$EI_{\tilde{x}\tilde{x}}(\omega) = \left(\int_{-1/2}^{1/2} |H^T(\alpha)|^2 \, d\alpha\right)^{-1} \int_{-1/2}^{1/2} |H^T(\omega-\alpha) - H^T(\omega) H^T(-\alpha)/H^T(0)|^2 s_{xx}(\alpha)\, d\alpha\,. \tag{4.10}$$

It may be shown that as T becomes large, (4.10) approaches $s_{xx}(\omega)$ under broad regularity conditions for $h(x)$ (Brillinger, 1975, Section 5.3). Hence the modified periodogram $I_{\tilde{x}\tilde{x}}(\omega)$ may be used to form periodogram-based spectrum estimates as described in Section 2.

We may similarly define the estimated autocovariance function for tapered data as

$$\hat{c}_{xx}(u) = \sum_{0 \le t, t+u < T} \tilde{x}(t+u)\tilde{x}(t) / \sum_{t=0}^{T-1} h^T(t)^2\,, \tag{4.11}$$

and proceed with the covariance-based spectrum estimation of Section 2. Note that we may still employ the FFT, described in Section 3, to rapidly compute (4.11).

The purpose of tapering is to produce functions $H^T(\alpha)$ with smaller sidelobes than that of $\sin \pi T\alpha / \sin \pi\alpha$ and to thereby reduce the periodogram bias due to 'leakage'. The price paid for reducing the magnitude of the sidelobes is that the main lobe will typically be broadened. Tapering therefore involves some loss of frequency resolution, e.g. if $x(t)$ were a sine wave at a Fourier frequency ω_k the untapered periodogram would assume a positive value at ω_k and would be zero at other Fourier frequencies, whereas with tapering the spike at ω_k would be blurred over adjacent Fourier frequencies. See Section 5.8 of Brillinger (1975) for further details. Tapering also increases the variance of the periodogram slightly, as discussed in the next section. These drawbacks are typically minor compared with the potential bias reduction obtained by tapering.

One tapering function in common use is

$$h_C(x;p)=\begin{cases}\frac{1}{2}\left(1+\cos\dfrac{\pi(x-p)}{p}\right), & 0\leqslant x\leqslant p,\\ 1 & p\leqslant x\leqslant 1-p,\\ \frac{1}{2}\left(1+\cos\dfrac{\pi(x-1+p)}{p}\right), & 1-p\leqslant x\leqslant 1,\end{cases}\tag{4.12}$$

with $h_C(x;p)=0$ for $x\leqslant 0$ or $x\geqslant 1$. A computational advantage of this tapering function is that the middle portion of the data need not be multiplied. Another advantage is that p may take values between 0 and $\frac{1}{2}$, a larger value of p taken to counter the more severe frequency leakage problem. Bingham, Godfrey and Tukey (1967) proposed this taper with $p=0.1$. For further discussion of tapering functions, see Brillinger (1975, Section 3.3), Brillinger (1981), Geçkinli and Yavuz (1978), Hannan (1970, Section 5.3), and Thomson (1977).

The question of whether or not to taper the data prior to spectrum estimation may be broadened to the question of which tapering function $h(x)$ to use, because 'not tapering' really amounts to the use of the uniform or rectangular taper

$$h_R(x)=\begin{cases}1, & 0\leqslant x\leqslant 1,\\ 0, & \text{otherwise}.\end{cases}\tag{4.13}$$

Padding the data consists of appending zeros to the data after centering and possibly tapering have been performed. A primary consequence of extending the data in this manner

$$x'(t)=\begin{cases}\tilde{x}(t), & 0\leqslant t<T,\\ 0, & T\leqslant t<S,\end{cases}\tag{4.14}$$

is that the fast Fourier transform applied to $x'(t)$ yields Fourier frequencies of the form k/S, $k=0, 1, 2, \ldots, S-1$ rather than j/T, $j=0, 1, 2, \ldots, T-1$. This finer mesh of Fourier frequencies allows greater flexibility when smoothing the periodogram $I_{x'x'}(\omega_k)$ and, by inverse transforming $I_{x'x'}(\omega_k)$, provides a rapid means of calculating the sample autocovariance function $\hat{c}_{xx}(u)$, $0\leqslant|u|\leqslant S-T$.

Another reason for padding the data is that the fast Fourier transform is most efficient when the length of the input sequence is a highly composite number (capable of being factored into many smaller numbers). If the length T

of the original data sequence is not highly composite, padding the data out to a highly composite number S can substantially reduce computing time.

For periodogram-based spectrum estimation, we recommend that the data be centered, i.e. have an estimate of the mean subtracted out, prior to any padding. (Note that our covariance-based estimates automatically use centered data.) This is because padding changes the set of Fourier frequencies so that the second term in (4.1) (periodogram bias due to the mean c_x) will no longer vanish, i.e. ω will no longer be of the form j/T.

5. Variance approximations

The large sample variance of the spectrum estimate $\hat{s}_{xx}(\alpha)$, $0 < \alpha < \frac{1}{2}$, may be approximated using the asymptotic variance

$$\operatorname{var} \hat{s}_{xx}(\alpha) \sim \frac{s_{xx}(\alpha)^2}{\beta_T T} \frac{\int_0^1 h(x)^4 \, \mathrm{d}x}{(\int_0^1 h(x)^2 \, \mathrm{d}x)^2} \int_{-\infty}^{\infty} W(\gamma)^2 \, \mathrm{d}\gamma . \tag{5.1}$$

The asymptotic variance is doubled at $\alpha = 0, \frac{1}{2}$ where, due to symmetry properties (2.14) and (2.16), half as many 'independent' periodograms are being averaged.

Using Schwarz's inequality, it may be seen that the use of a nonuniform data tapering function $h(x)$ increases the asymptotic variance (5.1). Similarly, when $W(\alpha)$ has bounded support, the usual case for periodogram-based estimates, the rectangular window $W_R(\alpha)$ minimizes the variance. The use of (nonuniform) tapers $h(x)$ and spectral windows $W(\alpha)$ is justified by their potential for bias-reduction (Brillinger, 1975, Sections 5.5 and 5.8). Thomson (1977) heavily tapers the data prior to spectrum estimation because bias, rather than variance, is the overriding concern. Similarly, Alekseev and Yaglom (1980) report substantial reductions in bias achieved in simulation studies using a special family of spectral windows.

A proof of (5.1) and related expressions may be found in Brillinger (1975, Theorem 7.7.1) for covariance-based estimates. This asymptotic variance may be similarly derived for periodogram-based estimates (Brillinger, 1975, Theorem 5.6.4). One possible approximation to (5.1) consists of replacing the squared spectrum $s_{xx}(\alpha)^2$ by the squared estimate $\hat{s}_{xx}(\alpha)^2$. Recall that we have normalized the spectral window $W(\alpha)$ so that

$$\int_{-\infty}^{\infty} W(\alpha) \, \mathrm{d}\alpha = 1 \tag{5.2}$$

which is equivalent to the normalization

$$w(0) = 1 \tag{5.3}$$

for lag window $w(x)$. The evaluation of (5.1) will be further simplified by the normalizations

$$\int_{-\infty}^{\infty} w(x)^2 \, dx = \int_{-\infty}^{\infty} W(\alpha)^2 \, d\alpha = 1 \tag{5.4}$$

and

$$\int_{-\infty}^{\infty} h(x)^2 \, dx = 1 . \tag{5.5}$$

As noted above, the symmetry properties of the periodogram account for the discontinuity of the asymptotic variance at frequencies 0, $\frac{1}{2}$. For a finite sample size T, these symmetry properties come into play for frequencies α near 0 or $\frac{1}{2}$. For periodogram-based estimates we may adjust for this using the approximation

$$\widehat{\operatorname{var}}\, \hat{s}_{xx}(\alpha) = \frac{\frac{1}{T}\sum_{t=0}^{T-1} h^T(t)^4}{\left(\frac{1}{T}\sum_{t=0}^{T-1} h^T(t)^2\right)^2} \frac{S}{T} \sum_{k=1}^{S-1} W_k(\alpha)^2 \hat{s}_{xx}(\alpha)^2 , \tag{5.6}$$

where the periodogram weights $W_k(\alpha)$, which sum to unity, are obtained from the spectral window $W(\alpha)$ as explained in Section 2, and where S is the overall length of the padded data (producing a spacing of $1/S$ between Fourier frequencies).

In the untapered, unpadded case ($S = T$, $h(x) = h_R(x) = 1$), the approximation (5.6) may be further refined to allow for the periodogram at Fourier frequency $\frac{1}{2}$ having double the asymptotic variance of periodograms at adjacent Fourier frequencies (Brillinger, 1975, Theorem 5.2.5). We thus obtain

$$\widehat{\operatorname{var}}\, \hat{s}_{xx}(\alpha) = \left[2W_{\bar{t}}(\alpha)^2 + \sum_{\substack{k=1 \\ k \neq t}}^{T-1} W_k(\alpha)^2\right] \hat{s}_{xx}(\alpha)^2 \tag{5.7}$$

for $h(x) = h_R(x) = 1$, $S = T$, T even, and $t = T/2$.

For covariance-based estimates (2.9) the expression

$$\widehat{\operatorname{var}}\, \hat{s}_{xx}(\alpha) = \frac{\Sigma_t h^T(t)^4}{(\Sigma_t h^T(t)^2)^2} \sum_u W(\beta_T u)^2 \hat{s}_{xx}(\alpha)^2 \tag{5.8}$$

may be a more convenient approximation to (5.1) than those presented so far.

A useful result in addition to that of (5.1) is the asymptotic normality of $\hat{s}_{xx}(\alpha)$ as $\beta_T T$ becomes large. Brillinger (1975, Theorem 5.6.3) proves this result for untapered, unpadded, periodogram-based estimates, but we may use normality as an approximate distribution in other cases as well. Because $\hat{s}_{xx}(\alpha)$ is approximately normally distributed and has variance proportional to $s_{xx}(\alpha)^2$, we may regard the logarithm of the estimate as a convenient variance-stabilizing transformation, i.e. $\ln \hat{s}_{xx}(\alpha)$ is approximately normally distributed with approximate variance (for $0<\alpha<\frac{1}{2}$)

$$\widehat{\text{var}} \ln \hat{s}_{xx}(\alpha) \sim \frac{1}{\beta_T T} \frac{\int_0^1 h(x)^4 \, dx}{(\int_0^1 h(x)^2 \, dx)^2} \int_{-\infty}^{\infty} W(\alpha)^2 \, d\alpha . \tag{5.9}$$

More commonly, the logarithm base 10 is taken, with

$$\log_{10} \hat{s}_{xx}(\alpha) = (\log_{10} e) \ln \hat{s}_{xx}(\alpha) \doteq (0.4343) \ln \hat{s}_{xx}(\alpha) ,$$

introducing a factor of 0.4343 for the corresponding variance approximation. In addition to stabilizing the variance, the logarithmic transformation is useful when plotting the spectrum estimates $\hat{s}_{xx}(\alpha_j)$. Detail is preserved in the log scale plot even when the computed spectrum assumes a wide range of values at different frequencies. The refinements to the approximation (5.9), particularly the doubling of the variance at $\alpha = 0, \frac{1}{2}$, parallel those for (5.1).

By taking square roots of the above variance approximations, we arrive at standard errors for spectrum estimates. We may also use the approximate normality of the estimates to construct approximate confidence limits (Brillinger, 1975, Section 5.7). Under the log transformation, (5.9) shows that the confidence intervals will be of approximately the same width for frequencies interior to the interval $(0, \frac{1}{2})$.

6. Bandwidths and degrees of freedom

We commonly attach a number of 'degrees of freedom' to a spectrum estimate as a rough index of its statistical variability. This terminology may be motivated as follows: suppose that $x(t)$ consists of independent and identically distributed normal random variables having zero mean, and that we refrain from either tapering or padding the data prior to Fourier transformation. Then for any frequency ω in the interval $0<\omega<\frac{1}{2}$, $J_x(\omega)$ is a mean-zero complex normal variable, i.e. the real and imaginary parts of $J_x(\omega)$ are independent normal variables having mean zero and common variance. Thus the periodogram

$$\begin{aligned} I_{xx}(\omega) &= \frac{1}{T} |J_x(\omega)|^2 \\ &= \frac{1}{T} \{[\text{Re}\, J_x(\omega)]^2 + [\text{Im}\, J_x(\omega)]^2\} \end{aligned} \tag{6.1}$$

is proportional to a variable having a chi-square distribution with two degrees of freedom. Moreover, the entire set of discrete Fourier transform values, $J_x(\omega_k)$, $0<\omega_k<\frac{1}{2}$, are independent complex normal variables having mean zero and common variance. If we estimate the spectrum of $x(t)$ by taking a simple average of K adjacent periodograms, viz.

$$\frac{1}{K}\sum_{k=j+1}^{j+K} I_{xx}(\omega_k) \tag{6.2}$$

with $0<\omega_{j+1}<\omega_{j+K}<\frac{1}{2}$, then our estimate is proportional to a chi-square variable having $2K$ degrees of freedom. In this simple case, our description of the spectrum estimate as having $2K$ degrees of freedom is in strict accordance with standard terminology.

The above periodogram average employs uniform weights, i.e. the rectangular spectral window (periodogram weighting function) $W_R(\alpha)$. Since the Fourier frequencies are spaced $1/T$ apart and the average is over K adjacent Fourier frequencies, the bandwidth used is $\beta = K/T$. We therefore obtain

$$\nu = 2\beta T \tag{6.3}$$

relating degrees of freedom ν to bandwidth β and data length T.

When $x(t)$ is not a sequence of independent, identically distributed normal random variables, but is a stationary mixing time series, the spectrum estimate (6.2) is proportional to a random variable having, as T becomes large, an asymptotic chi-square distribution with $2K$ degrees of freedom (Brillinger, 1975, Theorem 5.4.3). So the chi-square terminology remains consistent with standard statistical usage when uniform periodogram weights are used. When we estimate the spectrum using a weighted average of the form

$$\hat{s}_{xx}(\alpha) = \sum_k W_k(\alpha) I_{xx}(\omega_k) \tag{6.4}$$

with

$$\sum_k W_k(\alpha) = 1\,,$$

the asymptotic distribution will not in general be chi-square. However, we may still make a chi-square approximation to the asymptotic distribution using the first two moments of the spectrum estimate as follows (Brillinger, 1975, p. 145).

If X is proportional to a chi-square distribution with ν degrees of freedom, i.e.

$$X \sim \theta\chi^2_\nu\,,$$

then

$$EX = \theta\nu \quad \text{and} \quad \operatorname{var} X = 2\theta^2\nu\,,$$

so that

$$\nu = 2\frac{(EX)^2}{\operatorname{var} X}\,.$$

Using the approximations

$$E\hat{s}_{xx}(\alpha) \doteq s_{xx}(\alpha) \tag{6.5}$$

and

$$\operatorname{var} \hat{s}_{xx}(\alpha) \doteq \Big(\sum_k W_k(\alpha)^2\Big) s_{xx}(\alpha)^2 , \tag{6.6}$$

we obtain

$$\begin{aligned} \nu &= 2\frac{(E\hat{s}_{xx}(\alpha))^2}{\operatorname{var} \hat{s}_{xx}(\alpha)} \\ &\doteq \frac{2}{\sum_k W_k(\alpha)^2} . \end{aligned} \tag{6.7}$$

When we use (5.1) instead of (6.6) as the approximate variance of $\hat{s}_{xx}(\alpha)$ we obtain an approximate degrees of freedom appropriate for either periodogram-based or covariance-based estimates, providing T is large, bandwidth parameter β_T is small, and $0 < \alpha < \frac{1}{2}$. This is

$$\nu \doteq 2\beta_T T \frac{(\int_0^1 h(x)^2\,\mathrm{d}x)^2}{\int_0^1 h(x)^4\,\mathrm{d}x} \frac{1}{\int_{-\infty}^{\infty} W(\alpha)^2\,\mathrm{d}\alpha} \tag{6.8}$$

which agrees with (6.3) for untapered data ($h(x) = h_R(x) = 1$) with uniform periodogram weights ($W(\alpha) = W_R(\alpha)$).

We have so far used the 'bandwidth parameter' β_T as some indication of the width of the frequency interval in which periodograms were averaged, under the normalization

$$W(0) = \int_{-\infty}^{\infty} W(\alpha)\,\mathrm{d}\alpha = 1 . \tag{6.9}$$

From (5.1) and (6.8), however, we see that a given bandwidth parameter can yield different asymptotic variances, and consequently different degrees of freedom, depending on the tapering function $h(x)$ and spectral window $W(\alpha)$ used. One convention is to define 'the bandwidth' of a spectrum estimate so that it reflects not only the bandwidth parameter β_T, but the effects of $h(x)$ and $W(\alpha)$ as well. Using the discussions leading to (6.3) and (6.8) as a basis, we define

$$\begin{aligned} \text{bandwidth} &= \frac{1}{T}\frac{(E\hat{s}_{xx}(\alpha))^2}{\operatorname{var} \hat{s}_{xx}(\alpha)} \\ &= \frac{\nu}{2T} \\ &\doteq \beta_T \frac{(\int_0^1 h(x)^2\,\mathrm{d}x)^2}{\int_0^1 h(x)^4\,\mathrm{d}x} \frac{1}{\int_{-\infty}^{\infty} W(\alpha)^2\,\mathrm{d}\alpha} . \end{aligned} \tag{6.10}$$

We see that bandwidth parameter β_T is 'the bandwidth' when the data are untapered ($h(x) = h_R(x) = 1$) and when the transform pair $w(x)$, $W(\alpha)$ satisfy

$$\int_{-\infty}^{\infty} w(x)^2\,\mathrm{d}x = \int_{-\infty}^{\infty} W(\alpha)^2\,\mathrm{d}\alpha = 1 \tag{6.11}$$

in addition to (6.9). Clearly we may refine (6.8) and (6.10) parallel to the refinements of the asymptotic variance expression (5.1).

An advantage of the terminology we have discussed here is that it can assist us in the selection of weights, tapers, and bandwidth parameters. A narrow bandwidth estimate can reveal important detail, but a large number of degrees of freedom (hence a large bandwidth) is also desirable as it provides a more reliable estimate. Values for the bandwidth parameter are often most easily selected on the basis of the corresponding degrees of freedom. In the absence of other guidelines, we might initially estimate the spectrum at three bandwidths, with corresponding degrees of freedom $\nu = 8$, $3T^{1/3}$, $T^{2/3}$, approximately. The three plots of the spectrum estimate may then be used as a basis for further adjustment of parameters.

Several alternative definitions of the bandwidth of a spectrum estimate have been proposed. See, for example, Sections 3.3 and 5.8 of Brillinger (1975). Our definition essentially follows that of Blackman and Tukey (1959). No matter what approach is taken, one should remember that, under the broad stationarity and mixing assumptions about the distribution of $x(t)$ implicit in our discussion, standard errors, degrees of freedom, and bandwidths can only be rough indicators of the statistical behavior of the nonparametric spectrum estimate $\hat{s}_{xx}(\alpha)$.

7. Prefiltering and recoloring

From Section 2 we see that the expected values of both periodogram-based and covariance-based spectrum estimates are weighted averages of the spectrum of interest in a band of frequencies. The smoothing operation used to form the spectrum estimates thus reduces variance but also entails a loss of precision, i.e. a locally averaged spectrum rather than the spectrum at one frequency is being estimated.

An exception to the loss of precision occurs when the underlying spectrum is constant. Then a smoothed version of the true spectrum is identical to the true spectrum. A constant, i.e. flat, spectrum corresponds to a time series whose values are uncorrelated. Such a time series is called a noise series. Although rare in practice, it is an important concept.

A simple means of regaining some of the lost precision incurred by spectral smoothing is to form a linearly filtered version $y(t)$ of the input sequence $x(t)$ of the form

$$y(t) = a(u_1)x(t-u) + \cdots + a(u_K)x(t-u_K)\,, \tag{7.1}$$

to then estimate the spectrum $s_{yy}(\alpha)$ of $y(t)$, and finally to estimate the spectrum $s_{xx}(\alpha)$ of $x(t)$ by adjusting $\hat{s}_{yy}(\alpha)$.

When $y(t)$ is a linearly filtered version of $x(t)$, as given in (7.1), the spectra of y and x have the relation

$$s_{yy}(\alpha) = |A(\alpha)|^2 s_{xx}(\alpha)\,, \tag{7.2}$$

where $A(\alpha)$ is the frequency response function, or transfer function of the filter, and is given by

$$A(\alpha) = \sum_{k=1}^{K} a(u_k)\exp(-2\pi i\alpha u_k)\,. \tag{7.3}$$

The proposed estimate of $s_{xx}(\alpha)$ is then

$$\hat{s}_{xx}(\alpha) = |A(\alpha)|^{-2}\hat{s}_{yy}(\alpha)\,. \tag{7.4}$$

The idea of the prefiltering and recoloring procedure, given in (7.1) through (7.4), is to obtain filter coefficients $a(u)$ such that the spectrum of y will be much flatter than that of x. The loss of precision brought about by the smoothing operation, discussed in Section 2, used to form the spectrum estimate $\hat{s}_{yy}(\alpha)$ will therefore be diminished. The prefiltering operation is also called prewhitening and the adjustment to the computed spectrum is called recoloring. These names derive from the fact that white light exhibits a nearly flat spectrum as opposed, say, to red light having a spectrum concentrated at low frequencies.

An effective method, proposed by Parzen and by Tukey, of obtaining filter coefficients $a(u)$ is to fit an autoregressive scheme to $x(t)$ by minimizing

$$\sum_{t=m}^{T-1} [x(t) - \phi(1)x(t-1) - \cdots - \phi(m)x(t-m)]^2 \tag{7.5}$$

with the understanding that

$$a(u) = \begin{cases} 1, & u = 0\,, \\ -\phi(u), & 1 \leq u \leq m\,, \\ 0, & \text{otherwise}\,. \end{cases} \tag{7.6}$$

Note that the ϕ's constitute a set of regression coefficients of $x(t)$ upon its past (thus the name autoregression) and that the output sequence $y(t)$, obtained from (7.1), (7.5) and (7.6), represents the set of residuals from the autoregression.

There are several possible algorithms for obtaining the autoregressive coefficients ϕ. Durbin (1960) gives a recursive algorithm which is a computationally efficient version of the Levinson (1947) algorithm. The Durbin or Levinson algorithm, as it is commonly called, is also described in Bartlett (1955) and in Box and Jenkins (1976). It consists in solving the Yule–Walker system of equations

$$\hat{\boldsymbol{R}}\boldsymbol{\phi} = \hat{\boldsymbol{\rho}}, \tag{7.7}$$

where

$$\hat{\rho}_l = \frac{\hat{c}_{xx}(l)}{\hat{c}_{xx}(0)}, \quad 1 \leq l \leq m, \tag{7.8}$$

are the sample autocorrelations of the sequence $x(t)$ and where

$$\hat{R}_{jk} = \begin{cases} 1 & \text{if } j = k, \\ \hat{\rho}_{|j-k|} & \text{if } j \neq k, \end{cases} \tag{7.9}$$

for $1 \leq j$, $k \leq m$, are the elements of the sample autocorrelation matrix of $x(t)$. From (7.9) we see that R differs from a general correlation matrix in its Toeplitz structure, i.e. the main diagonal consists of unity, the first upper and lower diagonals consist of $\hat{\rho}_1$, and so on. The Levinson, or Durbin, algorithm takes advantage of this structure and calculates the solution of (7.7) for increasing orders m of the autoregressive scheme.

The recursive algorithm has been criticized for numerical instability, i.e. the numerical solution ϕ may be quite sensitive to small perturbations in the sample correlations if the matrix $\hat{\boldsymbol{R}}$ is ill conditioned. See Cybenko (1980), however. Hannan suggests calculating the residual variance as $\hat{\sigma}_m^2 = (1 - \hat{\phi}(m)^2)\hat{\sigma}_{m-1}^2$ to guarantee its nonnegativity, a slight modification to the Durbin algorithm. Pagano (1972) presents an alternative algorithm based upon the Cholesky decomposition of matrix $\hat{\boldsymbol{R}}$, the algorithm being a special case of that presented by Golub (1969) and known to be numerically stable. Voevodin, Boyle, et al. (1979) describe a subroutine library for the solution of general Toeplitz systems of equations, where recursive algorithms are again employed.

Two interesting variations on autoregressive prefiltering have recently been proposed. McClave (1975) provides an algorithm for finding an optimal subset of lags $u_1, \ldots, u_k$ under the constraints

$$0 < u_1 < u_2 < \cdots < u_k \leq U, \quad k \leq K,$$

where U and K are user-designated. The resulting filter is of the form

$$y(t) = x(t) - a(u_1)x(t - u_1) - \cdots - a(u_k)x(t - u_k).$$

The possible optimality criteria include AIC, BIC (Akaike, 1978), or CAT (Parzen, 1980). In another direction, Kleiner, Martin and Thomson (1979) robustly fit ϕ, and use the autoregressive filter to detect two subtle outliers.

Their robustly prefiltered and recolored spectrum estimate differs markedly from the initial spectrum estimate, and is in better agreement with other information about the data.

8. Trend removal

Many economic series, for example the consumer price index, display an upward trend over time. A different, cyclic type of pattern is sometimes apparent in both economic and meteorological data. For example, both unemployment and temperatures tend to be higher during summer months. We often wish to remove the effect of these trends or patterns by fitting some parametric model to the data. We may then calculate the spectrum of the residuals, or otherwise analyze the residuals, to see how the model might be refined, or to uncover subtle features in the data that were not initially apparent.

We begin our discussion with a consideration of seasonal adjustment, i.e. removing a periodic component from the data. As an example, the data might consist of monthly temperatures over several years, from which we can estimate and remove (subtract) the average temperature for each month of the year. We thus proposed the formation of a residual series

$$r(t) = x(t) - \hat{m}(t),$$

where $\hat{m}(t)$ is an estimate of a P-periodic trend $m(t)$ (in our example $P = 12$). When the total length T of the input sequence $x(t)$ is a multiple of P, this operation has a very simple effect; the Fourier transform of $r(t)$ is zero at the seasonal Fourier frequencies $\omega_k = k/T$ such that $\omega_k P$ is an integer (i.e. such that k is a multiple of T/P), and equals the transform of $x(t)$ at nonseasonal Fourier frequencies. Whereas the estimated spectrum of $x(t)$ may have peaks at the seasonal frequencies, that of $r(t)$ may have valleys at these frequencies if we do not allow for the seasonal adjustment. See Grether and Nerlove (1970).

A simple remedy is to estimate the spectrum using the periodogram-smoothing method, taking care to assign a weight of zero to periodogram values at seasonal frequencies. This procedure may be extended to the case where T is not a multiple of P by first truncating the sequence $r(t)$ to the nearest multiple of P and making sure that the number of padded zeros, if any, prior to Fourier transformation is also a multiple of P. See Thrall (1979) for further remarks.

For large values of T, the above refinement to the estimated spectrum is of less consequence, because each estimate $\hat{s}_{xx}(\alpha)$ is an average of a large number of periodogram values $I_{xx}(\omega_k)$, no single periodogram value having undue influence. (Recall that the Fourier frequencies $\omega_k = k/T$ become dense as T becomes large.) This observation is generalized in a theorem to the following effect: the fitting by least squares and subsequent removal of a parametric trend which is linear in the unknown parameters will result in consistent estimates, both of the parameters and of the residual spectrum, as T becomes large. A key condition here is that the trend components possess a spectral measure. See Brillinger (1975, Section 5.11) for a precise formulation of the theorem. In our example above the trend spectral measure is concentrated at the seasonal frequencies.

The theorem cited above is useful, but not all-encompassing. It does not include the case of a simple linear trend, for example. The removal of a linear, or polynomial trend deflates the spectrum at low frequencies, as one would expect. But there is no simple remedy for estimating the residual spectrum in the affected frequency range, unlike the case for seasonal adjustment. Hannan (1960, 1970) gives a detailed treatment of polynomial and other parametric trend removal. One of Tukey's suggestions in this area is that we forego the linear parameterization and instead remove low-frequency sinusoidal components from the data. Depending on how close the removed frequencies are to zero, this operation can have very much the same effect as linear trend removal.

The removal of low-frequency sinusoids from the data is akin to another operation, that of extracting the trend by passing the data through a lowpass filter. The residuals are then a highpass filtered version of the data. In general, filtering is a promising alternative to parametric trend removal in the absence of a clear choice for the functional form of the trend. An advantage to linear filtering as a means of trend removal is that the spectrum may be easily recolored, as described in Section 7. An important limitation to filtering is that one must either decide how to extend the original data or else accept the loss of some data at the beginning and end of the filtered sequence. Hamming (1977) gives a valuable introduction to linear filters.

Taking nonparametric trend removal a step further, Huber (1978), Mallows (1980a,b) and Velleman (1980), discuss the robustification (and nonlinearization) of linear filters, e.g. replacing a simple moving average by a moving median. Recoloring the residual spectrum is no longer a simple matter. However, given any nonlinear filter, Mallows obtains linear components for both the filter and its frequency response, leading us to the possibility of using Mallow's linear approximation to recolor as described in Section 7.

9. Missing values

Time-series data, like most types of data collected in real life, are occasionally incomplete for a variety of reasons. Wyzga (1978) analyzes the coefficient of haze and mortality in Philadelphia, where a few values for the haze are missed due to a failure in the recording equipment. Wu (1978) discusses an entomological experiment where daily counts were made for the treatment group of insects but counts of the control group were taken only about 80% of the time during the week and 10% of the time during weekends.

Generally speaking, there are two types of procedure commonly used to estimate the spectrum of a time series in the presence of missing values: (1) to replace the missing values as accurately as possible and then to proceed as though the series were fully observed, (2) to calculate the sample autocovariance function from observed values, to adjust the sample autocovariance function according to the pattern of missing values, and to Fourier transform the adjusted autocovariance function. In both types of procedure, we assume that the mechanism causing values to be missed, and the values of the time series are independent. Jones (1980) derives a parametric spectrum estimate for time series with missing values, thus providing a third type of procedure.

A number of methods for missing value replacement suggest themselves, including linear interpolation, or taking a local mean or median of observed values. Leneman and Lewis (1966) compare a number of these methods under the criterion of mean squared error for the replaced value. Jones (1980) fits an ARMA scheme to the data in the presence of missing values. While a spectrum estimate may be obtained directly from the fitted parametric model, we might also use the model to replace missing values.

The second type of procedure, that of Fourier transforming an adjusted autocovariance function, is obtained by Jones (1962), Parzen (1963) and Bloomfield (1970). Jones and Parzen assume that missing values occur in a fixed pattern, while Bloomfield assumes they follow a renewal process (e.g. sporadic failures in the recording equipment). Computationally, the three authors derive essentially the same technique.

For the sake of simplicity, assume that the time series of interest, $x(t)$, has zero mean. Let $z(t)$ denote the observed series, with missing values coded as zero, and let $y(t)$ denote a 0–1 valued series indicating the absence or presence, respectively, of an observation at time t. Thus

$$y(t) = \begin{cases} 1 & \text{if } x(t) \text{ is observed}, \\ 0 & \text{otherwise}, \end{cases}$$

and

$$z(t) = x(t)y(t). \tag{9.1}$$

In the absence of the assumption $Ex(t) = 0$, we may form $z(t)$ by taking the sample average of the observed values, subtracting the sample average from each of the observations, and replacing missing values by zero. Letting $m(\cdot)$ denote second-order moments, we have

$$\begin{aligned} m_{zz}(u) &= E\{z(t+u)z(t)\} \\ &= E\{x(t+u)x(t)\}E\{y(t+u)y(t)\} \\ &= m_{xx}(u)m_{yy}(u). \end{aligned} \tag{9.2}$$

Here we have made use of the assumption that $x(\cdot)$ and $y(\cdot)$ are independent. When $y(t)$ is nonrandom, $m_{yy}(u)$ is defined (and assumed to exist) by the limit

$$m_{yy}(u) = \lim_{T\to\infty} 1/T \sum_{0\leq t,t+u<T} y(t+u)y(t)\,. \tag{9.3}$$

Since we have assumed $Ex(t) = 0$, implying $Ez(t) = 0$, (9.2) may be written as

$$c_{zz}(u) = c_{xx}(u)m_{yy}(u)\,. \tag{9.4}$$

From (9.4) we see that the recovery of $c_{xx}(u)$ from $c_{zz}(u)$, $m_{yy}(u)$ is only possible if $m_{yy}(u) \neq 0$ for all lags u, clearly an important restriction. Supposing $m_{yy}(u) > \varepsilon$, for some $\varepsilon > 0$, we may estimate $c_{xx}(u)$ by

$$\hat{c}_{xx}(u) = (1 - |u|/T)\frac{\hat{c}_{zz}(u)}{\max(\varepsilon, \hat{m}_{yy}(u))} \tag{9.5}$$

where

$$\hat{c}_{zz}(u) = 1/T \sum_{0\leq t,t+u<T} z(t+u)z(t) \tag{9.6}$$

and

$$m_{yy}(u) = 1/T \sum_{0\leq t,t+u<T} y(t+u)y(t)\,. \tag{9.7}$$

When missing values occur in a predetermined pattern, we may calculate $m_{yy}(u)$ exactly and use it rather than $\max(\varepsilon, \hat{m}_{yy}(u))$ in (9.5). Note that when there is no pair of observations at lag u, $\hat{m}_{yy}(u)$, $\hat{c}_{zz}(u)$ and $\hat{c}_{xx}(u)$ are all zero. The factor $(1-|u|/T)$ has been inserted into (9.5) so that $\hat{c}_{xx}(u)$ is the usual positive-definite autocovariance estimate when $x(t)$ is fully observed, i.e. has no missing values. We may now use $\hat{c}_{xx}(u)$ to form a covariance-based spectrum estimate $\hat{s}_{xx}(\alpha)$ as given by (2.9).

There are, of course, other methods of spectrum estimation which are appropriate for special patterns of missing values. If the sequence $x(t)$ contains one or more fully observed subsequences of moderate length we could form a spectrum estimate for each subsequence, perhaps averaging them to a form a final estimate. In another special situation, Thrall (1980) provides a method appropriate for quasi-periodically missing values, i.e. the $y(t)$ are assumed to be independent, taking the value 0 or 1 according to periodic probabilities $p(t)$. Thrall applies the method to the Wu (1978) control group of insects mentioned above.

All of these methods depend heavily on underlying assumptions (particularly on the independence of $x(\cdot)$ and $y(\cdot)$) to produce consistent, or otherwise acceptable estimates of the spectrum. Even when the assumptions hold, the adjusted autocovariance procedure may produce negative spectrum estimates at some frequencies, and the missing value replacement procedure may seriously bias the spectrum estimate. A precise assessment of the replacement procedure would need to take into account the pattern of missing values, the spectrum of $x(t)$, and the actual method used to replace missing values. We are led to some obvious conclusions (1) for the purpose of spectrum estimation, it is almost always worthwhile to obtain a fully observed sequence if possible, and (2) if missing values are unavoidable, several methods of spectrum estimation are worth trying.

10. Other problems in spectrum estimation and computation

Lengthy series. Occasionally the data sequence $x(t)$ is too lengthy to be held in the available high-speed core storage of the computer so that some modification must be made in the methods described above. Welch (1967) discusses the formation of overlapping segments of data and the computation of tapered periodograms for each segment. The final spectrum estimate is obtained by averaging periodogram values, not at adjacent frequencies but rather across data segments. Zhurbenko (1979) examines the statistical properties of this spectrum estimate, and shows that near optimal mean square error is achieved when a data taper suggested by A. N. Kolmogorov is used. Welch's method offers a solution to the lengthy data problem, for we need only hold the current data segment, its Fourier transform, and the currently accumulated average of periodograms in core storage.

Another possibility is to compute the Fourier transform of the entire data sequence using external storage. Singleton (1967) gives an algorithm appropriate for magnetic tape or any other external storage device, while Brenner (1969) provides an algorithm appropriate for a disk, drum, or other direct access device. The Fourier transform, requiring the same amount of storage as the data, must also reside on an external storage device. Periodogram values and spectrum estimates could be computed, printed and plotted, by reading the transform into core in successive frequency intervals. This method typically requires many more input–output and computational operations than the Welch method. The advantage here is that the Fourier frequencies are very finely spaced, thus permitting high-resolution spectrum estimates.

Outliers have not been treated in our discussion. See R. D. Martin (Chapter 10). The primary approach to this problem has two stages: (1) robustify the estimation of autoregressive coefficients, using robust regression methods, and (2) modify the prefiltering and recoloring technique described in Section 7 by 'cleaning' the residuals (filtered output) of outliers. The approach may be automated so that the user need not identify outliers. Brillinger (1973) gives a simpler method that may be used when outliers are easily identified.

Locating a peak in the spectrum. Geophysicists concerned with the exact frequencies at which the earth 'rings' or freely vibrates (as evidenced by the earth's response to a seismic disturbance) have turned to autoregressive and other parametric spectrum estimates. The autoregressive spectrum estimate is similar to the prefiltered, recolored estimate of (7.4), but the residual spectrum estimate $\hat{s}_{yy}(\alpha)$ is replaced by a residual variance estimate $\hat{\sigma}_y^2$ (constant across frequencies α). See the chapters by E. Parzen and P. M. Robinson for detailed discussions.

Further references include Kleiner, Martin and Thomson (1979), Thomson (1977), Alekseev and Yaglom (1980) and Childers (1978). The first two papers extend Welch's spectrum estimate and further discuss robust spectrum estimation in the presence of outliers. Alekseev and Yaglom discuss parametric and nonparametric methods of spectrum estimation, and the Childers volume consists of selected papers in parametric spectrum estimation.

References

Akaike, H. (1978). Time series analysis and control through parametric models. In: D. F. Findley, ed., *Applied Time Series Analysis*. Academic Press, New York.

Alekseev, V. G. and Yaglom, A. M. (1980). Nonparametric and parametric spectrum estimation methods for stationary time series. In: O. D. Anderson, ed., *Time Series*. North-Holland, Amsterdam.

Bingham, C., Godfrey, M. D. and Tukey, J. W. (1967). Modern techniques in power spectrum estimation. *IEEE Trans. Audio Electroacoust.* **AU-15**, 56–66.

Blackman, R. B. and Tukey, J. W. (1959). *The Measurement of Power Spectra*. Dover, New York.

Bloomfield, P. (1970). Spectral analysis with randomly missing observations. *J. R. Statist. Soc.* **B32**, 369–380.

Bloomfield, P. (1976). *Fourier Analysis of Time Series*. Wiley, New York.

Box, G. E. P. and Jenkins, G. M. (1976). *Time Series Analysis, Forecasting and Control*. Holden-Day, San Francisco.

Brenner, N. M. (1969). Fast Fourier transform of externally stored data. *IEEE Trans. Audio and Electroacoust.* **AU-17**, 128–132.

Brillinger, D. R. (1973). A power spectral estimate which is insensitive to transients. *Technometrics* **15**, 559–562.

Brillinger, D. R. (1975). *Time Series: Data Analysis and Theory*. Holt, Rinehart and Winston, New York.

Brillinger, D. R. (1981). The key role of tapering in spectrum estimation. *IEEE Trans. Acoust., Speech, Signal Processing* **ASSP-29**, 1075–1076.

Childers, D. G. (ed.) (1978). *Modern Spectrum Analysis*. IEEE Press, New York.

Cooley, J. W., Lewis, P. A. W. and Welch, P. D. (1977). The fast Fourier transform and its application to time series analysis. In: K. Enslein et al., eds., *Statistical Methods for Digital Computers*. Wiley, New York.

Cooley, J. W. and Tukey, J. W. (1965). An algorithm for the machine calculation of complex Fourier series. *Math. Comp.* **19**, 297–301.

Cybenko, G. (1980). The numerical stability of the Levinson–Durbin algorithm for Toeplitz systems of equations. *SIAM J. on Sci. and Statist. Computing* **1**, 235–237.

Durbin, J. (1960). The fitting of time series models. *Rev. Int. Statist. Inst.* **28**, 233–244.

Geçkinli, N. C. and Yavuz, D. (1978). Some novel windows and a concise tutorial comparison of window families. *IEEE Trans. Audio Electrocoust.* **ASSP-26**, 6.

Golub, G. H. (1969). Matrix decompositions and statistical calculations. In: R. C. Milton and J. A. Nelder, eds., *Statistical Computation*, pp. 365–397. Academic Press, New York.

Grether, D. M. and Nerlove, M. (1970). Some properties of 'optimal' seasonal adjustment. *Econometrica* **38**, 682–703.

Hamming, R. W. (1977). *Digital Filters*. Prentice-Hall, Englewood Cliffs, N.J.

Hannan, E. J. (1960). *Time Series Analysis*. Methuen, London.

Hannan, E. J. (1970). *Multiple Time Series*. Wiley, New York.

Huber, P. J. (1979). Robust smoothing. In: R. L. Launer and G. N. Wilkinson, eds., *Robustness in Statistics*. Academic Press, New York.

Jones, R. H. (1962). Spectral analysis with regularly missed observations *Ann. Math. Statist.* **33**, 455–461.

Jones, R. H. (1980). Maximum likelihood fitting of ARMA models to time series with missing observations. *Technometrics* **22**(3) 389–395.

Kleiner, B., Martin, R. D. and Thomson, D. J. (1979). Robust estimation of power spectra. *J. R. Statist. Soc.* **B41**, 313–351.

Leneman, O. A. Z. and Lewis, J. B. (1966). Random sampling of random processes: mean square comparison of various interpolators. *IEEE Trans. Aut. Control* **AC-11**, 3.

Levinson, N. (1947). The Wiener RMS (root mean square) error criterion in filter design and predicton. *J. Math. and Physics* **25**(4), 261–278. Reprinted in *Time Series*, N. Wiener (1949) MIT Press, New York.

McClave, J. (1975). Subset autoregression. *Technometrics* **17**, 213–220.

Mallows, C. L. (1980a). Resistant smoothing. In: O. D. Anderson, ed., *Time Series*. North-Holland, Amsterdam.

Mallows, C. L. (1980b). Some theory of nonlinear smoothers. *Ann. of Statist.* **8**(4), 695–715.

Martin, R. D. (1983). Robust resistant spectrum estimation. Chapter 10 in this volume.

Pagano, M. (1972). An algorithm for fitting autoregressive schemes. *J. R. Statist. Soc.* **C21**, 274–281.

Parzen, E. (1963). On spectral analysis with missing observations and amplitude modulation. *Sankhya*. **A25**, 180–189.

Parzen, E. (1980). Time series modeling, spectral analysis, and forecasting. In: D. R. Brillinger and G. C. Tiao, eds., *Directions in Time Series*. Institute of Mathematical Statistics.

Parzen, E. (1983). Autoregressive spectral estimation. Chapter 11 in this volume.

Robinson, P. M. (1983). Review of various approaches to power spectrum estimation. Chapter 2 in this volume.

Singleton, R. C. (1967). A method for computing the fast Fourier transform with auxiliary memory and limited high-speed storage. *IEEE Trans. Audio and Electroacoust.* **AU-15**, 91–98.

Thomson, D. J. (1977). Spectrum estimation techniques for characterization and development of WT4 waveguide. *Bell Sys. Tech. J.* **56**, 1769–1815, 1983–2005.

Thrall, A. D. (1979). Spectral estimation following simple seasonal adjustment. BMDP Technical Report No. 60, Biomath. Dept., UCLA.

Thrall, A. D. (1980). A spectral analysis of a time series in which probabilities of observation are periodic. In: O. D. Anderson, ed., *Time Series*. North-Holland, Amsterdam.

Velleman, P. F. (1980). Definition and comparison of robust nonlinear data smoothing algorithms. *J. Amer. Statist. Assoc.* **75**, 609–615.

Voevodin, V. V., Boyle, J. M., et al. (1979). *The Toeplitz Package User's Guide*. Preliminary Version. Argonne National Laboratory.

Welch, P. D. (1967). The use of the fast Fourier transform for estimation of spectra: a method based on time averaging over short, modified periodograms. *IEEE Trans. Electr. Acoust.* **AU-15**, 70.

Wu, Y. C. (1978). An experimental theoretical study of population cycles of the blowfly, Phaenicia sericata (Calliphoridae), in a laboratory ecosystem. Ph.D. dissertation, University of California, Berkeley.

Wyzga, R. E. (1978). The effect of air pollution upon mortality: a consideration of distributed lag models. *J. Amer. Statist. Assoc.* **73**, 463–472.

Zhurbenko, I. G. (1979). On the efficiency of estimates of a spectral density. *Scand. J. Statist.* **6**, 49–56.

D. R. Brillinger and P. R. Krishnaiah, eds., *Handbook of Statistics, Vol. 3*
© Elsevier Science Publishers B.V. (1983) 439–476

20

Likelihood Ratio Tests on Covariance Matrices and Mean Vectors of Complex Multivariate Normal Populations and Their Applications in Time Series

P. R. Krishnaiah, J. C. Lee and T. C. Chang*

1. Introduction

It is known that certain estimates of spectral density matrices of the stationary and Gaussian multiple time series are distributed approximately as complex Wishart matrices. So, complex multivariate distributions are useful (e.g. see Brillinger, 1974; Hannan, 1970) in the area of inference on multiple time series. These distributions are useful in the area of nuclear physics (see Carmeli, 1974) also.

Wooding (1956) introduced the complex multivariate normal distribution. A complex random vector is said to be distributed as a complex multivariate normal if its real and imaginary parts are distributed jointly as a multivariate normal with a structured covariance matrix. Motivated by applications in time series, Goodman (1963a,b) made a systematic study of the complex multivariate normal distribution and complex Wishart matrix. Since then, James (1964), Khatri (1965), Krishnaiah (1976) and other workers in the field have investigated various aspects of complex multivariate distributions. For a review of the literature on complex multivariate distributions, the reader is referred to Krishnaiah (1976). In this paper, we review the literature on the likelihood ratio tests on mean vectors and covariance matrices of the complex multivariate normal populations as well as some of their applications in the area of inference on multiple time series in the frequency domain.

In Section 2 of this paper, we discuss the complex multivariate normal and complex Wishart matrix. The distribution of the determinant of the complex multivariate beta matrix is discussed in Section 3, whereas Section 4 is devoted to the likelihood ratio test procedure for testing the hypothesis of multiple independence of several sets of variables when their joint distribution is complex multivariate normal. Likelihood ratio tests for the hypothesis of sphericity and the hypothesis specifying the covariance matrix are discussed in Sections 5 and 6 respectively. In Section 7 we discuss the likelihood ratio test

*The work of this author is sponsored by the Air Force Office of Scientific Research under Contract F49620-82-K-0001. Reproduction in whole or in part is permitted for any purpose of the United States Government.

for the homogeneity of the covariance matrices, whereas the likelihood ratio test procedure for the homogeneity of several complex multivariate normal populations is discussed in Section 8. Likelihood ratio test procedure specifying the covariance matrix and mean vector is discussed in Section 9. Applications of some test procedures on the covariance matrices of the complex multivariate normal populations to the area of inference on multiple time series in the frequency domain are discussed in Section 10. Various tables useful in implementation of certain likelihood ratio test procedures are given in the Appendix. These tables are constructed by approximating a suitable power of the likelihood ratio statistics with Pearson's type I distribution by using the first four moments. The accuracy of these tables is found to be good.

2. Complex multivariate normal and complex Wishart distributions

Let $z = x + \mathrm{i}y$, where x and y are of order $p \times 1$ and (x', y') is distributed as $2p$-variate normal with mean vector (μ_1', μ_2') and covariance matrix

$$C = \begin{pmatrix} \Sigma_1 & \Sigma_2 \\ -\Sigma_2 & \Sigma_1 \end{pmatrix}, \tag{2.1}$$

where A' denotes the transpose of A. Then, the distribution of z is known to be the complex multivariate normal distribution with mean vector μ and covariance matrix Σ where $\mu = \mu_1 + \mathrm{i}\mu_2$ and $\Sigma = 2(\Sigma_1 - \mathrm{i}\Sigma_2)$. The probability density function (p.d.f.) of z is given by

$$f(z) = \frac{1}{\pi^p|\Sigma|} \exp\{-(\overline{z - \mu})'\Sigma^{-1}(z - \mu)\}, \tag{2.2}$$

whereas the characteristic function of z is

$$\phi(t) = \exp\{\mathrm{i}\bar{t}'\mu - \tfrac{1}{4}\bar{t}'\Sigma^{-1}t\}, \tag{2.3}$$

where $t = t_1 + \mathrm{i}t_2$ and $\bar{t}$ is the complex conjugate of t. Wooding (1956) derived expressions for the p.d.f. and characteristic function of z. The maximum likelihood estimates of μ and Σ based on a random sample $(z_1, \ldots, z_N)$ are known to be

$$\hat{\mu} = N^{-1}\sum_{j=1}^{N} z_j, \qquad \hat{\Sigma} = N^{-1}\sum_{j=1}^{N} (z_j - \hat{\mu})(\overline{z_j - \hat{\mu}})'. \tag{2.4}$$

Also, $\hat{\mu}$ and $\hat{\Sigma}$ are distributed independent of each other.

Next, let $S = N\hat{\Sigma}$. Then, the distribution of S is known to be a central complex Wishart matrix with $n = N - 1$ degrees of freedom. The probability density of S is known (Goodman, 1963b) to be

$$f(S) = \frac{|S|^{n-p} \operatorname{etr}\{-\Sigma^{-1}S\}}{\pi^{p(p-1)/2}\prod_{j=1}^{p} \Gamma(n-j+1)|\Sigma|^{n}}, \tag{2.5}$$

where etr B denotes the exponential of the trace of the matrix B.

3. Distribution of the determinant of the complex multivariate beta matrix

In this section we discuss the distribution of the determinant of the complex multivariate beta matrix. This distribution is useful for testing the *hypothesis* of the equality of several mean vectors and the equality of two covariance matrices when the underlying distributions are complex multivariate normal. It is also useful in testing the hypothesis H_1: $\Sigma_{12} = 0$, where Σ_{12} is the covariance between two sets of variables whose joint distribution is complex multivariate normal.

Let $A_1: p \times p$ and $A_2: p \times p$ be independently distributed as the central complex Wishart matrices with n and q degrees of freedom, and let $E(A_1/n) = E(A_2/q) = \Sigma$. Then $A_1(A_1 + A_2)^{-1}$ is known to be a (central) complex multivariate beta matrix. Now, let

$$U = |A_1(A_1 + A_2)^{-1}|. \tag{3.1}$$

The hth moment of U is known to be

$$E(U^h) = \prod_{j=1}^{p} \left[\frac{\Gamma(n+h-j+1)\Gamma(n+q-j+1)}{\Gamma(n-j+1)\Gamma(n+h+q-j+1)}\right]. \tag{3.2}$$

Using the first four moments of $U^{1/b}$, Lee, Krishnaiah and Chang (1975) have approximated the distribution of $U^{1/b}$ with the Pearson type I distribution, where b is a properly chosen integer. The constant b is chosen to be equal to 1 or 2 according as $M > 20$ or $M < 20$, where $M = n - p + 1$. By making use of this approximation, values of c_1 are computed, where

$$P[C_1 \leq c_1] = (1 - \alpha), \tag{3.3}$$

$c_1 = -(2n + q - p) \log U/\chi^2_{2pq,\alpha}$, and $\chi^2_{2pq,\alpha}$ is the upper $100\alpha\%$ value of χ^2 with $2pq$ degrees of freedom. The values of c_1 are computed for $\alpha = 0.005$, 0.01, 0.025, 0.05, 0.1, 0.90, 0.95, 0.99, 0.995, $M = 1(1)10(2)20$, 30, 60, 120, where $M = n - p + 1$. These values are given in a technical report by Lee, Krishnaiah and Chang (1975). The upper 5% and 1% points are reproduced in Table 7 of the Appendix. To check for the accuracy of the entries in Table 7, the above authors compared some of the values obtained by the Pearson type approximation with the corresponding exact values. These comparisons are given in Table 1.

Table 1
Comparison of the Pearson type approximation with exact expression for the distribution of C_1

	$p=2$ $q=3$			$p=2$ $q=20$		
M	α	L-K-C	Exact	α	L-K-C	Exact
1	0.05	1.286	1.289	0.05	1.928	1.932
1	0.01	1.350	1.349	0.01	2.085	2.080
5	0.05	1.029	1.029	0.05	1.243	1.243
5	0.01	1.033	1.033	0.01	1.262	1.262
9	0.05	1.010	1.011	0.05	1.128	1.128
9	0.01	1.012	1.012	0.01	1.137	1.137

The constant α in Table 1 is defined by (3.3). Also, the values under the column 'L-K-C' are the values of c_1 obtained by Lee, Krishnaiah and Chang (1975) using the Pearson type approximation, whereas the values under the column 'Exact' are the corresponding values given by Gupta (1971). Table 1 indicates that the accuracy of the Pearson type approximation is sufficient for practical purposes.

4. Test for independence of sets of variates

Let $z' = (z_1', \ldots, z_q')$ be distributed as a complex multivariate normal distribution with mean vector $\boldsymbol{\mu}' = (\boldsymbol{\mu}_1', \ldots, \boldsymbol{\mu}_q')$ and covariance matrix Σ. Also, let $E\{(z_i - \boldsymbol{\mu}_i)\overline{(z_j - \boldsymbol{\mu}_j)'}\} = \Sigma_{ij}$, and $E(z_i) = \boldsymbol{\mu}_i$. It is assumed that z_i is of order $p_i \times 1$ and $p_1 + \cdots + p_q = s$. In this section we discuss the problem of testing the hypothesis H_2, where

$$H_2: \quad \Sigma_{ij} = 0 \tag{4.1}$$

for $i \neq j = 1, \ldots, q$. Now let

$$A = \begin{bmatrix} A_{11} & A_{12} & \cdots & A_{1q} \\ A_{21} & A_{22} & \cdots & A_{2q} \\ \vdots & \vdots & \vdots\vdots\vdots & \vdots \\ A_{q1} & A_{q2} & \cdots & A_{qq} \end{bmatrix},$$

where

$$A_{gh} = \sum_{j=1}^{N} (z_{gj} - z_{g.})\overline{(z_{hj} - z_{h.})'}, \quad z_{g.} = N^{-1} \sum_{j=1}^{N} z_{gj}$$

and $(z_{1j}', \ldots, z_{qj}')$ is jth independent observation on $(z_1', \ldots, z_q')$. The likelihood ratio statistic for testing H_2 is

$$\lambda_2 = \frac{|A|}{\prod_{j=1}^{q} |A_{jj}|} . \tag{4.2}$$

In the analogous real case, Wilks (1935) derived the likelihood ratio test for multiple independence. The hth moment of λ_2 is

$$E(\lambda_2^h) = \left[\prod_{j=1}^{s} \frac{\Gamma(n+h-j+1)}{\Gamma(n-j+1)}\right]\left[\prod_{i=1}^{q}\prod_{\alpha=1}^{p_i} \frac{\Gamma(n-\alpha+1)}{\Gamma(n+h-\alpha+1)}\right], \tag{4.3}$$

where $n = N-1$ and $\Gamma(\cdot)$ is the complete gamma function. The distribution of $\lambda_2^{1/4}$ is approximated by Pearson's type I distribution with density

$$g(x) = \{\beta(\alpha+1, \varepsilon+1)(\sigma_1-\sigma_0)^{\alpha+\varepsilon+1}\}^{-1}(x-\sigma_0)^{\alpha}(\sigma_1-x)^{\varepsilon}, \tag{4.4}$$

where $\sigma_0 < x < \sigma_1$ and α and ε are some real numbers.

Approximate percentage points of the distribution of $\tilde{\lambda}_2 = -2\log\lambda_2$ are constructed by Krishnaiah, Lee and Chang (1975, 1976) for $p_i = p = 1, 2, 3$; $q = 3, 4, 5$; $\alpha = 0.01, 0.05, 0.10$; $M = 1(1)20(2)30$, where $M = n - s - 3$, and $\Pr[\tilde{\lambda}_2 \leq c_2 | H_2] = (1-\alpha)$. These percentage points are reproduced in Table 8 (see Appendix). Percentage points for $q = 2$ can be obtained from Table 7.

Now, consider a class of statistics $W (0 < W \leq 1)$ whose moments are of the form

$$E\{W^h\} = K \prod_{j=1}^{c} y_j^{y_j} \prod_{k=1}^{a} x_k^{x_k} \frac{\prod_{k=1}^{a} \Gamma[x_k(1+h)+\xi_k]}{\prod_{j=1}^{c} \Gamma[y_j(1+h)+y_j]}, \quad h = 0, 1, \ldots, \tag{4.5}$$

where K is a normalizing constant such that $E\{W^0\} = 1$ and $\Sigma_{k=1}^{a} x_k = \Sigma_{j=1}^{c} y_j$.

Table 2
Comparison of the Pearson type approximation with the asymptotic expansion for the distribution of $\tilde{\lambda}_2$

n	$q=3$ $p=1$			$q=4$ $p=2$			$q=5$ $p=1$		
	c_2	α_1	α_2	c_2	α_1	α_2	c_2	α_1	α_2
10	1.459	0.05	0.0499	—	—	—	4.011	0.05	0.0487
10	1.949	0.01	0.0100	—	—	—	4.811	0.01	0.0095
15	0.923	0.05	0.0500	5.733	0.05	0.0479	2.435	0.05	0.0497
15	1.233	0.01	0.0100	6.496	0.01	0.0093	2.914	0.01	0.0099
20	0.675	0.05	0.0500	3.958	0.05	0.0493	1.752	0.05	0.0499
20	0.902	0.01	0.0100	4.479	0.01	0.0098	2.096	0.01	0.0100
30	0.439	0.05	0.0501	2.455	0.05	0.0498	1.124	0.05	0.0500
30	0.587	0.01	0.0100	2.777	0.01	0.0099	1.344	0.01	0.0100

α_1 is the value of α if we use the Pearson type approximation and α_2 is the value of α if we use the asymptotic expression of order n^{-13}. From the table, we observe that the accuracy of the Pearson type approximation is sufficient for practical purposes.

The likelihood ratio test statistics considered in this chapter are special cases of the above class of statistics. Box (1949) gave explicitly the first few terms of an asymptotic expression for the distribution of a class of statistics whose moments are of the form (4.5). But the first few terms alone are not sufficient to get the desired degree of accuracy in a number of practical situations. So, Lee, Krishnaiah and Chang (1976) gave terms up to order n^{-15} explicitly. In Table 2 a comparison of the values obtained by using Pearson type approximation is made with the corresponding values obtained by the asymptotic expression of order n^{-13}.

5. Test for sphericity

The likelihood ratio statistic for testing H_3: $\Sigma = \sigma^2\Sigma_0$ is given by

$$\lambda_3 = \frac{|A\Sigma_0^{-1}|}{(\operatorname{tr} A\Sigma_0^{-1}/s)^s}, \tag{5.1}$$

where Σ_0 is known, A was defined in Section 4 and tr A denotes the trace of A. The hth moment of λ_3 is known to be

$$E(\lambda_3^h) = \frac{s^{hs}\Gamma(sn)}{\Gamma(sn+sh)} \prod_{j=1}^{s} \frac{\Gamma(n+h-j+1)}{\Gamma(n-j+1)}. \tag{5.2}$$

Mauchly (1940) derived the likelihood ratio statistic for testing the hypothesis of sphericity when the underlying distribution is real multivariate normal.

Table 3
Comparison of the Pearson type approximation with the asymptotic expression for the distribution of λ_3

	$s=5$			$s=8$		
n	c_3	α_1	α_2	c_3	α_1	α_2
15	2.763	0.05	0.0496	—	—	—
15	3.265	0.01	0.0099	—	—	—
21	1.895	0.05	0.0498	4.565	0.05	0.0490
21	2.238	0.01	0.0100	5.093	0.01	0.0097
41	0.928	0.05	0.0500	2.160	0.05	0.0499
41	1.095	0.01	0.0100	2.409	0.01	0.0100
51	0.739	0.05	0.0500	1.711	0.05	0.0499
51	0.873	0.01	0.0100	1.908	0.01	0.0100

α_1 is the value of α obtained by using the Pearson type approximation, whereas α_2 is the value of α obtained by using Box's asymptotic expansion of order n^{-13}. This table indicates that the accuracy of Pearson type approximation is sufficient for practical purposes.

The distribution of $\lambda_3^{1/b}$ is approximated by a Pearson type I distribution, where b is a suitably chosen integer. For $M \geqslant 22$, we took $b = 2$ and for $M < 22$, $b = 4$. Using the approximation described above, approximate upper percentage points of distribution of $\tilde{\lambda}_3 = -2 \log \lambda_3$ were constructed by Krishnaiah, Lee and Chang (1975, 1976) for $s = 2(1)10$, $\alpha = 0.01$, 0.05, $M = 1(1)20(2)30(5)50, 60$, where $M = n - s - 3$ and $P[\tilde{\lambda}_3 \leqslant c_3 | H_3] = (1 - \alpha)$. These values are reproduced in Table 9 (see Appendix).

In Table 3 we compare the values obtained by the Pearson type approximation with the corresponding values obtained by using Box's asymptotic expression of order n^{-13}. Upper percentage points of the distribution of $\tilde{\lambda}_3$ for $\alpha = 0.01$, 0.05 and $s = 3(1)6$ are also given by Nagarsenker and Das (1975).

6. Test specifying the covariance matrix

The modified likelihood ratio statistic for testing the hypothesis H_4: $\Sigma = \Sigma_0$ is given by

$$\lambda_4 = (e/n)^{sn} |A\Sigma_0^{-1}|^n \operatorname{etr}(-A\Sigma_0^{-1}) . \tag{6.1}$$

The modified likelihood ratio test statistic is obtained from the likelihood ratio test statistic by changing N to n. The moments of λ_4 are seen to be

$$E(\lambda_4^h) = (e/n)^{shn} |\Sigma_0|^{nh} |I + h\Sigma_0|^{-n(1+h)} \times \prod_{i=1}^{s} \{\Gamma(n + nh + 1 - i)/\Gamma(n + 1 - i)\} . \tag{6.2}$$

Anderson (1958) derived the likelihood ratio statistic for testing the hypothesis that the covariance matrix is equal to a specified matrix when the underlying distribution is real multivariate normal. The distribution of $\lambda_4^{1/b}$ can be approximated with a Pearson type I distribution using the first four moments, where b is a suitably chosen integer. Using the above approximation, Krishnaiah, Lee and Chang (1975, 1976) computed the percentage points of the distribution of $\tilde{\lambda}_4 = -2 \log \lambda_4$ for $s = 2(1)10$, $\alpha = 0.01$, 0.05, $M = 1(1)20(2)30$, where $M = n - s - 1$ and $P[\tilde{\lambda}_4 \leq c_4 | H_4] = (1 - \alpha)$. These percentage points are reproduced in Table 10 (see Appendix).

7. Test for multiple homogeneity of the covariance matrices

Let $z_1, \ldots, z_q$ be independently distributed as complex p-variate normal with mean vectors $\boldsymbol{\mu}_1, \ldots, \boldsymbol{\mu}_q$ and covariance matrices $\Sigma_{11}, \ldots, \Sigma_{qq}$, respectively. Also, let z_{ij} $(j = 1, \ldots, N_i)$ be the jth independent observation on z_i. In this section we study the Pearson type approximation to the distribution of the likelihood ratio statistic for testing H_5, where

$$H_5: \begin{cases} \Sigma_{11} = \cdots = \Sigma_{q_1 q_1}, \\ \Sigma_{q_1+1, q_1+1} = \cdots = \Sigma_{q_2^*, q_2^*}, \\ \vdots \qquad \vdots \qquad \vdots \\ \Sigma_{q_{d-1}^*+1, q_{d-1}^*+1} = \cdots = \Sigma_{qq}, \end{cases}$$

$q_0^* = 0$, $q_1^* = q$, $q_d^* = q$ and $q_j^* = q_1 + \cdots + q_j$. The modified likelihood ratio statistic (obtained by changing N_i to n_i in the likelihood ratio test statistic) for testing H_5, is given by

$$\lambda_5 = \frac{\prod_{i=1}^{q} |A_{ii}/n_i|^{n_i}}{\prod_{j=1}^{d} |\sum_{i=q_{j-1}^*+1}^{q_j^*} A_{jj}/n_j^*|^{n_j^*}}, \tag{7.1}$$

where $n_i = N_i - 1$, $n_j^* = \sum_{i=q_{j-1}^*+1}^{q_j^*} n_i$ and

$$A_{ii} = \sum_{j=1}^{N_i} (z_{ij} - z_{i.})(\overline{z_{ij} - z_{i.}})', \quad z_{i.} = \sum_{j=1}^{N_i} z_{ij}/N_i .$$

The moments of λ_5 are given by

$$E(\lambda_5^h) = \left[\frac{\prod_{\alpha=1}^{d} n_\alpha^{*phn_\alpha^*}}{\prod_{\alpha=1}^{q} (n_\alpha)^{phn_\alpha}}\right] \prod_{i=1}^{p} \prod_{\alpha=1}^{d} \left[\prod_{g=q_{\alpha-1}^*+1}^{q_\alpha^*} \frac{\Gamma(n_g + hn_g + 1 - i)}{\Gamma(n_g + 1 - i)}\right] \times \frac{\Gamma(n_\alpha^* + 1 - i)}{\Gamma(n_\alpha^* + hn_\alpha^* + 1 - i)}. \tag{7.2}$$

Using the first four moments of λ_5, the distribution of $\lambda_5^{1/b}$ can be approximated with a Pearson's type I distribution, where b is a suitably chosen integer. This approximation was used by Krishnaiah, Lee and Chang (1975, 1976) to compute approximate percentage points of the distribution of $\tilde{\lambda}_5 = -2 \log \lambda_5$ for $n_i = n_0$, $q = dk$ (i.e. there are k populations in each of the d groups). These points are reproduced in Table 11 (see Appendix) for $d = 1$.

In Tables 4 and 5, we compare the values obtained by the Pearson type approximation for the distribution function of $\tilde{\lambda}_5$ with the corresponding values obtained by using the Box's asymptotic expansion up to terms of order n^{-13}. In these tables, the constant c_5 is defined as

$$P[-2 \log \lambda_5 \leq c_5 | H_5] = (1 - \alpha). \tag{7.3}$$

Also, α_1 is the value of α if we use the Pearson type approximation, whereas α_2 is the value of α if we use the asymptotic expression of order n^{-13}. Tables 4 and 5 indicate that the accuracy of the Pearson type approximation is sufficient for practical purposes.

In the real case, Wilks (1932) derived the likelihood ratio statistic for testing the homogeneity of the covariance matrices, whereas Krishnaiah and Lee (1976)

Table 4
Comparison of the Pearson type approximation with the asymptotic expansion for the distribution of $\tilde{\lambda}_5$ when $d = 1$

		$p = 3$			$p = 4$		
n_0	q	c_5	α_1	α_2	c_5	α_1	α_2
10	2	19.82	0.05	0.0501	33.00	0.05	0.0502
10	2	25.40	0.01	0.0100	40.21	0.01	0.0101
10	6	69.68	0.05	0.0500	12.14	0.05	0.0492
10	6	79.11	0.01	0.0100	13.39	0.01	0.0098
15	2	18.72	0.05	0.0501	30.33	0.05	0.0500
15	2	23.99	0.01	0.0100	36.93	0.01	0.0100
15	6	66.70	0.05	0.0500	113.77	0.05	0.0498
15	6	75.69	0.01	0.0100	125.48	0.01	0.0099
20	2	18.23	0.05	0.0500	29.18	0.05	0.0500
20	2	23.35	0.01	0.0100	35.52	0.01	0.0100
20	6	65.34	0.05	0.0499	110.45	0.05	0.0499
20	6	74.13	0.01	0.0100	121.78	0.01	0.0100

discussed how certain tests of hypotheses on linear structure of the covariance matrices can be reduced to the problem of testing for the multiple homogeneity of the covariance matrices.

8. Simultaneous tests for the homogeneity of populations

In this section we discuss the likelihood ratio test for the homogeneity of complex multivariate normal populations. The hypothesis of the homogeneity of the q complex multivariate distributions defined in Section 7 is equivalent to the hypothesis H_6, where

$$H_6: \quad \begin{cases} \Sigma_{11} = \cdots = \Sigma_{qq}\,, \\ \boldsymbol{\mu}_1 = \cdots = \boldsymbol{\mu}_q\,. \end{cases} \tag{8.1}$$

The modified likelihood ratio statistic for testing H_6 is given by

$$\lambda_6 = \frac{n^{pn} \prod_{i=1}^{q} |G_i|^{n_i}}{\prod_{i=1}^{q} n_i^{pn_i} |G + \sum_{i=1}^{q} N_i(z_{i.} - z_{..})\overline{(z_{i.} - z_{..})}'|^n}\,, \tag{8.2}$$

where

$$n = \sum_{i=1}^{q} n_i\,, \quad n_i = N_i - 1\,,$$

$$N = \sum_{i=1}^{q} N_i\,, \qquad z_{..} = \frac{1}{N} \sum_i \sum_j z_{ij}\,, \qquad z_{i.} = \frac{1}{N_i} \sum_{j=1}^{N_i} z_{ij}\,,$$

$$G_i = \sum_{j=1}^{N_i} (z_{ij} - z_{i.})\overline{(z_{ij} - z_{i.})}'\,, \qquad G = \sum_{i=1}^{q} G_i\,.$$

Table 5
Comparison of the Pearson type approximation with asymptotic expansion for the distribution of $\bar{\lambda}_5$ when $d > 1$

			$p = 1$			$p = 2$			$p = 3$			$p = 4$		
n_0	q	d	c_5	α_1	α_2	c_5	α_1	α_2	c_5	α_1	α_2	c_5	α_1	α_2
10	6	3	8.01	0.05	0.0500	23.08	0.05	0.0500	46.97	0.05	0.0501	81.73	0.05	0.0503
10	6	3	11.62	0.01	0.0100	28.78	0.01	0.0100	55.03	0.01	0.0100	92.47	0.01	0.0101
10	6	2	9.70	0.05	0.0500	28.56	0.05	0.0500	58.66	0.05	0.0499	102.21	0.05	0.0500
10	6	2	13.57	0.01	0.0100	34.76	0.01	0.0100	67.44	0.01	0.0100	113.93	0.01	0.0100
20	6	3	7.91	0.05	0.0500	22.00	0.05	0.0500	43.22	0.05	0.0500	72.32	0.05	0.0500
20	6	3	11.49	0.01	0.0100	27.43	0.01	0.0100	50.60	0.01	0.0100	81.77	0.01	0.0100
20	6	2	9.59	0.05	0.0500	27.37	0.05	0.0500	54.49	0.05	0.0500	91.76	0.05	0.0500
20	6	2	13.42	0.01	0.0100	33.31	0.01	0.0100	62.63	0.01	0.0100	102.25	0.01	0.0100
30	6	3	7.88	0.05	0.0500	21.66	0.05	0.0500	42.12	0.05	0.0500	69.73	0.05	0.0500
30	6	3	11.44	0.01	0.0100	27.01	0.01	0.0100	49.31	0.01	0.0100	78.85	0.01	0.0100
30	6	2	9.56	0.05	0.0500	27.00	0.05	0.0500	53.26	0.05	0.0500	88.85	0.05	0.0500
30	6	2	13.38	0.01	0.0100	32.86	0.01	0.0100	61.22	0.01	0.0100	98.98	0.01	0.0100

The moments of λ_6 are given by

$$E(\lambda_6^h) = \left[n^{phn} / \prod_{i=1}^{q} n_i^{phn_i} \right] \prod_{i=1}^{p} \prod_{j=1}^{q} \frac{\Gamma(n_j + hn_j + 1 - i)}{\Gamma(n_j + 1 - i)} \frac{\Gamma(n + q - i)}{\Gamma(n + hn + q - i)} . \tag{8.3}$$

The distribution of $\lambda_6^{1/b}$ is approximated with a Pearson type I distribution, where b is a suitably chosen integer. Using this approximation, percentage points of the distribution of $\tilde{\lambda}_6 = -2 \log \lambda_6$ are computed by Chang, Krishnaiah and Lee (1975) for $\alpha = 0.01, 0.025, 0.05, 0.10$; $n_i = n_0$; $q = 2, 3, 4, 5$; $p = 1, 2, 3, 4$ and $M = n_0 - p = 1(1)20, 25, 30$. Table 12 (see Appendix) gives the value of c_6 for $\alpha = 0.05, 0.01$, where c_6 is given by

$$P[\tilde{\lambda}_6 \leq c_6 \mid H_6] = (1 - \alpha) . \tag{8.4}$$

To check for the accuracy of the entries in Table 12, we compared some of these values with the corresponding values obtained by using Box's asymptotic series of order n^{-13}. These comparisons are given in Table 6.

Table 6
Comparison of the Pearson type approximation with the asymptotic expansion

		$p=2$			$p=3$		
n_0	q	c_6	α_1	α_2	c_6	α_1	α_2
7	3	28.03	0.05	0.0499	50.49	0.05	0.0497
7	3	34.15	0.01	0.0100	58.80	0.01	0.0098
10	3	27.45	0.05	0.0500	48.06	0.05	0.0498
10	3	33.42	0.01	0.0100	55.92	0.01	0.0099
15	3	27.04	0.05	0.0500	46.46	0.05	0.0499
15	3	32.90	0.01	0.0100	54.04	0.01	0.0100
20	3	26.84	0.05	0.0500	45.73	0.05	0.0500
20	3	32.67	0.01	0.0100	53.18	0.01	0.0100

α_1 is the value of α obtained by approximating $\lambda_6^{1/b}$ with Pearson's type I distribution, whereas α_2 is the value of α obtained by using Box's asymptotic series. This table indicates that the accuracy of the values of α_1 is good.

9. Test specifying the values of the covariance matrix and mean vector

In this section we consider the distribution of the likelihood ratio statistic for testing the hypothesis H_7, where

$$H_7: \begin{cases} \Sigma_{11} = \Sigma_0 , \\ \boldsymbol{\mu}_1 = \boldsymbol{\mu}_0 , \end{cases}$$

and Σ_0 and $\boldsymbol{\mu}_0$ are known. The likelihood ratio statistic for testing H_7 is given by

$$\lambda_7 = (e/N_1)^{pN_1}|G_1\Sigma_0^{-1}|^{N_1} \operatorname{etr}[-\Sigma_0^{-1}\{G_1 + N_1(z_{1.} - \boldsymbol{\mu}_0)(\overline{z_{1.} - \boldsymbol{\mu}_0})'\}]. \quad (9.1)$$

The moments of λ_7 are given by

$$E(\lambda_7^h) = \left(\frac{e}{N_1}\right)^{phN_1} \frac{1}{(1+h)^{pN_1(1+h)}} \prod_{i=1}^{p} \frac{\Gamma(N_1 - i + N_1 h)}{\Gamma(N_1 - i)}. \quad (9.2)$$

Using the first four moments, Chang, Krishnaiah and Lee (1975, 1977) approximated the distribution of $\lambda_7^{1/b}$ with Pearson type I distribution, where b is a suitably chosen integer. This approximation is used to compute the values of c_7 for $M = n_1 - p - 1 = 1(1)20(2)30$, $n_1 = N_1 - 1$ and $p = 2, 3, 4, 5, 6$, where

$$P[\tilde{\lambda}_7 \leq c_7 | H_7] = (1 - \alpha).$$

and $\tilde{\lambda}_7 = -2 \log \lambda_7$. These values are given in Table 13 (see Appendix) for $\alpha = 0.05, 0.01, 0.025, 0.10$.

10. Applications in time series in the frequency domain

In this section we discuss as to how the likelihood ratio test procedures on the covariance matrices of the complex multivariate normal populations can be used in the area of inference on multiple time series.

Let $\boldsymbol{X}'(t) = (\boldsymbol{X}_1'(t), \ldots, \boldsymbol{X}_q'(t))$ $(t = 1, \ldots, T)$ form a Gaussian, stationary, multiple time series with zero means and covariance matrix $R(s) = (R_{jk}(s))$, where $R_{jk}(s) = E\{\boldsymbol{X}_j(t)\boldsymbol{X}_k'(t+s)\}$ and $\boldsymbol{X}_j(t)$ is of order $p_j \times 1$. The spectral density matrix of the above time series is given by $F(\omega) = (F_{lj}(\omega))$, where

$$F_{lj}(\omega) = \frac{1}{2\pi} \sum_{s=-\infty}^{\infty} e^{-i\omega s} R_{lj}(s). \quad (10.1)$$

A well-known estimate (e.g. see Brillinger (1974)) of $F(\omega)$ is $\hat{F}(\omega) = (\hat{F}_{lj}(\omega))$, where

$$\hat{F}_{lj}(\omega) = \frac{1}{(2m+1)} \sum_{r=-m}^{m} I_{lj}\left(\omega + \frac{2\pi r}{T}\right) \quad (10.2)$$

and m is a suitably chosen integer. In (10.2)

$$\boldsymbol{Z}_l(\lambda) = \frac{1}{\sqrt{2\pi T}} \sum_{t=1}^{T} \boldsymbol{X}_l(t) \exp(-it\lambda),$$

and

$$I_{lj}(\lambda) = Z_l(\lambda)\bar{Z}_l'(\lambda).$$

Goodman (1963b) and Wahba (1968) showed that $A(\omega) = (2m+1)\hat{F}(\omega)$ is approximately distributed as complex Wishart distribution with $(2m+1)$ degrees of freedom and $E(\hat{F}(\omega)) = F(\omega)$.

Now, let $H_2(\omega)$, $H_3(\omega)$ and $H_4(\omega)$ denote the following hypotheses:

$$\begin{aligned} H_2(\omega)&: \quad F_{lj}(\omega) = 0 \quad (l \neq j = 1, \ldots, q), \\ H_3(\omega)&: \quad F(\omega) = \sigma^2 F_0(\omega), \\ H_4(\omega)&: \quad F(\omega) = F_0(\omega), \end{aligned} \tag{10.3}$$

where σ^2 is unknown and $F_0(\omega)$ is known. Let the statistics $L_2(\omega)$, $L_3(\omega)$, and $L_4(\omega)$ be defined as follows:

$$L_2(\omega) = \frac{|A(\omega)|}{\prod_{i=1}^{q} |A_{ii}(\omega)|}, \tag{10.4}$$

$$L_3(\omega) = \frac{|A(\omega)F_0^{-1}(\omega)|}{\{\operatorname{tr} A(\omega)F_0^{-1}(\omega)/s\}^s}, \tag{10.5}$$

$$L_4(\omega) = (e/(2m+1))^{s(2m+1)}|A(\omega)F_0^{-1}(\omega)|^{(2m+1)} \operatorname{etr}(-A(\omega)F_0^{-1}(\omega)), \tag{10.6}$$

where $(2m+1)A_{ij}(\omega) = \hat{F}_{ij}(\omega)$. Also, let $\tilde{L}_i(\omega) = -2 \log L_i(\omega)$ for $i = 2, 3, 4$. The hypothesis $H_2(\omega)$ is accepted or rejected according as

$$\tilde{L}_2(\omega) \lesseqgtr d_2,$$

where

$$P[\tilde{L}_2(\omega) \leq d_2 \mid H_2(\omega)] = (1-\alpha). \tag{10.7}$$

We accept or reject $H_3(\omega)$ according as

$$\tilde{L}_3(\omega) \lesseqgtr d_3, \tag{10.8}$$

where

$$P[\tilde{L}_3(\omega) \leq d_3 \mid H_3(\omega)] = (1-\alpha). \tag{10.9}$$

Similarly, the hypothesis $H_4(\omega)$ is accepted or rejected according as

$$\tilde{L}_4(\omega) \lesseqgtr d_4,$$

where

$$P[\tilde{L}_4(\omega) \leq d_4 \mid H_4(\omega)] = (1-\alpha). \tag{10.10}$$

Since $A(\omega)$ is approximately distributed as the complex Wishart matrix, approximate values of d_2, d_3, and d_4 can be obtained from Table 8, Table 9, and Table 10, respectively (see Appendix).

Next, let $H_2 = \bigcap_{j=1}^{k} H_2(\omega_j)$, $H_3 = \bigcap_{j=1}^{k} H_3(\omega_j)$ and $H_4 = \bigcap_{j=1}^{k} H_4(\omega_j)$, where

$\omega_1, \ldots, \omega_k$ are widely separated. Then, we accept or reject H_2 according as

$$T_2 \lessgtr d_5, \tag{10.11}$$

where

$$P[T_2 \leq d_5 \mid H_2] = (1-\alpha) \tag{10.12}$$

and $T_2 = \max(\tilde{L}_2(\omega_1), \ldots, \tilde{L}_2(\omega_k))$. An alternative procedure is to accept or reject H_2 according as

$$T_2^* \lessgtr d_6, \tag{10.13}$$

where

$$P[T_2^* \leq d_6 \mid H_2] = (1-\alpha) \tag{10.14}$$

and

$$T_2^* = \prod_{j=1}^{k} \tilde{L}_2(\omega_j).$$

Since $\omega_1, \ldots, \omega_k$ are widely separated, $L_2(\omega_1), \ldots, L_2(\omega_k)$ are distributed independently. So the critical values d_5 and d_6 can be computed by using the methods discussed in this chapter. We can propose similar procedures to test H_3 and H_4.

Let H_5: $F(\omega_1) = \cdots = F(\omega_k)$, where the frequencies $\omega_1, \ldots, \omega_k$ are 'sufficiently' wide apart. Also, let $\tilde{L}_5(\omega) = -2 \log L_5(\omega)$, where

$$L_5 = \frac{\prod_{i=1}^{k} |\hat{F}(\omega_i)|^{(2m+1)}}{|\sum_{i=1}^{k} \hat{F}(\omega_i)/k|^{k(2m+1)}}.$$

Then, we accept or reject H_5 accordingly as $\tilde{L}_5 \lessgtr d_5^*$, where

$$P[\tilde{L}_5 \leq d_5^* \mid H_5] = (1-\alpha).$$

Since $\omega_1, \ldots, \omega_k$ are 'sufficiently' wide apart, $\hat{F}(\omega_1), \ldots, \hat{F}(\omega_k)$ are distributed independently. Also, $(2m+1)\hat{F}(\omega_i)$ is distributed approximately as the complex Wishart matrix with $(2m+1)$ degrees of freedom for $i = 1, \ldots, k$. Hence, the values of d_5^* can be obtained from Table 11 (see Appendix).

Next, let $\boldsymbol{X}_i'(t) = (\boldsymbol{X}_{i1}'(t), \ldots, \boldsymbol{X}_{iq}'(t))$, $(t = 1, \ldots, T_i)$ be a Gaussian, stationary, multiple time series with zero means and covariance matrix $R_i(s)$ and spectral density matrix $F_i(\omega)$, where $R_i(s) = (R_{iuv}(s))$ and $R_{iuv}(s) = E\{\boldsymbol{X}_{iu}(t)\boldsymbol{X}_{iv}'(t+s)\}$. Also, let $\boldsymbol{X}_1(t), \ldots, X_k(t)$ be distributed independently and $\boldsymbol{X}_i(t)$ be of order $p \times 1$ for $i = 1, \ldots, k$. Let the estimate $\hat{F}_i(\omega)$ of $F_i(\omega)$ be defined in a similar way as $\hat{F}(\omega)$. Here, $(2m_i + 1)\hat{F}_i(\omega)$ is distributed approximately as the complex Wishart matrix with $2m_i + 1$ degrees freedom. The hypothesis $H_6(\omega)$: $F_1(\omega) = \cdots = F_k(\omega)$ is tested as follows. We accept or reject H_6 accordingly as

$$\tilde{L}_6(\omega) \lessgtr d_6^*,$$

where

$$P[\tilde{L}_6(\omega) \leq d_6^* \mid H_6] = (1 - \alpha),$$

$$\tilde{L}_6(\omega) = -2 \log L_6(\omega),$$

$$L_6(\omega) = \frac{\prod_{i=1}^{k} |\hat{F}_i(\omega)|^{(2m_i+1)}}{|\sum_{i=1}^{k} (2m_i + 1)\hat{F}_i(\omega)/m_0|^{m_0}}$$

and $m_0 = 2(\sum_{i=1}^{k} m_i) + k$. The critical values d_6^* can be obtained from Table 11 when the m_i's are equal.

We will now illustrate the usefulness of some of the tables in this paper with vibration data on a C-5A transport aircraft.

Vibration measurements have been taken on the cargo deck of a C-5A transport aircraft to provide information about the dynamic environments that cargo must survive in transit and to understand better the distribution and transmission of vibrational energy throughout the aircraft structure. Measurements have been taken over certain periods by locating accelerometers at different locations on the cargo deck. We will treat each location as a variable. Data on the variables shown in Table 14 were taken.

Table 14

Variables	Longitudinal location	Lateral location	Directional orientation
1 (FRV)	Forward	Right	Vertical
2 (FRL)	Forward	Right	Lateral
3 (FLV)	Forward	Left	Vertical
4 (FLL)	Forward	Left	Lateral
5 (ARV)	Aft	Right	Vertical
6 (ARL)	Aft	Right	Lateral
7 (ALV)	Aft	Left	Vertical
8 (ALL)	Aft	Left	Lateral

The basic unit of measurement is the acceleration due to gravity ($1g = 980\ \text{cm/sec}^2$). Let the spectral density of the data on the above 8 variables at frequency ω be denoted by $\hat{F}(\omega)$ and let the corresponding population spectral density matrix be denoted by $F(\omega)$. The sample spectral density matrix at frequency ω_j is $\hat{F}(\omega_j) = S_{j0} + \mathrm{i}S_{j1}$, where $\omega_1 = 0.15907$ Hz, $\omega_a = a\omega_1$, $a = 2, 3, 4, 5$ and

$$S_{10} = \begin{bmatrix} 0.311200 & 0.0027030 & 0.3204000 & 0.0033170 & 0.4229000 & -0.0007585 & 0.4451000 & -0.0052180 \\ 0.0027030 & 0.0014000 & 0.0026990 & -0.0008533 & 0.0026720 & 0.0027410 & 0.0015100 & -0.0031180 \\ 0.3204000 & 0.0026990 & 0.3316000 & 0.0036660 & 0.4340000 & -0.0008463 & 0.4608000 & -0.0055720 \\ 0.0033170 & -0.0008533 & 0.0036660 & 0.0008898 & 0.0047820 & -0.0020820 & 0.0073780 & 0.0021640 \\ 0.4229000 & 0.0026720 & 0.4340000 & 0.0047820 & 0.6178000 & -0.0051030 & 0.6390000 & -0.0034100 \\ -0.0007585 & 0.0027410 & -0.0008463 & -0.0020820 & -0.0051030 & 0.0080290 & -0.0098130 & -0.0090220 \\ 0.4451000 & 0.0015100 & 0.4608000 & 0.0073780 & 0.6390000 & -0.0098130 & 0.7003000 & 0.0008291 \\ -0.0052180 & -0.0031180 & -0.0055720 & 0.0021640 & -0.0034100 & -0.0090220 & 0.0008291 & 0.0104800 \end{bmatrix},$$

$$S_{11} = \begin{bmatrix} 0.0000000 & -0.0044000 & -0.0104100 & 0.0004039 & 0.0915800 & -0.0152700 & 0.0657400 & 0.0159700 \\ 0.0044000 & 0.0000000 & 0.0049750 & 0.0003763 & 0.0061860 & -0.0005155 & 0.0102700 & 0.0001463 \\ 0.0104100 & -0.0049750 & 0.0000000 & 0.0007700 & 0.1081000 & -0.0172500 & 0.0805500 & 0.0181600 \\ -0.0004039 & -0.0003763 & -0.0007700 & 0.0000000 & 0.0013310 & -0.0007427 & -0.0010270 & 0.0010250 \\ -0.0915800 & -0.0061860 & -0.1081000 & -0.0013310 & 0.0000000 & -0.0191800 & -0.0460900 & 0.0216900 \\ 0.0152700 & 0.0005155 & 0.0172500 & 0.0007427 & 0.0191800 & 0.0000000 & 0.0296300 & -0.0011300 \\ -0.0657400 & -0.0102700 & -0.0805500 & 0.0010270 & 0.0460900 & -0.0296300 & 0.0000000 & 0.0331300 \\ -0.0159700 & -0.0001463 & -0.0181600 & -0.0010250 & -0.0216900 & 0.0011300 & -0.0331300 & 0.0000000 \end{bmatrix},$$

$$S_{20} = \begin{bmatrix} 0.0700800 & 0.0005669 & 0.0728700 & -0.0003783 & 0.1112000 & 0.0004678 & 0.1069000 & -0.0016200 \\ 0.0005689 & 0.0004357 & 0.0007168 & -0.0003547 & -0.0001567 & 0.0003975 & -0.0002877 & -0.0003964 \\ 0.0728700 & 0.0007163 & 0.0765300 & -0.0003989 & 0.1129000 & 0.0006873 & 0.1096000 & -0.0018970 \\ -0.0003783 & -0.0003547 & -0.0003989 & 0.0004464 & -0.0007892 & -0.0003550 & -0.0010090 & 0.0003924 \\ 0.1112000 & -0.0001607 & 0.1129000 & -0.0007892 & 0.2341000 & 0.0000914 & 0.2381000 & -0.0020400 \\ 0.0004678 & 0.0003975 & 0.0006873 & -0.0003550 & 0.0000914 & 0.0013360 & -0.0014520 & -0.0014630 \\ 0.1069000 & -0.0002877 & 0.1096000 & -0.0010090 & 0.2381000 & -0.0014520 & 0.2754000 & -0.0005648 \\ -0.0016200 & -0.0003964 & -0.0018970 & 0.0003924 & -0.0020400 & -0.0014630 & -0.0005648 & 0.0017120 \end{bmatrix},$$

$$S_{21} = \begin{bmatrix} 0.0000000 & 0.0010550 & -0.0024270 & -0.0024020 & 0.0413600 & 0.0023660 & 0.0465300 & -0.0027390 \\ -0.0010550 & 0.0000000 & -0.0011480 & 0.0000026 & -0.0022600 & -0.0001995 & -0.0020950 & -0.0002051 \\ 0.0024270 & 0.0011480 & 0.0000000 & -0.0025040 & 0.0468300 & 0.0022950 & 0.0524000 & -0.0026480 \\ 0.0024020 & -0.0000026 & 0.0025040 & 0.0000000 & 0.0041830 & -0.0000834 & 0.0038860 & 0.0000812 \\ -0.0413600 & 0.0022600 & -0.0465300 & -0.0041830 & 0.0000000 & 0.0042450 & 0.0130200 & -0.0050250 \\ -0.0023660 & -0.0001995 & -0.0022950 & 0.0000834 & -0.0042450 & 0.0000000 & -0.0029430 & -0.0000308 \\ -0.0465300 & 0.0020950 & -0.0524000 & -0.0038860 & -0.0130200 & 0.0029430 & 0.0000000 & -0.0038830 \\ 0.0027390 & 0.0002051 & 0.0026480 & -0.0000812 & 0.0051250 & 0.0000308 & 0.0038830 & 0.0000000 \end{bmatrix},$$

$$S_{30} = \begin{bmatrix} 0.0184600 & 0.0008811 & 0.0185700 & -0.0005107 & 0.0180400 & 0.0011710 & 0.0220500 & -0.0015350 \\ 0.0008811 & 0.0007630 & 0.0009133 & -0.0006139 & 0.0034630 & -0.0002562 & 0.0039880 & 0.0003266 \\ 0.0185700 & 0.0009133 & 0.0196800 & -0.0006396 & 0.0147100 & 0.0012600 & 0.0225900 & -0.0015980 \\ -0.0005107 & -0.0006139 & -0.0006396 & 0.0005513 & -0.0028510 & 0.0001858 & -0.0037240 & -0.0002224 \\ 0.0180400 & 0.0034630 & 0.0147100 & -0.0028510 & 0.1323000 & 0.0018900 & 0.1422000 & 0.0042980 \\ 0.0011710 & -0.0002562 & 0.0012600 & 0.0001858 & 0.0018900 & 0.0015540 & 0.0018380 & -0.0017730 \\ 0.0220500 & 0.0039880 & 0.0225900 & -0.0037240 & 0.1422000 & 0.0018380 & 0.1801000 & -0.0045580 \\ -0.0015350 & 0.0003266 & -0.0015980 & -0.0002224 & -0.0042980 & -0.0017730 & -0.0045580 & 0.0021400 \end{bmatrix},$$

$$S_{31} = \begin{bmatrix} 0.0000000 & -0.0006045 & 0.0012290 & 0.0004898 & -0.0022730 & 0.0014420 & 0.0030790 & -0.0016530 \\ 0.0006045 & 0.0000000 & 0.0002828 & 0.0000029 & 0.0021210 & 0.0006690 & 0.0011520 & -0.0007933 \\ -0.0012290 & -0.0002828 & 0.0000000 & 0.0001996 & 0.0021090 & 0.0013270 & 0.0077890 & -0.0015220 \\ -0.0004898 & -0.000029 & -0.0001996 & 0.0000000 & -0.0026050 & -0.0006137 & -0.0019060 & 0.0007250 \\ 0.0022730 & -0.0021210 & -0.0021090 & 0.0026050 & 0.0000000 & 0.0065620 & -0.0284800 & -0.0080510 \\ -0.0014420 & -0.0006690 & -0.0013270 & 0.0006137 & -0.0065620 & 0.0000000 & -0.0075690 & 0.0001471 \\ -0.0030790 & -0.0011520 & -0.0077890 & 0.0019060 & 0.0284800 & 0.0075690 & 0.0000000 & -0.0095390 \\ 0.0016530 & 0.0007933 & 0.0015220 & -0.0007250 & 0.0080510 & -0.0001471 & 0.0095390 & 0.0000000 \end{bmatrix},$$

$$S_{40} = \begin{bmatrix} 0.0178100 & 0.0009329 & 0.0173700 & -0.0009332 & -0.0019830 & -0.0020970 & -0.0035530 & 0.0023870 \\ 0.0009329 & 0.0010400 & 0.0011190 & -0.0010330 & 0.0036700 & 0.0001542 & 0.0027770 & -0.0001921 \\ 0.0173700 & 0.0011190 & 0.0177500 & -0.0011380 & -0.0043440 & -0.0016860 & -0.0056940 & 0.0018640 \\ -0.0009332 & -0.0010330 & -0.0011380 & 0.0011560 & -0.0039250 & 0.0000601 & -0.0026070 & -0.0000872 \\ -0.0019830 & 0.0036700 & -0.0043440 & -0.0039250 & 0.0891900 & 0.0027140 & 0.0974400 & -0.0030700 \\ -0.0020970 & 0.0001542 & -0.0016860 & 0.0000601 & 0.0027140 & 0.0018160 & 0.0031220 & -0.0020710 \\ -0.0035530 & 0.0027770 & -0.0056940 & -0.0026070 & 0.0974400 & 0.0031220 & 0.1220000 & -0.0037580 \\ 0.0023870 & -0.0001921 & 0.0018640 & -0.0000872 & -0.0030700 & -0.0020710 & -0.0037580 & 0.0024530 \end{bmatrix},$$

$$S_{41} = \begin{bmatrix} 0.0000000 & 0.0004232 & -0.0022300 & -0.0003321 & 0.0283700 & 0.0001471 & 0.0295500 & 0.0000247 \\ -0.0004232 & 0.0000000 & -0.0004769 & 0.0002420 & 0.0017030 & 0.0010000 & 0.0017070 & -0.0011890 \\ 0.0022300 & 0.0004769 & 0.0000000 & -0.0002708 & 0.0251500 & 0.0000198 & 0.0257100 & 0.0001436 \\ 0.0003321 & -0.0002420 & 0.0002708 & 0.0000000 & -0.0029520 & -0.0011320 & -0.0029710 & 0.0013630 \\ -0.0283700 & -0.0017030 & -0.0251500 & 0.0029520 & 0.0000000 & 0.0062310 & 0.0078030 & -0.0085380 \\ -0.0001471 & -0.0010000 & -0.0000198 & 0.0011320 & -0.0062310 & 0.0000000 & -0.0056680 & -0.0000241 \\ -0.0295500 & -0.0017070 & -0.0257100 & 0.0029710 & -0.0078030 & 0.0056680 & 0.0000000 & -0.0081600 \\ -0.0000247 & 0.0011890 & -0.0001436 & -0.0013630 & 0.0085380 & 0.0000241 & 0.0081600 & 0.0000000 \end{bmatrix},$$

$$S_{50} = \begin{bmatrix} 0.0177900 & -0.0002297 & 0.0194000 & 0.0007346 & -0.0168300 & 0.0010770 & -0.0202900 & -0.0010980 \\ -0.0002297 & 0.0006892 & -0.0003287 & -0.0006763 & -0.0011050 & -0.0004262 & -0.0007752 & 0.0004927 \\ 0.0194000 & -0.0003287 & 0.0219900 & 0.0010720 & -0.0203800 & 0.0020040 & -0.0240700 & -0.0021080 \\ 0.0007346 & -0.0006763 & 0.0010720 & 0.0007642 & -0.0000531 & 0.0006507 & -0.0006490 & -0.0007398 \\ -0.0168300 & -0.0011050 & -0.0203800 & -0.0000531 & 0.0916600 & -0.0017990 & 0.1052000 & 0.0009515 \\ 0.0010770 & -0.0004262 & 0.0020040 & 0.0006507 & -0.0017990 & 0.0019100 & -0.0011770 & -0.0020380 \\ -0.0202900 & -0.0007752 & -0.0240760 & -0.0006490 & 0.1052000 & -0.0011770 & 0.1295000 & 0.0001445 \\ -0.0010980 & 0.0004927 & -0.0021080 & -0.0007398 & 0.0009515 & -0.0020380 & 0.0001445 & 0.0022160 \end{bmatrix},$$

$$S_{51} = \begin{bmatrix} 0.0000000 & -0.0004046 & 0.0002579 & 0.0004528 & 0.0143100 & 0.0002641 & 0.0135000 & -0.0005494 \\ 0.0004046 & 0.0000000 & 0.0003417 & -0.0000041 & 0.0002183 & 0.0004148 & 0.0010520 & -0.0004537 \\ -0.0002579 & -0.0003417 & 0.0000000 & 0.0003401 & 0.0162500 & -0.0000514 & 0.0160100 & -0.0002829 \\ -0.0004523 & 0.0000041 & -0.0003401 & 0.0000000 & -0.0001052 & -0.0004866 & -0.0009598 & 0.0005184 \\ -0.0143100 & -0.0002183 & -0.0162500 & 0.0001052 & 0.0000000 & -0.0024870 & -0.0100500 & 0.0022780 \\ -0.0002641 & -0.0004148 & 0.0000514 & 0.0004866 & 0.0024870 & 0.0000000 & 0.0022010 & -0.0000928 \\ -0.0135000 & -0.0010520 & -0.0160100 & 0.0009598 & 0.0100500 & -0.0022010 & 0.0000000 & 0.0017320 \\ 0.0005494 & 0.0004537 & 0.0002829 & -0.0005184 & -0.0022780 & 0.0000928 & -0.0017320 & 0.0000000 \end{bmatrix}.$$

Let $\hat{F}(\omega)$ be partitioned as

$$\hat{F}(\omega) = \begin{bmatrix} \hat{F}_{11}(\omega) & \hat{F}_{12}(\omega) & \hat{F}_{13}(\omega) & \hat{F}_{14}(\omega) \\ \hat{F}_{21}(\omega) & \hat{F}_{22}(\omega) & \hat{F}_{23}(\omega) & \hat{F}_{24}(\omega) \\ \hat{F}_{31}(\omega) & \hat{F}_{32}(\omega) & \hat{F}_{33}(\omega) & \hat{F}_{34}(\omega) \\ \hat{F}_{41}(\omega) & \hat{F}_{42}(\omega) & \hat{F}_{43}(\omega) & \hat{F}_{44}(\omega) \end{bmatrix},$$

where $\hat{F}_{ii}(\omega)$ is of order 2×2 for $i = 1, 2, 3, 4$. We computed $\tilde{L}_2(\omega_j) = -2 \log L_2(\omega_j)$, where

$$L_2(\omega_j) = \frac{|(2m+1)\hat{F}(\omega_j)|}{\Pi_{i=1}^{4} |(2m+1)\hat{F}_{ii}(\omega_j)|}$$

and $(2m + 1) = 19$. The values of $\tilde{L}_2(\omega_j)$ in this case are found to be 47.760, 31,667, 33.684, 35.646 and 37.738, respectively. The value of the critical value d_2 for $n = 19$, $s = 8$ and $q = 4$ from Table 8 (see Appendix) is found to be 4.217 at 5% significance level. Since the computed values of $L_2(\omega_j)$ are greater than the value from the table, we conclude that the sets (1, 2), (3, 4), (5, 6), (7, 8) of variables are not independent for each of the five frequencies considered.

Next, we computed the value of $\tilde{L}_3(\omega_j) = -2 \log L_3(\omega_j)$, where

$$L_3(\omega_j) = \frac{|(2m+1)\hat{F}(\omega_j)|}{\{(2m+1) \operatorname{tr} \hat{F}(\omega_j)/s\}^s},$$

$s = 8$, $(2m + 1) = 19$. It is found that the values of $\tilde{L}_3(\omega_j)$ in this case are 78.321, 65.267, 63.226, 59.019 and 61.852 respectively. The critical value d_3 for $n = 19$ and $s = 8$ is found to be 5.142 at 5% level. So, we reject (individually) the hypotheses that $F(\omega_j) = \sigma^2 I_p$ for $j = 1, 2, 3, 4, 5$.

For the hypothesis H_5 we will consider only the first four variables. Let spectral density matrix of the data on the first four variables at frequency ω_i be denoted by $\hat{F}(\omega_i)$ and let the corresponding population spectral density matrix be denoted by $F(\omega_i)$. The sample spectral density matrices at frequencies 0.15907 Hz, 0.47721 Hz and 0.79535 Hz are $\hat{F}(0.15907) = S_{10} + \mathrm{i}S_{11}$, $F(0.47721) = S_{20} + \mathrm{i}S_{21}$ and $\hat{F}(0.79535) = S_{30} + \mathrm{i}S_{31}$, respectively, where

$$S_{10} = \begin{bmatrix} 0.3112000 & 0.0027030 & 0.3204000 & 0.0033170 \\ 0.0027030 & 0.0014000 & 0.0026990 & -0.0008533 \\ 0.3204000 & 0.0026990 & 0.3316000 & 0.0036660 \\ 0.0033170 & -0.0008533 & 0.0036660 & 0.0008898 \end{bmatrix},$$

$$S_{11} = \begin{bmatrix} 0.0000000 & -0.0044000 & -0.0104100 & 0.0004039 \\ 0.0044000 & 0.0000000 & 0.0049750 & 0.0003763 \\ 0.0104100 & -0.0049750 & 0.0000000 & 0.0007700 \\ -0.0004039 & -0.0003763 & -0.0007700 & 0.0000000 \end{bmatrix},$$

$$S_{20} = \begin{bmatrix} 0.0184600 & 0.0008811 & 0.0185700 & -0.0005107 \\ 0.0008811 & 0.0007630 & 0.0009133 & -0.0006139 \\ 0.0185700 & 0.0009133 & 0.0196800 & -0.0006396 \\ -0.0005107 & -0.0006139 & -0.0006396 & 0.0005513 \end{bmatrix},$$

$$S_{21} = \begin{bmatrix} 0.0000000 & -0.0006045 & 0.0012290 & 0.0004898 \\ 0.0006045 & 0.0000000 & 0.0002828 & 0.0000029 \\ -0.0012290 & -0.0002828 & 0.0000000 & 0.0001996 \\ -0.0004898 & -0.0000029 & -0.0001996 & 0.0000000 \end{bmatrix},$$

$$S_{30} = \begin{bmatrix} 0.0177900 & -0.0002297 & 0.0194000 & 0.0007346 \\ -0.0002297 & 0.0006892 & -0.0003287 & -0.0006763 \\ 0.0194000 & -0.0003287 & 0.0219900 & 0.0010720 \\ 0.0007346 & -0.0006763 & 0.0010720 & 0.0007642 \end{bmatrix},$$

$$S_{31} = \begin{bmatrix} 0.0000000 & -0.0004046 & 0.0002579 & 0.0004523 \\ 0.0004046 & 0.0000000 & 0.0003417 & -0.0000041 \\ -0.0002579 & -0.0003417 & 0.0000000 & 0.0003401 \\ -0.0004523 & 0.0000041 & -0.0003401 & 0.0000000 \end{bmatrix}.$$

We computed the value of $\tilde{L}_5 = -2 \log L_5$, where

$$L_5 = \frac{\prod_{i=1}^{3} |A(\omega_i)/(2m+1)|^{(2m+1)}}{|\sum_{i=1}^{3} A(\omega_i)/3(2m+1)|^{3(2m+1)}},$$

$\omega_1 = 0.15907$ Hz, $\omega_2 = 0.47721$ Hz, $\omega_3 = 0.79535$ Hz, $(2m+1) = 19$. The value of $\tilde{L}_5$ is found to be 224.72. The critical value d_5^* for $n_0 = 19$, $p = 4$ and $q = 3$ from Table 11 (see Appendix) is found to be 50.93 at the 5% significance level. Since $\tilde{L}_5$ is greater than the value from the table, we conclude that the spectral density matrices $F(\omega_1)$, $F(\omega_2)$, and $F(\omega_3)$ are significantly different from one another.

Appendix

Tables 7 through 13 give percentage points of various distributions discussed in this chapter. These tables are useful in implementation of the likelihood ratio test procedures. A description of these tables is given below.

Table 7. The entries in this table give the values of c_1 where

$$P[C_1 \leq c_1] = (1-\alpha),$$

$$C_1 = -\{(2n+q-p)\log U\}/\chi^2_{2pq,\alpha} \quad \text{and} \quad U = |A_1(A_1+A_2)^{-1}|.$$

Table 7
Percentage points of the distribution of the determinant of the complex multivariate beta matrix

	$p=3$	$q=3$	$p=3$	$q=4$	$p=3$	$q=5$	$p=3$	$q=6$	$p=3$	$q=7$	$p=3$	$q=8$
M α	0.050	0.010	0.050	0.010	0.050	0.010	0.050	0.010	0.050	0.010	0.050	0.010
1	1.310	1.378	1.343	1.414	1.377	1.453	1.412	1.491	1.446	1.528	1.478	1.564
2	1.132	1.153	1.154	1.177	1.177	1.202	1.201	1.228	1.225	1.253	1.249	1.278
3	1.076	1.087	1.092	1.104	1.109	1.122	1.127	1.141	1.145	1.160	1.163	1.179
4	1.050	1.057	1.062	1.070	1.075	1.083	1.089	1.098	1.103	1.113	1.117	1.128
5	1.036	1.041	1.045	1.050	1.055	1.061	1.066	1.073	1.078	1.085	1.089	1.097
6	1.027	1.030	1.034	1.038	1.043	1.047	1.051	1.056	1.061	1.066	1.071	1.077
7	1.021	1.024	1.027	1.030	1.034	1.037	1.041	1.045	1.049	1.054	1.057	1.062
8	1.017	1.019	1.022	1.024	1.028	1.030	1.034	1.037	1.041	1.044	1.048	1.052
9	1.014	1.016	1.018	1.020	1.023	1.025	1.029	1.031	1.034	1.037	1.040	1.044
10	1.011	1.013	1.015	1.017	1.019	1.021	1.024	1.026	1.029	1.032	1.035	1.037
12	1.008	1.009	1.011	1.012	1.015	1.016	1.018	1.020	1.022	1.024	1.026	1.028
14	1.007	1.007	1.009	1.010	1.011	1.012	1.014	1.015	1.017	1.019	1.021	1.022
16	1.005	1.006	1.007	1.008	1.009	1.010	1.011	1.012	1.014	1.015	1.017	1.018
18	1.004	1.005	1.006	1.006	1.007	1.008	1.009	1.010	1.011	1.012	1.014	1.015
20	1.003	1.004	1.005	1.005	1.006	1.007	1.008	1.008	1.010	1.010	1.012	1.012
30	1.002	1.002	1.002	1.002	1.003	1.003	1.004	1.004	1.005	1.005	1.006	1.006
60	1.000	1.000	1.001	1.001	1.001	1.001	1.001	1.001	1.001	1.001	1.002	1.002
120	1.000	1.000	1.000	1.000	1.000	1.000	1.000	1.000	1.000	1.000	1.001	1.001
$\chi^2_{2pq,\alpha}$	28.8693	34.8053	36.4150	42.9798	43.7730	50.8922	50.9985	58.6192	58.1240	66.2062	65.1708	73.6826

Table 7 (continued)

	$p=3$	$q=9$	$p=3$	$q=10$	$p=4$	$q=4$	$p=4$	$q=5$	$p=4$	$q=6$	$p=4$	$q=7$
M α	0.050	0.010	0.050	0.010	0.050	0.010	0.050	0.010	0.050	0.010	0.050	0.010
1	1.509	1.598	1.538	1.630	1.354	1.425	1.373	1.444	1.394	1.466	1.416	1.490
2	1.271	1.303	1.293	1.326	1.166	1.189	1.182	1.205	1.199	1.223	1.217	1.242
3	1.180	1.198	1.197	1.216	1.102	1.114	1.114	1.127	1.128	1.141	1.142	1.156
4	1.131	1.143	1.145	1.158	1.070	1.078	1.080	1.088	1.091	1.100	1.103	1.112
5	1.101	1.109	1.113	1.122	1.052	1.057	1.060	1.065	1.069	1.075	1.078	1.085
6	1.081	1.087	1.090	1.097	1.040	1.044	1.047	1.051	1.054	1.059	1.062	1.067
7	1.066	1.071	1.074	1.080	1.032	1.035	1.037	1.041	1.044	1.047	1.051	1.054
8	1.055	1.059	1.062	1.067	1.026	1.028	1.031	1.033	1.036	1.039	1.042	1.045
9	1.047	1.050	1.053	1.057	1.021	1.024	1.026	1.028	1.030	1.033	1.035	1.038
10	1.040	1.043	1.046	1.049	1.018	1.020	1.022	1.024	1.026	1.028	1.030	1.033
12	1.031	1.033	1.035	1.038	1.013	1.015	1.016	1.018	1.020	1.021	1.023	1.025
14	1.024	1.026	1.028	1.030	1.010	1.011	1.013	1.014	1.015	1.016	1.018	1.020
16	1.020	1.021	1.023	1.025	1.008	1.009	1.010	1.011	1.012	1.013	1.015	1.016
18	1.016	1.018	1.019	1.020	1.007	1.007	1.008	1.009	1.010	1.011	1.012	1.013
20	1.014	1.015	1.016	1.017	1.006	1.006	1.007	1.008	1.009	1.009	1.010	1.011
30	1.007	1.008	1.008	1.009	1.003	1.003	1.003	1.004	1.004	1.004	1.005	1.005
60	1.002	1.002	1.002	1.002	1.001	1.001	1.001	1.001	1.001	1.001	1.001	1.001
120	1.001	1.001	1.001	1.001	1.000	1.000	1.000	1.000	1.000	1.000	1.000	1.000
$\chi^2_{2pq,\alpha}$	72.1532	81.0688	79.0819	88.3794	46.1943	53.4858	55.7585	63.6907	65.1708	73.6826	74.4683	83.5134

Table 7 (continued)

	$p = 4$	$q = 8$	$p = 4$	$q = 9$	$p = 4$	$q = 10$	$p = 5$	$q = 5$	$p = 5$	$q = 6$	$p = 5$	$q = 7$
M α	0.050	0.010	0.050	0.010	0.050	0.010	0.050	0.010	0.050	0.010	0.050	0.010
1	1.440	1.514	1.463	1.539	1.485	1.562	1.378	1.448	1.390	1.458	1.403	1.472
2	1.235	1.261	1.254	1.280	1.272	1.299	1.191	1.214	1.203	1.226	1.216	1.240
3	1.157	1.171	1.172	1.187	1.186	1.202	1.123	1.135	1.133	1.145	1.144	1.157
4	1.115	1.124	1.127	1.137	1.139	1.149	1.087	1.095	1.096	1.104	1.105	1.114
5	1.088	1.095	1.098	1.106	1.108	1.116	1.066	1.072	1.073	1.079	1.081	1.087
6	1.070	1.076	1.079	1.084	1.088	1.094	1.052	1.056	1.058	1.062	1.065	1.069
7	1.058	1.062	1.065	1.070	1.073	1.077	1.042	1.045	1.047	1.051	1.053	1.057
8	1.048	1.052	1.055	1.058	1.061	1.065	1.035	1.037	1.039	1.042	1.044	1.047
9	1.041	1.044	1.047	1.050	1.053	1.056	1.029	1.031	1.033	1.035	1.038	1.040
10	1.035	1.038	1.040	1.043	1.046	1.048	1.025	1.027	1.028	1.031	1.032	1.034
12	1.027	1.029	1.031	1.033	1.035	1.037	1.019	1.020	1.022	1.023	1.025	1.026
14	1.021	1.023	1.025	1.026	1.028	1.030	1.015	1.016	1.017	1.018	1.020	1.021
16	1.017	1.018	1.020	1.021	1.023	1.024	1.012	1.013	1.014	1.015	1.016	1.017
18	1.014	1.015	1.017	1.018	1.019	1.020	1.010	1.011	1.011	1.012	1.013	1.014
20	1.012	1.013	1.014	1.015	1.016	1.017	1.008	1.009	1.010	1.010	1.011	1.012
30	1.006	1.006	1.007	1.008	1.008	1.009	1.004	1.004	1.005	1.005	1.006	1.006
60	1.002	1.002	1.002	1.002	1.002	1.003	1.001	1.001	1.001	1.001	1.002	1.002
120	1.001	1.001	1.001	1.001	1.001	1.001	1.000	1.000	1.000	1.000	1.000	1.000
$\chi^2_{2pq,\alpha}$	83.6753	93.2168	92.8083	102.8160	101.8790	112.3290	67.5048	76.1539	79.0819	88.3794	90.5312	100.4250

Table 7 (continued)

	$p=5$	$q=8$	$p=5$	$q=9$	$p=5$	$q=10$	$p=6$	$q=6$	$p=6$	$q=7$	$p=6$	$q=8$
M α	0.050	0.010	0.050	0.010	0.050	0.010	0.050	0.010	0.050	0.010	0.050	0.010
1	1.419	1.488	1.436	1.505	1.453	1.522	1.392	1.459	1.400	1.465	1.410	1.474
2	1.230	1.254	1.245	1.269	1.259	1.284	1.210	1.233	1.219	1.242	1.230	1.252
3	1.156	1.169	1.168	1.182	1.180	1.194	1.140	1.152	1.148	1.161	1.158	1.171
4	1.115	1.124	1.125	1.135	1.136	1.145	1.102	1.110	1.110	1.118	1.118	1.126
5	1.089	1.096	1.098	1.105	1.107	1.114	1.079	1.084	1.085	1.091	1.092	1.098
6	1.072	1.077	1.079	1.084	1.087	1.092	1.063	1.067	1.068	1.073	1.075	1.079
7	1.059	1.063	1.066	1.070	1.072	1.077	1.051	0.055	1.056	1.060	1.062	1.066
8	1.050	1.053	1.056	1.059	1.062	1.065	1.043	1.046	1.047	1.050	1.052	1.055
9	1.042	1.045	1.048	1.050	1.053	1.056	1.037	1.039	1.040	1.043	1.045	1.047
10	1.037	1.039	1.041	1.044	1.046	1.049	1.031	1.034	1.035	1.037	1.039	1.041
12	1.028	1.030	1.032	1.034	1.036	1.038	1.024	1.026	1.027	1.029	1.030	1.032
14	1.022	1.024	1.025	1.027	1.029	1.030	1.019	1.020	1.021	1.023	1.024	1.025
16	1.018	1.019	1.021	1.022	1.024	1.025	1.015	1.017	1.018	1.018	1.020	1.021
18	1.015	1.016	1.017	1.018	1.020	1.021	1.013	1.014	1.014	1.015	1.016	1.017
20	1.013	1.013	1.015	1.015	1.017	1.017	1.011	1.011	1.012	1.013	1.014	1.014
30	1.007	1.007	1.008	1.008	1.009	1.009	1.006	1.006	1.006	1.007	1.007	1.007
60	1.002	1.002	1.002	1.002	1.003	1.003	1.002	1.002	1.002	1.002	1.002	1.002
120	1.001	1.001	1.001	1.001	1.001	1.001	1.000	1.000	1.000	1.001	1.001	1.001
$\chi^2_{2pq,\alpha}$	101.8790	112.3290	113.1450	124.1160	124.3420	135.8070	92.8083	102.8160	106.3950	117.0570	119.8710	131.1410

Table 7 (continued)

	$p = 6$	$q = 9$	$p = 6$	$q = 10$	$p = 7$	$q = 7$	$p = 7$	$q = 8$
M α	0.050	0.010	0.050	0.010	0.050	0.010	0.050	0.010
1	1.421	1.485	1.433	1.497	1.401	1.464	1.406	1.467
2	1.241	1.264	1.253	1.276	1.225	1.247	1.232	1.254
3	1.168	1.180	1.178	1.191	1.154	1.167	1.162	1.174
4	1.126	1.135	1.135	1.144	1.115	1.124	1.122	1.130
5	1.100	1.106	1.107	1.114	1.090	1.096	1.096	1.102
6	1.081	1.086	1.088	1.093	1.073	1.077	1.078	1.083
7	1.068	1.071	1.074	1.078	1.060	1.064	1.065	1.069
8	1.057	1.061	1.063	1.066	1.051	1.054	1.055	1.058
9	1.049	1.052	1.054	1.057	1.044	1.046	1.047	1.050
10	1.043	1.045	1.047	1.050	1.038	1.040	1.041	1.043
12	1.033	1.035	1.037	1.039	1.029	1.031	1.032	1.034
14	1.027	1.028	1.030	1.031	1.023	1.025	1.026	1.027
16	1.022	1.023	1.025	1.026	1.019	1.020	1.021	1.022
18	1.018	1.019	1.021	1.022	1.016	1.017	1.018	1.019
20	1.016	1.016	1.018	1.018	1.014	1.014	1.015	1.015
30	1.008	1.009	1.009	1.010	1.007	1.007	1.008	1.008
60	1.002	1.003	1.003	1.003	1.002	1.002	1.002	1.003
120	1.001	1.001	1.001	1.001	10..01	1.001	1.001	1.001
$\chi^2_{2pq,\alpha}$	133.2570	145.0990	146.5670	158.9500	122.1080	133.4760	137.7010	149.7270

Here A_1 and A_2 are distributed independently as central $p \times p$ complex Wishart matrices with n and q degrees of freedom respectively. These entries in this table are reproduced from a technical report by Lee, Krishnaiah and Chang (1975).

Table 8. The entries in this table give the values of c_2, where

$$P[\tilde{\lambda}_2 \leq c_2] = (1 - \alpha),$$

$\tilde{\lambda}_2 = -2 \log \lambda_2$ and

$$\lambda_2 = \frac{|A|}{\prod_{i=1}^{q} |A_{ii}|}.$$

Here, $A = (A_{ij})$: $pq \times pq$ is the central complex Wishart matrix with n degrees of freedom and $E(A) = n\Sigma$, and $\Sigma = (\Sigma_{ij})$. Also, A_{ij} and Σ_{ij} are of order $p \times p$. In addition, $\Sigma_{ij} = 0$ for $i \neq j = 1, 2, \ldots, q$, $M = n - s - 3$ and $s = pq$.

Krishnaiah, Lee and Chang (1976) gave the values of c_2 for $\alpha = 0.05$, $M = 1(1)4(2)16$, 20, 24, 30, $p = 1$, 2, 3, and $q = 3$, 4, 5. These entries are included in Table 8 with the kind permission of Biometrika Trustees. The remaining entries in the table are reproduced from a technical report by Krishnaiah, Lee and Chang (1975).

Table 8
Percentage points of the distribution $\tilde{\lambda}_2$ for multiple independence

	$q = 3$								
	$p = 1$			$p = 2$			$p = 3$		
M α	10%	5%	1%	10%	5%	1%	10%	5%	1%
1	1.895	2.243	3.002	4.679	5.143	6.096	7.936	8.481	9.575
2	1.607	1.902	2.543	4.085	4.488	5.314	7.061	7.542	8.504
3	1.395	1.651	2.207	3.628	3.984	4.713	6.368	6.798	7.661
4	1.233	1.459	1.949	3.264	3.584	4.238	5.802	6.193	6.974
5	1.104	1.306	1.746	2.967	3.258	3.851	5.331	5.690	6.405
6	1.001	1.184	1.581	2.722	2.987	3.530	4.933	5.265	5.925
7	0.914	1.081	1.445	2.513	2.759	3.259	4.592	4.899	5.512
8	0.842	0.996	1.330	2.335	2.562	3.027	4.295	4.582	5.155
9	0.780	0.923	1.233	2.180	2.393	2.826	4.035	4.305	4.841

Table 8 (continued)

	q = 3								
	p = 1			p = 2			p = 3		
M α	10%	5%	1%	10%	5%	1%	10%	5%	1%
10	0.727	0.859	1.148	2.045	2.244	2.651	3.805	4.059	4.565
11	0.680	0.805	1.075	1.926	2.113	2.495	3.600	3.840	4.318
12	0.639	0.756	1.010	1.819	1.997	2.358	3.416	3.644	4.098
13	0.603	0.713	0.952	1.724	1.892	2.235	3.251	3.467	3.898
14	0.570	0.675	0.902	1.639	1.799	2.124	3.101	3.307	3.718
15	0.542	0.641	0.856	1.562	1.714	2.023	2.963	3.161	3.553
16	0.515	0.610	0.814	1.491	1.636	1.931	2.838	3.027	3.403
17	0.492	0.582	0.776	1.427	1.566	1.848	2.723	2.904	3.265
18	0.470	0.556	0.742	1.368	1.501	1.772	2.618	2.791	3.138
19	0.450	0.532	0.711	1.313	1.441	1.702	2.520	2.687	3.020
20	0.432	0.511	0.682	1.263	1.386	1.637	2.429	2.590	2.911
22	0.400	0.473	0.631	1.174	1.288	1.521	2.265	2.416	2.715
24	0.372	0.439	0.587	1.096	1.202	1.420	2.122	2.263	2.544
26	0.347	0.410	0.548	1.028	1.128	1.332	1.997	2.129	2.393
28	0.326	0.386	0.515	0.968	1.062	1.254	1.886	2.010	2.259
30	0.307	0.364	0.485	0.915	1.004	1.185	1.786	1.904	2.140
	q = 4								
1	2.982	3.384	4.234	7.374	7.909	8.983	12.428	13.049	14.276
2	2.563	2.908	3.635	6.542	7.012	7.956	11.237	11.794	12.891
3	2.248	2.551	3.186	5.886	6.306	7.149	10.271	10.775	11.768
4	2.003	2.272	2.837	5.352	5.733	6.496	9.466	9.927	10.836
5	1.807	2.049	2.558	4.909	5.259	5.955	8.783	9.210	10.049
6	1.645	1.866	2.328	4.536	4.858	5.500	8.196	8.593	9.373
7	1.511	1.713	2.138	4.217	4.514	5.111	7.684	8.057	8.785
8	1.397	1.583	1.976	3.939	4.217	4.774	7.236	7.584	8.269
9	1.298	1.472	1.837	3.697	3.958	4.479	6.837	7.166	7.812
10	1.213	1.376	1.716	3.483	3.729	4.219	6.480	6.792	7.404
11	1.139	1.291	1.610	3.293	3.525	3.988	6.161	6.457	7.037
12	1.073	1.216	1.517	3.122	3.342	3.782	5.872	6.153	6.705
13	1.014	1.149	1.434	2.969	3.178	3.596	5.608	5.878	6.404
14	0.961	1.090	1.359	2.830	3.029	3.427	5.369	5.625	6.129
15	0.914	1.036	1.292	2.704	2.894	3.273	5.148	5.395	5.878
16	0.871	0.987	1.231	2.588	2.770	3.133	4.946	5.182	5.646
17	0.831	0.943	1.176	2.482	2.656	3.005	4.758	4.986	5.432
18	0.796	0.902	1.125	2.384	2.552	2.886	4.585	4.805	5.234
19	0.763	0.865	1.079	2.294	2.455	2.777	4.424	4.636	5.050
20	0.733	0.831	1.036	2.211	2.366	2.676	4.274	4.479	4.878
22	0.680	0.770	0.960	2.061	2.205	2.493	4.004	4.194	4.568
24	0.633	0.717	0.894	1.930	2.064	2.335	3.765	3.944	4.297
26	0.593	0.672	0.838	1.814	1.941	2.195	3.553	3.723	4.054
28	0.557	0.631	0.787	1.712	1.832	2.071	3.364	3.525	3.839
30	0.525	0.595	0.742	1.621	1.734	1.961	3.195	3.346	3.645

Table 8 (continued)

	$q=5$								
	$p=1$			$p=2$			$p=3$		
M α	10%	5%	1%	10%	5%	1%	10%	5%	1%
1	4.169	4.617	5.543	10.270	10.860	12.030	17.180	17.862	19.194
2	3.623	4.011	4.811	9.222	9.746	10.784	15.717	16.330	17.534
3	3.206	3.549	4.253	8.379	8.852	9.787	14.506	15.072	16.165
4	2.877	3.183	3.813	7.683	8.115	8.968	13.483	14.002	15.012
5	2.609	2.886	3.457	7.098	7.496	8.280	12.603	13.086	14.024
6	2.388	2.642	3.163	6.599	6.967	7.694	11.837	12.289	13.166
7	2.202	2.435	2.914	6.167	6.510	7.188	11.162	11.587	12.412
8	2.042	2.259	2.703	5.789	6.110	6.745	10.563	10.965	11.743
9	1.904	2.106	2.521	5.456	5.758	6.355	10.028	10.407	11.144
10	1.784	1.973	2.361	5.159	5.444	6.009	9.547	9.906	10.606
11	1.678	1.856	2.221	4.893	5.164	5.699	9.109	9.452	10.119
12	1.585	1.752	2.096	4.654	4.912	5.420	8.713	9.041	9.677
13	1.500	1.659	1.985	4.438	4.683	5.167	8.349	8.663	9.273
14	1.425	1.575	1.885	4.241	4.475	4.938	8.015	8.316	8.901
15	1.357	1.500	1.794	4.061	4.285	4.727	7.708	7.997	8.559
16	1.295	1.431	1.713	3.895	4.111	4.534	7.423	7.702	8.242
17	1.238	1.369	1.638	3.743	3.950	4.357	7.159	7.428	7.949
18	1.186	1.312	1.569	3.602	3.801	4.192	6.914	7.173	7.676
19	1.139	1.259	1.506	3.472	3.663	4.041	6.685	6.935	7.420
20	1.095	1.210	1.448	3.350	3.536	3.900	6.471	6.713	7.183
22	1.016	1.124	1.344	3.132	3.305	3.645	6.082	6.309	6.750
24	0.948	1.048	1.254	2.940	3.102	3.421	5.738	5.952	6.367
26	0.889	0.983	1.176	2.771	2.924	3.224	5.430	5.633	6.026
28	0.836	0.925	1.106	2.620	2.764	3.048	5.155	5.347	5.720
30	0.790	0.873	1.045	2.485	2.622	2.891	4.906	5.089	5.444

Table 9. The entries in this table are the values of c_3 where

$$P[\tilde{\lambda}_3 \leqslant c_3] = (1-\alpha),$$

$$\tilde{\lambda}_3 = -2\log\{|A\Sigma_0^{-1}|/(\mathrm{tr}\, A\Sigma_0^{-1}/s)^s\},$$

and $M = n - s - 3$. Also, A: $s \times s$ is distributed as the central complex Wishart matrix with n degrees of freedom and $E(A) = n\sigma^2\Sigma_0$, where σ^2 is unknown and Σ_0 is known. Upper 5% points of the distribution of $\tilde{\lambda}_3$ are given in Krishnaiah, Lee and Chang (1976) for $s = 7(1)10$ and $M = 1(1)5$, 7, 10, 15, 20, 30(5)50, 60. These entries are reproduced in Table 9 with the kind permission of Biometrika Trustees. The remaining entries in the table are reproduced from a technical report by Krishnaiah, Lee and Chang (1975).

Table 9
Percentage points of the likelihood ratio statistic for the sphericity test of complex covariance matrix

M	$s=2$	$s=3$	$s=4$	$s=5$	$s=6$	$s=7$	$s=8$	$s=9$	$s=10$
				$\alpha = 0.05$					
1	1.496	2.648	3.866	5.157	6.512	7.921	9.375	10.870	12.399
2	1.255	2.257	3.338	4.499	5.731	7.022	8.366	9.755	11.183
3	1.080	1.968	2.939	3.993	5.122	6.314	7.562	8.860	10.199
4	0.949	1.745	2.626	3.592	4.633	5.740	6.906	8.122	9.383
5	0.846	1.568	2.374	3.265	4.231	5.264	6.357	7.502	8.693
6	0.763	1.423	2.166	2.993	3.895	4.863	5.891	6.973	8.100
7	0.696	1.303	1.992	2.763	3.608	4.519	5.490	6.514	7.586
8	0.639	1.201	1.844	2.567	3.362	4.222	5.142	6.115	7.137
9	0.590	1.115	1.717	2.397	3.147	3.962	4.836	5.762	6.736
10	0.549	1.040	1.606	2.248	2.959	3.732	4.565	5.448	6.381
11	0.512	0.975	1.509	2.116	2.791	3.528	4.322	5.168	6.061
12	0.481	0.916	1.423	1.999	2.642	3.346	4.104	4.915	5.773
13	0.453	0.866	1.346	1.895	2.508	3.181	3.908	4.686	5.511
14	0.429	0.820	1.277	1.801	2.387	3.031	3.730	4.478	5.272
15	0.406	0.779	1.215	1.715	2.277	2.896	3.567	4.287	5.053
16	0.386	0.741	1.158	1.638	2.177	2.772	3.419	4.113	4.851
17	0.368	0.707	1.107	1.567	2.086	2.659	3.281	3.952	4.666
18	0.352	0.676	1.060	1.503	2.001	2.554	3.155	3.803	4.494
19	0.336	0.648	1.016	1.443	1.924	2.457	3.038	3.665	4.335
20	0.322	0.622	0.977	1.387	1.852	2.367	2.930	3.537	4.186
22	0.298	0.576	0.906	1.289	1.724	2.207	2.735	3.306	3.918
24	0.277	0.536	0.845	1.204	1.612	2.066	2.564	3.103	3.683
26	0.258	0.501	0.791	1.129	1.513	1.943	2.413	2.925	3.474
28	0.242	0.471	0.744	1.063	1.427	1.833	2.280	2.765	3.287
30	0.228	0.444	0.702	1.005	1.349	1.735	2.160	2.622	3.120
35	0.199	0.388	0.616	0.883	1.188	1.531	1.909	2.322	2.769
40	0.177	0.345	0.548	0.787	1.061	1.370	1.711	2.084	2.488
45	0.159	0.311	0.494	0.710	0.959	1.239	1.550	1.891	2.259
50	0.144	0.282	0.450	0.647	0.875	1.132	1.417	1.730	2.069
60	0.122	0.239	0.381	0.549	0.744	0.964	1.209	1.478	1.771
				$\alpha = 0.01$					
1	2.174	3.440	4.746	6.112	7.531	8.995	10.501	12.041	13.612
2	1.823	2.930	4.095	5.326	6.620	7.967	9.361	10.794	12.265
3	1.570	2.554	3.602	4.724	5.912	7.158	8.455	9.797	11.176
4	1.378	2.263	3.218	4.247	5.346	6.505	7.716	8.976	10.277
5	1.229	2.033	2.908	3.859	4.880	5.962	7.101	8.286	9.517
6	1.109	1.845	2.653	3.537	4.491	5.506	6.579	7.700	8.866
7	1.010	1.689	2.439	3.265	4.160	5.117	6.129	7.192	8.302
8	0.927	1.557	2.258	3.032	3.875	4.779	5.739	6.750	7.807
9	0.857	1.445	2.102	2.831	3.627	4.484	5.397	6.360	7.369
10	0.797	1.348	1.966	2.655	3.409	4.223	5.093	6.013	6.979
11	0.745	1.263	1.847	2.499	3.216	3.992	4.823	5.703	6.629
12	0.699	1.188	1.741	2.361	3.044	3.785	4.579	5.423	6.313
13	0.658	1.121	1.647	2.238	2.889	3.598	4.360	5.170	6.025

Table 9 (continued)

				$\alpha = 0.01$					
M	$s=2$	$s=3$	$s=4$	$s=5$	$s=6$	$s=7$	$s=8$	$s=9$	$s=10$
14	0.622	1.062	1.563	2.126	2.750	3.429	4.160	4.940	5.764
15	0.590	1.009	1.486	2.026	2.624	3.276	3.979	4.730	5.525
16	0.560	0.960	1.418	1.934	2.508	3.135	3.813	4.537	5.304
17	0.534	0.917	1.355	1.850	2.402	3.007	3.660	4.359	5.101
18	0.510	0.877	1.297	1.774	2.305	2.888	3.520	4.195	4.913
19	0.488	0.840	1.244	1.703	2.216	2.779	3.389	4.043	4.738
20	0.468	0.806	1.195	1.638	2.133	2.677	3.267	3.901	4.576
22	0.432	0.746	1.108	1.522	1.985	2.495	3.050	3.646	4.283
24	0.402	0.694	1.034	1.421	1.856	2.337	2.860	3.423	4.025
26	0.375	0.650	0.968	1.333	1.743	2.197	2.691	3.225	3.797
28	0.352	0.610	0.910	1.255	1.643	2.072	2.542	3.049	3.593
30	0.331	0.575	0.859	1.186	1.554	1.962	2.409	2.892	3.410
35	0.289	0.503	0.753	1.042	1.368	1.731	2.129	2.561	3.025
40	0.256	0.447	0.671	0.929	1.222	1.549	1.908	2.298	2.719
45	0.230	0.402	0.604	0.838	1.104	1.401	1.729	2.085	2.469
50	0.209	0.366	0.550	0.764	1.007	1.280	1.580	1.907	2.261
60	0.177	0.309	0.466	0.649	0.857	1.090	1.348	1.630	1.936

Table 10. The entries in this table give the values of c_4, where

$$P[\tilde{\lambda}_4 \leqslant c_4] = (1-\alpha),$$
$$\lambda_4 = (e/n)^{sn}|A\Sigma_0^{-1}|^n \operatorname{etr}(-A\Sigma_0^{-1}) \text{ and } \tilde{\lambda}_4 = -2\log\lambda_4 .$$

In the above equation, A: $s \times s$ is distributed as the central complex Wishart matrix with n degrees of freedom and $E(A) = n\Sigma_0$, where Σ_0 is known. Krishnaiah, Lee and Chang (1976) gave upper 5% points of the distribution of $\tilde{\lambda}_4$ for $s = 2(1)7$, $M = 1(1)5, 7, 10, 15, 20, 30$, where $M = n - s - 1$. These percentage points are reproduced in Table 10 with the kind permission of the Biometrika Trustees. The remaining entries in the table are reproduced from a technical report by Krishnaiah, Lee and Chang (1975).

Table 10
Percentage points of the likelihood ratio statistic for specifying the covariance matrix

				$\alpha = 0.05$					
M	$s=2$	$s=3$	$s=4$	$s=5$	$s=6$	$s=7$	$s=8$	$s=9$	$s=10$
1	11.31	21.68	35.43	52.70	73.55	98.03	126.19	158.02	193.65
2	10.86	20.57	33.44	49.60	69.16	92.18	118.72	148.79	182.49
3	10.58	19.88	32.17	47.60	66.26	88.25	113.62	142.40	174.68
4	10.40	19.42	31.29	46.18	64.19	85.40	109.87	137.65	168.85
5	10.27	19.07	30.64	45.12	62.62	83.22	106.98	134.00	164.28

Table 10 (continued)

$\alpha = 0.05$									
M	$s = 2$	$s = 3$	$s = 4$	$s = 5$	$s = 6$	$s = 7$	$s = 8$	$s = 9$	$s = 10$
6	10.18	18.82	30.14	44.30	61.40	81.50	104.69	131.04	160.60
7	10.10	18.61	29.75	43.64	60.40	80.11	102.83	128.62	157.57
8	10.04	18.45	29.43	43.11	59.58	78.96	101.28	126.61	155.01
9	9.99	18.32	29.16	42.66	58.91	77.98	99.96	124.87	152.84
10	9.95	18.21	28.94	42.28	58.32	77.15	98.82	123.41	150.97
11	9.91	18.11	28.74	41.95	57.82	76.43	97.84	122.11	149.30
12	9.88	18.02	28.58	41.67	57.38	75.80	96.98	120.98	147.87
13	9.85	17.95	28.44	41.42	57.00	75.24	96.21	119.97	146.57
14	9.83	17.89	28.31	41.21	56.67	74.75	95.54	119.06	145.44
15	9.81	17.83	28.20	41.01	56.36	74.31	94.92	118.27	144.39
16	9.79	17.79	28.10	40.84	56.09	73.91	94.38	117.55	143.44
17	9.78	17.74	28.01	40.68	55.84	73.55	93.88	116.87	142.60
18	9.77	17.70	27.92	40.54	55.61	73.23	93.43	116.29	141.81
19	9.75	17.66	27.85	40.41	55.41	72.93	93.01	115.74	141.11
20	9.74	17.63	27.78	40.29	55.23	72.66	92.64	115.22	140.44
22	9.72	17.57	27.66	40.08	54.90	72.17	91.96	114.32	139.27
24	9.70	17.52	27.56	39.90	54.61	71.76	91.39	113.54	138.25
26	9.68	17.48	27.47	39.75	54.38	71.40	90.88	112.86	137.36
28	9.67	17.44	27.40	39.61	54.17	71.09	90.45	112.27	136.61
30	9.66	17.41	27.33	39.49	53.97	70.81	90.05	111.73	135.91
$\alpha = 0.01$									
1	15.90	27.97	43.50	62.62	85.36	111.79	141.94	175.84	213.42
2	15.23	26.46	40.91	58.73	80.00	104.77	133.08	165.01	200.54
3	14.84	25.53	39.29	56.25	76.51	100.11	127.13	157.63	191.59
4	14.58	24.91	38.18	54.51	74.02	96.77	122.81	152.21	184.99
5	14.39	24.47	37.37	53.23	72.15	94.24	119.52	148.06	179.91
6	14.25	24.13	36.74	52.24	70.71	92.25	116.89	144.72	175.77
7	14.14	23.86	36.25	51.44	69.54	90.63	114.77	142.02	172.38
8	14.05	23.65	35.85	50.80	68.59	89.30	112.99	139.70	169.53
9	13.98	23.47	35.52	50.26	67.79	88.18	111.48	137.78	167.15
10	13.92	23.33	35.24	49.80	67.11	87.22	110.20	136.15	165.05
11	13.88	23.20	35.00	49.41	66.51	86.39	109.09	134.71	163.21
12	13.83	23.10	34.80	49.08	66.01	85.67	108.12	133.44	161.62
13	13.80	23.00	34.63	48.78	65.56	85.03	107.25	132.28	160.20
14	13.76	22.92	34.47	48.52	65.17	84.48	106.50	131.28	158.91
15	13.73	22.85	34.33	48.29	64.82	83.96	105.82	130.42	157.78
16	13.71	22.78	34.20	48.08	64.49	83.52	105.20	129.58	156.76
17	13.68	22.72	34.09	47.89	64.21	83.11	104.64	128.86	155.81
18	13.67	22.67	33.99	47.73	63.95	82.74	104.13	128.18	154.94
19	13.65	22.62	33.90	47.57	63.71	82.39	103.66	127.58	154.14
20	13.63	22.58	33.81	47.43	63.50	82.09	103.24	126.99	153.41
22	13.60	22.51	33.67	47.18	63.12	81.53	102.48	126.00	152.15
24	13.58	22.44	33.54	46.97	62.79	81.06	101.83	125.13	151.03
26	13.55	22.39	33.43	46.79	62.51	80.65	101.27	124.37	150.03
28	13.54	22.34	33.34	46.63	62.27	80.30	100.77	123.72	149.20
30	13.52	22.30	33.26	46.49	62.05	79.98	100.33	123.14	148.44

Table 11. The entries in this table are the values of c_5 where

$$P[\tilde{\lambda}_5 \leq c_5] = (1 - \alpha),$$

$\tilde{\lambda}_5 = -2 \log \lambda_5$ and

$$\lambda_5 = \frac{\Pi_{i=1}^{q} |A_{ii}/n_i|^{n_i}}{|\Sigma_{i=1}^{q} A_{ii}/n|^{n}},$$

where $A_{11}, \ldots, A_{qq}$ are distributed independently as central $p \times p$ complex Wishart matrices with $n_1, \ldots, n_q$ degrees of freedom respectively, $E(A_{11}/n_1) = \cdots = E(A_{qq}/n_q)$ and $n = n_1 + \cdots + n_q$. Upper 5% points of the distribution of $\tilde{\lambda}_5$ are given in Krishnaiah, Lee and Chang (1976) for $p = 3, 4$; $q = 2(1)6, 8$; $n_0 = 5(1)20, 25, 30$, where $n_1 = \cdots = n_q = n_0$. These points are reproduced in Table 11 with the kind permission of Biometrika Trustees. The remaining entries in this table are reproduced from a technical report by Krishnaiah, Lee and Chang (1975).

Table 11
Percentage points of likelihood ratio statistic for the homogeneity of the covariance matrices

				$\alpha = 0.10$	$p = 2$		
n_0	$q = 2$	$q = 3$	$q = 4$	$q = 5$	$q = 6$	$q = 7$	$q = 8$
4	10.07	16.80	22.99	28.93	34.72	40.41	46.01
5	9.49	15.94	21.89	27.60	33.17	38.63	44.02
6	9.15	15.43	21.23	26.80	32.22	37.56	42.82
7	8.92	15.08	20.78	26.26	31.60	36.84	42.01
8	8.76	14.84	20.47	25.87	31.15	36.33	41.43
9	8.63	14.65	20.23	25.58	30.81	35.93	41.00
10	8.54	14.51	20.04	25.36	30.54	35.63	40.66
11	8.46	14.40	19.89	25.18	30.33	35.39	40.39
12	8.40	14.30	19.77	25.03	30.16	35.19	40.17
13	8.35	14.23	19.67	24.91	30.02	35.03	39.98
14	8.30	14.16	19.59	24.80	29.89	34.89	39.82
15	8.27	14.10	19.51	24.71	29.79	34.77	39.69
16	8.23	14.05	19.45	24.64	29.70	34.67	39.57
17	8.21	14.01	19.39	24.57	29.62	34.57	39.47
18	8.18	13.97	19.34	24.51	29.54	34.49	39.38
19	8.16	13.94	19.30	24.46	29.48	34.42	39.29
20	8.14	13.91	19.26	24.41	29.43	34.36	39.23
25	8.06	13.79	19.11	24.23	29.22	34.12	38.96
30	8.02	13.72	19.02	24.11	29.08	33.96	38.77

Table 11 (continued)

n_0	$q=2$	$q=3$	$q=4$	$q=5$	$q=6$	$q=7$	$q=8$
			$\alpha=0.10$ $p=3$				
5	21.08	35.76	49.54	62.91	76.03	88.97	101.79
6	19.56	33.49	46.58	59.30	71.77	84.09	96.30
7	18.63	32.08	44.75	57.04	69.12	81.04	92.85
8	18.00	31.12	43.49	55.50	67.31	78.97	90.51
9	17.54	30.42	42.57	54.39	65.98	77.44	88.79
10	17.20	29.89	41.88	53.54	64.98	76.28	87.48
11	16.92	29.47	41.33	52.87	64.19	75.37	86.46
12	16.71	29.14	40.90	52.32	63.56	74.63	85.62
13	16.52	28.86	40.53	51.88	63.03	74.03	84.93
14	16.38	28.63	40.23	51.50	62.59	73.52	84.36
15	16.25	28.43	39.97	51.18	62.20	73.09	83.87
16	16.14	28.26	39.74	50.91	61.88	72.71	83.44
17	16.05	28.11	39.55	50.67	61.60	72.38	83.08
18	15.96	27.98	39.38	50.46	61.34	72.10	82.75
19	15.89	27.87	39.23	50.27	61.13	71.85	82.46
20	15.82	27.77	39.09	50.11	60.93	71.62	82.22
25	15.58	27.39	38.59	49.49	60.20	70.78	81.27
30	15.42	27.14	38.26	49.09	59.73	70.23	80.64
			$\alpha=0.10$ $p=4$				
6	36.34	62.39	87.08	111.15	134.85	158.32	181.59
7	33.48	58.07	81.43	104.21	126.66	148.87	170.92
8	31.68	55.33	77.82	99.78	121.41	142.82	164.09
9	30.43	53.42	75.31	96.68	117.74	138.61	159.27
10	29.52	52.01	73.45	94.39	115.04	135.46	155.73
11	28.82	50.93	72.02	92.62	112.94	133.05	153.00
12	28.27	50.07	70.89	91.23	111.28	131.13	150.80
13	27.82	49.37	69.96	90.08	109.92	129.54	149.05
14	27.45	48.80	69.19	89.14	108.78	128.24	147.57
15	27.15	48.31	68.54	88.33	107.85	127.16	146.32
16	26.88	47.89	67.99	87.65	107.04	126.23	145.26
17	26.66	47.53	67.51	87.06	106.33	125.42	144.35
18	26.45	47.22	67.09	86.54	105.71	124.69	143.54
19	26.28	46.94	66.73	86.09	105.17	124.08	142.83
20	26.13	46.70	66.41	85.68	104.70	123.53	142.20
25	25.55	45.80	65.21	84.19	102.92	121.47	139.90
30	25.19	45.22	64.43	83.24	101.79	120.16	138.41
			$\alpha=0.10$ $p=5$				
7	56.07	97.04	136.08	174.28	211.94	249.31	286.40
8	51.49	90.09	126.96	163.03	198.64	233.96	269.00
9	48.53	85.58	121.00	155.68	189.91	223.87	257.52
10	46.47	82.38	116.75	150.43	183.69	216.64	249.45
11	44.93	79.98	113.59	146.52	179.02	211.24	243.26
12	43.75	78.14	111.12	143.46	175.38	207.09	238.56
13	42.79	76.66	109.14	141.02	172.47	203.72	234.68
14	42.03	75.44	107.53	139.00	170.11	200.93	231.59
15	41.39	74.43	106.18	137.33	168.11	198.61	228.94

Table 11 (continued)

n_0	$q=2$	$q=3$	$q=4$	$q=5$	$q=6$	$q=7$	$q=8$
			$\alpha = 0.10$ $p = 5$				
16	40.85	73.57	105.04	135.91	166.41	196.68	226.71
17	40.39	72.83	104.06	134.69	164.96	194.99	224.83
18	39.98	72.20	103.20	133.62	163.69	193.53	223.18
19	39.64	71.64	102.46	132.70	162.58	192.22	221.69
20	39.33	71.14	101.79	131.87	161.61	191.08	220.42
25	38.20	69.35	99.39	128.89	158.03	186.95	215.70
30	37.49	68.21	97.87	126.98	155.77	184.34	212.71
			$\alpha = 0.05$ $p = 2$				
4	12.30	19.51	26.08	32.34	38.42	44.36	50.21
5	11.59	18.51	24.82	30.84	36.68	42.39	48.02
6	11.16	17.91	24.07	29.94	35.64	41.21	46.69
7	10.88	17.51	23.56	29.34	34.94	40.42	45.81
8	10.68	17.23	23.20	28.90	34.44	39.86	45.18
9	10.53	17.01	22.93	28.58	34.07	39.43	44.70
10	10.42	16.85	22.72	28.33	33.77	39.10	44.34
11	10.32	16.71	22.55	28.13	33.54	38.83	44.04
12	10.25	16.60	22.41	27.96	33.38	38.61	43.79
13	10.18	16.51	22.30	27.82	33.18	38.43	43.59
14	10.13	16.43	22.20	27.71	33.05	38.28	43.42
15	10.08	16.37	22.12	27.61	32.93	38.15	43.27
16	10.04	16.31	22.05	27.52	32.83	38.03	43.14
17	10.01	16.26	21.98	27.44	32.75	37.93	43.03
18	9.98	16.22	21.92	27.38	32.67	37.84	42.94
19	9.95	16.18	21.88	27.32	32.60	37.76	42.84
20	9.93	16.14	21.83	27.26	32.54	37.69	42.77
25	9.83	16.01	21.66	27.06	32.30	37.43	42.47
30	9.77	15.92	21.55	26.93	32.15	37.26	42.27
			$\alpha = 0.05$ $p = 3$				
5	24.35	39.80	54.17	68.05	81.62	94.99	108.18
6	22.57	37.24	50.91	64.10	77.02	89.73	102.28
7	21.48	35.66	48.88	61.66	74.16	86.46	98.62
8	20.75	34.58	47.50	59.99	72.19	84.23	96.11
9	20.22	33.80	46.50	58.77	70.78	82.59	94.28
10	19.82	33.21	45.74	57.85	69.68	81.35	92.88
11	19.51	32.75	45.14	57.12	68.84	80.37	91.79
12	19.26	32.37	44.66	56.53	68.15	79.59	90.89
13	19.04	32.06	44.26	56.04	67.58	78.93	90.18
14	18.87	31.81	43.92	55.64	67.10	78.39	89.55
15	18.72	31.59	43.64	55.30	66.70	77.94	89.03
16	18.60	31.40	43.40	55.00	66.35	77.54	88.58
17	18.49	31.23	43.18	54.74	66.05	77.18	88.18
18	18.39	31.09	43.00	54.51	65.78	76.88	87.85
19	18.30	30.96	42.83	54.31	65.54	76.61	87.55
20	18.23	30.85	42.68	54.13	65.34	76.37	87.28
25	17.95	30.43	42.13	53.46	64.56	75.47	86.26
30	17.76	30.15	41.78	53.03	64.04	74.89	85.60

Table 11 (continued)

$\alpha = 0.05 \quad p = 4$

n_0	$q=2$	$q=3$	$q=4$	$q=5$	$q=6$	$q=7$	$q=8$
6	40.71	67.84	93.36	118.15	142.48	166.51	190.31
7	37.47	63.08	87.24	110.70	133.75	156.49	179.04
8	35.43	60.08	83.34	105.96	128.17	150.09	171.85
9	34.03	57.99	80.62	102.65	124.27	145.63	166.81
10	33.00	56.45	78.63	100.19	121.39	142.31	163.04
11	32.21	55.26	77.09	98.31	119.16	139.76	160.18
12	31.59	54.33	75.87	96.82	117.40	137.74	157.89
13	31.09	53.57	74.88	95.60	115.97	136.11	156.04
14	30.68	52.94	74.05	94.59	114.77	134.73	154.51
15	30.33	52.42	73.36	93.74	113.77	133.56	153.18
16	30.03	51.96	72.77	93.02	112.92	132.58	152.06
17	29.78	51.57	72.25	92.38	112.18	131.73	151.10
18	29.56	51.23	71.81	91.84	111.53	130.98	150.25
19	29.36	50.93	71.41	91.35	110.95	130.32	149.52
20	29.18	50.67	71.06	90.92	110.45	129.73	148.86
25	28.54	49.69	69.77	89.34	108.57	127.58	146.43
30	28.14	49.05	68.94	88.32	107.37	126.20	144.85

$\alpha = 0.05 \quad p = 5$

n_0	$q=2$	$q=3$	$q=4$	$q=5$	$q=6$	$q=7$	$q=8$
7	61.61	103.97	144.09	183.19	221.69	259.84	297.64
8	56.50	96.44	134.32	171.26	207.67	243.68	279.40
9	53.23	91.55	127.95	163.48	198.48	233.10	267.46
10	50.94	88.10	123.45	157.95	191.93	225.56	258.89
11	49.25	85.53	120.08	153.81	187.05	219.92	252.55
12	47.94	83.54	117.47	150.58	183.21	215.52	247.58
13	46.90	81.96	115.37	148.00	180.19	212.01	243.60
14	46.05	80.65	113.65	145.90	177.68	209.14	240.32
15	45.34	79.57	112.22	144.12	175.60	206.71	237.59
16	44.75	78.64	111.01	142.65	173.82	204.64	235.31
17	44.24	77.86	109.98	141.36	172.28	202.89	233.28
18	43.80	77.17	109.07	140.24	170.97	201.38	231.56
19	43.42	76.57	108.27	139.27	169.82	200.07	230.08
20	43.07	76.04	107.56	138.40	168.80	198.86	228.72
25	41.84	74.12	105.02	135.25	165.04	194.56	223.82
30	41.06	72.90	103.41	133.25	162.66	191.82	220.72

Table 12. The entries in this table are the values of c_6, where

$$P[\tilde{\lambda}_6 \leqslant c_6 \mid H_6] = (1-\alpha),$$

$\tilde{\lambda}_6 = -2\log\lambda_6$ and λ_6 is the likelihood ratio statistic for testing the hypothesis H_6 of the homogeneity of q complex p-variate normal populations. The statistic λ_6 is defined in Section 8. In the table, q denotes the number of populations $M = N_0 - p$ and N_0 is the common size of various groups. Chang, Krishnaiah and Lee (1977) gave upper 5% points of the distribution of $\tilde{\lambda}_6$ for $q = 2(1)5$ and

Table 12
Percentage points of the likelihood ratio test statistic for the homogeneity of complex multivariate normal populations

	$\alpha = 0.05$ $p = 1$				$\alpha = 0.05$ $p = 2$			
M	$q = 2$	$q = 3$	$q = 4$	$q = 5$	$q = 2$	$q = 3$	$q = 4$	$q = 5$
1	8.12	12.47	16.36	20.06	19.87	31.66	42.59	53.10
2	8.03	12.49	16.50	20.30	18.41	29.83	40.42	50.62
3	7.97	12.51	16.59	20.45	17.68	28.92	39.37	49.43
4	7.94	12.52	16.64	20.55	17.25	28.39	38.75	48.74
5	7.92	12.53	16.68	20.62	16.96	28.03	38.35	48.29
6	7.91	12.53	16.71	20.67	16.75	27.78	38.06	47.97
7	7.89	12.54	16.73	20.71	16.60	27.60	37.85	47.74
8	7.89	12.55	16.75	20.74	16.47	27.45	37.69	47.56
9	7.88	12.55	16.77	20.77	16.38	27.33	37.55	47.41
10	7.87	12.56	16.78	20.79	16.30	27.24	37.44	47.30
11	7.87	12.56	16.79	20.81	16.23	27.16	37.36	47.21
12	7.86	12.56	16.80	20.82	16.18	27.09	37.28	47.12
13	7.86	12.56	16.81	20.84	16.13	27.04	37.22	47.05
14	7.86	12.56	16.81	20.85	16.09	26.99	37.16	47.00
15	7.85	12.57	16.82	20.86	16.05	26.94	37.12	46.94
16	7.85	12.56	16.83	20.87	16.02	26.90	37.07	46.90
17	7.85	12.57	16.83	20.88	15.99	26.87	37.04	46.86
18	7.85	12.57	16.84	20.89	15.97	26.84	37.01	46.82
19	7.85	12.57	16.84	20.89	15.94	26.81	36.97	46.79
20	7.85	12.57	16.85	20.90	15.92	26.79	36.95	46.76
25	7.84	12.57	16.86	20.92	15.84	26.69	36.85	46.65
30	7.84	12.58	16.87	20.94	15.79	26.63	36.78	46.57
	$\alpha = 0.05$ $p = 3$				$\alpha = 0.05$ $p = 4$			
1	36.46	59.39	80.89	101.75	58.06	95.91	131.68	166.49
2	32.90	54.56	74.91	94.65	51.79	87.16	120.63	153.23
3	31.04	52.04	71.80	90.97	48.40	82.37	114.58	145.98
4	29.90	50.49	69.88	88.70	46.26	79.35	110.73	141.37
5	29.13	49.42	68.57	87.16	44.77	77.23	108.05	138.13
6	28.56	48.65	67.62	86.05	43.68	75.67	106.08	135.78
7	28.13	48.06	66.90	85.21	42.86	74.46	104.57	133.95
8	27.79	47.61	66.34	84.55	42.17	73.51	103.36	132.51
9	27.52	47.24	65.89	84.02	41.63	72.74	102.39	131.33
10	27.30	46.94	65.52	83.57	41.19	72.10	101.57	130.37
11	27.11	46.68	65.19	83.21	40.81	71.56	100.89	129.54
12	26.96	46.46	64.94	82.90	40.50	71.09	100.29	128.84
13	26.82	46.27	64.71	82.63	40.22	70.70	99.80	128.22
14	26.70	46.11	64.50	82.40	39.97	70.35	99.36	127.70
15	26.60	45.97	64.33	82.20	39.76	70.03	98.97	127.23
16	26.51	45.84	64.17	82.01	39.57	69.77	98.62	126.83
17	26.42	45.73	64.04	81.85	39.41	69.52	98.31	126.45
18	26.35	45.63	63.92	81.71	39.25	69.30	98.04	126.13
19	26.28	45.54	63.80	81.58	39.12	69.10	97.79	125.82
20	26.22	45.45	63.70	81.46	38.99	68.92	97.55	125.54
25	25.99	45.14	63.31	81.01	38.51	68.22	96.67	124.50
30	25.83	44.92	63.05	80.70	38.18	67.74	96.07	123.76

Table 12 (continued)

	$\alpha = 0.01$ $p = 1$				$\alpha = 0.01$ $p = 2$			
M	$q = 2$	$q = 3$	$q = 4$	$q = 5$	$q = 2$	$q = 3$	$q = 4$	$q = 5$
1	11.78	16.56	20.99	25.06	25.93	38.78	50.55	61.80
2	11.65	16.69	21.15	25.34	23.94	36.42	47.84	58.76
3	11.57	16.71	21.25	25.52	22.96	35.27	46.55	57.32
4	11.53	16.72	21.32	25.63	22.38	34.59	45.79	56.49
5	11.50	16.73	21.37	25.72	21.99	34.15	45.30	55.96
6	11.48	16.74	21.40	25.78	21.72	33.83	44.95	55.58
7	11.45	16.75	21.43	25.83	21.51	33.60	44.70	55.30
8	11.45	16.75	21.46	25.87	21.35	33.42	44.49	55.08
9	11.44	16.76	21.47	25.90	21.23	33.27	44.34	54.92
10	11.43	16.76	21.49	25.92	21.12	33.16	44.21	54.78
11	11.42	16.77	21.51	25.95	21.04	33.05	44.10	54.66
12	11.42	16.77	21.52	25.96	20.97	32.98	44.02	54.57
13	11.41	16.77	21.52	25.98	20.90	32.90	43.94	54.49
14	11.41	16.77	21.53	26.00	20.84	32.84	43.87	54.42
15	11.40	16.78	21.54	26.01	20.80	32.79	43.82	54.36
16	11.40	16.78	21.55	26.02	20.76	32.74	43.77	54.31
17	11.40	16.78	21.55	26.03	20.72	32.70	43.72	54.26
18	11.40	16.78	21.56	26.04	20.69	32.67	43.68	54.22
19	11.39	16.78	21.57	26.05	20.66	32.63	43.64	54.18
20	11.39	16.78	21.57	26.06	20.63	32.60	43.62	54.14
25	11.38	16.79	21.59	26.09	20.53	32.48	43.49	54.02
30	11.38	16.79	21.60	26.11	20.45	32.40	43.41	53.93
	$\alpha = 0.01$ $p = 3$				$\alpha = 0.01$ $p = 4$			
1	45.07	69.62	92.43	114.42	69.34	109.42	146.98	183.36
2	40.46	63.71	85.31	106.12	61.52	99.01	134.17	168.24
3	38.10	60.66	81.65	101.88	57.35	93.43	127.28	160.05
4	36.66	58.80	79.40	99.26	54.75	89.89	122.89	154.88
5	35.68	57.53	77.88	97.51	52.95	87.44	119.87	151.30
6	34.98	56.62	76.79	96.23	51.63	85.65	117.65	148.66
7	34.45	55.92	75.95	95.27	50.62	84.27	115.94	146.65
8	34.03	55.38	75.31	94.52	49.82	83.18	114.59	145.05
9	33.70	54.95	74.78	93.93	49.18	82.29	113.48	143.74
10	33.41	54.59	74.36	93.44	48.64	81.56	112.57	142.68
11	33.19	54.29	73.99	93.01	48.21	80.94	111.80	141.76
12	32.99	54.04	73.69	92.67	47.82	80.41	111.16	140.99
13	32.82	53.82	73.43	92.37	47.49	79.95	110.58	140.31
14	32.67	53.62	73.20	92.10	47.20	79.56	110.09	139.73
15	32.55	53.46	73.00	91.87	46.95	79.20	109.66	139.21
16	32.43	53.31	72.83	91.67	46.72	78.90	109.28	138.77
17	32.33	53.18	72.68	91.48	46.53	78.62	108.94	138.36
18	32.24	53.05	72.53	91.32	46.34	78.36	108.62	138.00
19	32.16	52.95	72.40	91.18	46.18	78.15	108.35	137.66
20	32.09	52.86	72.29	91.05	46.03	77.94	108.09	137.37
25	31.80	52.48	71.85	90.55	45.46	77.14	107.10	136.20
30	31.60	52.23	71.54	90.19	45.07	76.60	106.42	135.40

$M = 1(1)20$, 25, 30. These percentage points are reproduced in Table 12 with the kind permission of the North-Holland Publishing Company. The remaining entries in this table are reproduced from a technical report by Chang, Krishnaiah and Lee (1975).

Table 13. The entries in this table are the values of c_7, where

$$P[\tilde{\lambda}_7 \leq c_7 | H_7] = (1-\alpha), \quad \tilde{\lambda}_7 = -2 \log \lambda_7$$

and λ_7, defined in Section 9, is the likelihood ratio statistic for testing the hypothesis H_7, where

$$H_7: \quad \Sigma_{11} = \Sigma_0, \quad \boldsymbol{\mu}_1 = \boldsymbol{\mu}_0 .$$

Here $\boldsymbol{\mu}_1$ and Σ_{11} are respectively the mean vector and covariance matrix of

Table 13
Percentage points of the likelihood ratio test statistic for $\Sigma_{11} = \Sigma_0$ and $\boldsymbol{\mu}_1 = \boldsymbol{\mu}_0$

	$\alpha = 0.05$					$\alpha = 0.01$				
p	2	3	4	5	6	2	3	4	5	6
M										
1	20.33	34.19	51.51	72.40	96.91	26.52	42.19	61.37	84.17	110.61
2	19.20	32.18	48.40	68.00	91.05	24.99	39.58	57.48	78.79	103.56
3	18.51	30.91	46.39	65.10	87.12	24.05	37.94	54.98	75.27	98.92
4	18.03	30.02	44.97	63.02	84.25	23.41	36.82	53.24	72.79	95.57
5	17.69	29.37	43.91	61.45	82.09	22.96	36.00	51.95	70.93	93.03
6	17.43	28.87	43.09	60.22	80.36	22.61	35.37	50.95	69.47	91.02
7	17.22	28.47	42.43	59.23	78.96	22.34	34.87	50.15	68.30	89.41
8	17.06	28.15	41.89	58.41	77.80	22.12	34.47	49.50	67.35	88.07
9	16.92	27.88	41.44	57.73	76.84	21.94	34.14	48.96	66.54	86.96
10	16.81	27.65	41.06	57.15	76.00	21.79	33.85	48.50	65.85	85.99
11	16.71	27.46	40.73	56.64	75.29	21.66	33.62	48.11	65.27	85.17
12	16.63	27.29	40.45	56.20	74.64	21.55	33.41	47.77	64.76	84.44
13	16.55	27.15	40.20	55.82	74.09	21.46	33.23	47.48	64.31	83.81
14	16.49	27.02	39.98	55.48	73.61	21.37	33.07	47.22	63.92	83.26
15	16.44	26.91	39.79	55.17	73.15	21.30	32.93	46.98	63.56	82.74
16	16.38	26.81	39.61	54.90	72.76	21.23	32.81	46.77	63.25	82.29
17	16.34	26.72	39.45	54.65	72.40	21.17	32.69	46.58	62.96	81.89
18	16.30	26.63	39.31	54.43	72.08	21.12	32.59	46.41	62.70	81.51
19	16.26	26.56	39.18	54.22	71.78	21.07	32.50	46.26	62.46	81.16
20	16.23	26.49	39.06	54.04	71.51	21.03	32.41	46.11	62.24	80.86
22	16.17	26.37	38.85	53.71	71.02	20.95	32.26	45.87	61.86	80.31
24	16.12	26.27	38.57	53.43	70.60	20.88	32.14	45.65	61.53	79.83
26	16.07	26.18	38.52	53.18	70.25	20.83	32.03	45.47	61.25	79.42
28	16.04	26.10	38.39	52.98	69.93	20.78	31.94	45.31	61.00	79.07
30	16.01	26.03	38.27	52.79	69.65	20.74	31.85	45.17	60.79	78.74

Table 13 (continued)

	$\alpha = 0.025$					$\alpha = 0.10$				
p	2	3	4	5	6	2	3	4	5	6
M										
1	23.05	37.74	55.91	77.68	103.07	17.46	30.39	46.75	66.65	90.15
2	21.75	35.47	52.47	72.85	96.69	16.51	28.65	44.00	62.71	84.84
3	20.95	34.04	50.24	69.68	92.44	15.92	27.53	42.21	60.09	81.25
4	20.41	33.06	48.68	67.42	89.37	15.52	26.76	40.93	58.19	78.62
5	20.01	32.32	47.51	65.72	87.03	15.23	26.18	39.98	56.76	76.61
6	19.72	31.77	46.62	64.39	85.19	15.01	25.74	39.24	55.64	75.02
7	19.49	31.33	45.89	63.32	83.69	14.83	25.39	38.65	54.73	73.73
8	19.30	30.97	45.30	62.44	82.46	14.69	25.11	38.16	53.98	72.66
9	19.14	30.67	44.82	61.70	81.42	14.58	24.87	37.76	53.36	71.76
10	19.01	30.42	44.40	61.08	80.52	14.48	24.67	37.41	52.82	71.00
11	18.90	30.21	44.05	60.54	79.76	14.39	24.50	37.12	52.37	70.33
12	18.80	30.02	43.74	60.06	79.09	14.32	24.35	36.86	51.96	69.74
13	18.72	29.86	43.46	59.65	78.49	14.26	24.22	36.64	51.61	69.22
14	18.65	29.72	43.23	59.29	77.96	14.21	24.11	36.44	51.29	68.76
15	18.58	29.60	43.02	58.96	77.50	14.16	24.01	36.26	51.02	68.35
16	18.53	29.48	42.83	58.66	77.08	14.12	23.92	36.10	50.76	67.98
17	18.48	29.38	42.65	58.40	76.69	14.08	23.84	35.96	50.53	67.64
18	18.43	29.29	42.50	58.16	76.35	14.04	23.76	35.83	50.33	67.34
19	18.39	29.21	42.36	57.94	76.02	14.01	23.70	35.71	50.14	67.07
20	18.35	29.13	42.23	57.74	75.75	13.98	23.64	35.60	49.97	66.81
22	18.28	29.00	42.01	57.39	75.22	13.93	23.53	35.41	49.67	66.36
24	18.23	28.89	41.81	57.09	74.78	13.89	23.44	35.25	49.41	65.98
26	18.18	28.79	41.64	56.82	74.39	13.85	23.36	35.11	49.18	65.65
28	18.14	28.71	41.49	56.60	74.06	13.82	23.29	34.99	48.99	65.35
30	18.10	28.63	41.36	56.39	73.77	13.79	23.23	34.88	48.81	65.09

p-variate complex normal distribution and $\boldsymbol{\mu}_0$ and Σ_0 are known. Also, $M = N_1 - p - 2$ and N_1 is the sample size. Chang, Krishnaiah and Lee (1977) gave the values of c_7 for $\alpha = 0.05$, $M = 1(1)20(2)30$ and $p = 2(1)6$. These values are reproduced in Table 13 with the kind permission of North-Holland Publishing Company. The remaining entries are reproduced from the technical report by Chang, Krishnaiah and Lee (1975).

Some of the entries in Tables 2–5 are reproduced from Krishnaiah, Lee and Chang (1976) with the kind permission of the Biometrika Trustees, whereas some of the entries in Table 6 are reproduced from Chang, Krishnaiah and Lee (1977) with the kind permission of North-Holland Publishing Company.

References

Anderson, T. W. (1958). *An Introduction to Multivariate Statistical Analysis.* Wiley, New York.

Box, G. E. P. (1949). A general distribution theory for a class of likelihood criteria. *Biometrika* **36**, 317–346.

Brillinger, D. R. (1974). *Time Series: Data Analysis and Theory.* Holt, Rinehart and Winston, New York.

Carmeli, M. (1974). Statistical theory of energy levels and random matrices in physics. *J. Statist. Phy.* **10**, 259–297.

Chang, T. C., Krishnaiah, P. R. and Lee, J. C. (1975). Approximations to the distributions of the likelihood ratio statistics for testing the hypotheses on covariance matrices and mean vectors simultaneously. Aerospace Research Laboratories TR 75-0176.

Chang, T. C., Krishnaiah, P. R. and Lee, J. C. (1977). Approximations to the distributions of the likelihood ratio statistics for testing the hypotheses on covariance matrices and mean vectors simultaneously. In: P. R. Krishnaiah, ed., *Applications of Statistics*, 97–108. North-Holland, Amsterdam.

Goodman, N. R. (1963a). Statistical analysis based on a certain multivariate complex Gaussian distribution (an introduction). *Ann. Math. Statist.* **34**, 152–176.

Goodman, N. R. (1963b). The distribution of the determinant of a complex Wishart distributed matrix. *Ann. Math. Statist.* **34**, 178–180.

Gupta, A. K. (1971). Distribution of Wilks' likelihood-ratio criterion in the complex case. *Ann. Inst. Statist. Math.* **23**, 77–87.

Hannan, E. J. (1970). *Multiple Time Series.* Wiley, New York.

James, A. T. (1964). Distributions of matrix variates and latent roots derived from normal samples. *Ann. Math. Statist.* **35**, 475–501.

Khatri, C. G. (1965). Classical statistical analysis based on a certain multivariate complex Gaussian distribution. *Ann. Math. Statist.* **36**, 98–114.

Krishnaiah, P. R. (1976). Some recent developments on complex multivariate distributions. *J. Multivariate Anal.* **6**, 1–30.

Krishnaiah, P. R. and Lee, J. C. (1976). On covariance structure. *Sankhyā, Series A*, **38**, 357–371.

Krishnaiah, P. R., Lee, J. C. and Chang, T. C. (1975). On the distributions of the likelihood ratio statistics for tests of certain covariance structures of complex multivariate normal populations. Aerospace Research Laboratories TR 75-0169.

Krishnaiah, P. R., Lee, J. C. and Chang, T. C. (1976). The distributions of the likelihood ratio statistics for tests of certain covariance structures of complex multivariate normal populations. *Biometrika* **63**, 543–549.

Lee, J. C., Krishnaiah, P. R. and Chang, T. C. (1975). Approximations to the distributions of the determinants of real and complex multivariate beta matrices. Aerospace Research Laboratories TR 75-1068.

Lee, J. C., Krishnaiah, P. R. and Chang, T. C. (1976). On the distribution of the likelihood ratio test statistic for compound symmetry. *S. African Statist. J.* **10**, 49–62.

Mauchly, J. W. (1940). Significance test for sphericity of a normal n-variate distribution. *Ann. Math. Statist.* **11**, 204–209.

Nagarsenker, B. N. and Das, M. M. (1975). Exact distribution of sphericity criterion in the complex case and its percentage points. *Comm. Statist.* **4**, 363–375.

Wahba, G. (1968). On the distributions of some statistics useful in the analysis of jointly stationary time series. *Ann. Math. Statist.* **39**, 1849–1862.

Wilks, S. S. (1932). Certain generalization in the analysis of variance. *Biometrika* **24**, 471–494.

Wilks, S. S. (1935). On the independeence of K sets of normally distributed statistical variables. *Econometrica* **3**, 309–325.

Wooding, R. A. (1956). The multivariate distribution of complex normal variables. *Biometrika* **43**, 212–215.

Subject Index

Handbook of Statistics
Contents of Previous Volumes

Volume 1. Analysis of Variance
Edited by P. R. Krishnaiah
1980 xviii + 1002 pp.

Volume 2. Classification, Pattern Recognition and Reduction of Dimensionality
Edited by P. R. Krishnaiah and L. N. Kanal
1982 xxii + 903 pp.